ENVIRONMENTAL ENGINEER'S MATHEMATICS HANDBOOK

Frank R. Spellman and
Nancy E. Whiting

CRC Press
Taylor & Francis Group
Boca Raton London New York

CRC Press is an imprint of the
Taylor & Francis Group, an **informa** business

CRC Press
Taylor & Francis Group
6000 Broken Sound Parkway NW, Suite 300
Boca Raton, FL 33487-2742

First issued in paperback 2020

© 2000 by Taylor & Francis Group, LLC
CRC Press is an imprint of Taylor & Francis Group, an Informa business

No claim to original U.S. Government works

ISBN 13: 978-0-367-57823-7 (pbk)
ISBN 13: 978-1-56670-681-0 (hbk)

Visit the Taylor & Francis Web site at
http://www.taylorandfrancis.com

and the CRC Press Web site at
http://www.crcpress.com

Library of Congress Cataloging-in-Publication Data

Spellman, Frank R.
 Environmental engineer's mathematics handbook / by Frank R. Spellman, Nancy Whiting.
  p. cm.
 Includes bibliographical references and index.
 ISBN 1-56670-681-5 (alk. paper)
  1. Environmental engineering--Mathematics--Handbooks, manuals, etc. I. Whiting, Nancy E. II. Title.

TD145.S676 2004
629.8'95--dc22

                        2004051872

Library of Congress Card Number 2004051872

Preface

Environmental Engineer's Mathematics Handbook brings together and integrates in a single text the more practical math operations of environmental engineering for air, water, wastewater, biosolids and stormwater. Taking an unusual approach to the overall concept of environmental engineering math concepts, this offers the reader an approach that emphasizes the relationship between the principles in natural processes and those employed in engineered processes.

The text covers in detail the engineering principles, practices, and math operations involved in the design and operation of conventional environmental engineering works and presents engineering modeling tools and environmental algorithm examples. The arrangement of the material lends itself to several different specific environmental specialties and several different formal course formats.

Major subjects covered in this book include:

- Math concepts review
- Modeling
- Algorithms
- Air pollution control calculations
- Water assessment and control calculations
- Stormwater engineering math calculations

In our approach, we emphasize concepts, definitions, descriptions, and derivations, as well as a touch of common sense. This book is intended to be a combination textbook and reference tool for practitioners involved in the protection of the three environmental media: air, water, and land resources.

Frank R. Spellman
Norfolk, Virginia

Nancy E. Whiting
Columbia, Pennsylvania

Acknowledgments

This text would not have been possible without the tireless efforts of Mimi Williams. We appreciate her astute sense of sensibility and correctness. Thanks.

Contents

PART I

Fundamental Computation and Modeling

CHAPTER 1

Conversion Factors and SI Units

1.1 INTRODUCTION

The units most commonly used by environmental engineering professionals are based on the complicated English System of Weights and Measures. However, bench work is usually based on the metric system or the International System of Units (SI) because of the convenient relationship among milliliters (mL), cubic centimeters (cm^3), and grams (g).

The SI is a modernized version of the metric system established by international agreement. The metric system of measurement was developed during the French Revolution and was first promoted in the U.S. in 1866. In 1902, proposed congressional legislation requiring the U.S. government to use the metric system exclusively was defeated by a single vote. Although we use both systems in this text, SI provides a logical and interconnected framework for all measurements in engineering, science, industry, and commerce. The metric system is much simpler to use than the existing English system because all its units of measurement are divisible by 10.

Before we list the various conversion factors commonly used in environmental engineering, we describe the prefixes commonly used in the SI system. These prefixes are based on the power 10. For example, a "kilo" means 1000 g, and a "centimeter" means 1/100 of 1 m. The 20 SI prefixes used to form decimal multiples and submultiples of SI units are given in Table 1.1.

Note that the kilogram is the only SI unit with a prefix as part of its name and symbol. Because multiple prefixes are not used, in the case of the kilogram the prefix names of Table 1.1 are used with the unit name "gram" and the prefix symbols are used with the unit symbol "g." With this exception, any SI prefix may be used with any SI unit, including the degree Celsius and its symbol °C.

Example 1.1

10^{-6} kg = 1 mg (1 milligram), but not 10^{-6} kg = 1 μkg (1 microkilogram)

Example 1.2

Consider the height of the Washington Monument. We may write h_w = 169,000 mm = 16,900 cm = 169 m = 0.169 km, using the millimeter (SI prefix "milli," symbol "m"); centimeter (SI prefix "centi," symbol "c"); or kilometer (SI prefix "kilo," symbol "k").

1.2 CONVERSION FACTORS

Conversion factors are given in alphabetical order in Table 1.2 and in unit category listing order in Table 1.3.

3

Table 1.1　SI Prefixes

Factor	Name	Symbol
10^{24}	Yotta	Y
10^{21}	Zetta	Z
10^{18}	Exa	E
10^{15}	Peta	P
10^{12}	Tera	T
10^{9}	Giga	G
10^{6}	Mega	M
10^{3}	Kilo	k
10^{2}	Hecto	h
10^{1}	Deka	da
10^{-1}	Deci	d
10^{-2}	Centi	c
10^{-3}	Milli	m
10^{-6}	Micro	m
10^{-9}	Nano	n
10^{-12}	Pico	p
10^{-15}	Femto	f
10^{-18}	Atto	a
10^{-21}	Zepto	z
10^{-24}	Yocto	y

Table 1.2　Alphabetical Listing of Conversion Factors

Factors	Metric (SI) or English conversions
1 atm (atmosphere) =	1.013 bar
	10.133 N/cm^2 (newtons per square centimeter)
	33.90 ft of H_2O (feet of water)
	101.325 kPa (kilopascals)
	1013.25 mbar (millibars)
	13.70 psia (pounds per square inch — absolute)
	760 torr
	760 mm Hg (millimeters of mercury)
1 bar =	0.987 atm (atmospheres)
	1×10^6 dyn/cm^2 (dynes per square centimeter)
	33.45 ft of H_2O (feet of water)
	1×10^5 Pa [N/m^2] (pascals; newtons per square meter)
	750.06 torr
	750.06 mm Hg (millimeters of mercury)
1 Bq (becquerel) =	1 radioactive disintegration per second
	2.7×10^{-11} Ci (curie)
	2.7×10^{-8} mCi (millicurie)
1 Btu (British thermal unit) =	252 cal (calories)
	1055.06 J (joules)
	10.41 L–atm (liter–atmospheres)
	0.293 Wh (watt–hours)
1 cal (calories) =	3.97×10^{-3} Btu (British thermal units)
	4.18 J (joules)
	0.0413 L–atm (liter–atmospheres)
	1.163×10^{-3} Wh (watt–hours)
1 cm (centimeters) =	0.0328 ft (feet)
	0.394 in. (inches)
	10,000 μm (microns/micrometers)
	100,000,000 Å = 10^8 Å (angstroms)

Table 1.2 Alphabetical Listing of Conversion Factors (continued)

Factors	Metric (SI) or English conversions
1 cm³ (cubic centimeter) =	3.53×10^{-5} ft³ (cubic feet)
	0.061 in.³ (cubic inches)
	2.64×10^{-4} gal (gallons)
	52.18 L (liters)
	52.18 mL (milliliters)
1 ft³ (cubic foot) =	28.317 cm³ (cubic centimeters)
	1728 in.³ (cubic inches)
	0.0283 m³ (cubic meters)
	7.48 gal (gallons)
	28.32 L (liters)
	29.92 qt (quarts)
1 in.³ (cubic inch) =	16.39 cm³ (cubic centimeters)
	16.39 mL (milliliters)
	5.79×10^{-4} ft³ (cubic feet)
	1.64×10^{-5} m³ (cubic meters)
	4.33×10^{-3} gal (gallons)
	0.0164 L (liters)
	0.55 fl oz (fluid ounces)
1 m³ (cubic meter) =	1,000,000 cm³ = 10^6 cm³ (cubic centimeters)
	33.32 ft³ (cubic feet)
	61,023 in³ (cubic inches)
	264.17 gal (gallons)
	1000 L (liters)
1 yd³ (cubic yard) =	201.97 gal (gallons)
	764.55 L (liters)
1 Ci (curie) =	3.7×10^{10} radioactive disintegrations per second
	3.7×10^{10} Bq (becquerel)
	1000 mCi (millicurie)
1 day =	24 h (hours)
	1440 min (minutes)
	86,400 sec (seconds)
	0.143 weeks
	2.738×10^{-3} yr (years)
1°C (expressed as an interval) =	1.8°F = [9/5]°F (degrees Fahrenheit)
	1.8°R (degrees Rankine)
	1.0 K (degrees Kelvin)
°C (degree Celsius) =	[(5/9)(°F − 32°)]
1°F (expressed as an interval) =	0.556°C = [5/9]°C (degrees Celsius)
	1.0°R (degrees Rankine)
	0.556 K (degrees Kelvin)
°F (degree Fahrenheit) =	[(9/5)(°C) + 32°]
1 dyn (dyne) =	1×10^{-5} N (newton)
1 eV (electron volt) =	1.602×10^{-12} ergs
	1.602×10^{-19} J (joules)
1 erg =	1 dyn–cm (dyne–centimeter)
	1×10^{-7} J (joules)
	2.78×10^{-11} Wh (watt–hours)
1 ft/sec (feet per second) =	1.097 km/h (kilometers per hour)
	0.305 m/sec (meters per second)
	0.01136 mi/h (miles per hour)
1 ft (foot) =	30.48 cm (centimeters)
	12 in. (inches)

Table 1.2   Alphabetical Listing of Conversion Factors (continued)

Factors	Metric (SI) or English conversions
1 gal (gallon) =	0.3048 m (meters)
	1.65×10^{-4} Nmi (nautical miles)
	1.89×10^{-4} mi (statute miles)
	3785 cm^3 (cubic centimeters)
	0.134 ft^3 (cubic feet)
	231 in.^{3} (cubic inches)
	3.785 L (liters)
1 g (gram)	0.001 kg (kilogram)
	1000 mg (milligrams)
	$1,000,000$ ng = 10^6 ng (nanograms)
	2.205×10^{-3} lb (pounds)
1 g/cm^3 (grams per cubic centimeters) =	62.43 lb/ft^3 (pounds per cubic foot)
	0.0361 lb/in.^{3} (pounds per cubic inch)
	8.345 lb/gal (pounds per gallon)
1 Gy (gray) =	1 J/kg (joules per kilogram)
	100 rad
	1 Sv (sievert) (unless modified through division by an appropriate factor, such as Q and/or N)
1 hp (horsepower) =	745.7 J/sec (joules per sec)
1 h (hour) =	0.0417 D (day)
	60 min (minutes)
	3600 sec (seconds)
	5.95×10^{-3} weeks
	1.14×10^{-4} yr (years)
1 in. (inch) =	2.54 cm (centimeters)
	1000 mils
1 in. (inch) of water =	1.86 mm Hg (millimeters of mercury)
	249.09 Pa (pascals)
	0.0361 psi (pounds per square inch)
1 J (joule) =	9.48×10^{-4} Btu ((British thermal units)
	0.239 cal (calories)
	$10,000,000$ ergs = 1×10^7 ergs
	9.87×10^{-3} L-atm liter–atmospheres
	1.0 N–m (newton–meter)
1 kcal (kilocalories) =	3.97 Btu (British thermal units)
	1000 cal (calories)
	4186.8 J (joules)
1 kg (kilogram) =	1000 g (grams)
	2205 lb (pounds)
1 km (kilometer) =	3280 ft (feet)
	0.54 Nmi (nautical miles)
	0.6214 mi (statute miles)
1 kW (kilowatt) =	56.87 Btu/min (British thermal units per minute)
	1.341 hp (horsepower)
	1000 J/sec (joules per second)
1 kWh (kilowatt–hour) =	3412.14 Btu (British thermal units)
	3.6×10^6 J (joules)
	859.8 kcal (kilocalories)
1 L (liter) =	1000 cm^3 (cubic centimeters)
	1 dm^3 (cubic decimeters)
	0.0353 ft^3 (cubic feet)

Table 1.2   Alphabetical Listing of Conversion Factors (continued)

Factors	Metric (SI) or English conversions
	61.02 in.^{3} (cubic inches)
	0.264 gal (gallons)
	1000 ml (milliliters)
	1.057 qt (quarts)
1 m (meter) =	1 × 10^{10} Å (angstroms)
	100 cm (centimeters)
	3.28 ft (feet)
	39.37 in. (inches)
	1 × 10^{-3} km (kilometers)
	1000 mm (millimeters)
	1,000,000 µm = 1 × 10^6 µm (micrometers)
	1 × 10^9 nm (nanometers)
1 m/sec (meters per second) =	196.9 ft/min (feet per minute)
	3.6 km/h (kilometers per hour)
	2.237 mi/h (miles per hour)
1 mi/h (mile per hour) =	88 ft/min (feet per minute)
	1.61 km/h (kilometers per hour)
	0.447 m/sec (meter per second)
1 Nmi (nautical mile) =	6076.1 ft (feet)
	1.852 Km (kilometers)
	1.15 mi (statute miles)
	2025.4 yd (yards)
1 mi (statute mile) =	5280 ft (feet)
	1.609 km (kilometers)
	1609.3 m (meters)
	0.869 Nmi (nautical miles)
	1760 yd (yards)
1 mCi (millicurie) =	0.001 Ci (curie)
	3.7 ×10^{10} radioactive disintegrations per second
	3.7 × 10^{10} Bq (becquerel)
1 mm Hg (mm of mercury) =	1.316 × 10^{-3} atm (atmosphere)
	0.535 in. H$_2$O (inches of water)
	1.33 mbar (millibars)
	133.32 Pa (pascals)
	1 torr
	0.0193 psia (pounds per square inch — absolute)
1 min (minute) =	6.94 × 10^{-4} days
	0.0167 h (hour)
	60 sec (seconds)
	9.92 × 10^{-5} weeks
	1.90 × 10^{-6} yr (years)
1 N (newton) =	1 × 10^5 dyn (dynes)
1 N-m (newton–meter) =	1.00 J (joules)
	2.78 × 10^{-4} Wh (watt–hours)
1 ppm (parts per million–volume) =	1.00 mL/m^3 (milliliters per cubic meter)
1 ppm [wt] (parts per million–weight) =	1.00 mg/kg (milligrams per kilograms)
1 Pa (pascal) =	9.87 × 10^{-6} atm (atmospheres)
	4.015 × 10^{-3} in. H$_2$O (inches of water)
	0.01 mbar (millibars)
	7.5 × 10^{-3} mm Hg (milliliters of mercury)

Table 1.2   Alphabetical Listing of Conversion Factors (continued)

Factors	Metric (SI) or English conversions
1 lb (pound) =	453.59 g (grams)
	16 oz (ounces)
l lb/ft³ (pounds per cubic foot) =	16.02 g/L (grams per liter)
1 lb/in.³ (pounds per cubic inch) =	27.68 g/cm³ (grams per cubic centimeter)
	1728 lb/ft³ (pounds per cubic feet)
1 psi (pounds per square inch) =	0.068 atm (atmospheres)
	27.67 in. H_2O (inches of water)
	68.85 mbar (millibars)
	51.71 mm Hg (millimeters of mercury)
	6894.76 Pa (pascals)
1 qt (quart) =	946.4 cm³ (cubic centimeters)
	57.75 in.³ (cubic inches)
	0.946 L (liters)
1 rad =	100 ergs/g (gram)
	0.01 Gy (gray)
	1 rem (unless modified through division by an appropriate factor, such as Q and/or N)
1 rem	1 rad (unless modified through division by an appropriate factor, such as Q and/or N)
1 Sv (sievert) =	1 Gy (gray) (unless modified through division by an appropriate factor, such as Q and/or N)
1 cm² (square centimeter) =	1.076×10^{-3} ft² (square feet)
	0.155 in.² (square inches)
	1×10^{-4} m² (square meters)
1 ft² (square foot) =	2.296×10^{-5} acres
	9.296 cm² (square centimeters)
	144 in.² (square inches)
	0.0929 m² (square meter)
1 m² (square meter) =	10.76 ft² (square feet)
	1550 in.² (square inches)
1 mi² (square mile) =	640 acres
	2.79×10^{7} ft² (square feet)
	2.59×10^{6} m² (square meters)
1 torr =	1.33 mbar (millibars)
1 watt =	3.41 Btu/h (British thermal units per hour)
	1.341×10^{-3} hp (horsepower)
	52.18 J/sec (joules per second)
1 Wh (watt–hour)=	3.412 Btu (British thermal units)
	859.8 cal (calories)
	3600 J (joules)
	35.53 L–atm (liter–atmospheres)
1 week =	7 days
	168 h (hours)
	10,080 min (minutes)
	6.048×10^{5} sec (seconds)
	0.0192 yr (years)
1 yr (year) =	365.25 days
	8766 h (hours)
	5.26×10^{5} min (minutes)
	3.16×10^{7} sec (seconds)
	52.18 weeks

Table 1.3 Conversion Factors by Unit Category

Units of length

1 cm (centimeter) =	0.0328 ft (feet)
	0.394 in. (inches)
	10,000 μm (microns or micrometers)
	100,000,000 Å = 10^8 Å (angstroms)
1 ft (foot) =	30.48 cm (centimeters)
	12 in. (inches)
	0.3048 m (meter)
	1.65×10^{-4} Nmi (nautical miles)
	1.89×10^{-4} mi (statute miles)
1 in. (inch) =	2.54 cm (centimeters)
	1000 mils
1 km (kilometer) =	3280.8 ft (feet)
	0.54 Nmi (nautical mile)
	0.6214 mi (statute mile)
1 m (meter) =	1×10^{10} Å (angstroms)
	100 cm (centimeters)
	3.28 ft (feet)
	39.37 in. (inches)
	1×10^{-3} km (kilometers)
	1000 mm (millimeters)
	1,000,000 μm = 1×10^6 μm (micrometers)
	1×10^9 nm (nanometers)
1 Nmi (nautical mile) =	6076.1 ft (feet)
	1.852 km (kilometers)
	1.15 mi (statute miles)
	2025.4 yd (yards)
1 mi (statute mile) =	5280 ft (feet)
	1609 km (kilometers)
	1690.3 m (meters)
	0.869 Nmi (nautical mile)
	1760 yd (yards)

Units of area

1 cm² (square centimeter) =	1.076×10^{-3} ft² (square feet)
	0.155 in.² (square inches)
	1×10^{-4} m² (square meters)
1 ft² (square foot) =	2.296×10^{-5} acres
	929.03 cm² (square centimeters)
	144 in.² (square inches)
	0.0929 m² (square meters)
1 m² (square meter) =	10.76 ft² (square feet)
	1550 in.² (square inches)
1 mi² (square mile) =	640 acres
	2.79×10^7 ft² (square feet)
	2.59×10^6 m² (square meters)

Units of volume

1 cm³ (cubic centimeter) =	3.53×10^{-5} ft³ (cubic feet)
	0.061 in.³ (cubic inches)
	2.64×10^{-4} gal (gallons)
	0.001 L (liter)
	1.00 mL (milliliter)
1 ft³ (cubic foot) =	28,317 cm³ (cubic centimeters)
	1728 in.³ (cubic inches)
	0.0283 m³ (cubic meter)
	7.48 gal (gallons)

Table 1.3   Conversion Factors by Unit Category (continued)

	28.32 L (liters)
	29.92 qt (quarts)
1 in.^{3} (cubic inch) =	16.39 cm^3 (cubic centimeters)
	16.39 mL (milliliters)
	5.79×10^{-4} ft^3 (cubic feet)
	1.64×10^{-5} m^3 (cubic meters)
	4.33×10^{-3} gal (gallons)
	0.0164 L (liter)
	0.55 fl oz (fluid ounce)
1 m^3 (cubic meter) =	1,000,000 cm^3 = 10^6 cm^3 (cubic centimeters)
	35.31 ft^3 (cubic feet)
	61,023 in.^{3} (cubic inches)
	264.17 gal (gallons)
	1000 L (liters)
1 yd^3 (cubic yards) =	201.97 gal (gallons)
	764.55 L (liters)
1 gal (gallon) =	3785 cm^3 (cubic centimeters)
	0.134 ft^3 (cubic feet)
	231 in.^{3} (cubic inches)
	3.785 L (liters)
1 L (liter) =	1000 cm^3 (cubic centimeters)
	1 dm^3 (cubic decimeters)
	0.0353 ft^3 (cubic foot)
	61.02 in.^{3} (cubic inches)
	0.264 gal (gallon)
	1000 mL (milliliters)
	1.057 qt (quarts)
1 qt (quart) =	946.4 cm^3 (cubic centimeters)
	57.75 in.^{3} (cubic inches)
	0.946 L (liter)

Units of mass

1 g (gram) =	0.001 kg (kilogram)
	1000 mg (milligrams)
	1,000,000 mg = 10^6 ng (nanograms)
	2.205×10^{-3} lb (pounds)
1 kg (kilogram) =	1000 g (grams)
	2.205 lb (pounds)
1 lb (pound) =	453.59 g (grams)
	16 oz (ounces)

Units of time

1 day =	24 h (hours)
	1440 min (minutes)
	86,400 sec (seconds)
	0.143 weeks
	2.738×10^{-3} yr (years)
1 h (hour) =	0.0417 days
	60 min (minutes)
	3600 sec (seconds)
	5.95×10^{-3} yr (years)
	1.14×10^{-4} yr (years)
1 min (minutes) =	6.94×10^{-4} days
	0.0167 h (hour)
	60 sec (seconds)
	9.92×10^{-5} weeks
	1.90×10^{-6} yr (years)

Table 1.3 Conversion Factors by Unit Category (continued)

1 week =	7 days
	168 h (hours)
	10,080 min (minutes)
	6.048×10^5 sec (seconds)
	0.0192 yr (years)
1 yr (year) =	365.25 days
	8766 h (hours)
	5.26×10^5 min (minutes)
	3.16×10^7 sec (seconds)
	52.18 weeks

Units of the measure of temperature

°C (degrees Celsius) =	[(5/9)(°F− 32°)]
1°C (expressed as an interval) =	1.8°F = [9/5]°F (degrees Fahrenheit)
	1.8°R (degrees Rankine)
	1.0 K (degrees Kelvin)
°F (degree Fahrenheit) =	[(9/5)(°C) + 32°]
1°F (expressed as an interval) =	0.556°C = [5/9]°C (degrees Celsius)
	1.0°R (degrees Rankine)
	0.556 K (degrees Kelvin)

Units of force

1 dyn (dyne) =	1×10^{-5} N (newtons)
1 N (newton) =	1×10^5 dyn (dynes)

Units of work or energy

1 Btu (British thermal unit) =	252 cal (calories)
	1055.06 J (joules)
	10.41 L–atm (liter–atmospheres)
	0.293 Wh (watt–hour)
1 cal (calories) =	3.97×10^{-3} Btu (British thermal units)
	4.18 J (joules)
	0.0413 L–atm (liter–atmosphere)
	1.163×10^{-3} Wh (watt–hours)
1 eV (electron volt) =	1.602×10^{-12} ergs
	1.602×10^{-19} J (joules)
1 erg =	1 dyn–cm (dyne–centimeter)
	1×10^{-7} J (joules)
	2.78×10^{-11} Wh (watt–hours)
1 J (joule) =	9.48×10^{-4} Btu (British thermal units)
	0.239 cal (calorie)
	10,000,000 ergs = 1×10^7 ergs
	9.87×10^{-3} L–atm (liter–atmospheres)
	1.00 N–m (newton–meters)
1 kcal (kilocalorie) =	3.97 Btu (British thermal units)
	1000 cal (calories)
	4186.8 J (joules)
1 kWh (kilowatt–hour) =	3412.14 Btu (British thermal units)
	3.6×10^6 J (joules)
	859.8 kcal (kilocalories)
1 N–m (newton–meter) =	1.00 J (joule)
	2.78×10^{-4} Wh (watt–hours)
1 Wh (watt–hour) =	3.412 Btu (British thermal units)
	859.8 cal (calories)
	3600 J (joules)
	35.53 L–atm (liter–atmospheres)

Table 1.3   Conversion Factors by Unit Category (continued)

Units of power

1 hp (horsepower) =	745.7 J/sec (joules per second)
1 kW (kilowatt) =	56.87 Btu/min (British thermal units per minute)
	1.341 hp (horsepower)
	1000 J/sec (joules per second)
1 W (watt) =	3.41 Btu/h (British thermal units per hour)
	1.341×10^{-3} hp (horsepower)
	1.00 J/sec (joules per second)

Units of pressure

1 atm (atmosphere) =	1.013 bar
	10.133 N/cm^2 (newtons per square centimeter)
	33.90 ft of H_2O (feet of water)
	101.325 kPa (kilopascals)
	14.70 psia (pounds per square inch — absolute)
	760 torr
	760 mm Hg (millimeters of mercury)
1 bar =	0.987 atm (atmospheres)
	1×10^6 dyn/cm^2 (dynes per square centimeters)
	33.45 ft of H_2O (feet of water)
	1×10^5 Pa (pascals) [N/m^2] (newtons per square meter)
	750.06 torr
	750.06 mm Hg (millimeters of mercury)
1 in. (inch) of water =	1.86 mm Hg (millimeters of mercury)
	249.09 Pa (pascals)
	0.0361 psi (pounds per square inch)
1 mm Hg (millimeter of merc.) =	1.316×10^{-3} atm (atmospheres)
	0.535 in. H_2O (inch of water)
	1.33 mbar (millibars)
	133.32 Pa (pascals)
	1 torr
	0.0193 psia (pounds per square inch — absolute)
1 Pa (pascal) =	9.87×10^{-6} atm (atmospheres)
	4.015×10^{-3} in. H_2O (inches of water)
	0.01 mbar (millibars)
	7.5×10^{-3} mm Hg (millimeters of mercury)
1 psi (pounds per square inch) =	0.068 atm (atmospheres)
	27.67 in. H_2O (inches of water)
	68.85 mbar (millibars)
	51.71 mm Hg (millimeters of mercury)
	6894.76 Pa (pascals)
1 torr =	1.33 mbar (millibars)

Units of velocity or speed

1 fps (foot per second) =	1.097 kmph (kilometer per hour)
	0.305 mps (meter per second)
	0.01136 mph (mile per hour)
1 mps (meter per second) =	196.9 fpm (feet per minute)
	3.6 kmph (kilometers per hour)
	2.237 mph (miles per hour)
1 mph (mile per hour) =	88 fpm (feet per minute)
	1.61 kmph (kilometers per hour)
	0.447 mps (meter per second)

Units of density

1 g/cm^3 (gram per cubic centimeters) =	62.43 lb/ft^3 (pounds per cubic foot)
	0.0361 $lb/in.^3$ (pound per cubic inch)

Table 1.3 Conversion Factors by Unit Category (continued)

	8.345 lb/gal (pounds per gallon)
1 lb/ft³ (pound per cubic foot) =	16.02 g/L (grams per liter)
1 lb/in.² (pound per square inch) =	27.68 g/c³ (grams per cubic centimeter)
	1.728 lb/ft³ (pounds per cubic foot)

Units of concentration

1 ppm (part per million–volume) =	1.00 mL/m³ (milliliter per cubic meter)
1 ppm (part per million–wt) =	1.00 mg/kg (milligram per kilogram)

Radiation and dose-related units

1 Bq (becquerel) =	1 radioactive disintegration per second
	2.7×10^{-11} Ci (curies)
	2.7×10^{-8} mCi (millicurie)
1 Ci (curie) =	3.7×10^{10} radioactive disintegrations per second
	3.7×10^{10} Bq (becquerel)
	1000 mCi (millicuries)
1 Gy (gray) =	1 J/kg (joule per kilogram)
	100 rad
	1 Sv (sievert) (unless modified through division by an appropriate factor, such as Q and/or N)
1 mCi (millicurie) =	0.001 Ci (curie)
	3.7×10^{10} radioactive disintegrations per second
	3.7×10^{10} Bq (becquerels)
1 rad =	100 ergs/g (gram)
	0.01 Gy (gray)
	1 rem (unless modified through division by an appropriate factor, such as Q and/or N)
1 rem =	1 rad (unless modified through division by an appropriate factor, such as Q and/or N)
1 Sv (sievert) =	1 Gy (gray) (unless modified through division by an appropriate factor, such as Q and/or N)

Example 1.3

Problem:

Find degrees in Celsius of water at 72°F.

Solution:

$$°C = (°F - 32) \times \frac{5}{9} = (72 - 32) \times \frac{5}{9} = 22.2$$

1.3 CONVERSION FACTORS: PRACTICAL EXAMPLES

Sometimes we must convert between different units. Suppose that a 60-in. piece of pipe is attached to an existing 6-ft piece of pipe. Joined together, how long are they? Obviously, we cannot find the answer to this question by adding 60 to 6. Why? — because the two lengths are given in different units. Before we can add the two lengths, we must convert one of them to the units of the other. Then, when we have two lengths in the same units, we can add them.

To perform this conversion, we need a *conversion factor*. In this case, we need to know how many inches make up a foot — that is, 12 in. is 1 ft. Knowing this, we can perform the calculation in two steps:

1. 60 in. is really 60/12 = 5 ft
2. 5 ft + 6 ft = 11 ft

From this example, we see that a conversion factor changes known quantities in one unit of measure to an equivalent quantity in another unit of measure.

In making the conversion from one unit to another, we must know two things:

- The exact number that relates the two units
- Whether to multiply or divide by that number

When conversions are necessary, confusion over whether to multiply or divide is common; however, the number that relates the two units is usually known and thus is not a problem. Understanding the proper methodology — the "mechanics" — to use for various operations requires practice and common sense.

Along with using the proper mechanics (and practice and common sense) in making conversions, probably the easiest and fastest method of converting units is to use a conversion table. The simplest conversions require that the measurement be multiplied or divided by a constant value. For instance, if the depth of wet cement in a form is 0.85 ft, multiplying by 12 in./ft converts the measured depth to inches (10.2 in.). Likewise, if the depth of the cement in the form is measured as 16 in., dividing by 12 in./ft converts the depth measurement to feet (1.33 ft).

1.3.1   Weight, Concentration, and Flow

Using Table 1.4 to convert from one unit expression to another and vice versa is good practice. However, in making conversions to solve process computations in water treatment operations, for example, we must be familiar with conversion calculations based upon a relationship among weight, flow or volume, and concentration. The basic relationship is

$$\text{Weight} = \text{Concentration} \times \text{Flow or Volume} \times \text{Factor} \qquad (1.1)$$

Table 1.5 summarizes weight, volume, and concentration calculations. With practice, many of these calculations become second nature to users.

Table 1.4   Conversion Table

To convert (to get)	Multiply by (divide by)	To get (to convert)
Feet	12	Inches
Yards	3	Feet
Yards	36	Inches
Inches	2.54	Centimeters
Meters	3.3	Feet
Meters	100	Centimeters
Meters	1000	Millimeters
Square yards	9	Square feet
Square feet	144	Square inches
Acres	43,560	Square feet
Cubic yards	27	Cubic feet
Cubic feet	1728	Cubic inches
Cubic feet (water)	7.48	Gallons
Cubic feet (water)	62.4	Pounds
Acre–feet	43,560	Cubic feet
Gallons (water)	8.34	Pounds
Gallons (water)	3.785	Liters
Gallons (water)	3785	Milliliters
Gallons (water)	3785	Cubic centimeters
Gallons (water)	3785	Grams
Liters	1000	Milliliters
Days	24	Hours

Table 1.4   Conversion Table (continued)

To convert (to get)	Multiply by (divide by)	To get (to convert)
Days	1440	Minutes
Days	86,400	Seconds
Million gallons per day	1,000,000	Gallons per day
Million gallons per day	1.55	Cubic feet per second
Million gallons per day	3.069	Acre-feet per day
Million gallons per day	36.8	Acre-inches per day
Million gallons per day	3785	Cubic meters per day
Gallons per minute	1440	Gallons per day
Gallons per minute	63.08	Liters per minute
Pounds	454	Grams
Grams	1000	Milligrams
Pressure, psi	2.31	Head, ft (water)
Horsepower	33,000	Foot-pounds per minute
Horsepower	0.746	Kilowatts

Table 1.5   Weight, Volume, and Concentration Calculations

To calculate	Formula
Pounds	Concentration, mg/L × tank vol. MG (million gallons) × 8.34 lb/MG/mg/L
Pounds per day	Concentration, mg/L × flow, MGD × 8.34 lb/MG/mg/L
Million gallons per day	$\dfrac{\text{Quantity, pounds per day}}{(\text{Conc., mg/L} \times 8.34\ \text{lb/mg/L/MG})}$
Milligrams per liter	$\dfrac{\text{Quantity, pounds}}{(\text{Tank vol., MG} \times 8.34\ \text{lb/mg/L/MG})}$
Kilograms per liter	Conc., mg/L × vol., MG × 3.785 lb/MG/mg/L
Kilograms per day	Conc., mg/L × flow, MGD × 3.785 lb/MG/mg/L
Pounds per dry ton (d.t.)	Conc. Mg/kg × 0.002 lb/d.t./mg/kg

The following conversion factors are used extensively in environmental engineering (water and wastewater operations):

- 7.48 gal/ft^3
- 3.785 L/gal
- 454 g/lb
- 1000 mL/L
- 1000 mg/g
- 1 ft^3/sec (cfs) = 0.6465 MGD (million gallons per day)

Key point: Density (also called specific weight) is mass per unit volume and may be registered as pounds per cubic foot; pounds per gallon; grams per milliliter; or grams per cubic meter. If we take a fixed volume container, fill it with a fluid, and weigh it, we can determine density of the fluid (after subtracting the weight of the container).

- 8.34 lb/gal (water) — (density = 8.34 lb/gal)
- 1 mL of water weighs 1 g — (density = 1 g/mL)
- 62.4 lb/ft^3 (water) — (density = 8.34 lb/gal)
- 8.34 lb/gal = milligrams per liter (converts dosage in milligrams per liter into pounds per day per million gallons per day)

Example: 1 mg/L × 10 MGD × 8.3 = 83.4 lb/day

- 1 psi = 2.31 ft of water (head)
- 1 foot head = 0.433 psi
- °F = 9/5(°C + 32)
- °C = 5/9(°F − 32)
- Average water usage: 100 gal/capita/day (gpcd)
- Persons per single-family residence: 3.7

1.3.2   Water/Wastewater Conversion Examples

Use Table 1.4 and Table 1.5 to make the conversions indicated in the following example problems. Other conversions are presented in appropriate sections of the text.

Example 1.4

Convert cubic feet to gallons.

$$\text{Gallons} = \text{Cubic Feet, } ft^3 \times gal/ft^3$$

Sample problem:
   How many gallons of biosolids can be pumped to a digester that has 3600 ft³ of volume available?

$$\text{Gallons} = 3600 \ ft^3 \times 7.48 \ gal/ft^3 = 26{,}928 \ gal$$

Example 1.5

Convert gallons to cubic feet.

$$\text{Cubic Feet} = \frac{gal}{7.48 \ gal/ft^3}$$

Sample problem:
   How many cubic feet of biosolids are removed when 18,200 gal are withdrawn?

$$\text{Cubic Feet} = \frac{18{,}200 \ gal}{7.48 \ gal/ft^3} = 2433 \ ft^3$$

Example 1.6

Convert gallons to pounds.

$$\text{Pounds, lb} = gal \times 8.34 \ lb/gal$$

Sample problem:
   If 1650 gal of solids are removed from the primary settling tank, how many pounds of solids are removed?

$$\text{Pounds} = 1650 \ gal \times 8.34/gal = 13{,}761 \ lb$$

Example 1.7

Convert pounds to gallons.

$$\text{Gallons} = \frac{\text{lb}}{8.34 \text{ lb/gal}}$$

Sample problem:
How many gallons of water are required to fill a tank that holds 7540 lb of water?

$$\text{Gallons} = \frac{7540 \text{ lb}}{8.34 \text{ lb/gal}} = 904 \text{ gal}$$

Example 1.8

Convert milligrams per liter to pounds.

> *Key point*: For plant operations, concentrations in milligrams per liter or parts per million determined by laboratory testing must be converted to quantities of pounds, kilograms, pounds per day, or kilograms per day.

$$\text{Pounds} = \text{Concentration, mg/L} \times \text{Volume, MG} \times 8.34 \text{ lb/mg/L/MG}$$

Sample problem:
The solids concentration in the aeration tank is 2580 mg/L. The aeration tank volume is 0.95 MG. How many pounds of solids are in the tank?

$$\text{Pounds} = 2580 \text{ mg/L} \times 0.95 \text{ MG} \times 8.34 \text{ lb/mg/L/MG} = 20,441.3 \text{ lb}$$

Example 1.9

Convert milligrams per liter to pounds per day.

$$\text{Pounds/Day} = \text{Concentration, mg/L} \times \text{Flow, MGD} \times 8.34 \text{ lb/mg/L/MG}$$

Sample problem:
How many pounds of solids are discharged per day when the plant effluent flow rate is 4.75 MGD and the effluent solids concentration is 26 mg/L?

$$\text{Pounds/Day} = 26 \text{ mg/L} \times 4.75 \text{ MGD} \times 8.34 \text{ lb/mg/L/MG} = 1030 \text{ lb/day}$$

Example 1.10

Convert milligrams per liter to kilograms per day.

$$\text{Kg/Day} = \text{Concentration, mg/L} \times \text{Volume, MG} \times 3.785 \text{ kg/mg/L/MG}$$

Sample problem:
   The effluent contains 26 mg/L of BOD_5. How many kilograms per day of BOD_5 are discharged when the effluent flow rate is 9.5 MGD?

$$Kg/Day = 26 \text{ mg/L} \times 9.5 \text{ MG} \times 3.785 \text{ kg/mg/L/MG} = 934 \text{ kg/day}$$

Example 1.11

Convert pounds to milligrams per liter.

$$Concentration, \text{ mg/L} = \frac{Quantity, \text{ lb}}{Volume, \text{ MG} \times 8.34 \text{ lb/mg/L/MG}}$$

Sample problem:
   The aeration tank contains 89,990 lb of solids. The volume of the aeration tank is 4.45 MG. What is the concentration of solids in the aeration tank in milligrams per liter?

$$Concentration, \text{ mg/L} = \frac{89,990 \text{ lb}}{4.45 \text{ MG} \times 8.34 \text{ lb/mg/L/MG}} = 2425 \text{ mg/L}$$

Example 1.12

Convert pounds per day to milligrams per liter.

$$Concentration, \text{ mg/L} = \frac{Quantity, \text{ lb/day}}{Volume, \text{ MG} \times 8.34 \text{ lb/mg/L/MG}}$$

Sample problem:
   The disinfecting process uses 4820 lb per day of chlorine to disinfect a flow of 25.2 MGD. What is the concentration of chlorine applied to the effluent?

$$Concentration, \text{ mg/L} = \frac{4820}{25.2 \text{ MGD} \times 8.34 \text{ lb/mg/L/MG}} = 22.9 \text{ mg/L}$$

Example 1.13

Convert pounds to flow in million gallons per day.

$$Flow = \frac{Quantity, \text{ lb/day}}{Concentration, \text{ mg/L} \times 8.34 \text{ lb/mg/L/MG}}$$

Sample problem:
   Per day, 9640 lb of solids must be removed from the activated biosolids process. The waste-activated biosolids concentration is 7699 mg/L. How many million gallons per day of waste-activated biosolids must be removed?

$$\text{Flow} = \frac{9640 \text{ lb}}{7699 \text{ mg/L} \times 8.34 \text{ lb/MG/mg/L}} = 0.15 \text{ MGD}$$

Example 1.14

Convert million gallons per day to gallons per minute (gpm).

$$\text{Flow} = \frac{\text{Flow, MGD} \times 1{,}000{,}000 \text{ gal/MG}}{1440 \text{ min/day}}$$

Sample problem:
   The current flow rate is 5.55 MGD. What is the flow rate in gallons per minute?

$$\text{Flow} = \frac{5.55 \text{ MGD} \times 1{,}000{,}000 \text{ gal/MG}}{1440 \text{ min/day}} = 3854 \text{ gpm}$$

Example 1.15

Convert million gallons per day to gallons per day (gpd).

$$\text{Flow} = \text{Flow, MGD} \times 1{,}000{,}000 \text{ gal/MG}$$

Sample problem:
   The influent meter reads 28.8 MGD. What is the current flow rate in gallons per day?

$$\text{Flow} = 28.8 \text{ MGD} \times 1{,}000{,}000 \text{ gal/MG} = 28{,}800{,}000 \text{ gpd}$$

Example 1.16

Convert million gallons per day to cubic feet per second (cfs).

$$\text{Flow, cfs} = \text{Flow, MGD} \times 1.55 \text{ ft}^3/\text{sec/MGD}$$

Sample problem:
   The flow rate entering the grit channel is 2.89 MGD. What is the flow rate in cubic feet per second?

$$\text{Flow} = 2.89 \text{ MGD} \times 1.55 \text{ ft}^3/\text{sec/MGD} = 4.48 \text{ ft}^3/\text{sec}$$

Example 1.17

Convert gallons per minute to million gallons per day.

$$\text{Flow, MGD} \frac{\text{Flow, gpm} \times 1440 \text{ min/day}}{1{,}000{,}000 \text{ gal/MG}}$$

Sample problem:

The flow meter indicates that the current flow rate is 1469 gpm. What is the flow rate in million gallons per day?

$$\text{Flow, MGD } \frac{1469 \text{ gpm} \times 1440 \text{ min/day}}{1,000,000 \text{ gal/MG}} = 2.12 \text{ MGD (rounded)}$$

Example 1.18

Convert gallons per day to million gallons per day.

$$\text{Flow, MGD } \frac{\text{Flow, gal/day}}{1,000,000 \text{ gal/MG}}$$

Sample problem:

The totalizing flow meter indicates that 33,444,950 gal of wastewater have entered the plant in the past 24 h. What is the flow rate in million gallons per day?

$$\text{Flow, MGD } \frac{33,444,950 \text{ gal/day}}{1,000,000 \text{ gal/MG}} = 33.44 \text{ MGD}$$

Example 1.19

Convert flow in cubic feet per second to million gallons per day.

$$\text{Flow, MGD } \frac{\text{Flow, ft}^3/\text{sec}}{1.55 \text{ ft}^3/\text{sec/MG}}$$

Sample problem:

The flow in a channel is determined to be 3.89 ft³/sec. What is the flow rate in million gallons per day?

$$\text{Flow, MGD } \frac{3.89 \text{ ft}^3/\text{sec}}{1.55 \text{ ft}^3/\text{sec/MG}} = 2.5 \text{ MGD}$$

Example 1.20

Problem:

The water in a tank weighs 675 lb. How many gallons does the tank hold?

Solution:

Water weighs 8.34 lb/gal. Therefore:

$$\frac{675 \text{ lb}}{8.34 \text{ lb/gal}} = 80.9 \text{ gallons}$$

Example 1.21

Problem:
A liquid chemical weighs 62 lb/ft³. How much does a 5-gal can of it weigh?

Solution:
Solve for specific gravity; get pounds per gallon; multiply by 5.

$$\text{Specific Gravity} = \frac{\text{wt. channel}}{\text{wt. water}}$$

$$\frac{62 \text{ lb/ft}^3}{62.4 \text{ lb/ft}^3} = .99$$

$$\text{Specific Gravity} = \frac{\text{wt. channel}}{\text{wt. water}}$$

$$.99 = \frac{\text{wt. chemical}}{8.34 \text{ lb/gal}}$$

$$8.26 \text{ lb/gal} = \text{wt. chemical}$$

$$8.26 \text{ lb/gal} \times 5 \text{ gal} = 41.3 \text{ lb}$$

Example 1.22

Problem:
A wooden piling with a diameter of 16 in. and a length of 16 ft weighs 50 lb/ft³. If it is inserted vertically into a body of water, what vertical force is required to hold it below the water surface?

Solution:
If this piling had the same weight as water, it would rest just barely submerged. Find the difference between its weight and that of the same volume of water. This is the weight needed to keep it down.

$$
\begin{array}{r}
62.4 \text{ lb/ft}^3 \text{ (water)} \\
-50.0 \text{ lb/ft}^3 \text{ (piling)} \\
\hline
12.4 \text{ lb/ft}^3 \text{ difference}
\end{array}
$$

$$\text{Volume of piling} = .785 \times 1.33^2 \times 16 \text{ ft} = 22.21 \text{ ft}^3$$

$$12.4 \text{ lb/ft}^3 \times 22.21 \text{ ft}^3 = 275.4 \text{ lb (needed to hold piling below water surface)}$$

Example 1.23

Problem:

A liquid chemical with a specific gravity (SG) of 1.22 is pumped at a rate of 40 gpm. How many pounds per day is the pump delivering?

Solution:

Solve for pounds pumped per minute; change to pounds per day.

$$8.34 \text{ lb/gal water} \times 1.22 \text{ SG liquid chemical} = 10.2 \text{ lb/gal liquid}$$

$$40 \text{ gpm} \times 10.2 \text{ lb/gal} = 408 \text{ lb/min}$$

$$408 \text{ lb/min} \times 1440 \text{ min/day} = 587,520 \text{ lb/day}$$

Example 1.24

Problem:

A cinder block weighs 70 lb in air. When immersed in water, it weighs 40 lb. What is the volume and specific gravity of the cinder block?

Solution:

The cinder block displaces 30 lb of water; solve for cubic feet of water displaced (equivalent to volume of cinder block).

$$\frac{30 \text{ lb water displaced}}{62.4 \text{ lb/ft}^3} = .48 \text{ ft}^3 \text{ water displaced}$$

Cinder block volume = 0.48 ft³; this weighs 70 lb.

$$\frac{70 \text{ lb}}{.48 \text{ ft}^3} = 145.8 \text{ lb/ft}^3 \text{ density of cinder block}$$

$$\text{Specific Gravity} = \frac{\text{density of cinder block}}{\text{density of water}}$$

$$= \frac{\text{density of cinder block}}{\text{density of water}} = 2.34$$

1.3.3 Temperature Conversions

Two methods are commonly used to make temperature conversions. We have already demonstrated the following method:

- °F = 9/5(°C + 32)
- °C = 5/9(°F − 32)

Example 1.25

Problem:

At a temperature of 4°C, water is at its greatest density. What is the degree in Fahrenheit?

Solution:

$$°F = 9/5(°C + 32)$$

$$= 4 \times 9/5 + 32$$

$$= 7.2 + 32$$

$$= 39.2$$

However, the difficulty arises when one tries to recall these formulae from memory. Probably the easiest way to recall them is to remember three basic steps for Fahrenheit and Celsius conversions:

- Add 40°.
- Multiply by the appropriate fraction (5/9 or 9/5).
- Subtract 40°.

Obviously, the only variable in this method is the choice of 5/9 or 9/5 in the multiplication step. To make the proper choice, you must be familiar with the two scales. The freezing point of water is 32° on the Fahrenheit scale and 0° on the Celsius scale. The boiling point of water is 212° on the Fahrenheit scale and 100° on the Celsius scale.

> *Key point*: Note, for example, that at the same temperature, higher numbers are associated with the Fahrenheit scale and lower numbers with the Celsius scale. This important relationship helps you decide whether to multiply by 5/9 or 9/5.

Now look at a few conversion problems to see how the three-step process works.

Example 1.26

Suppose that we wish to convert 240°F to Celsius. Using the three-step process, we proceed as follows:

Step 1. Add 40°:

$$240° + 40° = 280°$$

Step 2. 280° must be multiplied by 5/9 or 9/5. Because the conversion is to the Celsius scale, we will be moving to a number *smaller* than 280. Through reason and observation, obviously, if 280 were multiplied by 9/5, the result would be almost the same as multiplying by 2, which would double 280 rather than make it smaller. If we multiply by 5/9, the result will be about the same as multiplying by ½, which would cut 280 in half. Because in this problem we wish to move to a smaller number, we should multiply by 5/9:

$$(5/9)(280°) = 156.0°C$$

Step 3. Now subtract 40°:

$$156.0°C - 40.0°C = 116.0°C$$

Therefore, $240°F = 116.0°C$.

Example 1.27

Convert $22°C$ to Fahrenheit.

Step 1. Add 40°:

$$22° + 40° = 62°$$

Step 2. Because we are converting from Celsius to Fahrenheit, we are moving from a smaller to a larger number, so 9/5 should be used in the multiplication:

$$(9/5)(62°) = 112°$$

Step 3. Subtract 40°:

$$112° - 40° = 72°$$

Thus, $22°C = 72°F$.

Obviously, knowing how to make these temperature conversion calculations is useful. However, in practical *in situ* or non-*in situ* operations, you may wish to use a temperature conversion table.

1.4 CONVERSION FACTORS: AIR POLLUTION MEASUREMENTS

The recommended units for reporting air pollutant emissions are commonly stated in metric system whole numbers. If possible, the reported units should be the same as those that are actually measured. For example, weight should be recorded in grams; volume of air should be recorded in cubic meters. When the analytical system is calibrated in one unit, the emissions should also be reported in the units of the calibration standard. For example, if a gas chromatograph is calibrated with a 1-ppm (parts per million) standard of toluene in air, then the emissions monitored by the system should also be reported in parts per million. Finally, if the emission standard is defined in a specific unit, the monitoring system should be selected to monitor in that unit.

The preferred reporting units for the following types of emissions should be:

- Nonmethane organic and volatile organic compound emissions ppm; ppb
- Semivolatile organic compound emissions $\mu g/m^3$; mg/m^3
- Particulate matter (TSP/PM-10) emissions $\mu g/m^3$
- Metal compound emissions ng/m^3

1.4.1 Conversion from Parts per Million to Micrograms per Cubic Meter

Often, environmental engineers must be able to convert from parts per million to micrograms per cubic meter. Following is an example of how to perform that conversion, using sulfur dioxide (SO_2) as the monitored constituent.

Example 1.28

The expression *parts per million* is without dimensions (no units of weight or volume are specifically designated). Using the format of other units, the expression may be written:

$$\frac{parts}{million\ parts}$$

"Parts" are not defined. If cubic centimeters replace parts, we obtain:

$$\frac{cubic\ centimeters}{million\ cubic\ centimeters}$$

Similarly, we might write pounds per million pounds, tons per million tons, or liters per million liters. In each expression, identical units of weight or volume appear in the numerator and the denominator and may be canceled out, leaving a dimensionless term. An analog of parts per million is the more familiar term *percent*. Percent can be written:

$$\frac{parts}{hundred\ parts}$$

To convert from parts per million by volume, ppm (μL/L), to micrograms per cubic meter (μg/m^3) at the Environmental Protection Agency (EPA) standard temperature (25°C) and standard pressure (760 mm Hg) STP, we must know the molar volume at the given temperature and pressure and the molecular weight of the pollutant. At 25°C and 760 mm Hg, 1 mol of any gas occupies 24.46 L.

Problem:
The atmospheric concentration was reported as 2.5 ppm by volume of sulfur dioxide (SO_2). What is this concentration in micrograms (μg) per cubic meter (m^3) at 25°C and 760 mmHg? What is the concentration in micrograms per cubic meter at 37°C and 752 mmHg?

> *Note*: The following example problem points out the need for reporting temperature and pressure when the results are present on a weight-to-volume basis.

Solution:
Let parts per million equal microliters per liter; then, 2.5 ppm = 2.5 μL/L. The molar volume at 25°C and 760 mm Hg is 24.46 L and the molecular weight of SO_2 is 64.1 g/mol.

Step 1. 25°C and 760 mm Hg:

$$\frac{2.5\ \mu L}{L} \times \frac{1\ \mu mol}{24.46\ \mu L} \times \frac{64.1\ \mu g}{\mu mol} \times \frac{1000\ L}{m^3} = \frac{6.6 \times 10^3\ \mu g}{m^3}\ \text{at STP}$$

Step 2. 37°C and 752 mm Hg:

$$24.46\ \mu L \left(\frac{310\ K}{298\ K}\right)\left(\frac{760\ mm\ Hg}{752\ mm\ Hg}\right) = 25.72\ \mu L$$

$$\frac{2.5\ \mu L}{L} \times \frac{1\ \mu mol}{25.72\ \mu L} \times \frac{64.1\ \mu g}{\mu mol} \times \frac{1000\ L}{m^3} = \frac{6.2 \times 10^3 \mu g}{m^3}\ \text{at } 37°, 752\ \text{mm Hg}$$

1.4.2   Conversion Tables for Common Air Pollution Measurements

To assist the environmental engineer in converting from one set of units to another, we supply the following conversion factors for common air pollution measurements and other useful information. The conversion tables provide factors for:

- Atmospheric gases
- Atmospheric pressure
- Gas velocity
- Concentration
- Atmospheric particulate matter

Following is a list of conversions from parts per million to micrograms per cubic meters (at 25°C and 760 mmHg) for several common air pollutants:

ppm SO_2 × 2620 = $\mu g/m^3$ SO_2 (sulfur dioxide)
ppm CO × 1150 = $\mu g/m^3$ CO (carbon monoxide)
ppm CO_x × 1.15 = mg/m^3 CO (carbon dioxide)
ppm CO_2 × 1.8 = mg/m^3 CO_2 (carbon dioxide)
ppm NO × 1230 = $\mu g/m^3$ NO (nitrogen oxide)
ppm NO_2 × 1880 = $\mu g/m^3$ NO_2 (nitrogen dioxide)
ppm O_2 × 1960 = $\mu g/m^3$ O_3 (ozone)
ppm CH_4 × 655 = $\mu g/m^3$ CH_4 (methane)
ppm CH_4 × 655 = mg/m^3 CH_4 (methane)
ppm CH_3SH × 2000 = $\mu g/m^3$ CH_3SH (methyl mercaptan)
ppm C_3H_8 × 1800 = $\mu g/m^3$ C_3H_8 (propane)
ppm C_3H_8 × 1.8 = mg/m^3 C_3H_8 (propane)
ppm F^- × 790 = $\mu g/m^3$ F^- (fluoride)
ppm H_2S × 1400 = $\mu g/m^3$ H_2S (hydrogen sulfide)
ppm NH_3 × 696 = $\mu g/m^3$ NH_3 (ammonia)
ppm $HCHO$ × 1230 = $\mu g/m^3$ $HCHO$ (formaldehyde)

Table 1.6 through Table 1.10 show various conversion calculations.

1.5   SOIL TEST RESULTS CONVERSION FACTORS

Soil test results can be converted from parts per million to pounds per acre by multiplying parts per million by a conversion factor based on the depth to which the soil was sampled. Because a slice of soil 1 acre in area and 3 in. deep weighs approximately 1 million lb, we can use the conversion factors given in Table 1.11.

1.6   CONCLUSION

The conversions presented here are equations that will be used constantly — daily. Although conversion tables, graphs, and charts are useful tools for efficiency, knowing the factors behind the charts is essential to understanding what one is doing.

Table 1.6 Atmospheric Gases

To convert from	To	Multiply by
Milligram per cubic meter	Micrograms per cubic meter	1000.0
	Micrograms per liter	1.0
	Parts per million by volume (20°C)	$\dfrac{24.04}{M}$
	Parts per million by weight	0.8347
	Pounds per cubic foot	62.43×10^{-9}
Micrograms per cubic foot	Milligrams per cubic foot	0.001
	Parts per million by volume (20°C)	$\dfrac{0.02404}{M}$
	Parts per million by weight	834.7×10^{-6}
		62.43×10^{-12}
Micrograms per liter	Milligrams per cubic meter	1.0
	Micrograms per cubic meter	1000.0
	Parts per million by volume (20°C)	$\dfrac{24.04}{M}$
	Parts per million by weight	0.8347
	Pounds per cubic foot	62.43×10^{-9}
Parts per million by volume (20°C)	Milligrams per cubic meter	$\dfrac{M}{24.04}$
	Micrograms per cubic meter	$\dfrac{M}{0.02404}$
	Micrograms per liter	$\dfrac{M}{24.04}$
	Parts per million by weight	$\dfrac{M}{28.8}$
	Pounds per cubic foot	$\dfrac{M}{385.1 \times 10^{6}}$
Parts per million by weight	Milligrams per cubic meter	1.198
	Micrograms per cubic meter	1.198×10^{3}
	Micrograms per liter	1.198
	Parts per million by volume (20°C)	$\dfrac{28.8}{M}$
	Pounds per cubic foot	7.48×10^{-6}
Pounds per cubic foot	Milligrams per cubic meter	16.018×10^{6}
	Micrograms per cubic meter	16.018×10^{9}
	Micrograms per liter	16.018×10^{6}
	Parts per million by volume (20°C)	$\dfrac{385.1 \times 10^{6}}{M}$
	Parts per million by weight	133.7×10^{3}

Table 1.7   Atmospheric Pressure

To convert from	To	Multiply by
Atmospheres	Millimeters of mercury	760.0
	Inches of mercury	29.92
	Millibars	1013.2
Millimeters of mercury	Atmospheres	1.316×10^{-3}
	Inches of mercury	39.37×10^{-3}
	Millibars	1.333
Inches of mercury	Atmospheres	0.03333
	Millimeters of mercury	25.4005
	Millibars	33.35
Millibars	Atmospheres	0.000987
	Millimeters of mercury	0.75
	Inches of mercury	0.30

Sampling Pressures

Millimeters of mercury	Inches of water (60°C)	0.5358
	(0°C)	
Inches of mercury	Inches of water (60°C)	13.609
	(0°C)	
Inches of water	Millimeters of mercury (0°C)	1.8663
	Inches of mercury (0°C)	73.48×10^{-2}

Table 1.8   Velocity

To convert from	To	Multiply by
Meters per second	Kilometers per hour	3.6
	Feet per second	3.281
	Miles per hour	2.237
Kilometers per hour	Meters per second	0.2778
	Feet per second	0.9113
	Miles per hour	0.6241
Feet per hour	Meters per second	0.3048
	Kilometers per hour	1.0973
	Miles per hour	0.6818
Miles per hour	Meters per second	0.4470
	Kilometers per hour	1.6093
	Feet per second	1.4667

Table 1.9   Atmospheric Particulate Matter

To convert from	To	Multiply by
Milligrams per cubic meter	Grams per cubic foot	283.2×10^{-6}
	Grams per cubic meter	0.001
	Micrograms per cubic meter	1000.0
	Monograms per cubic foot	28.32
	Pounds per 1000 ft^3	62.43×10^{-6}
Grams per cubic foot	Milligrams per cubic meter	35.3145×10^{3}
	Grams per cubic meter	35.314
	Micrograms per cubic meter	35.314×10^{3}
	Micrograms per cubic foot	1.0×10^{6}
	Pounds per 1000 ft^3	2.2046

Table 1.10   Concentration

To convert from	To	Multiply by
Grams per cubic meter	Milligrams per cubic meter	1000.0
	Grams per cubic foot	0.02832
	Micrograms per cubic foot	1.0×10^6
	Pounds per 1000 cubic foot	0.06243
Micrograms per cubic meter	Milligrams per cubic meter	0.001
	Grams per cubic foot	28.43×10^{-9}
	Grams per cubic meter	1.0×10^{-6}
	Micrograms per cubic foot	0.02832
	Pounds per 1000 ft^3	62.43×10^{-9}
Micrograms per cubic foot	Milligrams per cubic meter	35.314×10^{-3}
	Grams per cubic foot	1.0×10^{-6}
	Grams per cubic meter	35.314×10^{-6}
	Micrograms	35.314
	Pounds per 1000 ft^3	2.2046×10^{-6}
Pounds per 1000 ft^3	Milligrams per cubic meter	16.018×10^3
	Grams per cubic foot	0.35314
	Micrograms per cubic meter	16.018×10^6
	Grams per cubic meter	16.018
	Micrograms per cubic foot	353.14×10^2

Table 1.11   Soil Test Conversion Factors

Soil sample depth (inches)	Multiply parts per million by
3	1
6	2
7	2.33
8	2.66
9	3
10	3.33
12	4

CHAPTER **2**

Basic Math Operations

2.1 INTRODUCTION

Most calculations required by environmental engineers (as with many others) start with the basics, such as addition, subtraction, multiplication, division, and sequence of operations. Although many of the operations are fundamental tools within each environmental engineer's toolbox, using these tools on a consistent basis is important in order to remain sharp in their use. Engineers should master basic math definitions and the formation of problems; daily operations require calculation of percentage; average; simple ratio; geometric dimensions; threshold odor number; force; pressure; and head, as well as the use of dimensional analysis and advanced math operations.

2.2 BASIC MATH TERMINOLOGY AND DEFINITIONS

The following basic definitions will aid in understanding the material in this chapter.

> *Integer* or *integral number*: a whole number. Thus 1, 2, 3, 4, 5, 6, 7, 8, 9, 10, 11, and 12 are the first 12 positive integers.
> *Factor* or *divisor* of a whole number: any other whole number that exactly divides it. Thus, 2 and 5 are factors of 10.
> *Prime number:* a number that has no factors except itself and 1. Examples of prime numbers are 1, 3, 5, 7, and 11.
> *Composite number:* a number that has factors other than itself and 1. Examples of composite numbers are 4, 6, 8, 9, and 12.
> *Common factor* or *common divisor* of two or more numbers: a factor that will exactly divide each of the numbers. If this factor is the largest factor possible, it is called the *greatest common divisor.* Thus, 3 is a common divisor of 9 and 27, but 9 is the greatest common divisor of 9 and 27.
> *Multiple* of a given number: a number that is exactly divisible by the given number. If a number is exactly divisible by two or more other numbers, it is their common multiple. The least (smallest) such number is called the *lowest common multiple.* Thus, 36 and 72 are common multiples of 12, 9, and 4; however, 36 is the lowest common multiple.
> *Even number:* a number exactly divisible by 2. Thus, 2, 4, 6, 8, 10, and 12 are even integers.
> *Odd number:* an integer that is not exactly divisible by 2. Thus, 1, 3, 5, 7, 9, and 11 are odd integers.
> *Product:* the result of multiplying two or more numbers together. Thus, 25 is the product of 5×5, and 4 and 5 are factors of the product 20.
> *Quotient:* the result of dividing one number by another. For example, 5 is the quotient of 20 divided by 4.
> *Dividend:* a number to be divided; a *divisor* is a number that divides. For example, in $100 \div 20 = 5$, 100 is the dividend, 20 is the divisor, and 5 is the quotient.

Area: the area of an object, measured in square units — the amount of surface an object contains or the amount of material required to cover the surface.

Base: a term used to identify the bottom leg of a triangle, measured in linear units.

Circumference: the distance around an object, measured in linear units. When determined for other than circles, it may be called the *perimeter* of the figure, object, or landscape.

Cubic units: measurements used to express volume: cubic feet, cubic meters, and so on.

Depth: the vertical distance from the bottom of the unit to the top. This is normally measured in terms of liquid depth and given in terms of sidewall depth (SWD), measured in linear units.

Diameter: the distance from one edge of a circle to the opposite edge, passing through the center, measured in linear units.

Height: the vertical distance from the base or bottom of a unit to the top or surface.

Linear units: measurements used to express distances: feet, inches, meters, yards, and so on.

Pi, (π): a number in the calculations involving circles, spheres, or cones: $\pi = 3.14$.

Radius: the distance from the center of a circle to the edge, measured in linear units.

Sphere: a container shaped like a ball.

Square units: measurements used to express area: square feet, square meters, acres, and so on.

Volume: the capacity of the unit (how much it will hold) measured in cubic units (cubic feet, cubic meters) or in liquid volume units (gallons, liters, million gallons).

Width: the distance from one side of the unit to the other, measured in linear units.

Key words:

- *Of* means to multiply.
- *And* means to add.
- *Per* means to divide.
- *Less than* means to subtract.

2.3   SEQUENCE OF OPERATIONS

Mathematical operations such as addition, subtraction, multiplication, and division are usually performed in a certain order or sequence. Typically, multiplication and division operations are done prior to addition and subtraction operations. In addition, mathematical operations are also generally performed from left to right, using this hierarchy. Parentheses are commonly used to set apart operations that should be performed in a particular sequence.

Note: We assume that the reader has a fundamental knowledge of basic arithmetic and math operations. Thus, the purpose of the following subsection is to provide a brief review of the mathematical concepts and applications frequently employed by environmental engineers.

2.3.1   Sequence of Operations — Rules

Rule 1: In a series of additions, the terms may be placed in any order and grouped in any way. Thus,

$$4 + 3 = 7 \text{ and } 3 + 4 = 7; (4 + 3) + (6 + 4) = 17, (6 + 3) + (4 + 4) = 17,$$

$$\text{and } [6 + (3 + 4) + 4] = 17.$$

Rule 2: In a series of subtractions, changing the order or the grouping of the terms may change the result. Thus,

$$100 - 30 = 70, \text{ but } 30 - 100 = -70; (100 - 30) - 10 = 60, \text{ but } 100 - (30 - 10) = 80.$$

Rule 3: When no grouping is given, the subtractions are performed in the order written, from left to right. Thus,

$$100 - 30 - 15 - 4 = 51; \text{ or by steps, } 100 - 30 = 70, 70 - 15 = 55, 55 - 4 = 51.$$

Rule 4: In a series of multiplications, the factors may be placed in any order and in any grouping. Thus,

$$[(2 \times 3) \times 5] \times 6 = 180 \text{ and } 5 \times [2 \times (6 \times 3)] = 180$$

Rule 5: In a series of divisions, changing the order or the grouping may change the result. Thus,

$$100 \div 10 = 10, \text{ but } 10 \div 100 = 0.1; (100 \div 10) \div 2 = 5, \text{ but } 100 \div (10 + 2) = 20.$$

Again, if no grouping is indicated, the divisions are performed in the order written, from left to right. Thus, $100 \div 10 \div 2$ is understood to mean $(100 \div 10) \div 2$.

Rule 6: In a series of mixed mathematical operations, the convention is that whenever no grouping is given, multiplications and divisions are to be performed in the order written, and then additions and subtractions in the order written.

2.3.2 Sequence of Operations — Examples

In a series of additions, the terms may be placed in any order and grouped in any way. Examples:

$$3 + 6 = 10 \text{ and } 6 + 4 = 10$$

$$(4 + 5) + (3 + 7) = 19, (3 + 5) + (4 + 7) = 19, \text{ and } [7 + (5 + 4)] + 3 = 19$$

In a series of subtractions, changing the order or the grouping of the terms may change the result. Examples:

$$100 - 20 = 80, \text{ but } 20 - 100 = -80$$

$$(100 - 30) - 20 = 50, \text{ but } 100 - (30 - 20) = 90$$

When no grouping is given, the subtractions are performed in the order written — from left to right. Example:

$$100 - 30 - 20 - 3 = 47$$

or by steps,

$$100 - 30 = 70, 70 - 20 = 50, 50 - 3 = 47$$

In a series of multiplications, the factors may be placed in any order and in any grouping. Example:

$$[(3 \times 3) \times 5] \times 6 = 270 \text{ and } 5 \times [3 \times (6 \times 3)] = 270$$

In a series of divisions, changing the order or the grouping may change the result. Examples:

$$100 \div 10 = 10, \text{ but } 10 \div 100 = 0.1$$

$$(100 \div 10) \div 2 = 5, \text{ but } 100 \div (10 \div 2) = 20$$

If no grouping is indicated, the divisions are performed in the order written — from left to right. Example:

$$100 \div 5 \div 2 \text{ is understood to mean } (100 \div 5) \div 2$$

In a series of mixed mathematical operations, the rule of thumb is that whenever no grouping is given, multiplications and divisions are performed in the order written, and then additions and subtractions in the order written.

2.4   PERCENT

The word "percent" means "by the hundred." Percentage is often designated by the symbol "%." Thus, 15% means 15 percent or 15/100 or 0.15. These equivalents may be written in the reverse order: 0.15 = 15/100 = 15%. In environmental engineering (water/wastewater treatment, for example), percent is frequently used to express plant performance and for control of biosolids treatment processes. When working with percent, the following key points are important:

- Percents are another way of expressing a part of a whole.
- As mentioned, percent means "by the hundred," so a percentage is the number out of 100. To determine percent, divide the quantity to be expressed as a percent by the total quantity, then multiply by 100:

$$\text{Percent } (\%) = \frac{\text{Part}}{\text{Whole}} \qquad (2.1)$$

   For example, 22 percent (or 22%) means 22 out of 100, or 22/100. Dividing 22 by 100 results in the decimal 0.22:

$$22\% = \frac{22}{100} = 0.22$$

- When using percentage in calculations (for example, to calculate hypochlorite dosages when the percent of available chlorine must be considered), the percentage must be converted to an equivalent decimal number; this is accomplished by dividing the percentage by 100.
   For example, calcium hypochlorite (HTH) contains 65% available chlorine. What is the decimal equivalent of 65%? Because 65% means 65 per 100, divide 65 by 100: 65/100 is 0.65.
- Decimals and fractions can be converted to percentages. The fraction is first converted to a decimal, and then the decimal is multiplied by 100 to get the percentage.
   For example, If a 50-ft high water tank has 26 ft of water in it, how full is the tank in terms of percentage of its capacity?

$$\frac{26 \text{ ft}}{50 \text{ ft}} = 0.52 \text{ (decimal equivalent)}$$

$$0.52 \times 100 = 52$$

The tank is 52% full.

Example 2.1

Problem:

The plant operator removes 6500 gal of biosolids from the settling tank. The biosolids contain 325 gal of solids. What is the percent solids in the biosolids?

Solution:

$$\text{Percent} = \frac{325 \text{ gal}}{6500 \text{ gal}} \times 100 = 5\%$$

Example 2.2

Problem:

Convert 65% to decimal percent.

Solution:

$$\text{Decimal percent} = \frac{\text{Percent}}{100}$$

$$\frac{65}{100} = 0.65$$

Example 2.3

Problem:

Biosolids contains 5.8% solids. What is the concentration of solids in decimal percent?

Solution:

$$\text{Decimal percent} = \frac{5.8\%}{100} = 0.058$$

Key point: Unless otherwise noted, all calculations in the text using percent values require the percent to be converted to a decimal before use.

Key point: To determine what quantity a percent equals, first convert the percent to a decimal, then multiply by the total quantity.

$$\text{Quantity} = \text{Total} \times \text{Decimal Percent} \qquad (2.2)$$

Example 2.4

Problem:

Biosolids drawn from the settling tank are 5% solids. If 2800 gal of biosolids are withdrawn, how many gallons of solids are removed?

Solution:

$$\frac{5\%}{100} \times 2800 \text{ gal} = 140 \text{ gal}$$

Example 2.5

Problem:
   Convert 0.55 to percent.

Solution:

$$0.55 = \frac{55}{100} = 0.55 = 55\%$$

In converting 0.55 to 55%, we simply moved the decimal point two places to the right.

Example 2.6

Problem:
   Convert 7/22 to a percent.

Solution:

$$\frac{7}{22} = 0.318 = 0.318 \times 100 = 31.8\%$$

Example 2.7

Problem:
   What is the percentage of 3 ppm?

   Key point: Because 1 L of water weighs 1 kg (1000 g = 1,000,000 mg), milligrams per liter is parts per million (ppm).

Solution:
   Since 3 ppm = 3 mg/L,

$$3 \text{ mg/L} = \frac{3 \text{ mg}}{1 \text{ L} \times 1,000,000 \text{ mg/L}} \times 100\%$$

$$\frac{3}{10,000}\% = 0.0003\%$$

Example 2.8

Problem:
   How many milligrams per liter is a 1.4% solution?

Solution:

$$1.4 = \frac{1.4}{100}$$

Since the weight of 1 L water is 10^6 mg,

$$\frac{1.4}{100} \times 1,000,000 \text{ mg/L} = 14,000 \text{ mg/L}$$

Example 2.9

Problem:
Calculate pounds per MG (million gallons) for 1 ppm (1 mg/L) of water.

Solution:
Since 1 gal of water = 8.34 lb,

$$1 \text{ ppm} \frac{1 \text{ gal}}{10^6 \text{ gal}}$$

$$\frac{1 \text{ gal} \times 8.34 \text{ lb/gal}}{\text{MG}} = 8.34 \text{ lb/MG}$$

Example 2.10

Problem:
How many pounds of activated carbon (AC) should be added to 42 lb of sand for a mixture that contains 26% of AC?

Solution:
Let x be the weight of AC:

$$\frac{x}{42 + x} = 0.26$$

$$x = 0.26(42 + x)$$

$$x = 10.92 + 0.26x$$

$$(1 - 0.26)x = 10.92$$

$$x = \frac{10.92}{0.74} = 14.76 \text{ lb}$$

Example 2.11

Problem:
   A pipe is laid at a rise of 140 mm in 22 m. What is the grade?

Solution:

$$\text{Grade} = \frac{140 \text{ mm}}{22 \text{ m}} \times 100(\%)$$

$$= \frac{140 \text{ mm}}{22 \times 1000 \text{ mm}} \times 100\% = 0.64\%$$

Example 2.12

Problem:
   A motor is rated as 40 hp. However, the output horsepower of the motor is only 26.5 hp. What is the efficiency of the motor?

Solution:

$$\text{Efficiency} = \frac{\text{hp output}}{\text{hp input}} \times 100\%$$

$$= \frac{26.5 \text{ hp}}{40 \text{ hp}} \times 100\% = 66\%$$

2.5 SIGNIFICANT DIGITS

When rounding numbers, remember the following key points:

- Numbers are rounded to reduce the number of digits to the right of the decimal point. This is done for convenience, not for accuracy.
- Rule: a number is rounded off by dropping one or more numbers from the right and adding zeros if necessary to place the decimal point. If the last figure dropped is 5 or more, increase the last retained figure by 1. If the last digit dropped is less than 5, do not increase the last retained figure. If the digit 5 is dropped, round off the preceding digit to the nearest *even* number.

Example 2.13

Problem:
   Round off the following to one decimal: 34.73; 34.77; 34.75; 34.45; 34.35.

Solution:

   34.73 = 34.7
   34.77 = 34.8
   34.75 = 34.8
   34.45 = 34.4
   34.35 = 34.4

Example 2.14

Problem:
   Round off 10,546 to 4, 3, 2, and 1 significant figures.

Solution:

   10,546 = 10,550 to four significant figures
   10,546 = 10,500 to three significant figures
   10,546 = 11,000 to two significant figures
   10,546 = 10,000 to one significant figure

   In determining significant figures, remember the following key points:

- The concept of significant figures is related to rounding.
- Significant figures can be used to determine where to round off.
     Key point: No answer can be more accurate than the least accurate piece of data used to calculate the answer.
- Rule: significant figures are those numbers known to be reliable. The position of the decimal point does not determine the number of significant figures.

Example 2.15

Problem:
   How many significant figures are in a measurement of 1.35 in.?

Solution:
   Three significant figures: 1, 3, and 5.

Example 2.16

Problem:
   How many significant figures are in a measurement of 0.000135?

Solution:
   Again, three significant figures: 1, 3, and 5. The three zeros are used only to place the decimal point.

Example 2.17

Problem:
   How many significant figures are in a measurement of 103.500?

Solution:
   Four significant figures: 1, 0, 3, and 5. The remaining two zeros are used to place the decimal point.

Example 2.18

Problem:
   How many significant figures are in 27,000.0?

Solution:

Six significant figures: 2, 7, 0, 0, 0, 0. In this case, the.0 means that the measurement is precise to 1/10 unit. The zeros indicate measured values and are not used solely to place the decimal point.

2.6   POWERS AND EXPONENTS

In working with powers and exponents, important key points include:

- Powers are used to identify area (as in square feet) and volume (as in cubic feet).
- Powers can also be used to indicate that a number should be squared, cubed, etc. This latter designation is the number of times a number must be multiplied times itself. For example, when several numbers are multiplied together, such as $4 \times 5 \times 6 = 120$, the numbers, 4, 5, and 6 are the *factors*; 120 is the *product*.
- If all the factors are alike, such as $4 \times 4 \times 4 \times 4 = 256$, the product is called a *power*. Thus, 256 is a power of 4, and 4 is the *base* of the power. A power is a product obtained by using a base as a factor for a certain number of times.
- Instead of writing $4 \times 4 \times 4 \times 4$, it is more convenient to use an *exponent* to indicate that the factor 4 is used as a factor four times. This exponent (a small number placed above and to the right of the base number) indicates how many times the base is to be used as a factor. Using this system of notation, the multiplication $4 \times 4 \times 4 \times 4$ is written as 4^4. The "4" is the *exponent*, showing that 4 is to be used as a factor four times.
- These same considerations apply to letters (*a, b, x, y*, etc.) as well. For example:

$$z^2 \; = \; (z)\,(z) \; \text{ or } \; z^4 \; = \; (z)\,(z)\,(z)\,(z)$$

- When a number or letter does not have an exponent, it is considered to have an exponent of one.

The powers of 1:

$1^0 = 1$
$1^1 = 1$
$1^2 = 1$
$1^3 = 1$
$1^4 = 1$

The powers of 10:

$10^0 = 1$
$10^1 = 10$
$10^2 = 100$
$10^3 = 1000$
$10^4 = 10,000$

Example 2.19

Problem:

How is the term 2^3 written in expanded form?

Solution:

The power (exponent) of 3 means that the base number (2) is multiplied by itself three times:

$$2^3 \; = \; (2)(2)(2)$$

Example 2.20

Problem:
  How is the term $(3/8)^2$ written in expanded form?

Solution:

  Key point: When parentheses are used, the exponent refers to the entire term within the parentheses. Thus, in this example, $(3/8)^2$ means:

$$(3/8)^2 = (3/8)(3/8)$$

  Key point: When a negative exponent is used with a number or term, the number can be re-expressed using a positive exponent:

$$6^{-3} = 1/6^3$$

  Another example is

$$11^{-5} = 1/11^5$$

Example 2.21

Problem:
  How is the term 8^{-3} written in expanded form?

Solution:

$$8^{-3} = \frac{1}{8^3} = \frac{1}{(8)(8)(8)}$$

  Key point: Any number or letter such as 3^0 or X^0 does not equal 3×1 or $X \times 1$, but simply 1.

2.7 AVERAGES (ARITHMETIC MEAN)

Whether we speak of harmonic mean, geometric mean, or arithmetic mean, each is designed to find the "center," or the "middle," of a set of numbers. They capture the intuitive notion of a central tendency that may be present in the data. In statistical analysis, an average of data is a number that indicates the middle of the distribution of data values.

  An *average* is a way of representing several different measurements as a single number. Although averages can be useful by telling approximately how much or how many, they can also be misleading, as we demonstrate next. Two kinds of averages can be found in environmental engineering calculations: the *arithmetic* mean (or simply *mean*) and the *median*.

Example 2.22

Problem:
  When working with averages, the mean (again, what we usually refer to as an average) is the total of values of a set of observations divided by the number of observations. We simply add up

all of the individual measurements and divide by the total number of measurements taken. For example, the operator of a waterworks or wastewater treatment plant takes a chlorine residual measurement every day; and part of his or her operating log is shown in Table 2.1. Find the mean.

Table 2.1   Daily Chlorine Residual Results

Day	Chlorine residual (mg/L)
Monday	0.9
Tuesday	1.0
Wednesday	0.9
Thursday	1.3
Friday	1.1
Saturday	1.4
Sunday	1.2

Solution:

Add up the seven chlorine residual readings: 0.9 + 1.0 + 0.9 + 1.3 + 1.1 + 1.4 + 1.2. = 7.8. Next, divide by the number of measurements, in this case seven: 7.8 ÷ 7 = 1.11. The mean chlorine residual for the week was 1.11 mg/L.

Example 2.23

Problem:

A water system has four wells with the following capacities: 115 gpm; 100 gpm; 125 gpm; and 90 gpm. What is the mean?

Solution:

$$\frac{115 \text{ gpm} + 100 \text{ gpm} + 125 \text{ gpm} + 90 \text{ gpm}}{4} = \frac{430}{4} = 107.5 \text{ gpm}$$

Example 2.24

Problem:

A water system has four storage tanks. Three of them have a capacity of 100,000 gal each, while the fourth has a capacity of 1 million gal. What is the mean capacity of the storage tanks?

Solution:

The mean capacity of the storage tanks is:

$$\frac{100,000 + 100,000 + 100,000 + 1,000,000}{4} = 325,000 \text{ gal}$$

Notice that no tank in this example has a capacity anywhere close to the mean.

Example 2.25

Problem:

Effluent BOD test results for the treatment plant during the month of August are shown below. What is the average effluent BOD for the month?

Test 1: 22 mg/L
Test 2: 33 mg/L
Test 3: 21 mg/L
Test 4: 13 mg/L

Solution:

$$\text{Average} = \frac{22 \text{ mg/L} + 33 \text{ mg/L} + 21 \text{ mg/L} + 13 \text{ mg/L}}{4} = 22.3 \text{ mg/L}$$

Example 2.26

Problem:

For the primary influent flow, the composite-sampled solids concentrations were recorded for the week in the following table. What is the average sampled solids (SS)?:

Monday	310 mg/L SS
Tuesday	322 mg/L SS
Wednesday	305 mg/L SS
Thursday	326 mg/L SS
Friday	313 mg/L SS
Saturday	310 mg/L SS
Sunday	320 mg/L SS
Total	2206 mg/L SS

Solution:

$$\text{Average SS} = \frac{\text{Sum of All Measurements}}{\text{Number of Measurements Used}}$$

$$= \frac{2206 \text{ mg/L SS}}{7} = 315.1 \text{ mg/L SS}$$

2.8 RATIO

A ratio is the established relationship between two numbers; it is simply one number divided by another number. For example, if someone says, "I'll give you four to one the Redskins over the Cowboys in the Super Bowl," what does that person mean? Four to one, or 4:1, is a ratio. If someone gives you 4 to 1, it is his or her $4 to your $1.

As another more pertinent example, if an average of 3 ft³ of screenings are removed from each million gallons of wastewater treated, the ratio of screenings removed (cubic feet) to treated wastewater (million gallons) is 3:1. Ratios are normally written using a colon (such as 2:1), or written as a fraction (such as 2/1).

When working with ratio, remember the following key points:

- One occasion when fractions are used in calculations is when ratios are used, such as calculating solutions.
- A ratio is usually stated in the form "A is to B as C is to D" and we can write it as two fractions that are equal to each other:

$$\frac{A}{B} = \frac{C}{D}$$

- Cross-multiplying solves ratio problems; that is, we multiply the left numerator (A) by the right denominator (D) and say that it is equal to the left denominator (B) times the right numerator (C):

$$A \times D = B \times C$$

$$AD = BC$$

- If one of the four items is unknown, dividing the two known items that are multiplied together by the known item that is multiplied by the unknown solves the ratio. For example, If 2 lb of alum are needed to treat 500 gal of water, how many pounds of alum will we need to treat 10,000 gal? We can state this as a ratio: 2 lb of alum is to 500 gal of water as x (unknown) pounds of alum is to 10,000 gal.

  This is set up in this manner:

$$\frac{1 \text{ lb alum}}{500 \text{ gal water}} = \frac{x \text{ lb alum}}{10,000 \text{ gal water}}$$

Cross-multiplying:

$$(500)(x) = (1) \times (10,000)$$

Transposing:

$$x = \frac{1 \times 10,000}{500}$$

$$x = 20 \text{ lb alum}$$

- For calculating proportion, for example, 5 gal of fuel costs $5.40. How much does 15 gal cost?

$$\frac{5 \text{ gal}}{\$5.40} = \frac{15 \text{ gal}}{\$y} =$$

$$5 \times y = 15 \times 5.40 = 81$$

$$y = \frac{81}{5} \times = \$16.20$$

Example 2.27

Problem:

  If a pump will fill a tank in 20 h at 4 gpm, how long will it take a 10-gpm pump to fill the same tank?

Solution:

First, analyze the problem. Here, the unknown is some number of hours. Should the answer be larger or smaller than 20 h? If a 4-gpm pump can fill the tank in 20 h, a larger pump (10 gpm) should be able to complete the filling in less than 20 h. Therefore, the answer should be less than 20 h.

Now set up the proportion:

$$\frac{x\ h}{20\ h} = \frac{4\ gpm}{10\ gpm} =$$

$$x = \frac{(4)(20)}{10} = 8h$$

Example 2.28

Problem:

Solve for x in the proportion problem given below.

Solution:

$$\frac{36}{180} = \frac{x}{4450} =$$

$$\frac{(4450)(36)}{180} = x$$

$$x = 890$$

Example 2.29

Problem:

Solve for the unknown value x in the problem given below.

Solution:

$$\frac{3.4}{2} = \frac{6}{x}$$

$$(3.4)\,(x) = (2)\,(6)$$

$$x = \frac{(2)(6)}{3.40}$$

$$x = 3.53$$

Example 2.30

Problem:

One pound of chlorine is dissolved in 65 gal of water. To maintain the same concentration, how many pounds of chlorine would have to be dissolved in 150 gal of water?

Solution:

$$\frac{1 \text{ lb}}{65 \text{ gal}} = \frac{x \text{ lb}}{150 \text{ gal}}$$

$$(65)(x) = (1)(150)$$

$$x = \frac{(1)(150)}{65}$$

$$= 2.3 \text{ lbs}$$

Example 2.31

Problem:

It takes five workers 50 h to complete a job. At the same rate, how many hours would it take eight workers to complete the job?

Solution:

$$\frac{5 \text{ workers}}{8 \text{ workers}} = \frac{x \text{ h}}{50 \text{ h}}$$

$$x = \frac{(5)(50)}{8}$$

$$x = 31.3 \text{ h}$$

Example 2.32

Problem:

If 1.6 L of activated sludge (biosolids) with volatile suspended solids (VSS) of 1900 mg/L is mixed with 7.2 L of raw domestic wastewater with BOD of 250 g/L, what is the F/M (food-to-microorganisms) ratio?

Solution:

$$\frac{F}{M} = \frac{\text{amount of BOD}}{\text{amount of VSS}}$$

$$= \frac{250 \text{ mg/L} \times 7.2 \text{ L}}{1900 \text{ mg/L} \times 1.6 \text{ L}}$$

$$= \frac{0.59}{1}$$

$$= 0.59$$

2.9 DIMENSIONAL ANALYSIS

Dimensional analysis is a problem-solving method that uses the fact that one, without changing its value, can multiply any number or expression. It is a useful technique to check whether a problem is set up correctly. In using dimensional analysis to check a math setup, we work with the dimensions (units of measure) only — not with numbers.

An example of dimensional analysis common to everyday life is the unit pricing found in many hardware stores. A shopper can purchase a 1-lb box of nails for $0.98 in one store, whereas a warehouse store sells a 5-lb bag of the same nails for $3.50. The shopper will analyze this problem almost without thinking about it. The solution calls for reducing the problem to the price per pound. The pound is selected as the unit common to both stores. A shopper will pay $0.70/lb for nails in the warehouse store or $0.98/lb in the local hardware store. Knowing the unit price, which is expressed in dollars per pound ($/lb), is implicit in the solution to this problem.

To use the dimensional analysis method, we must know how to perform three basic operations:

Note: Unit factors may be made from any two terms that describe the same or equivalent "amounts" of the objects of interest. For example, we know that 1 in. = 2.54 cm.

1. *Basic operation*: to complete a division of units, always ensure that all units are written in the same format; it is best to express a horizontal fraction (such as gallons per square foot) as a vertical fraction. Horizontal to vertical:

$$\text{gal/ft}^3 \text{ to } = \frac{\text{gal}}{\text{ft}^3}$$

$$\text{psi to } \frac{\text{lb}}{\text{in.}^2}$$

The same procedures are applied in the following examples:

$$\text{ft}^3/\text{min becomes } \frac{\text{ft}^3}{\text{min}}$$

$$\text{s/in.}^2 \text{ becomes } \frac{\text{s}}{\text{in.}^2}$$

2. *Basic operation*: we must know how to divide by a fraction. For example,

$$\frac{\dfrac{\text{lb}}{\text{d}}}{\dfrac{\text{min}}{\text{d}}} \text{ becomes } \frac{\text{lb}}{\text{d}} \times \frac{\text{d}}{\text{min}}$$

In the preceding example, notice that the terms in the denominator were inverted before the fractions were multiplied. This is a standard rule that must be followed when dividing fractions. Another example is

$$\frac{\dfrac{\text{mm}^2}{\text{mm}^2}}{\text{m}^2} \text{ becomes } \text{m}^2 \times \frac{\text{m}^2}{\text{mm}^2}$$

3. *Basic operation*: we must know how to cancel or divide terms in the numerator and denominator of a fraction. After fractions have been rewritten in the vertical form and division by the fraction has been re-expressed as multiplication as shown earlier, the terms can be canceled (or divided) out. *Key point*: For every term that is canceled in the numerator of a fraction, a similar term must be canceled in the denominator and vice versa, as shown below:

$$\frac{\text{Kg}}{\text{d}} \times \frac{\text{d}}{\text{min}} = \frac{\text{Kg}}{\text{min}}$$

$$\text{mm}^2 \times \frac{\text{m}^2}{\text{mm}^2} = \text{m}^2$$

$$\frac{\text{gal}}{\text{min}} \times \frac{\text{ft}^3}{\text{gal}} = \frac{\text{ft}^3}{\text{min}}$$

How are units that include exponents calculated?

When written with exponents (ft^3, for example), a unit can be left as it is, or put in expanded form (ft)(ft)(ft), depending on other units in the calculation. However, it is necessary to ensure that square and cubic terms are expressed uniformly, as sq ft, cu ft, or as ft^2, ft^3. For dimensional analysis, the latter system is preferred.

For example, if we wish to convert 1400 ft^3 volume to gallons and will use 7.48 gal/ft^3 in the conversions, dimensional analysis can be used to determine whether we multiply or divide by 7.48. To determine if the math setup is correct, only the dimensions are used.

First, try dividing the dimensions:

$$\frac{\text{ft}^3}{\text{gal/ft}^3} = \frac{\text{ft}^3}{\dfrac{\text{gal}}{\text{ft}^3}}$$

Then, the numerator and denominator are multiplied to get

$$= \frac{ft^6}{gal}$$

Thus, by dimensional analysis we determine that if we divide the two dimensions (ft^3 and gal/ft^3), the units of the answer are ft^6/gal, not gal. Clearly, division is not the right way to go in making this conversion.

What would have happened if we had multiplied the dimensions instead of dividing?

$$(ft^3)(gal/ft^3) = (ft^3) = \left(\frac{gal}{ft^3}\right)$$

Then, multiply the numerator and denominator to obtain

$$= \frac{(ft^3)(gal)}{ft^3}$$

and cancel common terms to obtain

$$= \frac{(\cancel{ft^3})(gal)}{\cancel{ft^3}}$$

$$= gal$$

Obviously, by multiplying the two dimensions (ft^3 and gal/ft^3), the answer will be in gallons, which is what we want. Thus, because the math setup is correct, we would then multiply the numbers to obtain the number of gallons.

$$(1400\ ft^3)(7.48\ gal/ft^3) = 10,472\ gal$$

Now try another problem with exponents. We wish to obtain an answer in square feet. If we are given the two terms — 70 ft^3/sec and 4.5 ft/sec — is the following math setup correct?

$$(70\ ft^3/sec)(4.5\ ft/sec)$$

First, only the dimensions are used to determine if the math setup is correct. Multiplying the two dimensions yields:

$$(ft^3/sec)(ft/sec) = \left(\frac{ft^3}{sec}\right) = \left(\frac{ft}{sec}\right)$$

Then, the terms in the numerators and denominators of the fraction are multiplied:

$$\frac{(ft^3)}{(sec)} = \frac{(ft)}{(sec)}$$

$$= \frac{(ft^4)}{(sec^2)}$$

Obviously, the math setup is incorrect because the dimensions of the answer are not square feet. Therefore, if we multiply the numbers as shown previously, the answer will be wrong.

Let us try division of the two dimensions instead:

$$ft^3/sec = \frac{\dfrac{ft^3}{sec}}{\dfrac{ft}{sec}}$$

Invert the denominator and multiply to get

$$= \left(\frac{ft^3}{sec}\right) = \left(\frac{sec}{ft}\right)$$

$$= \frac{(ft)(ft)(ft)\ (sec)}{(sec)(ft)}$$

$$= ft^2$$

Because the dimensions of the answer are square feet, this math setup is correct. Therefore, by dividing the numbers as we did with units, the answer will also be correct:

$$\frac{70\ ft^3/sec}{4.5\ ft/sec} = 15.56$$

Example 2.33

Problem:

We are given two terms — 5 m/sec and 7 m² — and the answer to be obtained is in cubic meters per second (m³/sec). Is multiplying the two terms the correct math setup?

Solution:

$$(m/sec)\ (m^2)\ = \frac{m^2}{sec} \times m^2$$

Multiply the numerators and denominator of the fraction:

$$= \frac{(m)\ (m^2)}{sec}$$

$$= \frac{m^3}{sec}$$

Because the dimensions of the answer are cubic meters per second (m³/sec), the math setup is correct. Therefore, multiply the numbers to get the correct answer:

$$5 \text{ (m/sec)} (7 \text{ m}^2) = 35 \text{ m}^3/\text{sec}$$

Example 2.34

Problem:

Solve the following problem: Given that the flow rate in a water line is 2.3 ft³/sec, what is the flow rate expressed as gallons per minute?

Solution:

Set up the math problem:

$$(2.3 \text{ ft}^3/\text{sec}) (7.48 \text{ gal/ft}^3) (60 \text{ sec/min})$$

Then, use dimensional analysis to check the math setup:

$$(\text{ft}^3/\text{sec}) (\text{gal/ft}^3) (\text{sec/min}) = \left(\frac{\text{ft}^3}{\text{sec}}\right)\left(\frac{\text{gal}}{\text{ft}^3}\right)\left(\frac{\text{sec}}{\text{min}}\right)$$

$$= \frac{\text{ft}^3}{\text{sec}} \frac{\text{gal}}{\text{ft}^3} \frac{\text{sec}}{\text{min}}$$

$$= \frac{\text{gal}}{\text{min}}$$

The math setup is correct as shown above. Therefore, this problem can be multiplied out to get the answer in correct units.

$$(2.3 \text{ ft}^3/\text{sec}) (7.48 \text{ gal/ft}^3) (60 \text{ sec/min}) = 1032.24 \text{ gal/min}$$

Example 2.35

Problem:

During an 8-h period, a water treatment plant treated 3.2 million gal of water. What is the plant total volume treated per day, assuming the same treatment rate?

Solution:

$$= \frac{3.2 \text{ mil gal}}{8 \text{ h}} \times \frac{24 \text{ h}}{\text{day}}$$

$$= \frac{3.2 \times 24}{8} \text{ MGD}$$

$$= 9.6 \text{ MGD}$$

Example 2.36

Problem:

How many cubic feet per second (cfs; ft³/sec) are equal to 1 MGD?

Solution:

$$1 \text{ MGD} = \frac{10^6}{1 \text{ day}}$$

$$= \frac{10^6 \text{ gal} \times 0.1337 \text{ ft}^3/\text{gal}}{1 \text{ day} \times 86,400 \text{ sec/day}}$$

$$= \frac{133,700 \text{ ft}^3}{86,400 \text{ sec}}$$

$$= 1.547 \text{ cfs (ft}^3/\text{sec})$$

Example 2.37

Problem:

A 10-gal empty tank weighs 4.6 lb. What is the total weight of the tank filled with 6 gal of water?

Solution:

$$\text{Weight of water} = 6 \text{ gal} \times 8.34 \text{ lb/gal}$$

$$= 50.04 \text{ lb}$$

$$\text{Total weight} = 50.04 + 4.6 \text{ lb}$$

$$= 54.6 \text{ lb}$$

Example 2.38

Problem:

The depth of biosolids applied to the biosolids drying bed is 10 in. What is the depth in centimeters (2.54 cm = 1 in.)?

Solution:

$$10 \text{ in.} = 10 \times 2.54 \text{ cm}$$

$$= 25.4 \text{ cm}$$

2.10   THRESHOLD ODOR NUMBER (TON)

The environmental engineer responsible for water supplies soon discovers that taste and odor are the most common customer complaint. Odor is typically measured and expressed in terms of a threshold odor number (TON), the ratio by which the sample must be diluted for the odor to become virtually unnoticeable. In 1989, the USEPA issued a "secondary maximum contaminant level" (SMCL) of 3 TON for odor.

> *Note*: Secondary standards are parameters not related to health.

When a dilution is used, a number can be devised to describe clarifying odor.

$$\text{TON (threshold odor number)} = \frac{V_T + V_P}{V_T} \tag{2.3}$$

where:
V_T = volume tested
V_P = volume of dilution with odor-free distilled water

For $V_P = 0$, TON = 1 (lowest value possible)
For $V_P = V_T$, TON = 2
For $V_P = 2V_T$, TON = 3, etc.

Example 2.39

Problem:
The first detectable odor is observed when a 50-mL sample is diluted to 200 mL with odor-free water. What is the TON of the water sample?

Solution:

$$\text{TON} = \frac{200}{V_T} = \frac{200 \text{ mL}}{50 \text{ mL}} = 4$$

2.11   GEOMETRICAL MEASUREMENTS

Water/wastewater treatment plants consist of a series of tanks and channels. Proper design and operational control require the engineer and operator to perform several process control calculations. Many of these calculations include parameters such as the circumference or perimeter; area; or the volume of the tank or channel as part of the information necessary to determine the result. Many process calculations require computation of surface areas. To aid in performing these calculations, we provide the following definitions and relevant equations used to calculate areas and volumes for several geometric shapes.

- *Area*: the area of an object, measured in square units; the amount of surface an object contains or the amount of material it takes to cover the surface.
- *Base*: the bottom leg of a triangle, measured in linear units.
- *Circumference*: the distance around an object, measured in linear units. When determined for other than circles, it may be called the perimeter of the figure, object, or landscape.

- *Cubic units*: measurements used to express volume: cubic feet, cubic meters, etc.
- *Depth*: the vertical distance from the bottom of the tank to the top; normally measured in terms of liquid depth and given in terms of side-wall depth (SWD), measured in linear units.
- *Diameter*: the distance from one edge of a circle to the opposite edge passing through the center, measured in linear units.
- *Height*: the vertical distance from one end of an object to the other, measured in linear units.
- *Length*: the distance from one end of an object to the other, measured in linear units.
- *Linear units*: measurements used to express distances: feet, inches, meters, yards, etc.
- *Pi, π:* a number in the calculations involving circles, spheres, or cones ($\pi = 3.14$).
- *Radius:* the distance from the center of a circle to the edge, measured in linear units.
- *Sphere:* a container shaped like a ball.
- *Square units:* measurements used to express area: square feet, square meters, acres, etc.
- *Volume:* the capacity of the unit (how much it will hold), measured in cubic units (cubic feet, cubic meters) or in liquid volume units (gallons, liters, million gallons).
- *Width:* the distance from one side of the tank to the other, measured in linear units.

Relevant geometric equations

Circumference of a circle	$C = \pi d = 2\pi r$
Perimeter of a square with side a	$P = 4a$
Perimeter of a rectangle with sides a and b	$P = 2a + 2b$
Perimeter of a triangle with sides a, b, and c	$P = a + b + c$
Area A of a circle with radius r ($d = 2r$)	$A = \pi d^2/4 = \pi r^2$
Area of duct in square feet when d is in inches	$A = 0.005454d^2$
Area of A of a triangle with base b and height h	$A = 0.5bh$
Area of A of a square with sides a	$A = a^2$
Area of A of a rectangle with sides a and b	$A = ab$
Area A of an ellipse with major axis a and minor axis b	$A = \pi ab$
Area A of a trapezoid with parallel sides a and b and height h	$A = 0.5(a + b)h$
Area A of a duct in square feet when d is in inches	$A = \pi d^2/576$
	$= 0.005454d^2$
Volume V of a sphere with a radius r ($d = 2r$)	$V = 1.33\pi r^3$
	$= 0.1667\pi d^3$
Volume V of a cube with sides a	$V = a^3$
Volume V of a rectangular solid (sides a and b and height c)	$V = abc$
Volume V of a cylinder with a radius r and height H	$V = \pi r^2 h$
	$= \pi d^2 h/4$
Volume V of a pyramid	$V = 0.33$

2.11.1 Geometrical Calculations

2.11.1.1 *Perimeter and Circumference*

On occasion, determining the distance around grounds or landscapes may be necessary. To measure the distance around property, buildings, and basin-like structures, it is necessary to determine perimeter or circumference. The *perimeter* is the distance around an object — a border or outer boundary. *Circumference* is the distance around a circle or circular object, such as a clarifier. Distance is a linear measurement that defines the distance (or length) along a line. Standard units of measurement like inches, feet, yards, and miles, as well as metric units like centimeters, meters, and kilometers are used.

The perimeter of a rectangle (a four-sided figure with four right angles) is obtained by adding the lengths of the four sides (see Figure 2.1):

$$\text{Perimeter} = L_1 + L_2 + L_3 + L_4 \qquad (2.4)$$

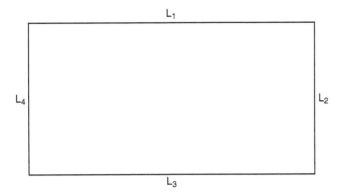

Figure 2.1   Perimeter.

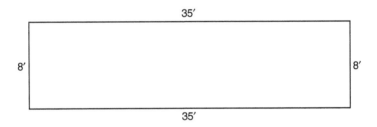

Figure 2.2   See Example 2.40.

Example 2.40

Problem:
   Find the perimeter of the rectangle shown in Figure 2.2

Solution:

$$P = 35' + 8' + 35' + 8'$$

$$P = 86'$$

Example 2.41

Problem:
   What is the perimeter of a rectangular field if its length is 100 ft and its width is 50 ft?

Solution:

$$\text{Perimeter} = (2 \times \text{length}) + (2 \times \text{width})$$

$$= (2 \times 100 \text{ ft}) + (2 \times 50 \text{ ft})$$

$$= 200 \text{ ft} + 100 \text{ ft}$$

$$= 300 \text{ ft}$$

Example 2.42

Problem:
   What is the perimeter of a square whose side is 8 in.?

Solution:

$$\text{Perimeter} = 2 \times \text{length} + 2 \times \text{width}$$

$$= 2 \times 8 \text{ in.} + 2 \times 8 \text{ in.}$$

$$= 16 \text{ in.} + 16 \text{ in.}$$

$$= 32 \text{ in.}$$

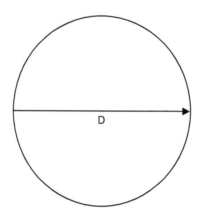

Figure 2.3   Diameter of circle.

   The circumference is the length of the outer border of a circle. The circumference is found by multiplying pi (π) times the diameter (*D*; diameter is a straight line passing through the center of a circle — the distance across the circle; see Figure 2.3).

$$C = \pi D \tag{2.5}$$

where
C = circumference
π = Greek letter pi = 3.1416
D = diameter

Use this calculation, for example, to determine the circumference of a circular tank.

Example 2.43

Problem:
   Find the circumference of a circle with a diameter of 25 ft (π = 3.14)

Solution:

$$C = \pi \times 25 \text{ ft}$$

$$C = 3.14 \times 25 \text{ ft}$$

$$C = 78.5 \text{ ft}$$

Example 2.44

Problem:

A circular chemical holding tank has a diameter of 18 m. What is the circumference of this tank?

Solution:

$$C = \pi \ 18 \text{ m}$$

$$C = (3.14)(18 \text{ m})$$

$$C = 56.52 \text{ m}$$

Example 2.45

Problem:

An influent pipe inlet opening has a diameter of 6 ft. What is the circumference of the inlet opening in inches?

Solution:

$$C = \pi \times 6 \text{ ft}$$

$$C = 3.14 \times 6 \text{ ft}$$

$$C = 18.84 \text{ ft}$$

2.11.1.2 Area

For area measurements in water/wastewater operations, three basic shapes are particularly impor-tant: circles, rectangles, and triangles. Area is the amount of surface an object contains or the amount of material needed to cover the surface. The area on top of a chemical tank is called the *surface area*. The area of the end of a ventilation duct is called the *cross-sectional area* (the area at right angles to the length of ducting). Area is usually expressed in square units, such as square inches (in.²) or square feet (ft²). Land may also be expressed in terms of square miles (sections) or acres (43,560 ft²) or, in the metric system, as *hectares*.

A *rectangle* is a two-dimensional box. The area of a rectangle is found by multiplying the length (*L*) times width (*W*); see Figure 2.4.

$$\text{Area} = \text{L} \times \text{W} \qquad\qquad (2.6)$$

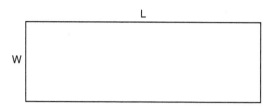

Figure 2.4   Rectangle.

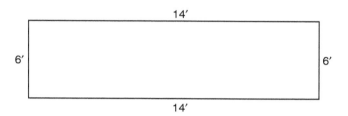

Figure 2.5   See Example 2.46.

Example 2.46

Problem:
   Find the area of the rectangle shown in Figure 2.5.

Solution:

$$\text{Area} = L \times W$$

$$= 14\ \text{ft} \times 6\ \text{ft}$$

$$= 84\ \text{ft}^2$$

To find the area of a circle, we need to introduce one new term: the *radius,* which is represented by *r.* In Figure 2.6, we have a circle with a radius of 6 ft. The radius is any straight line that radiates from the center of the circle to some point on the circumference. By definition, all radii (plural of radius) of the same circle are equal. The surface area of a circle is determined by multiplying π times the radius squared.

$$\text{Area of circle} = \pi r^2 \tag{2.7}$$

where
π = pi (3.14)
r = radius of circle radius is one half the diameter

Example 2.47

Problem:
   What is the area of the circle shown in Figure 2.6?

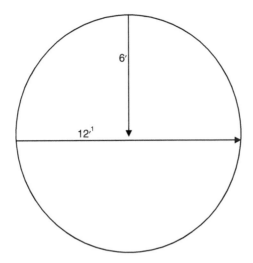

Figure 2.6 See Example 2.47.

Solution:

$$\text{Area of circle} = \pi r^2$$

$$= \pi 6^2$$

$$= 3.14 \times 36$$

$$= 113 \text{ ft}^2$$

If we were assigned to paint a water storage tank, we would need to know the surface area of the walls of the tank in order to know how much paint would be required. That is, we need to know the area of a circular or cylindrical tank. To determine the tank's surface area, we need to visualize the cylindrical walls as a rectangle wrapped around a circular base. The area of a rectangle is found by multiplying the length by the width; in this case, the width of the rectangle is the height of the wall, and the length of the rectangle is the distance around the circle — the circumference.

Thus, the area of the side walls of the circular tank is found by multiplying the circumference of the base ($C = \pi \times D$) times the height of the wall (H):

$$A = \pi \times D \times H \tag{2.8}$$

$$A = \pi \times 20 \text{ ft} \times 25 \text{ ft}$$

$$A = 3.14 \times 20 \text{ ft} \times 25 \text{ ft}$$

$$A = 1570 \text{ ft}^2$$

To determine the amount of paint needed, remember to add the surface area of the top of the tank, which is 314 ft². Thus, the amount of paint needed must cover 1570 ft² + 314 ft² = 1885 ft². If the tank floor should be painted, add another 314 ft².

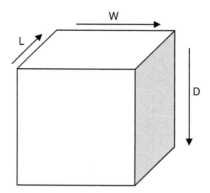

Figure 2.7   Volume.

2.11.1.3   Volume

The amount of space occupied by or contained in an object — volume (see Figure 2.7) — is expressed in cubic units, such as cubic inches (in.³), cubic feet (ft³), acre-feet (1 acre-ft = 43,560 ft³), etc. The volume of a rectangular object is obtained by multiplying the length times the width times the depth or height.

$$V = L \times W \times H \tag{2.9}$$

where
L = length
W = width
D or H = depth or height

Example 2.48

Problem:
    A unit rectangular process basin has a length of 15 ft, a width of 7 ft, and depth of 9 ft. What is the volume of the basin?

Solution:

$$V = L \times W \times H$$

$$= 15\ \text{ft} \times 7\ \text{ft} \times 9\ \text{ft}$$

$$= 945\ \text{ft}^3$$

For engineers and water/wastewater operators, representative surface areas are most often rectangles, triangles, circles, or a combination of these. Table 2.2 provides practical volume formulae used in water/wastewater calculations.

In determining the volume of round pipe and round surface areas, the following examples are helpful.

Table 2.2 Volume Formulae

Sphere volume	$= (\pi/6) \, (\text{diameter})^3$
Cone volume	$= 1/3 \, (\text{volume of a cylinder})$
Rectangular tank volume	$= (\text{area of rectangle}) \, (D \text{ or } H)$
	$= (LW) \, (D \text{ or } H)$
Cylinder volume	$= (\text{area of cylinder}) \, (D \text{ or } H)$
	$= \pi^2 \, (D \text{ or } H)$
Triangle volume	$= (\text{area of triangle}) \, (D \text{ or } H)$
	$= (bh/2) \, (D \text{ or } H)$

Example 2.49

Problem:

Find the volume of a 3-in. round pipe that is 300-ft long.

Solution:

Step 1. Change the diameter of the duct from inches to feet by dividing by 12.

$$D = 3 \div 12 = 0.25 \text{ ft}$$

Step 2. Find the radius by dividing the diameter by 2.

$$R = 0.25 \text{ ft} \div 2 = 0.125$$

Step 3. Find the volume.

$$V = L \times \pi r^2$$

$$V = 300 \text{ ft} \times 3.14 \times 0.0156$$

$$V = 14.72 \text{ ft}^2$$

Example 2.50

Problem:

Find the volume of a smokestack that is 24 in. in diameter (entire length) and 96 in. tall. Find the radius of the stack. The radius is one half the diameter.

$$24 \text{ in.} \div 2 = 12 \text{ in.}$$

Find the volume.

Solution:

$$V = H \times \pi r^2$$

$$V = 96 \text{ in.} \times \pi (12 \text{ in.})^2$$

$$V = 96 \text{ in.} \times \pi(144 \text{ in.}^2)$$

$$V = 43,407 \text{ in.}^3$$

To determine the volume of a cone and sphere, we use the following equations and examples.

$$\text{Volume of cone} = \frac{\pi}{12} \times \text{Diameter} \times \text{Diameter} \times \text{Height} \qquad (2.10)$$

$$\frac{\pi}{12} = \frac{3.14}{12} = 0.262$$

Key point: The diameter used in the formula is the diameter of the base of the cone.

Example 2.51

Problem:

The bottom section of a circular settling tank is cone shaped. How many cubic feet of water are contained in this section of the tank, if the tank has a diameter of 120 ft and the cone portion of the unit has a depth of 6 ft?

Solution:

$$\text{Volume, ft}^3 = 0.262 \times 120 \text{ ft} \times 120 \text{ ft} \times 6 \text{ ft} = 22,637 \text{ ft}^3$$

The volume of a sphere may be calculated as follows:

$$\text{Volume of sphere} = \frac{3.14}{6} \times \text{Diameter} \times \text{Diameter} \times \text{Diameter} \qquad (2.11)$$

$$\frac{\pi}{6} = \frac{3.14}{6} = 0.524$$

Example 2.52

Problem:

What is the volume in cubic feet of a spherical gas storage container with a diameter of 60 ft?

Solution:

$$\text{Volume, ft}^3 = 0.524 \times 60 \text{ ft} \times 60 \text{ ft} \times 60 \text{ ft} = 113,184 \text{ ft}^3$$

Circular process and various water/chemical storage tanks are commonly found in water/wastewater treatment. A circular tank consists of a circular floor surface with a cylinder rising above it (see Figure 2.8). The volume of a circular tank is calculated by multiplying the surface area times the height of the tank walls.

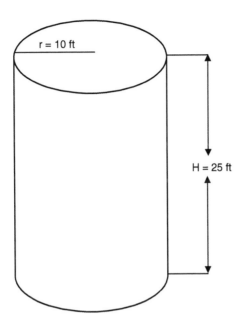

Figure 2.8 Circular or cylindrical water tank.

Example 2.53

Problem:

If a tank is 20 ft in diameter and 25 ft deep, how many gallons of water will it hold?

- *Hint*: In this type of problem, calculate the surface area first, multiply by the height, and then convert to gallons.

Solution:

$$r = D \div 2 = 20 \text{ ft} \div 2 = 10 \text{ ft}$$

$$A = \pi \times r^2$$

$$A = \pi \times 10 \text{ ft} \times 10 \text{ ft}$$

$$A = 314 \text{ ft}^2$$

$$V = A \times H$$

$$V = 314 \text{ ft}^2 \times 25 \text{ ft}$$

$$V = 7850 \text{ ft}^3 \times 7.5 \text{ gal/ft}^3 = 58{,}875 \text{ gal}$$

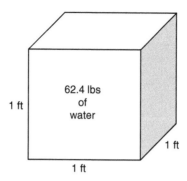

Figure 2.9   One cubic foot of water weighs 62.4 lb.

2.12   FORCE, PRESSURE, AND HEAD CALCULATIONS

Before we review calculations involving force, pressure, and head, we first define these terms:

- *Force:* the push exerted by water on any confined surface. Force can be expressed in pounds, tons, grams, or kilograms.
- *Pressure:* the force per unit area. The most common way of expressing pressure is in pounds per square inch (psi).
- *Head:* the vertical distance or height of water above a reference point. Head is usually expressed in feet. In the case of water, head and pressure are related.

2.12.1   Force and Pressure

Figure 2.9 helps to illustrate these terms. A cubical container measuring 1 ft on each side can hold 1 ft³ of water. A basic fact of science states that 1 ft³ of water weights 62.4 lb and contains 7.48 gal. The force acting on the bottom of the container would be 62.4 lb/ft³. The area of the bottom in square inches is:

$$1\ \text{ft}^2\ =\ 12\ \text{in.}\ \times\ 12\ \text{in.}\ =\ 144\ \text{in}^2$$

Therefore the pressure in pounds per square inch (psi) is:

$$\frac{62.4\ \text{lb/ft}^3}{1\ \text{ft}^2}\ =\ \frac{62.4\ \text{lb/ft}^3}{144\ \text{in}^2/\text{ft}^2} = 0.433\ \text{lb/in}^2\ \text{(psi)}$$

If we use the bottom of the container as our reference point, the head would be 1 ft. From this, we see that 1 ft of head is equal to 0.433 psi — an important parameter to remember. Figure 2.10 illustrates some other important relationships between pressure and head.

> *Important point*: Force acts in a particular direction. Water in a tank exerts force down on the bottom and out on the sides. Pressure, however, acts in all directions. A marble at a water depth of 1 ft would have 0.433 psi of pressure acting inward over the entire surface.

Using the preceding information, we can develop Equation 2.12 and Equation 2.13 for calculating pressure and head.

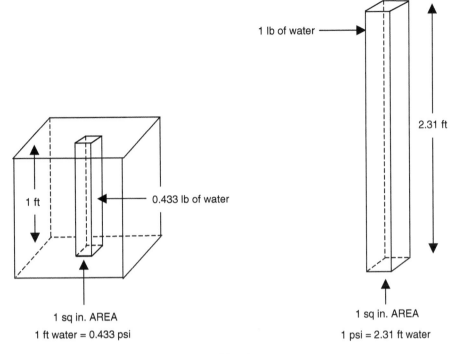

1 lb of water ——→

2.31 ft

1 ft

0.433 lb of water

1 sq in. AREA
1 ft water = 0.433 psi

1 sq in. AREA
1 psi = 2.31 ft water

Figure 2.10 The relationship between pressure and head.

$$\text{Pressure (psi)} = 0.433 \times \text{Head (ft)} \tag{2.12}$$

$$\text{Head (ft)} = 2.31 \times \text{Pressure (psi)} \tag{2.13}$$

2.12.2 Head

Head is the vertical distance that water must be lifted from the supply tank or unit process to the discharge. The total head includes the vertical distance the liquid must be lifted (static head); the loss to friction (friction head); and the energy required to maintain the desired velocity (velocity head).

$$\text{Total Head} = \text{Static Head} + \text{Friction Head} + \text{Velocity Head} \tag{2.14}$$

2.12.2.1 Static Head

Static head is the actual vertical distance the liquid must be lifted.

$$\text{Static Head} = \text{Discharge Elevation} - \text{Supply Elevation} \tag{2.15}$$

Example 2.54

Problem:
    The supply tank is located at elevation 108 ft. The discharge point is at elevation 205 ft. What is the static head in feet?

Solution:

$$\text{Static Head, ft} = 205 \text{ ft} - 108 \text{ ft} = 97 \text{ ft}$$

2.12.2.2 Friction Head

Friction head is the equivalent distance of the energy needed to overcome friction. Engineering references include tables showing the equivalent vertical distance for various sizes and types of pipes, fittings, and valves. The total friction head is the sum of the equivalent vertical distances for each component.

$$\text{Friction Head, ft} = \text{Energy Losses due to Friction} \qquad (2.16)$$

2.12.2.3 Velocity Head

Velocity head is the equivalent distance of the energy consumed in achieving and maintaining the desired velocity in the system.

$$\text{Velocity Head, ft} = \text{Energy Losses to Maintain Velocity} \qquad (2.17)$$

2.12.2.4 Total Dynamic Head (Total System Head)

$$\text{Total Head} = \text{Static Head} + \text{Friction Head} + \text{Velocity Head} \qquad (2.18)$$

2.12.2.5 Pressure/Head

The pressure exerted by water/wastewater is directly proportional to its depth or head in the pipe, tank, or channel. If the pressure is known, the equivalent head can be calculated.

$$\text{Head, ft} = \text{Pressure, psi} \times 2.31 \text{ ft/psi} \qquad (2.19)$$

Example 2.55

Problem:

The pressure gauge on the discharge line from the influent pump reads 75.3 psi. What is the equivalent head in feet?

Solution:

$$\text{Head, ft} = 75.3 \times 2.31 \text{ ft/psi} = 173.9 \text{ ft}$$

2.12.2.6 Head/Pressure

If the head is known, the equivalent pressure can be calculated by:

$$\text{Pressure, psi} \; \frac{\text{Head, ft}}{2.31 \text{ ft/psi}} \qquad (2.20)$$

Example 2.56

Problem:
The tank is 15 ft deep. What is the pressure in psi at the bottom of the tank when it is filled with wastewater?

Solution:

$$\text{Pressure, psi } \frac{15 \text{ ft}}{2.31 \text{ ft/psi}} = 6.49 \text{ psi}$$

Before we look at a few example problems, we review the key points related to force, pressure, and head:

- By definition, water weighs 62.4 lb/ft³.
- The surface of any one side of the cube contains 144 in.² (12 in. × 12 in. = 144 in.²). Therefore, the cube contains 144 columns of water 1 ft tall and 1 in. square.
- The weight of each of these pieces can be determined by dividing the weight of the water in the cube by the number of square inches.

$$\text{Weight} = \frac{62.4 \text{ lbs}}{144 \text{ in}^2} = 0.433 \text{ lb/in.}^2 \text{ or } 0.433 \text{ psi}$$

- Because this is the weight of one column of water 1 ft tall, the true expression would be 0.433 lb/in.²/ft of head or 0.433 psi/ft.

  Key point: 1 ft of head = 0.433 psi.

In addition to remembering the important parameter, 1 ft of head = 0.433 psi, it is necessary to understand the relationship between pressure and feet of head — in other words, how many feet of head 1 psi represents. This is determined by dividing 1 by 0.433:

$$\text{Feet of head} = \frac{1 \text{ ft}}{0.433 \text{ psi}} = 2.31 \text{ ft/psi}$$

If a pressure gauge reads 12 psi, the height of the water necessary to represent this pressure is 12 psi × 2.31 ft/psi =27.7 ft.

  Key point: Both the preceding conversions are commonly used in water/wastewater treatment calculations. However, the most accurate conversion is 1 ft = 0.433 psi. We use this conversion throughout this text.

Example 2.57

Problem:
Convert 40 psi to feet head.

Solution:

$$\frac{40 \text{ psi}}{1} \times \frac{\text{ft}}{0.433 \text{ psi}} = 92.4 \text{ ft}$$

Example 2.58

Problem:

Convert 40 ft to pounds per square inch.

Solution:

$$40 \, \frac{\text{ft}}{1} \times \frac{0.433 \text{ psi}}{1 \text{ ft}} = 17.32 \text{ psi}$$

As the preceding examples demonstrate, when attempting to convert pounds per square inch to feet, we divide by 0.433, and when attempting to convert feet to pounds per square inch, we multiply by 0.433. The preceding process can be most helpful in clearing up the confusion on whether to multiply or divide. Another way, however, may be more beneficial and easier for many operators to use. Notice that the relationship between pounds per square inch and feet is almost two to one. It takes slightly more than 2 ft to make 1 psi. Therefore, when looking at a problem in which the data are in pressure, the result should be in feet and the answer will be at least twice as large as the starting number. For instance, if the pressure were 25 psi, we intuitively know that the head is over 50 ft. We divide by 0.433 to obtain the correct answer.

Example 2.59

Problem:

Convert a pressure of 45 psi to feet of head.

Solution:

$$45 \, \frac{\text{psi}}{1} \times \frac{1 \text{ ft}}{0.433 \text{ psi}} = 104 \text{ ft}$$

Example 2.60

Problem:

Convert 15 psi to feet.

Solution:

$$15 \, \frac{\text{psi}}{1} \times \frac{1 \text{ ft}}{0.433 \text{ psi}} = 34.6 \text{ ft}$$

Example 2.61

Problem:

Between the top of a reservoir and the watering point, the elevation is 125 ft. What will the static pressure be at the watering point?

Solution:

$$125\ \frac{psi}{1} \times \frac{1\ ft}{0.433\ psi} = 288.7\ ft$$

Example 2.62

Problem:
Find the pressure (pounds per square inch) in a 12-ft deep tank at a point 5 ft below the water surface.

Solution:

$$Pressure\ (psi) = 0.433 \times 5\ ft$$

$$= 2.17\ psi$$

Example 2.63

Problem:
A pressure gauge at the bottom of a tank reads 12.2 psi. How deep is the water in the tank?

Solution:

$$Head\ (ft) = 2.31 \times 12.2\ psi$$

$$28.2\ ft$$

Example 2.64

Problem:
What is the pressure (static pressure) 4 miles beneath the ocean surface?

Solution:
Change miles to feet and then to pounds per square inch.

$$5280\ ft/mile \times 4 = 21,120\ ft$$

$$\frac{21,120\ ft}{2.31\ ft/psi} = 9143\ psi$$

Example 2.65

Problem:
A 150-ft diameter cylindrical tank contains 2.0 MG water. What is the water depth? At what pressure would a gauge at the bottom read in pounds per square inch?

Solution:

Step 1. Change MG to cubic feet:

$$\frac{2,000,000 \text{ gal}}{7.48} = 267,380 \text{ cu ft}$$

Step 2. Using volume, solve for depth:

$$\text{Volume} = .785 \times D^2 \times \text{depth}$$

$$267,380 \text{ cu ft} = .785 \times (150)^2 \times \text{depth}$$

$$\text{Depth} = 15.1 \text{ ft}$$

Example 2.66

Problem:

The pressure in a pipe is 70 psi. What is the pressure in feet of water? What is the pressure in pounds per square foot?

Solution:

Step 1. Convert pressure to feet of water:

$$70 \text{ psi} \times 2.31 \text{ ft/psi} = 161.7 \text{ ft of water}$$

Step 2. Convert pounds per square inch to pounds per square foot:

$$70 \text{ psi} \times 144 \text{ in.}^2/\text{ft}^2 = 10,080 \text{ psf}$$

Example 2.67

Problem:

The pressure in a pipeline is 6476 psf. What is the head on the pipe?

Solution:

$$\text{Head on pipe} = \text{ft of pressure}$$

$$\text{Pressure} = \text{Weight} \times \text{Height}$$

$$6476 \text{ psf} = 62.4 \text{ lbs/ft}^3 \times \text{height}$$

$$\text{Height} = 104 \text{ ft}$$

2.13 REVIEW OF ADVANCED ALGEBRA KEY TERMS AND CONCEPTS

Advanced algebraic operations (linear, linear differential, and ordinary differential equations) have in recent years become an essential part of the mathematical background required by environmental engineers, among others. Although we do not intend to provide complete coverage of the topics (engineers are normally well grounded in these critical foundational areas), we review the key terms and concepts germane to the topics.

Key definitions include:

Algebraic multiplicity of an eigenvalue: the algebraic multiplicity of an eigenvalue c of a matrix A is the number of times the factor $(t - c)$ occurs in the characteristic polynomial of A.

Basis for a subspace: a basis for a subspace W is a set of vectors $\{\mathbf{v}_1, ..., \mathbf{v}_k\}$ in which W is such that:
  $\{\mathbf{v}_1, ..., \boldsymbol{v}_k\}$ is linearly independent
  $\{\mathbf{v}_1, ..., \mathbf{v}_k\}$ spans W.

Characteristic polynomial of a matrix: the characteristic polynomial of a n-by-n matrix A is the polynomial in t given by the formula $\det(A - tI)$.

Column space of a matrix: the subspace spanned by the columns of the matrix considered as a set of vectors (also see row space).

Consistent linear system: a system of linear equations is consistent if it has at least one solution.

Defective matrix: a matrix A is defective if A has an eigenvalue whose geometric multiplicity is less than its algebraic multiplicity.

Diagonalizable matrix: a matrix is diagonalizable if it is similar to a diagonal matrix.

Dimension of a subspace: the dimension of a subspace W is the number of vectors in any basis of W. (If W is the subspace $\{0\}$, we say that its dimension is 0.)

Echelon form of a matrix: a matrix is in row echelon form if:
  All rows that consist entirely of zeros are grouped together at the bottom of the matrix.
  The first (counting left to right) nonzero entry in each nonzero row appears in a column to the right of the first nonzero entry in the preceding row (if there is a preceding row).

Eigenspace of a matrix: the eigenspace associated with the eigenvalue c of a matrix A is the null space of $A - cI$.

Eigenvalue of a matrix: an eigenvalue of a matrix A is a scalar c in which $Ax = cx$ holds for some nonzero vector $\boldsymbol{x}$.

Eigenvector of a matrix: an eigenvector of a square matrix A is a nonzero vector $\mathbf{x}$ in which $A\mathbf{x} = c\mathbf{x}$ holds for some scalar c.

Elementary matrix: a matrix that is obtained by performing an elementary row operation on an identity matrix.

Equivalent linear systems: two systems of linear equations in n unknowns are equivalent if they have the same set of solutions.

Geometric multiplicity of an eigenvalue: when an eigenvalue c of a matrix A is the dimension of the eigenspace of c.

Homogeneous linear system: a system of linear equations $A\mathbf{x} = \mathbf{b}$ is homogeneous if $\mathbf{b} = \mathbf{0}$.

Inconsistent linear system: a system of linear equations with no solutions.

Inverse of a matrix: the matrix B is an inverse for the matrix A if $AB = BA = 1$.

Invertible matrix: a matrix is invertible if it has no inverse.

Least squares solution of a linear system: a solution to a system of linear equations $A\mathbf{x} = \mathbf{b}$ is a vector $\mathbf{x}$ that minimizes the length of the vector $A\mathbf{x} - \mathbf{b}$.

Linear combination of vectors: a vector $\mathbf{v}$ is a linear combination of the vectors $\mathbf{v}_1, ..., \mathbf{v}_k$ if there exist scalars $a_1, ..., a_k$ in which $\mathbf{v} = a_1\mathbf{v}_1 + ... + a_k\mathbf{v}_k$.

Linear dependence relation for a set of vectors: a relation for the set of vectors $\{\mathbf{v}_1, ..., \mathbf{v}_k\}$ is an equation of the form $a_1\mathbf{v}_1 + ... + a_k\mathbf{v}_k = \mathbf{0}$, where the scalars $a_1, ..., a_k$ are zero.

Linearly dependent set of vectors: the set of vectors $\{\mathbf{v}_1, ..., \mathbf{v}_k\}$ is linearly dependent if the equation $a_1\mathbf{v}_1 + ... + a_k\mathbf{v}_k = \mathbf{0}$ has a solution where not all the scalars $a_1, ..., a_k$ are zero (i.e., if $\{\mathbf{v}_1, ..., \mathbf{v}_k\}$ satisfies a linear dependence relation).

Linearly independent set of vectors: the set of vectors $\{\mathbf{v}_1, ..., \mathbf{v}_k\}$ is linearly independent if the only solution to the equation $a_1\mathbf{v}_1 + ... + a_k\mathbf{v}_k = 0$ is the solution where all the scalars $a_1, ..., a_k$ are zero (i.e., if $\{\mathbf{v}_1, ..., \mathbf{v}_k\}$ does not satisfy any linear dependence relation).

Linear transformation: a transformation from V to W is a function T from V to W in which:

$T(\mathbf{u} + \mathbf{v}) = T(\mathbf{u}) + T(\mathbf{v})$ for all vectors $\mathbf{u}$ and $\mathbf{v}$ in V.

$T(a\mathbf{v}) = aT(\mathbf{v})$ for all vectors $\mathbf{v}$ in V and all scalars a.

Nonsingular matrix: a square matrix A is nonsingular if the only solution to the equation $A\mathbf{x} = \mathbf{0}$ is $\mathbf{x} = \mathbf{0}$.

Null space of a matrix: the null space of an m by n matrix A is the set of all vectors $\mathbf{x}$ in R^n such that $A\mathbf{x} = \mathbf{0}$.

Null space of a linear transformation: for a linear transformation, T is the set of vectors $\mathbf{v}$ in its domain such that $T(\mathbf{v}) = \mathbf{0}$.

Nullity of a matrix: the dimension of its null space.

Nullity of a linear transformation: the dimension of its null space.

Orthogonal complement of a subspace: the orthogonal complement of a subspace S of R^n is the set of all vectors $\mathbf{v}$ in R^n so that $\mathbf{v}$ is orthogonal to every vector in S.

Orthogonal set of vectors: a set of vectors in R^n is orthogonal if the product of any two of them is 0.

Orthogonal matrix: a matrix A is orthogonal if A is invertible and its inverse equals its transpose; i.e., $A^{-1} = A^T$.

Orthogonal linear transformation: a linear transformation T from V to W is orthogonal if $T(\mathbf{v})$ has the same length as $\mathbf{v}$ for all vectors $\mathbf{v}$ in V.

Orthonormal set of vectors: a set of vectors in R^n is orthonormal if it is an orthogonal set and each vector has length 1.

Range of a linear transformation: the range of a linear transformation T is the set of all vectors $T(\mathbf{v})$, where $\mathbf{v}$ is any vector in its domain.

Rank of a matrix: the rank of a matrix A is the number of nonzero rows in the reduced row echelon form of A, i.e., the dimension of the row space of A.

Rank of a linear transformation: the rank of a linear transformation (and thus of any matrix regarded as a linear transformation) is the dimension of its range. Note that a theorem tells us that the two definitions of rank of a matrix are equivalent.

Reduced row echelon form of a matrix: a matrix is in reduced row echelon form if:

The matrix is in row echelon form.

The first nonzero entry in each nonzero row is the number 1.

The first nonzero entry in each nonzero row is the only nonzero entry in its column.

Row equivalent matrices: two matrices are row equivalent if one can be obtained from the other by a sequence of elementary row operations.

Row operations: elementary row operations performed on a matrix:

Interchange two rows

Multiply a row by a nonzero scalar

Add a constant multiple of one row to another

Row space of a matrix: the row space of a matrix is the subspace spanned by the rows of the matrix considered as a set of vectors.

Similar matrices: matrices A and B are similar if a square invertible matrix S is an equivalent to $S^{-1}AS = B$.

Singular matrix: a square matrix A is singular if the equation $A\mathbf{x} = \mathbf{0}$ has a nonzero solution for $\mathbf{x}$.

Span of a set of vectors: the span of the set of vectors $\{\mathbf{v}_1, \ldots, \mathbf{v}_k\}$ is the subspace V consisting of all linear combinations of $\mathbf{v}_1, \ldots, \mathbf{v}_k$. One also says that the subspace V is spanned by the set of vectors $\{\mathbf{v}_1, \ldots, \mathbf{v}_k\}$ and that this set of vectors spans V.

Subspace: a subset W of R^n is a subspace of R^n if:

The zero vector is in W.

$\mathbf{x} + \mathbf{y}$ is in W whenever $\mathbf{x}$ and $\mathbf{y}$ are in W.

$a\mathbf{x}$ is in W whenever $\mathbf{x}$ is in W and a is any scalar.

Symmetric matrix: a matrix A is symmetric if it equals its transpose; i.e., $A = A^T$.

CHAPTER **3**

Environmental Modeling

3.1 INTRODUCTION

Interest in the field of environmental monitoring and quantitative assessment of environmental problems is growing. For some years now, the results of environmental models and assessment analyses have been influencing environmental regulation and policies. These results are widely cited by politicians in forecasting consequences of greenhouse gas emissions like carbon dioxide (CO_2) and in advocating dramatic reductions of energy consumption at local, state, national, and international levels. For this reason and because environmental modeling is often based on extreme conceptual and numerical intricacy and uncertain validity, environmental modeling has become one of the most controversial topics of applied mathematics.

Having said this, environmental modeling continues to be widely used in environmental engineering, with its growth only limited by the imagination of the modelers. Environmental engineering problem-solving techniques incorporating modeling are widely used in watershed mapping; surface water information; flood hazard mapping; climate modeling; groundwater modeling; and others. Keep in mind that the end product produced on any modeling system will be, at least in part, a reflection of the modeling system — sometimes more than a reflection of actual conditions.

In this chapter, we do not provide a complete treatment of environmental modeling. (For the reader who desires such a treatment, we highly recommend Nirmalakhandan's *Modeling Tools for Environmental Engineers and Scientists*, 2002. Much of the work presented in this chapter is modeled after his work.) We present an overview of quantitative operations implicit to environmental modeling processes.

3.2 MEDIA MATERIAL CONTENT

Media material content is a measure of the material contained in a bulk medium, quantified by the ratio of the amount of material present to the amount of the medium. The terms *mass*, *moles*, or *volume* can be used to quantify the amounts. Thus, the ratio can be expressed in several forms, such as mass or moles of material per volume of medium (resulting in mass or molar concentration); moles of material per mole of medium (resulting in mole fraction); and volume of material per volume of medium (resulting in volume fraction).

When dealing with mixtures of materials and media, the use of different forms of measures in the ratio to quantify material content may become confusing. With mixtures, the ratio can be expressed in concentration units. The concentration of a chemical substance (liquid, gaseous, or solid) expresses the amount of substance present in a mixture. Concentration can be expressed in many different ways.

Chemists use the term *solute* to describe the substance of interest and the term *solvent* to describe the material in which the solute is dissolved. For example, in a can of soft drink (a solution of sugar in carbonated water), approximately 12 tablespoons of sugar (the solute) are dissolved in the carbonated water (the solvent). In general, the component present in the greatest amount is termed the solvent.

Some of the more common concentration units are:

- *Mass per unit volume.* Some concentrations are expressed in milligrams per milliliter (mg/mL) or milligrams per cubic centimeter (mg/cm³). Note that "1 mL = 1 cm³" is sometimes denoted as a "cc." Mass per unit volume is handy when discussing how soluble a material is in water or a particular solvent. For example, "the solubility of substance X is 4 grams per liter."
- *Percent by mass.* Also called weight percent or percent by weight, this is the mass of the solute divided by the total mass of the solution and multiplied by 100%:

$$\text{Percent by mass} = \frac{\text{Mass of component}}{\text{Mass of solution}} = 100\% \tag{3.1}$$

The mass of the solution is equal to the mass of the solute plus the mass of the solvent. For example, a solution consisting of 30 g of sodium chloride and 70 g of water would be 30% sodium chloride by mass: (30 g NaCl)/(30 g NaCl + 70 g water) × 100% = 30%. To avoid confusion over whether a solution is percent by weight or percent by volume, the symbol "w/w" (for weight to weight) is often used after the concentration: "10% potassium iodide solution in water (w/w)."

- *Percent by volume.* Also called volume percent or percent by volume, this is typically only used for mixtures of liquids and is the volume of the solute divided by the sum of the volumes of the other components, multiplied by 100%. If we mix 30 mL of ethanol and 70 mL of water, the percent ethanol by volume will be 30%; however, the total volume of the solution will NOT be 100 mL (although it will be close) because ethanol and water molecules interact differently with each other than they do with themselves. To avoid confusion over whether we have a percent by weight or percent by volume solution, we could label this as "30% ethanol in water (v/v)" where v/v stands for "volume to volume."
- *Molarity.* This is the number of moles of solute dissolved in one liter of solution. For example, if we have 90 g of glucose (molar mass = 180 g/mol), this is (90 g)/(180 g/mol) = 0.50 mol of glucose. If we place this in a flask and add water until the total volume = 1 L, we would have a 0.5 molar solution. Molarity is usually denoted with a capital, italicized "M" — a 0.50-*M* solution. Recognize that molarity is moles of solute per liter of solution, not per liter of solvent. Also recognize that molarity changes slightly with temperature because the volume of a solution changes with temperature.
- *Molality* (*m*, used for calculations of colligative properties). Molality is the number of moles of solute dissolved in 1 kg of solvent. Notice two key differences between molarity and molality: molality uses mass rather than volume, and solvent instead of solution.

$$\text{Molality} = \frac{\text{Moles of Solute}}{\text{Kilograms of solution}} \tag{3.2}$$

Unlike molarity, molality is independent of temperature because mass does not change with temperature. If we place 90 g of glucose (0.50 mol) in a flask, then add 1 kg of water, we have a 0.50 molal solution. Molality is usually denoted with a small, italicized "m" — a 0.50-*m* solution.

- *Parts per million (ppm).* Parts per million works like percent by mass, but is more convenient when only a small amount of solute is present. It is defined as the mass of the component in solution, divided by the total mass of the solution, multiplied by 10⁶ (1 million):

$$\text{Parts per million} = \frac{\text{Mass of component}}{\text{Mass of solution}}(1,000,000) \qquad (3.3)$$

A solution with a concentration of 1 ppm has 1 g of substance for every million grams of solution. Because the density of water is 1 g/mL and we are adding such a tiny amount of solute, the density of a solution at such a low concentration is approximately 1 g/mL. Therefore, in general, 1 ppm implies 1 mg of solute per liter of solution. Finally, recognize that 1% = 10,000 ppm. Therefore, something that has a concentration of 300 ppm could also be said to have a concentration of (300 ppm)/(10,000 ppm/percent) = 0.03% percent by mass.

- *Parts per billion (ppb)*. This works like parts per million, but we multiply by 1 billion (10^9). (Caution: the word *billion* has different meanings in different countries.) A solution with 1 ppb of solute has 1 μg (10^{-6}) of material per liter.
- *Parts per trillion (PPT)*. Again, this works like parts per million and parts per billion, except that we multiply by 1 trillion (10^{12}). Few, if any, solutes are harmful at concentrations as low as 1 ppt.

The following notation and examples can help in standardizing these different forms; subscripts for components are $i = 1, 2, 3, \ldots N$; and subscripts for phases are g = gas; a = air; l = liquid; w = water; and s = solids and soil.

3.2.1 Material Content: Liquid Phases

Mass concentration, molar concentration, or mole fraction can be used to quantify material content in liquid phases.

$$\text{Mass conc. of component } i \text{ in water} = p_{i,\,w} \frac{\text{Mass of material, } i}{\text{Volume of water}} \qquad (3.4)$$

$$\text{Mass conc. of component } i \text{ in water} = C_{i,\,w} \frac{\text{Moles of material, } i}{\text{Volume of water}} \qquad (3.5)$$

Because moles of material = mass/molecular weight (MW), mass concentrations, $p_{i,w}$, are related as:

$$C_{i,\,w} \frac{p_{i,\,w}}{MW_i} \qquad (3.6)$$

For molarity M, [X] is molar concentration of "X."

Mole fraction, X, of a single chemical in water can be expressed as follows:

$$\text{Mole fraction, } X = \frac{\text{Moles of component/chemical}}{\text{Total moles of sol. (Moles of chemical + Moles of water)}} \qquad (3.7)$$

For dilute solutions, the moles of chemical in the denominator of the preceding can be ignored in comparison to the moles of water, n_w, and can be approximated by:

$$X = \frac{\text{Moles of chemical}}{\text{Moles of water}} \qquad (3.8)$$

If X is less than 0.02, an aqueous solution can be considered dilute. On a mass basis, similar expressions can be formulated to yield mass fractions. Mass fractions can also be expressed as a percentage or as other ratios, such as parts per million or parts per billion.

The mole fraction of a component in a solution is simply the number of moles of that component divided by the total moles of all the components. We use the mole fraction because the sum of the individual fractions should equal 1. This constraint can reduce the number of variables when modeling mixtures of chemicals. Mole fractions are strictly additive; the sum of the mole fractions of all components is equal to one. Mole fraction, X_i, of component i in an N-component mixture is defined as follows:

$$X_i = \frac{\text{Moles of i}}{\left(\sum_1^N n_i\right) + n_w} \tag{3.9}$$

$$\text{The sum of all the mole fractions} = \left(\sum_1^N X_w\right) = 1 \tag{3.10}$$

For dilute solutions of multiple chemicals (as in the case of single chemical systems), mole fraction X_i of component i in an N-component mixture can be approximated by:

$$X = \frac{\text{Moles of i}}{n_w} \tag{3.11}$$

Note that the preceding ratio is known as an intensive property because it is independent of the system and the mass of the sample. An *intensive* property is any property that can exist at a point in space. Temperature, pressure, and density are good examples. An *extensive* property is any property that depends on the size (or extent) of the system under consideration — volume, for example. If we double the length of all edges of a solid cube, the volume increases by a factor of eight. Mass is another extensive property. The same cube undergoes an eightfold increase mass when the length of the edges is doubled.

Note: The material content in solid and gas phases is different from those in liquid phases. For example, the material content in solid phases is often quantified by a ratio of masses and expressed as parts per million or parts per billion. The material content in gas phases is often quantified by a ratio of moles or volumes and expressed as parts per million or parts per billion. Reporting gas phase concentrations at standard temperature and pressure (*STP* — 0°C and 769 mm Hg or 273 K and 1 atm) is the preferred form.

Example 3.1

Problem:

A certain chemical has a molecular weight of 80. Derive the conversion factors to quantify the following:

1. 1 ppm (volume/volume) of the chemical in air in molar and mass concentration form
2. 1 ppm (mass ratio) of the chemical in water in mass and molar concentration form
3. 1 ppm (mass ratio) of the chemical in soil in mass ratio form

Solution:

1. Gas phase. The volume ratio of 1 ppm can be converted to the mole or mass concentration form using the assumption of ideal gas, with a molar volume of 22.4 L/g mol at STP conditions (273 K and 1.0 atm).

$$1 \text{ ppm}_v = \frac{1 \text{ m}^3 \text{ chemical}}{1,000,000 \text{ m}^3 \text{ of air}}$$

$$1 \text{ ppm}_v \equiv \frac{1 \text{ m}^3 \text{ chemical}}{1,000,000 \text{ m}^3 \text{ of air}} \left(\frac{\text{mol}}{22.4 \text{ L}}\right)\left(\frac{1000 \text{ L}}{\text{m}^3}\right) \equiv 4.46 \times 10^{-5} \frac{\text{mol}}{\text{m}^3}$$

$$\equiv 4.46 \times 10^{-5} \frac{\text{mol}}{\text{m}^3}\left(\frac{80 \text{ g}}{\text{gmol}}\right) \equiv 0.0035 \frac{\text{g}}{\text{m}^3} \equiv 3.5 \frac{\text{mg}}{\text{m}^3} \equiv 3.5 \frac{\mu\text{g}}{\text{L}}$$

The general relationship is 1 ppm = (MW/22.4) mg/m³.

2. Water phase. The mass ratio of 1 ppm can be converted to mole or mass concentration form using the density of water, which is 1 g/cm³ at 4°C and 1 atm.

$$1 \text{ ppm} = \frac{1 \text{ g chemical}}{1,000,000 \text{ g water}}$$

$$1 \text{ ppm} \equiv \frac{1 \text{ g chemical}}{1,000,000 \text{ g water}} \left(1\frac{\text{g}}{\text{cm}^3}\right)\left(\frac{100^3 \text{cm}^3}{\text{m}^3}\right) \equiv 1\frac{\text{g}}{\text{m}^3} \equiv 1\frac{\text{mg}}{\text{L}}$$

$$\equiv 1\frac{\text{g}}{\text{m}^3}\left(\frac{\text{mol}}{80 \text{ g}}\right) \equiv 0.0125 \equiv \frac{\text{mol}}{\text{m}^3}$$

3. Soil phase. The conversion is direct

$$1 \text{ ppm} = \frac{1 \text{ g chemical}}{1,000,000 \text{ g soil}}$$

$$1 \text{ ppm} = \frac{1 \text{ g chemical}}{1,000,000 \text{ g soil}} \left(\frac{1000 \text{ g}}{\text{kg}}\right)\left(\frac{1000 \text{ mg}}{\text{g}}\right) = 1\frac{\text{mg}}{\text{kg}}$$

Example 3.2

Problem:

Analysis of a water sample from a pond gave the following results: volume of sample = 2 L; concentration of suspended solids in the sample = 15 mg/L; concentration of dissolved chemical = 0.01 mol/L; and concentration of the chemical adsorbed onto the suspended solids = 400 μg/g solids. If the molecular weight of the chemical is 125, determine the total mass of the chemical in the sample.

Solution:

Dissolved concentration = molar concentration × MW

$$= \ 0.001 \frac{\text{Mol}}{\text{L}} \left(\frac{125 \ \text{g}}{\text{gmol}} \right) = 0.125 \frac{\text{g}}{\text{L}}$$

Dissolved mass in sample = dissolved concentration × volume

$$= \left(0.125 \frac{\text{g}}{\text{L}} \right) \times (2\text{L}) = 0.25\text{g}$$

Mass of solids in sample = concentration of solids × volume

$$= \left(25 \frac{\text{mg}}{\text{L}} \right) \times (2\text{L}) = 50 \ \text{mg} = 0.05 \ \text{g}$$

Adsorbed mass in sample = adsorbed concentration × mass of solids

$$= \left(400 \frac{\mu\text{g}}{\text{g}} \right) \times (0.05 \ \text{g}) \left(\frac{\text{g}}{10^6 \mu\text{g}} \right) = 0.00020 \ \text{g}$$

Thus, total mass of chemical in the sample = 0.25 g + 0.00020 g = 0.25020 g.

3.3 PHASE EQUILIBRIUM AND STEADY STATE

The concept of phase equilibrium (balance of forces) is an important one in environmental modeling. In the case of mechanical equilibrium, consider the following example. A cup sitting on a table top remains at rest because the downward force exerted by the Earth's gravity action on the cup's mass (this is what is meant by the "weight" of the cup) is exactly balanced by the repulsive force between atoms that prevents two objects from simultaneously occupying the same space, acting in this case between the table surface and the cup. If one picks up the cup and raises it above the tabletop, the additional upward force exerted by the arm destroys the state of equilibrium as the cup moves upward. If one wishes to hold the cup at rest above the table, it is necessary to adjust the upward force to balance the weight of the cup exactly, thus restoring equilibrium.

For more pertinent examples (chemical equilibrium, for example) consider the following. Chemical equilibrium is a dynamic system in which chemical changes are taking place in such a way that no overall change occurs in the composition of the system. In addition to partial ionization, equilibrium situations include simple reactions — for example, when the air in contact with a liquid is saturated with the liquid's vapor, meaning that the rate of evaporation is equal to the rate of condensation. When a solution is saturated with a solute, the dissolving rate is just equal to the precipitation rate from solution. In each of these cases, both processes continue. The equality of rate creates the illusion of static conditions, and no reaction actually goes to completion.

Equilibrium is best described by the principle of Le Chatelier, which sums up the effects of changes in any of the factors influencing the position of equilibrium. It states that a system in equilibrium, when subjected to a stress resulting from a change in temperature, pressure, or concentration, and causing the equilibrium to be upset, will adjust its position of equilibrium to relieve the stress and re-establish equilibrium.

What is the difference between steady state and equilibrium? Steady state implies no changes with the passage of time; similarly, equilibrium can also imply no change of state with passage of time. In many situations, the system not only is at steady state, but also is at equilibrium. However, this is not always the case. Sometimes, when flow rates are steady but the phase contents, for example, are not being maintained at the "equilibrium values," the system is at steady state but not at equilibrium.

3.4 MATH OPERATIONS AND LAWS OF EQUILIBRIUM

Earlier we observed that no chemical reaction goes to completion; the qualitative consequences of this insight go beyond the purpose of this text. However, in this text we describe and use the basic quantitative aspects of equilibria. The chemist usually starts with the chemistry of the reaction and fully uses chemical intuition before resorting to mathematical techniques. That is, science should always precede mathematics in the study of physical phenomena. Note, however, that most chemical problems do not need an exact, closed-form solution, and the direct application of mathematics to a problem can lead to an impasse.

Several basic math operations and fundamental laws from physical chemistry and thermodynamics serve as the tools, blueprints, and foundational structures of mathematical models. They can be used and applied to environmental systems under certain conditions, serving to solve various problems. Many laws serve as important links between the state of the system, chemical properties, and their behavior. In the following sections, we review some of the basic math operations used to solve basic equilibrium problems, as well as laws essential for modeling the fate and transport of chemicals in natural and engineered environmental systems.

3.4.1 Solving Equilibrium Problems

In the following math operations, we provide examples of the various forms of hydrogen combustion to yield water to demonstrate the solution of equilibrium problems. The example reactions are represented by the following equation. *Note*: Consider the reaction at 1000.0 K where all constituents are in the gas phase and the equilibrium constant is 1.15×10^{10} atm^{-1}:

$$2 H_2(g) + O_2(g) = 2 H_2O(g)$$

and the equilibrium constant expression:

$$K = [H_2O]^2/[H_2]^2[O_2] \tag{3.12}$$

where concentrations are given as partial pressures in atmospheres. Note that K is very large, and consequently the concentration of water is large and/or the concentration of at least one of the reactants is very small.

Example 3.3

Problem:

Consider a system at 1000.0 K in which 4.00 atm of oxygen is mixed with 0.500 atm of hydrogen and no water is initially present. Note that oxygen is in excess and hydrogen is the limiting reagent. Because the equilibrium constant is very large, virtually all the hydrogen is converted to water yielding $[H_2O] = 0.500$ atm and $[O_2] = 4.000 - 0.5(0.500) = 3.750$ atm. The final concentration of hydrogen, a small number, is an unknown — the only unknown.

Solution:

Using the equilibrium constant expression, we obtain:

$$1.15 \times 10^{10} = (0.500)^2/[H_2]^2(3.750)$$

from which we determine that $[H_2] = 2.41 \times 10^{-6}$ atm. Because this is a small number, our initial approximation is satisfactory.

Example 3.4

Problem:

Again, consider a system at 1000.0 K, where 0.250 atm of oxygen is mixed with 0.500 atm of hydrogen and 2000 atm of water.

Solution:

Again, the equilibrium constant is very large and the concentration of least reactants must be reduced to a very small value.

$$[H_2O] = 2.000 + 0.500 = 2.500 \text{ atm}$$

In this case, oxygen and hydrogen are present in a 1:2 ratio, the same ratio given by the stoichiometric coefficients. Neither reactant is in excess, and the equilibrium concentrations of both will be very small values. We have two unknowns, but they are related by stoichiometry. Because neither product is in excess and one molecule of oxygen is consumed for two of hydrogen, the ratio $[H_2]/[O_2] = 2/1$ is preserved during the entire reaction and $[H_2] = 2[O_2]$.

$$1.15 \times 10^{10} = 2.500^2/(2[O_2])^2[O_2]$$

$$[O_2] = 5.14 \times 10^{-4} \text{ atm and } [H_2] = 2[O_2] = 1.03 \times 10^{-3} \text{ atm}$$

3.4.2 Laws of Equilibrium

Some of the laws essential for modeling the fate and transport of chemicals in natural and engineered environmental system include:

- Ideal gas law
- Dalton's law
- Raoults' law
- Henry's law

3.4.2.1 Ideal Gas Law

An ideal gas is defined as one in which all collisions between atoms or molecules are perfectly elastic and have no intermolecular attractive forces. One can visualize it as collections of perfectly hard spheres, which collide but otherwise do not interact with each other. In such a gas, all the internal energy is in the form of kinetic energy, and any change in the internal energy is accompanied by a change in temperature. An ideal gas can be characterized by three state variables: absolute pressure (P); volume (V); and absolute temperature (T). The relationship between them may be deduced from kinetic theory and is called the ideal gas law:

$$PV = nRT = NkT \tag{3.13}$$

where:

n = number of moles
R = universal gas constant = 8.3145 J/mol K or 0.821 L-atm/de-mol
N = number of molecules
k = Boltzmann constant = 1.38066×10^{-23} J/K
   = R/N_A = where N_A = Avogadros number = 6.0221×10^{23}

Note: At standard temperature and pressure (STP), the volume of 1 mol of ideal gas is 22.4 L, a volume called the *molar volume of a gas*.

Example 3.5

Problem:
   Calculate the volume of 0.333 mol of gas at 300 K under a pressure of 0.950 atm.

Solution:

$$V = \frac{nRT}{P} = \frac{0.333 \text{ mol} \times 0.0821 \text{ L atm/K mol} \times 300 \text{ K}}{0.959 \text{ atm}} = 8.63 \text{ L}$$

Most gases in environmental systems can be assumed to obey this law. The ideal gas law can be viewed as arising from the kinetic pressure of gas molecules colliding with the walls of a container in accordance with Newton's laws. However, a statistical element is also present in the determination of the average kinetic energy of those molecules. The temperature is taken to be proportional to this average kinetic energy; this invokes the idea of kinetic temperature.

3.4.2.2 Dalton's Law

Dalton's law states that the pressure of a mixture of gases is equal to the sum of the pressures of all of the constituent gases alone. Mathematically, this can be represented as:

$$P_{Total} = P_1 + P_2 \ldots P_n \tag{3.14}$$

where:
P_{Total} = total pressure
P_1 = partial pressure and

$$\text{Partial P} = \frac{njRT}{V} \tag{3.15}$$

where n_j is the number of moles of component j in the mixture.

Note: Although Dalton's law explains that the total pressure is equal to the sum of all of the pressures of the parts, this is only absolutely true for ideal gases, although the error is small for real gases.

Example 3.6

Problem:

The atmospheric pressure in a lab is 102.4 kPa. The temperature of the water sample is 25°C, with pressure as 23.76 torr. If we use a 250-mL beaker to collect hydrogen from the water sample, what is the pressure of the hydrogen, and the moles of hydrogen using the ideal gas law?

Solution:

Step 1. Make the following conversions: a torr is 1 mm Hg at standard temperature. In kilopascals, that would be 3.17 (1 mm Hg = 7.5 kPa). Convert 250 mL to 0.250 L and 25°C to 298 L.
Step 2. Use Dalton's law to find the hydrogen pressure.

$$P_{Total} = P_{Water} + P_{Hydrogen}$$

$$102.4 \text{ kPa} = 3.17 \text{ kPa} + P_{Hydrogen}$$

$$P_{Hydrogen} = 99.23 \text{ kPa or pp.2 kPa}$$

Step 3. Recall that the ideal gas law is:

$$PV = nRT$$

where:
P is pressure
V is volume
n is moles
R is the ideal gas constant (0.821 L-atm/mol-K or 8.31 L-kPa/mol-K)
T is temperature

Therefore,

$$99.2 \text{ kPa} \times .250 \text{ L} = n \times 8.31 \text{ L-kPa/mol} \times 298 \text{ K}$$

rearranged:

$$n = 99.2 \text{ kPa} \times .250 \text{ L}/8.31 \text{ L-kPa/mol-K}/298 \text{ K}$$

$$n = .0100 \text{ mol or } 1.00 \times 10^{-2} \text{ mol H}$$

3.4.2.3 Raoult's Law

Raoult's law states that the vapor pressure of mixed liquids is dependent on the vapor pressures of the individual liquids and the molar fraction of each present. Accordingly, for concentrated solutions in which the components do not interact, the resulting vapor pressure (p) of component a in equilibrium with other solutions can be expressed as

$$P = x_a P_a \tag{3.16}$$

where:
p = resulting vapor pressure
x = mole fraction of component a in solution
P_a = vapor pressure of pure a at the same temperature and pressure as the solution

3.4.2.4 Henry's Law

Henry's law states that the mass of a gas that dissolves in a definite volume of liquid is directly proportional to the pressure of the gas, provided the gas does not react with the solvent. A formula for Henry's law is:

$$p = Hx \tag{3.17}$$

where:
x is the solubility of a gas in the solution phase
H is Henry's constant
p is the partial pressure of a gas above the solution

Hemond and Fechner–Levy (2000) point out that the Henry's law constant, H,

> ... is a partition coefficient usually defined as the ratio of a chemical's concentration in air to its concentration in water at equilibrium. Henry's law constants generally increase with increased temperature, primarily due to the significant temperature dependency of chemical vapor pressures; solubility is much less affected by the changes in temperature normally found in the environment.

H can be expressed in a dimensionless form or with units. Table 3.1 lists Henry's law constants for some common environmental chemicals.

3.5 CHEMICAL TRANSPORT SYSTEMS

In environmental modeling, environmental engineers must have a fundamental understanding of the phenomena involved with the transport of certain chemicals through the various components of the environment. The primary transport mechanism at the macroscopic level (referred to as *dispersive* transport) is by molecular diffusion driven by concentration gradients. Mixing and bulk movement (referred to as *advective* transport) of the medium are primary transport mechanisms at the macroscopic level.

Advective and dispersive transports are driven by fluid element; for example, advection is the movement of dissolved solute with flowing groundwater. The amount of contaminant transported is a function of its concentration in the groundwater and the quantity of groundwater flowing, and

Table 3.1   Henry's Law Constants (H)

Chemical	Henry's law constant (atm × m³/mol)	Henry's law constant (dimensionless)
Aroclor 1254	2.7×10^{-3}	1.2×10^{-1}
Aroclor 1260	7.1×10^{-3}	3.0×10^{-1}
Atrazine	3×10^{-9}	1×10^{-7}
Benzene	5.5×10^{-3}	2.4×10^{-1}
Benz[a]anthracene	5.75×10^{-6}	2.4×10^{-4}
Carbon tetrachloride	2.3×10^{-2}	9.7×10^{-1}
Chlorobenzene	3.7×10^{-3}	1.65×10^{-1}
Chloroform	4.8×10^{-3}	2.0×10^{-1}
Cyclohexane	0.18	7.3
1,1-Dichloroethane	6×10^{-3}	2.4×10^{-1}
1,2-Dichloroethane	10^{-3}	4.1×10^{-2}
cis-1,2-Dichloroethene	3.4×10^{-3}	0.25
trans-1,2-Dichlorethene	6.7×10^{-3}	0.23
Ethane	4.9×10^{-1}	20
Ethanol	6.3×10^{-6}	
Ethylbenzene	8.7×10^{-3}	3.7×10^{-1}
Lindane	4.8×10^{-7}	2.2×10^{-5}
Methane	0.66	27
Methylene chloride	3×10^{-3}	1.3×10^{-1}
n-Octane	2.95	121
Pentachlorophenol	3.4×10^{-6}	1.5×10^{-4}
n-Pentane	1.23	50.3
Perchloroethane	8.3×10^{-3}	3.4×10^{-1}
Phenanthrene	3.5×10^{-5}	1.5×10^{-3}
Toluene	6.6×10^{-3}	2.8×10^{-1}
1,1,1-Trichloroethane (TCA)	1.8×10^{-2}	7.7×10^{-1}
Trichloroethene (TCE)	1×10^{-2}	4.2×10^{-1}
o-Xylene	5.1×10^{-3}	2.2×10^{-1}
Vinyl chloride	2.4	99

Source: Adapted from Lyman, W.J., Reehl, W.F., and Rosenblatt, D.H., 1990, *Handbook of Chemical Property Estimation Methods*, 2nd printing, American Chemical Society, Washington, D.C.

advection will transport contaminants at different rates in each stratum. Diffusive transport is the process by which a contaminant in water moves from an area of greater concentration toward an area of less concentration. Diffusion occurs as long as a concentration gradient exists, even if the fluid is not moving; as a result, a contaminant may spread away from the place where it is introduced into a porous medium.

3.6   A FINAL WORD ON ENVIRONMENTAL MODELING

In this chapter we have provided survey coverage of some of the basic math and science involved in environmental modeling. In today's computer age, environmental engineers have the advantage of choosing from a plethora of available mathematical models. These models enable environmental engineers and students with minimal computer programming skills to develop computer-based mathematical models for natural and engineered environmental systems. Commercially available syntax-free authoring software can be adapted to create customized, high-level models of environmental phenomena in groundwater, air, soil, aquatic, and atmospheric systems.

We highly recommend that aspiring environmental engineering students take full advantage of college level computer modeling courses. Without such a background, the modern environmental engineer's technical toolbox is missing a vital tool.

REFERENCES

Hemond, F.H. and Fechner–Levy, E.J. (2000). *Chemical Fate and Transport in the Environment*, 2nd ed. San Diego: Academic Press.

Lyman, W.J., Reehl, W.R., and Rosenblatt, D.H. (1990). *Handbook of Chemical Property Estimation Methods*, 2nd printing. Washington, D.C.: American Chemical Society.

Nirmalakhandan, N. (2002). *Modeling Tools for Environmental Engineers and Scientists*. Boca Raton, FL: CRC Press.

CHAPTER 4

Algorithms and Environmental Engineering

4.1 INTRODUCTION

In Chapter 3, we pointed out that environmental modeling has become an important tool within the well-equipped environmental engineer's toolbox. Using an analogy, we can say that if a typical skilled handyperson's toolbox contains a socket set ratchet and several differently sized wrench attachments, then the well-equipped environmental engineer's toolbox includes a number of environmental models (socket set ratchets) with a varying set of algorithms (socket wrench attachments). Although a complete treatment or discussion of algorithms is beyond the scope of this book, we do provide basic underlying explanations of what algorithms are and examples of their applications in cyber space (to the real world?). For those interested in a more complete discussion of algorithms, many excellent texts are available on the general topic. We list several of these resources in the recommended reading section at the chapter's end.

4.2 ALGORITHMS: WHAT ARE THEY?

An *algorithm* is a specific mathematical calculation procedure. More specifically, according to Cormen et al. (2002), "an algorithm is any well-defined computational procedure that takes some value, or set of values, as input and produces some value, or set of values, as output." In other words, an algorithm is a recipe for an automated solution to a problem. A computer model may contain several algorithms. According to Knuth (1973), the word "algorithm" is derived from the name "al-Khowarizmi," a ninth-century Persian mathematician.

Algorithms should not be confused with computations. An algorithm is a systematic method for solving problems, and computer science is the study of algorithms (although the algorithm was developed and used long before any device resembling a modern computer was available); the act of executing an algorithm — that is, manipulating data in a systematic manner — is called *computation*.

For example, the following algorithm (attributed to Euclid, circa 300 B.C., and thus known for millennia) for finding the greatest common divisor of two given whole numbers may be stated as follows:

- Set a and b to the values A and B, respectively.
- Repeat the following sequence of operations until b has value 0:
  - Let r take the value of $a \bmod b$.
  - Let a take the value of b.
  - Let b take the value of r.
- The greatest common divisor of A and B is final value of a.

Note: The operation $a \bmod b$ gives the 'remainder' obtained on dividing a by b.

The problem — that of finding the greatest common divisor of two numbers — is specified by stating what is to be computed; the problem statement does not require that any particular algorithm be used to compute the value. Such method-independent specifications can be used to define the meaning of algorithms: the meaning of an algorithm is the value that it computes.

Several methods can be used to compute the required value; Euclid's method is just one. The chosen method assumes a set of standard operations (such as basic operations on the whole number and a means to repeat an operation) and combines these operations to form an operation that computes the required value. Also, something not at all obvious to the vast majority of people is that the proposed algorithm does actually compute the required value. One reason why a study of algorithms is important is to develop methods that can be used to establish what a proposed algorithm achieves.

4.3 EXPRESSING ALGORITHMS

Again, although an in-depth discussion is beyond the scope of this text, we point out that the analysis of algorithms often requires us to draw upon a body of mathematical operations. Some of these operations are as simple as high-school algebra, but others may be less familiar to the average environmental engineer. Along with learning how to manipulate asymptotic notations and solving recurrences, an environmental engineer must learn several other concepts and methods in order to analyze algorithms.

For example, methods for evaluating bounding summations are important. Bounding summations are algorithms that contain an iterative control construct such as a *while* or *for* loop; the algorithm running time can be expressed as the sum of the times spent on each execution of the body of the loop. These occur frequently in the analysis of algorithms. Many of the formulae commonly used in analyzing algorithms can be found in any calculus text. In addition, in order to analyze many algorithms, we must be familiar with the basic definitions and notations for sets, relations, functions, graphs, and trees. A basic understanding of elementary principles of counting (permutations, combinations, and the like) is important as well. Most algorithms used in environmental engineering require no probability for their analysis; however, a familiarity with these operations can be useful.

Because mathematical and scientific analysis (and many environmental engineering functions) is so heavily based upon numbers, people tend to associate computation with numbers. However, this need not be the case: algorithms can be expressed using any formal manipulation system, i.e., any system that defines a set of entities and a set of unambiguous rules for manipulating those entities. For example, the system called the SKI calculus consists of three combinators (entities) called (coincidentally) *S*, *K*, and *I*. The computation rules for the calculus are:

$$Sfgx = fx\,(gx)$$

$$Kxy = x$$

$$Ix = x$$

where *f, g, x,* and *y* are strings of the three entities. The SKI calculus is computationally complete; that is, any computation that can be performed using any formal system can be performed using the SKI calculus (equivalently, all algorithms can be expressed using the SKI calculus). Not all systems of computation are equally as powerful, however; some problems that can be solved using one system cannot be solved using another. Furthermore, we know that some problems cannot be solved using any formal computation system.

4.4 GENERAL ALGORITHM APPLICATIONS

Practical applications of algorithms are ubiquitous. All computer programs are expressions of algorithms, in which the instructions are expressed in the computer language used to develop the program. Computer programs are described as expressions of algorithms because an algorithm is a general technique for achieving some purpose that can be expressed in a number of different ways. Algorithms exist for many purposes and are expressed in many different ways. Examples of algorithms include recipes in cook books; servicing instructions in a computer's hardware manual; knitting patterns; digital instructions to a welding robot as to where each weld should be made; or cyber-speak to any system working cyberspace.

Algorithms can be used in sorting operations, for example, to reorder a list into some defined sequence. Expressing the same algorithms as instructions to a human with a similar requirement to reorder some list (for example, to sort a list of tax record reports into a sequence determined by the date of birth on the record) is possible. These instructions could employ the *insertion sort* algorithm, or the *bubble sort* algorithm or one of many other available algorithms. Thus, an algorithm as a general technique for expressing the process of completing a defined task is independent of the precise manner in which it is expressed.

Sorting is by no means the only application for which algorithms have been developed. Examples of practical applications of algorithms include:

- Internet routing: single-source shortest paths
- Search engines: string matching
- Public-key cryptography and digital signatures: number-theoretic algorithms
- Allocating scarce resources in the most beneficial way: linear programming

Algorithms are at the core of most technologies used in contemporary computers:

- Hardware design uses algorithms.
- The design of any GUI relies on algorithms.
- Routing in networks relies heavily on algorithms.
- Compilers, interpreters, or assemblers make extensive use of algorithms.

A frequently used example (because of its general usefulness and because the algorithm is easy to explain to almost anyone) of how an algorithm can be applied to real world situations is known as the *traveling salesman* problem. This problem is the most notorious NP-complete problem (no polynomial-time algorithm has yet been discovered for an NP-complete problem, nor has anyone yet been able to prove that no polynomial-time algorithm can exist for any one of them).

The traveling salesman problem: imagine a traveling salesman who must visit each of a given set of cities by car. Using Figure 4.1A, find the cycle of minimum cost (Figure 4.1B), visiting all of the vertices exactly once.

We have pointed out some of the functions that algorithms can perform. The question arises: "Can every problem be solved algorithmically?" The simple and complex answer is NO. For example, problems exist for which no generalized algorithmic solution can possibly exist (unsolvable). Also, problems exist for which no efficient solution is known — e.g., NP-complete problems — that is, whether or not efficient algorithms exist for NP-complete problems is unknown; if an efficient algorithm exists for any one of them, then efficient algorithms exist for all of them (e.g.,

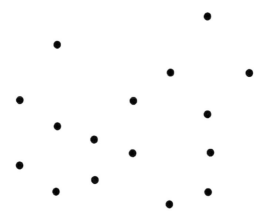

Figure 4.1a   Input.

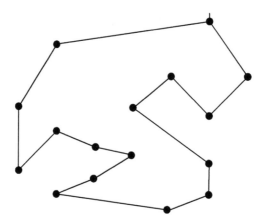

Figure 4.1b   Output.

traveling salesman problem). Finally, some problems exist that we do not know how to solve algorithmically.

From the preceding discussion, it should be apparent that computer science is not just about word processing and spreadsheets; it can include applications and not just software applications.

4.5   ENVIRONMENTAL ENGINEERING ALGORITHM APPLICATIONS

Although algorithms can be used in transportation applications (traveling salesman problem), many of their most important applications are applied to engineering functions. For example, consider a robot arm assigned to weld all the metal parts on an automobile in an assembly line. The shortest tour that visits each weld point exactly once defines the most efficient path for the robot. A similar application arises in minimizing the amount of time taken by a design engineer or draftsperson to draw a given structure.

Algorithms have found widespread application in all branches of engineering. In environmental engineering, for example, the USEPA uses computer models that employ various algorithms to monitor chemical-spill and ultimate-fate data. In the following section, we provide selected model summary descriptions of applications used in dispersion modeling. Specifically, we discuss how the USEPA (and others) employ preferred/recommended models (refined models recommended for a specific type of regulatory application) in monitoring air quality, ambient pollutant concentrations, and their temporal and spatial distribution. Further information on this important topic can be accessed at USEPA's Technology Transfer Network Support Center for Regulatory Air Models at www.epa.gov/scram001/tt22.htm Web site.

4.6 DISPERSION MODELS

The following algorithm-based models are currently listed in Appendix A of the *Guidelines on Air Quality Models* (published as Appendix W of 40 CFR Part 51):

- BLP (buoyant line and point source model) — a Gaussian plume dispersion model designed to handle unique modeling problems associated with aluminum reduction plants and other industrial sources in which plume rise and downwash effects from stationary line sources are important.
- CALINE3 — a steady-state Gaussian dispersion model designed to determine air pollution concentrations at receptor locations downwind of "at-grade," "fill," "bridge," and "cut section" highways located in relatively uncomplicated terrain.
- CALPUFF — a multilayer, multispecies non-steady-state puff dispersion model that simulates the effects of time- and space-varying meteorological conditions on pollution transport, transformation and removal. CALPUFF can be applied on scales of tens to hundreds of kilometers. It includes algorithms for subgrid scale effects (such as terrain impingement), as well as longer range effects (including pollutant removal from wet scavenging and dry deposition, chemical transformation, and visibility effects of particulate matter concentrations).
- CTDMPLUS (complex terrain dispersion model plus algorithms for unstable situations) — a refined point-source Gaussian air quality model for use in all stability conditions for complex terrain (terrain exceeding the height of the stack being modeled as contrasted with simple terrain — an area in which terrain features are lower in elevation than the top of the stack of the source). The model contains, in its entirety, the technology of CTDM for stable and neutral conditions.
- ISC3 (industrial source complex model) — a steady-state Gaussian plume model that can be used to assess pollutant concentrations from a wide variety of sources associated with an industrial complex. This model can account for the following: settling and dry deposition of particles; downwash; point, area, line, and volume sources; plume rise as a function of downwind distance; separation of point sources; and limited terrain adjustment. ISC3 operates in long- and short-term modes.
- OCD (offshore and coastal dispersion model) — a straight-line Gaussian model developed to determine the impact of offshore emissions from point, area, or line sources on the air quality of coastal regions. OCD incorporates overwater plume transport and dispersion as well as changes that occur as the plume crosses the shoreline. Hourly meteorological data are needed from offshore as well as onshore locations.

4.7 SCREENING TOOLS

Screening tools are relatively simple analysis techniques used to determine if a given source is likely to pose a threat to air quality. Concentration estimates from screening techniques precede a refined modeling analysis and are conservative.

- CAL3QHC/CAL3QHCR (CALINE3 with queuing and hot spot calculations) — a CALINE3-based CO model with a traffic model to calculate delays and queues that occur at signalized intersections; CAL3QHCR requires local meteorological data.
- COMPLEX 1 — a multiple-point source screening technique with terrain adjustment that incorporates the plume impaction algorithm of the VALLEY model.
- CTSCREEN (complex terrain screening model) — a Gaussian plume dispersion model designed as a screening technique for regulatory application to plume impaction assessments in complex terrain. CTSCREEN is a screening version of the CTDMPLUS model.

- LONGZ — a steady-state Gaussian plume formulation for urban and rural areas in flat or complex terrain to calculate long-term (seasonal and/or annual) ground-level ambient air concentrations attributable to emissions from up to 14,000 arbitrarily placed sources (stack, building, and area sources).
- SCREEN3 — a single source Gaussian plume model that provides maximum ground-level concentrations for point, area, flare, and volume sources, as well as concentrations in the cavity zone, and concentrations from inversion break-up and shoreline fumigation. SCREEN3 is a screening version of the ISC3 model.
- SHORTZ — a steady-state bivariate Gaussian plume formulation for urban and rural areas in flat or complex terrain to calculate ground-level ambient air concentrations. It can calculate 1-, 2-, 3-h, etc. average concentrations from emissions from stacks, buildings, and area sources for up to 300 arbitrarily placed sources.
- VALLEY — a steady-state, complex terrain, univariate Gaussian plume dispersion algorithm designed for estimating 24-h or annual concentrations resulting from emissions from up to 50 (total) point and area sources.
- VISCREEN — calculates the potential impact of a plume of specified emissions for specific transport and dispersion conditions.

REFERENCES

Cormen, T.H., Leiserson, C. E., Rivest, R.L., and Stein, C. (2002). *Introduction to Algorithms*, 2nd ed. New Delhi: Prentice–Hall of India.

Knuth, D.E. (1973). *Fundamental Algorithms,* Vol. 1 of *The Art of Computer Programming.* New York: Addison-Wesley.

USEPA (2003). Technology Transfer Network Support Center for Regulatory Air Models. Accessed at www.epa.gov/scram001/tt22.htm.

USEPA (1993). Guideline on Air Quality Models (revised Appendix W of 40 CFR part 51 [supplement b]). Accessed at http://www.osti.gov/energycitations/products.biblio.jsp?osti_id=5948584 on November 11, 2003.

SUGGESTED READING

Gusfield, D. (1997). *Algorithms on Strings, Trees, and Sequences: Computer Science and Computational Biology.* New York: Cambridge University Press.

Lafore, R. (2002). *Data Structures and Algorithms in Java*, 2nd ed. Indianapolis, IN: Sams.

Mitchell, T.M. (1997). *Machine Learning.* New York: McGraw-Hill.

Poynton, C. (2003). *Digital Video and HDTV Algorithms and Interfaces.* San Francisco: Morgan Kaufmann.

PART II

Fundamental Science and Statistics Review

CHAPTER 5

Fundamental Chemistry and Hydraulics

5.1 INTRODUCTION

It is not sufficient for a future working environmental engineer to understand the causes and effects of environmental problems in qualitative terms only. He or she must also be able to express the perceived problem and its potential solution in quantitative terms. To do this, environmental engineers must be able to draw on basic sciences such as chemistry, physics, and hydrology (as well as others) to predict the fate of pollutants in the environment and to design effective treatment systems to reduce impact. In this chapter, we discuss fundamental chemistry and basic hydraulics for environmental engineers.

5.2 FUNDAMENTAL CHEMISTRY

> The chemists are a strange class of mortals, impelled by an almost insane impulse to seek their pleasure among smoke and vapor, soot and flame, poisons and poverty; yet among all these evils I seem to live so sweetly that I may die if I would change places with the Persian King.
>
> **Johann Joachim Becher**

All matter on Earth consists of chemicals. This simplified definition may shock those who think chemistry is what happens in a test tube or between men and women. Chemistry is much more; it is the science of materials that make up the physical world. Chemistry is so complex that no one person could expect to master all aspects of such a vast field; thus, it has been found convenient to divide the subject into specialty areas. For example:

- *Organic chemists* study compounds of carbon. Atoms of this element can form stable chains and rings, giving rise to very large numbers of natural and synthetic compounds.
- *Inorganic chemists* are interested in all elements, particularly in metals, and are often involved in the preparation of new catalysts.
- *Biochemists* are concerned with the chemistry of the living world.
- *Physical chemists* study the structures of materials, and rates and energies of chemical reactions.
- *Theoretical chemists* use mathematics and computational techniques to derive unifying concepts to explain chemical behavior.
- *Analytical chemists* develop test procedures to determine the identity, composition, and purity of chemicals and materials. New analytical procedures often discover the presence of previously unknown compounds.

Why should we care about chemistry? Is it not enough to know that we do not want unnecessary chemicals in or on our food or harmful chemicals in our air, water, or soil?

Chemicals are everywhere in our environment. The vast majority of these chemicals are natural. Chemists often copy from nature to create new substances superior to and cheaper than natural materials. Our human nature makes us work to make nature serve us. Without chemistry (and the other sciences), we are at nature's mercy. To control nature, we must learn its laws and then use them.

Environmental engineers must also learn to use the laws of chemistry; however, they must know even more. Environmental engineers must know the ramifications of chemistry out of control. Chemistry properly used can perform miracles, but, out of control, chemicals and their effects can be devastating. In fact, many current environmental regulations dealing with chemical safety and emergency response for chemical spills resulted because of catastrophic events involving chemicals.

5.2.1 Density and Specific Gravity

When we say that iron is heavier than aluminum, we mean that iron has greater density than aluminum. In practice, what we are really saying is that a given volume of iron is heavier than the same volume of aluminum. Density (p) is the mass (weight) per unit volume of a substance at a particular temperature, although density generally varies with temperature. The weight may be expressed in terms of pounds, ounces, grams, kilograms, etc. The volume may be liters, milliliters, gallons, cubic feet, etc. Table 5.1 shows the relationship between the temperature, specific weight, and density of fresh water.

Table 5.1 Water Properties (Temperature, Specific Weight and Density)

Temperature (°F)	Specific weight (lb/ft³)	Density (slugs/ft³)	Temperature (°F)	Specific weight (lb/ft³)	Density (slugs/ft³)
32	62.4	1.94	130	61.5	1.91
40	62.4	1.94	140	61.4	1.91
50	62.4	1.94	150	61.2	1.90
60	62.4	1.94	160	61.0	1.90
70	62.3	1.94	170	60.8	1.89
80	62.2	1.93	180	60.6	1.88
90	62.1	1.93	190	60.4	1.88
100	62.0	1.93	200	60.1	1.87
110	61.9	1.92	210	59.8	1.86
120	61.7	1.92			

Source: Spellman, F.R., 2003, *Handbook of Water and Wastewater Treatment Plant Operations*. Boca Raton, FL: Lewis Publishers.

Suppose we had a tub of lard and a large box of crackers, each with a mass of 600 g. The density of the crackers would be much less than the density of the lard because the crackers occupy a much larger volume than the lard occupies. The density of an object can be calculated by using the formula:

$$\text{Density} = \frac{\text{Mass}}{\text{Volume}} \qquad (5.1)$$

In water/wastewater operations, perhaps the most common measures of density are pounds per cubic foot (lb/ft³) and pounds per gallon (lb/gal).

- 1 ft³ of water weighs 62.4 lb — density = 62.4 lb/ft³
- 1 gal of water weighs 8.34 lb — density = 8.34 lb/gal

The density of a dry material (e.g., cereal, lime, soda, or sand) is usually expressed in pounds per cubic foot. The densities of plain and reinforced concrete are 144 and 150 lb/ft^3, respectively. The density of a liquid (liquid alum, liquid chlorine, or water) can be expressed as pounds per cubic foot or as pounds per gallon. The density of a gas (chlorine gas, methane, carbon dioxide, or air) is usually expressed in pounds per cubic foot.

As shown in Table 5.1, the density of a substance like water changes slightly as the temperature of the substance changes. This occurs because substances usually increase in volume (size — they expand) as they become warmer. Because of this expansion with warming, the same weight is spread over a larger volume, so the density is lower when a substance is warm than when it is cold.

Specific gravity is defined as the weight (or density) of a substance compared to the weight (or density) of an equal volume of water. (The specific gravity of water is 1.) This relationship is easily seen when a cubic foot of water (62.4 lb) is compared to a cubic foot of aluminum (178 lb). Aluminum is 2.7 times as heavy as water. Finding the specific gravity of a piece of metal is not difficult. We weigh the metal in air, then weigh it under water. Its loss of weight is the weight of an equal volume of water. To find the specific gravity, divide the weight of the metal by its loss of weight in water.

$$\text{Specific Gravity} = \frac{\text{Weight of Substance}}{\text{Weight of Equal Volume of Water}} \qquad (5.2)$$

Example 5.1

Problem:
Suppose a piece of metal weighs 150 lb in air and 85 lb under water. What is the specific gravity?

Solution:

Step 1. 150 lb – 85 lb = 65 lb-loss of weight in water
Step 2.

$$\text{Specific Gravity} = \frac{150}{65} = 2.3$$

Note that in a calculation of specific gravity, the densities *must* be expressed in the same units.

As stated earlier, the specific gravity of water is one, which is the standard — the reference to which all other liquid or solid substances are compared. Specifically, any object that has a specific gravity greater than one will sink in water (rocks, steel, iron, grit, floc, sludge). Substances with a specific gravity of less than one will float (wood, scum, and gasoline). Because the total weight and volume of a ship is less than one, its specific gravity is less than one; therefore, it can float,

The most common use of specific gravity in water/wastewater treatment operations is in gallons-to-pounds conversions. In many cases, the liquids handled have a specific gravity of 1.00 or very nearly 1.00 (between 0.98 and 1.02), so 1.00 may be used in the calculations without introducing significant error. However, in calculations involving a liquid with a specific gravity of less than 0.98 or greater than 1.02, the conversions from gallons to pounds must consider the exact specific gravity. The technique is illustrated in the following example.

Example 5.2

Problem:
A basin holds 1455 gal of a certain liquid. If the specific gravity of the liquid is 0.94, how many pounds of liquid are in the basin?

Solution:

If the substance's specific gravity were between 0.98 and 1.02, we would use the factor 8.34 lb/gal (the density of water) for a conversion from gallons to pounds. However, in this instance the substance has a specific gravity outside this range, so the 8.34 factor must be adjusted.

Step 1. Multiply 8.34 lb/gal by the specific gravity to obtain the adjusted factor:

$$(8.34 \text{ lb/gal}) (0.94) = 7.84 \text{ lb/gal (rounded)}$$

Step 2. Convert 1455 gal to pounds using the corrected factor:

$$(1455 \text{ gal}) (7.84 \text{ lb/gal}) = 11,407 \text{ lb (rounded)}$$

Example 5.3

Problem:

The specific gravity of a liquid substance is 0.96 at 64°F. What is the weight of 1 gal of the substance?

Solution:

$$\text{Weight} = \text{specific gravity} \times \text{weight of water}$$

$$= 0.96 \times 8.34 \text{ lb/gal}$$

$$= 8.01 \text{ lb}$$

Example 5.4

Problem:

A liquid has a specific gravity of 1.15. How many pounds is 66 gal of the liquid?

Solution:

$$\text{Weight} = 66 \text{ gal} \times 8.34 \text{ lb/gal} \times 1.15$$

$$= 633 \text{ lb}$$

Example 5.5

Problem:

If a solid in water has a specific gravity of 1.30, what percent heavier is it than water?

Solution:

$$\text{Percent heavier} = \frac{\text{sp gr of solid} - \text{sp gr of water}}{\text{sp gr of water}}$$

$$\text{Percent heavier} = \frac{1.30 - 1.0}{1.0}$$

$$= 30$$

5.2.2   Water Chemistry Fundamentals

Whenever we add a chemical substance to another chemical substance (adding sugar to tea or adding hypochlorite to water to make it safe to drink), we are performing the work of chemists because we are working with chemical substances; how they react is important to success. Environmental engineers involved with water treatment operations, for example, may be required to determine the amount of chemicals or chemical compounds to add (dosing) to various unit processes. Table 5.2 lists some of the chemicals and their common applications in water treatment operations.

Table 5.2   Chemicals and Chemical Compounds Used in Water Treatment

Name	Common application	Name	Common application
Activated carbon	Taste and odor control	Aluminum sulfate	Coagulation
Ammonia	Chloramine disinfection	Ammonium sulfate	Coagulation
Calcium hydroxide	Softening	Calcium hypochlorite	Disinfection
Calcium oxide	Softening	Carbon dioxide	Recarbonation
Copper sulfate	Algae control	Ferric chloride	Coagulation
Ferric sulfate	Coagulation	Magnesium hydroxide	Defluoridation
Oxygen	Aeration	Potassium permanganate	Oxidation
Sodium aluminate	Coagulation	Sodium bicarbonate	pH adjustment
Sodium carbonate	Softening	Sodium chloride	Ion exchanger regeneration
Sodium fluoride	Fluoridation	Sodium fluosilicate	Fluoridation
Sodium hexametaphosphate	Corrosion control	Sodium hydroxide	pH adjustment
Sodium hypochlorite	Disinfection	Sodium silicate	Coagulation aid
Sodium thiosulfate	Dechlorination	Sulfur dioxide	Dechlorination
Sulfuric acid	pH adjustment		

Source: Spellman, F.R., 2003, *Handbook of Water and Wastewater Treatment Plant Operations*. Boca Raton, FL: Lewis Publishers.

5.2.2.1   The Water Molecule

Just about everyone knows that water is a chemical compound of two simple and abundant elements: H_2O. Yet scientists continue to argue the merits of rival theories on the structure of water. The fact is that we still know little about water — for example, we do not know how water works.

The reality is that water is very complex, with many unique properties that are essential to life and determine its environmental chemical behavior. The water molecule is different. Two hydrogen atoms (the two in the H_2 part of the water formula) *always* come to rest at an angle of approximately 105° from each other. The hydrogens tend to be positively charged, and the oxygen tends to be negatively charged. This arrangement gives the water molecule an electrical polarity; that is, one end is positively charged and one end negatively charged. This 105° relationship makes water lopsided, peculiar, and eccentric; it breaks all the rules (Figure 5.1).

In the laboratory, pure water contains no impurities, but in nature water contains a lot of materials besides water. The environmental professional tasked with maintaining the purest or cleanest water possible must always consider those extras that ride along in water's flow. Water is often called the *universal solvent*, a fitting description when you consider that given enough time and contact, water will dissolve anything and everything on Earth.

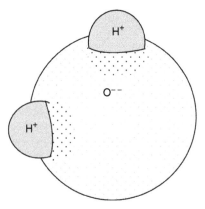

Figure 5.1 A molecule of water. (From Spellman, F.R., 2003, *Handbook of Water and Wastewater Treatment Plant Operations.* Boca Raton, FL: Lewis Publishers.)

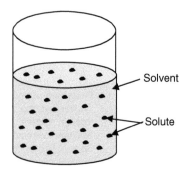

Figure 5.2 Solution with two components: solvent and solute. (From Spellman, F.R., 2003, *Handbook of Water and Wastewater Treatment Plant Operations.* Boca Raton, FL: Lewis Publishers.)

5.2.2.2 Water Solutions

A *solution* is a condition in which one or more substances are uniformly and evenly mixed or dissolved. In other words, a solution is a homogenous mixture of two or more substances. Solutions can be solids, liquids, or gases (drinking water, seawater, or air, for example). We focus primarily on liquid solutions.

A solution has two components: a *solvent* and a *solute* (see Figure 5.2). The solvent is the component that does the dissolving. Typically, the solvent is the substance present in the greater quantity. The solute is the component that is dissolved. When water dissolves substances, it creates solutions with many impurities.

Generally, a solution is usually transparent, not cloudy, and visible to longer wavelength ultraviolet light. Because water is colorless (hopefully), the light necessary for photosynthesis can travel to considerable depths. However, a solution may be colored when the solute remains uniformly distributed throughout the solution and does not settle with time.

When molecules dissolve in water, the atoms making up the molecules come apart (dissociate) in the water. This dissociation in water is called *ionization*. When the atoms in the molecules come apart, they do so as charged atoms (negatively charged and positively charged) called *ions*. The positively charged ions are called *cations* and the negatively charged ions are called *anions*.

Examples of Ionization

$CaCO_3 \leftrightarrow$ Calcium carbonate	Ca^{++} Calcium ion (cation)	+	CO_3^{-2} Carbonate ion (anion)
$NaCl \leftrightarrow$ Sodium chloride	Na^+ Sodium ion (cation)	+	Cl^- Chloride ion (anion)

Some of the common ions found in water are:

Ion	Symbol
Hydrogen	H^+
Sodium	Na^+
Potassium	K^+
Chloride	Cl^-
Bromide	Br^-
Iodide	I^-
Bicarbonate	HCO_3^-

Solutions serve as a vehicle to (1) allow chemical species to come into close proximity so that they can react; (2) provide a uniform matrix for solid materials, such as paints, inks, and other coatings, so that they can be applied to surfaces; and (3) dissolve oil and grease so that they can be rinsed away.

Water dissolves polar substances better than nonpolar substances. Polar substances (mineral acids, bases and salts) are easily dissolved in water. Nonpolar substances (oils and fats and many organic compounds) do not dissolve easily in water.

5.2.2.3 Concentrations

Because the properties of a solution depend largely on the relative amounts of solvent and solute, the concentrations of each must be specified.

> *Key point*: Chemists use relative terms such as saturated and unsaturated, as well as more exact concentration terms, such as weight percentages, molarity, and normality.

Although polar substances dissolve better than nonpolar substances in water, polar substances dissolve in water only to a point; that is, only so much solute will dissolve at a given temperature. When that limit is reached, the resulting solution is saturated. At that point, the solution is in equilibrium — no more solute can be dissolved. A liquid/solids solution is supersaturated when the solvent actually dissolves more than an equilibrium concentration of solute (usually when heated).

Specifying the relative amounts of solvent and solute, or specifying the amount of one component relative to the whole, usually gives the exact concentrations of solutions. Solution concentrations are sometimes specified as weight percentages.

$$\% \text{ of Solute} = \frac{\text{Mass of Solute}}{\text{Mass of Solute}} \times 100 \tag{5.3}$$

To understand the concepts of *molarity*, *molality*, and *normality*, we must first understand the concept of a *mole.* The mole is the amount of a substance that contains exactly the *same number of items* (i.e., atoms, molecules, or ions) as 12 g of Carbon-12. By experiment, Avagodro determined this number to be 6.02×10^{23} (to three significant figures).

If 1 mol of C atoms equals 12 g, how much is the mass of 1 mol of H atoms?

- Note that carbon is 12 times as heavy as hydrogen.
- Therefore, we need only 1/12 the weight of H to equal the same number of atoms of C.

   Key point: 1 mol of H equals 1 g.

By the same principle:

- 1 mol of CO_2 = 12 + 2(16) = 44 g
- 1 mol of Cl^- = 35.5 g
- 1 mol of Ra = 226 g

In other words, we can calculate the mass of a mole if we know the formula of the "item."

Molarity (M) is defined as the number of moles of solute per liter of solution. The volume of a solution is easier to measure in the lab than its mass.

$$M = \frac{\text{No. of moles of solute}}{\text{No. of liters of solution}} \tag{5.4}$$

Molality (m) is defined as the number of moles of solute per kilogram of solvent.

$$M = \frac{\text{No. of moles of solute}}{\text{No. of kilograms of solution}} \tag{5.5}$$

   Key point: Molality is not as frequently used as molarity, except in theoretical calculations.

Especially for acids and bases, the normality (N) rather than the molarity of a solution is often reported — the number of equivalents of solute per liter of solution (1 equivalent of a substance reacts with 1 equivalent of another substance).

$$N = \frac{\text{No. of equivalents of solute}}{\text{No. of liters of solution}} \tag{5.6}$$

In acid/base terms, an equivalent (or gram equivalent weight) is the amount that will react with 1 mol of H^+ or OH^-. For example,

1 mol of HCl will generate 1 mol of H^+
   Therefore, 1 mol HCl = 1 equivalent
1 mol of $Mg(OH)_2$ will generate 2 mol of OH^-
   Therefore, 1 mol of $Mg(OH)_2$ = 2 equivalents

$$HCl \Rightarrow H^+ + Cl^- \qquad\qquad Mg(OH)^{+2} \Rightarrow Mg^{+2} + 2OH^-$$

By the same principle:

A 1-M solution of H_3PO_4 = 3 N
A 2-N solution of H_2SO_4 = 1 M
A 0.5-N solution of NaOH = 0.5 M
A 2-M solution of HNO_3 = 2 N

Chemists titrate acid/base solutions to determine their normality. An endpoint indicator is used to identify the point at which the titrated solution is neutralized.

> *Key point*: If 100 mL of 1 N HCl neutralizes 100 mL of NaOH, then the NaOH solution must also be 1 N.

5.2.2.4 Predicting Solubility

Predicting solubility is difficult, but as a general rule of thumb: *like dissolves like*.

Liquid–Liquid Solubility

Liquids with similar structure and hence similar intermolecular forces are completely miscible. For example, we would correctly predict that methanol and water are completely soluble in any proportion.

Liquid–Solid Solubility

Solids *always* have limited solubilities in liquids because of the difference in magnitude of their intermolecular forces. Therefore, the closer the temperature comes to the melting point of a particular solid, the better the match is between a solid and a liquid.

> *Key point*: At a given temperature, lower melting solids are more soluble than higher melting solids. Structure is also important; for example, nonpolar solids are more soluble in nonpolar solvents.

Liquid–Gas Solubility

As with solids, the more similar the intermolecular forces, the higher the solubility. Therefore, the closer the match is between the temperature of the solvent and the boiling point of the gas, the higher the solubility is. When water is the solvent, an additional hydration factor promotes solubility of charged species. Other factors that can significantly affect solubility are temperature and pressure. In general, raising the temperature typically increases the solubility of solids in liquids.

> *Key point*: Dissolving a solid in a liquid is usually an *endothermic process* (i.e., heat is absorbed), so raising the temperature "fuels" this process. In contrast, dissolving a gas in a liquid is usually an *exothermic* process (it emits heat). Therefore lowering the temperature generally increases the solubility of gases in liquids.

> *Interesting point*: "Thermal" pollution is a problem because of the decreased solubility of O_2 in water at higher temperatures.

Pressure has an appreciable effect only on the solubility of gases in liquids. For example, carbonated beverages like soda water are typically bottled at significantly higher atmospheres. When the beverage is opened, the decrease in the pressure above the liquid causes the gas to bubble out of solution. When shaving cream is used, dissolved gas comes out of solution, bringing the liquid with it as foam.

5.2.2.5 Colligative Properties

Some properties of a solution depend on the concentrations of the solute species rather than their identity:

- Lowering vapor pressure
- Raising boiling point
- Decreasing freezing point
- Osmotic pressure

True colligative properties are directly proportional to the concentration of the solute but entirely independent of its identity.

Lowering Vapor Pressure

With all other conditions identical, the vapor pressure of water above the pure liquid is higher than that above sugar water. The vapor pressure above a 0.2-m sugar solution is the same as that above a 0.2-m urea solution. The lowering of vapor pressure above a 0.4-m sugar solution is twice as great as that above a 0.2-m sugar solution. Solutes lower vapor pressure because they lower the concentration of solvent molecules. To remain in equilibrium, the solvent vapor concentration must decrease (thus the vapor pressure decreases).

Raising the Boiling Point

A solution containing a nonvolatile solute boils at a higher temperature than the pure solvent. The increase in boiling point is directly proportional to the increase in solute concentration in dilute solutions. This phenomenon is explained by the lowering of vapor pressure already described.

Decreasing the Freezing Point

At low solute concentrations, solutions generally freeze or melt at lower temperatures than the pure solvent.

> *Key point*: The presence of dissolved "foreign bodies" tends to interfere with freezing, so solutions can only be frozen at temperatures below that of the pure solvent.
> *Key point*: We add antifreeze to the water in a radiator to lower its freezing point and increase its boiling point.

Osmotic Pressure

Water moves spontaneously from an area of high vapor pressure to an area of low vapor pressure. If allowed to continue, in the end all of the water would move to the solution. A similar process occurs when pure water is separated from a concentrated solution by a semipermeable membrane (one that only allows the passage of water molecules). The osmotic pressure is the pressure just adequate to prevent osmosis. In dilute solutions, the osmotic pressure is directly proportional to the solute concentration and is independent of its identity. The properties of electrolyte solutions follow the same trends as nonelectrolyte solutions, but are also dependent on the *nature* of the electrolyte as well as its concentration.

$$NaCl \quad Na_2SO_4 \quad CaCl_2 \quad MgSO_4$$

5.2.2.6 Colloids/Emulsions

A solution is a homogenous mixture of two or more substances (seawater, for example). A *suspension* is a brief comingling of solvent and undissolved particles (sand and water, for example). A *colloidal suspension* is a comingling of particles not visible to the naked eye but larger than individual molecules.

Table 5.3   Types of Colloids

Name	Dispersing medium	Dispersed phase
Solid sol	Solid	Solid
Gel	Solid	Liquid
Solid form	Solid	Gas
Sol	Liquid	Solid
Emulsion	Liquid	Liquid
Foam	Liquid	Gas
Solid aerosol	Gas	Solid
Aerosol	Liquid	Aerosol

Source: Adapted from Types of Colloids. Accessed @ http://www.ch.bris.ac.uk/webprojects2002/pdavies/types.html, December 18, 2002.

Key point: Colloidal particles do not settle out by gravity alone.

Colloidal suspensions can consist of:

- Hydrophilic "solutions" of macromolecules (proteins, for example) that spontaneously form in water
- Hydrophobic suspensions, which gain stability from their repulsive electrical charges
- Micelles — special colloids with charged hydrophilic "heads" and long hydrophobic "tails"

Colloids are usually classified according to the original states of their constituent parts (see Table 5.3). The stability of colloids can be primarily attributed to hydration and surface charge, which help to prevent contact and subsequent coagulation.

Key point: In many cases, water-based emulsions have been used to replace organic solvents (in paints and inks, for example), even though the compounds are not readily soluble in water.

In wastewater treatment, the elimination of colloidal species and emulsions is achieved by various means, including:

- Agitation
- Heat
- Acidification
- Coagulation (adding ions)
- Flocculation (adding bridging groups)

5.2.2.7   *Water Constituents*

Natural water can contain a number of substances — what we call impurities or *constituents*. When a particular constituent can affect the good health of the water user, it is called a *contaminant* or *pollutant*. These contaminants are the elements that the environmental practitioner works to prevent from entering or to remove from the water supply.

Solids

Other than gases, all water's contaminants contribute to the solids content. Natural waters carry a lot of dissolved solids as well as solids that are not dissolved. The undissolved solids are nonpolar substances and relatively large particles of materials — for example, silt, which will not dissolve. Classified by their size and state, by their chemical characteristics, and by their size distribution, solids can be dispersed in water in suspended and dissolved forms.

Size classifications for solids in water include:

* Suspended solids
* Settleable solids
* Colloidal solids
* Dissolved solids

Total solids are suspended and dissolved solids that remain behind when the water is removed by evaporation. Solids are also characterized as volatile or nonvolatile.

> *Key point*: Although not technically accurate from a chemical point of view because some finely suspended material can actually pass through the filter, suspended solids are defined as those that can be filtered out in the suspended solids laboratory test. Material that passes through the filter is defined as dissolved solids. Colloidal solids are extremely fine suspended solids (particles) of less than 1 μm in diameter; they are so small they will not settle even if allowed to sit quietly for days or weeks, although they may make water cloudy.

Turbidity

One of the first characteristics people notice about water is its clarity. Turbidity is a condition in water caused by the presence of suspended matter, resulting in the scattering and absorption of light rays. In plain English, turbidity is a measure of the light-transmitting properties of water. Natural water that is very clear (low turbidity) allows one to see images at considerable depths. High-turbidity water appears cloudy. Even water with low turbidity, however, can still contain dissolved solids because they do not cause light to be scattered or absorbed; thus, the water looks clear. High turbidity causes problems for the waterworks operator because the components that cause high turbidity can cause taste and odor problems and will reduce the effectiveness of disinfection.

Color

Water can be colored, but often the color of water can be deceiving. For example, color is considered an aesthetic quality of water, one with no direct health impact. Many of the colors associated with water are not "true" colors but the result of colloidal suspension (apparent color). This apparent color can be attributed to dissolved tannin extracted from decaying plant material. True color is the result of dissolved chemicals, usually organics that cannot be seen.

Dissolved Oxygen (DO)

Gases, including oxygen, carbon dioxide, hydrogen sulfide, and nitrogen, can be dissolved in water. Gases dissolved in water are important. For example, carbon dioxide plays an important role in pH and alkalinity. Carbon dioxide is released into the water by microorganisms and consumed by aquatic plants. Dissolved oxygen (DO) in water is most important to waterworks operators as an indicator of water quality. We stated earlier that solutions could become saturated with solute. Water can become saturated with oxygen. The amount of oxygen that can be dissolved at saturation depends upon the temperature of the water. However, in the case of oxygen, the effect is just the opposite of other solutes. The higher the temperature is, the lower the saturation level; the lower the temperature is, the higher the saturation level.

Metals

Metals are common constituents or impurities often carried by water. At normal levels, most metals are not harmful; however, a few metals can cause taste and odor problems in drinking water. Some

Table 5.4   Common Metals Found in Water

Metal	Health hazard
Barium	Circulatory system effects and increased blood pressure
Cadmium	Concentration in the liver, kidneys, pancreas, and thyroid
Copper	Nervous system damage and kidney effects; toxic to humans
Lead	Nervous system damage and kidney effects; toxic to humans
Mercury	Central nervous system disorders
Nickel	Central nervous system disorders
Selenium	Central nervous system disorders
Silver	Turns skin gray
Zinc	Causes taste problems, but not a health hazard

Source: Spellman, F.R., 2003, *Handbook of Water and Wastewater Treatment Plant Operations*. Boca Raton, FL: Lewis Publishers.

metals may be toxic to humans, animals, and microorganisms. Most metals enter water as part of compounds that ionize to release the metal as positive ions. Table 5.4 lists some metals commonly found in water and their potential health hazards.

Organic Matter

Organic matter or organic compounds contain the element carbon and are derived from material that was once alive (plants and animals):

- Fats
- Dyes
- Soaps
- Rubber products
- Woods
- Fuels
- Cotton
- Proteins
- Carbohydrates

Organic compounds in water are usually large, nonpolar molecules that do not dissolve well in water. They often provide large amounts of energy to animals and microorganisms.

Inorganic Matter

Inorganic matter or inorganic compounds are carbon free, not derived from living matter, and easily dissolved in water; they are of mineral origin. The inorganics include acids, bases, oxides, salts, and so forth. Several inorganic components are important in establishing and controlling water quality.

Acids

An *acid* is a substance that produces hydrogen ions (H^+) when dissolved in water. Hydrogen ions are hydrogen atoms that have been stripped of their electrons. A single hydrogen ion is nothing more than the nucleus of a hydrogen atom. Lemon juice, vinegar, and sour milk are acidic or contain acid. The common acids used in treating water are hydrochloric acid (HCl); sulfuric acid (H_2SO_4); nitric acid (HNO_3); and carbonic acid (H_2CO_3). Note that in each of these acids, hydrogen (H) is one of the elements. The relative strengths of acids in water, listed in descending order of strength, are classified in Table 5.5.

Table 5.5   Relative Strengths of Acids in Water

Acid	Chemical symbol
Perchloric acid	$HClO_4$
Sulfuric acid	H_2SO_2
Hydrochloric acid	HCl
Nitric acid	HNO_3
Phosphoric acid	H_3PO_4
Nitrous acid	HNO_2
Hydrofluoric acid	HF
Acetic acid	CH_3COOH
Carbonic acid	H_2CO_3
Hydrocyanic acid	HCN
Boric acid	H_3BO_3

Source: Spellman, F.R., 2003, *Handbook of Water and Wastewater Treatment Plant Operations*. Boca Raton, FL: Lewis Publishers.

Bases

A base is a substance that produces hydroxide ions (OH⁻) when dissolved in water. Lye or common soap (bitter things) contains bases. Bases used in waterworks operations are calcium hydroxide $(Ca(OH)_2)$, sodium hydroxide $(NaOH)$, and potassium hydroxide (KOH). Note that the hydroxyl group (OH) is found in all bases. Certain bases also contain metallic substances, such as sodium (Na), calcium (Ca), magnesium (Mg), and potassium (K). These bases contain the elements that produce alkalinity in water.

Salts

When acids and bases chemically interact, they neutralize each other. The compound other than water that forms from the neutralization of acids and bases is called a *salt*. Salts constitute, by far, the largest groups of inorganic compounds. A common salt used in waterworks operations, copper sulfate, is used to kill algae in water.

pH

pH is a measure of the hydrogen ion (H⁺) concentration. Solutions range from very acidic (having a high concentration of H⁺ ions) to very basic (having a high concentration of OH⁻ ions). The pH scale ranges from 0 to 14 with 7 as the neutral value (see Figure 5.3).

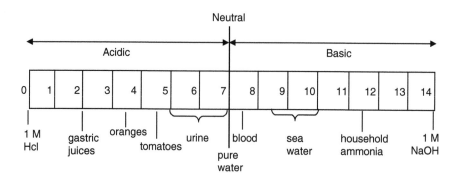

Figure 5.3   pH of selected liquids. (From Spellman, F.R., 2003, *Handbook of Water and Wastewater Treatment Plant Operations*. Boca Raton, FL: Lewis Publishers.)

The pH of water is important to the chemical reactions that take place within water, and pH values that are too high or low can inhibit the growth of microorganisms. High pH values are considered basic and low pH values are considered acidic. Stated another way, low pH values indicate a high level of H^+ concentration, while high pH values indicate a low H^+ concentration. Because of this inverse logarithmic relationship, H^+ concentrations have a 10-fold difference. The pH is the logarithm of the reciprocal of the molar concentration of the hydrogen ion. In mathematical form, it is:

$$pH = \log \frac{1}{H^+}$$ (5.7)

Natural water varies in pH depending on its source. Pure water has a neutral pH, with an equal number of H^+ and OH^-. Adding an acid to water causes additional "+" ions to be released so that the H^+ ion concentration goes up and the pH value goes down:

$$HCl \rightarrow H^+ + Cl^-$$

Changing the hydrogen ion activity in solution can shift the chemical equilibrium of water. Thus, pH adjustment is used to optimize coagulation, softening, and disinfection reactions, and for corrosion control. To control water coagulation and corrosion, the waterworks operator must test for hydrogen ion concentration of the water to get pH. In coagulation tests, as more alum (acid) is added, the pH value is lowered. If more lime (alkali — base) is added, the pH value is raised. This relationship is important; if good floc is formed, the pH should then be determined and maintained at that pH value until a change occurs in the new water.

Example 5.6

Problem:
A 0.1-*M* solution of acetic acid has a hydrogen ion concentration of 1.3×10^{-3} *M*. Find the pH of the solution.

Solution:

$$pH = \log \frac{1}{H^+}$$

$$= \log \frac{1}{1.3 \times 10^{-3}}$$

$$= \log \frac{10^{-3}}{1.3} \log 10^{-3} - 1.3$$

$$= 3 - \log 1.3$$

$$pH = 3 - 0.11 = 2.89$$

Alkalinity

Alkalinity is defined as the capacity of water to accept protons (positively charged particles); it can also be defined as a measure of water's ability to neutralize an acid. Alkalinity is a measure of water's capacity to absorb hydrogen ions without significant pH change (to neutralize acids).

Bicarbonates, carbonates, and hydrogen cause alkalinity compounds in a raw or treated water supply. Bicarbonates are the major components because of carbon dioxide action on "basic" materials of soil; borates, silicates, and phosphates may be minor components. Alkalinity of raw water may also contain salts formed from organic acids — humic acid, for example. Alkalinity in water acts as a buffer that tends to stabilize and prevent fluctuations in pH. Significant alkalinity in water is usually beneficial because it tends to prevent quick changes in pH, which interfere with the effectiveness of common water treatment processes. Low alkalinity also contributes to the corrosive tendencies of water. When alkalinity is below 80 mg/L, it is considered low.

Hardness

Hardness may be considered a physical or chemical parameter of water. It represents the total concentration of calcium and magnesium ions, reported as calcium carbonate. Hardness causes soaps and detergents to be less effective and contributes to scale formation in pipes and boilers, but is not considered a health hazard. However, lime precipitation or ion exchange often softens water that contains hardness. Low hardness contributes to the corrosive tendencies of water. Hardness and alkalinity often occur together because some compounds can contribute alkalinity as well as hardness ions. Hardness is generally classified as shown in Table 5.6.

Table 5.6   Water Hardness

Classification	mg/L CaCo$_3$
Soft	0–75
Moderately hard	75–150
Hard	150–300
Very hard	Over 300

Source: Spellman, F.R., 2003, *Handbook of Water and Wastewater Treatment Plant Operations.* Boca Raton, FL: Lewis Publishers.

Example 5.7

Problem:

22.4 g of Na_2CO_3 are dissolved in water, and the solution is made up to 500 mL. Find the molarity and normality of the solution.

Solution:

The molecular weight of Na_2CO_3 is 106.
Molarity is number of moles per volume in liters.
Number of moles is actual weight/molecular weight.
The net positive valence of Na_2CO_3 is $1 \times 2 = 2$.
The equivalent weight of Na_2CO_3 is $106/2 = 53$.

$$\text{Molarity} = \frac{\text{Actual weight/molecular weight}}{\text{Volume in liters}}$$

$$= \frac{\text{Actual weight}}{\text{molecular weight} \times \text{Volume in liters}}$$

$$\text{Molarity} = \frac{22.4}{106 \times 0.500} = 0.42\text{M}$$

$$\text{Normality} = \frac{\text{Actual weight}}{\text{equivalent weight} \times \text{Volume in liters}}$$

$$\frac{22.4}{53 \times 0.500} = 0.85 \text{ N}$$

Example 5.8

Problem:

Find the molarity of 0.03 mol NaOH in 90 mL of solution.

Solution:

$$\frac{0.03}{0.090} = 0.33 \text{ M}$$

Example 5.9

Problem:

Find the normality of 0.3 mol $CaCl_2$ in 300 mL of solution.

Solution:

$$\frac{2 \times 0.3}{0.300} = 2.0 \text{ N}$$

Example 5.10

Problem:

Find the molality of 5.0 g of NaOH in 500 g of water.

Solution:

$$\frac{5.0}{50 \times 0.500} = 0.2 \text{ m}$$

5.2.2.8 Simple Solutions and Dilutions

A simple dilution is one in which a unit volume of a liquid material of interest is combined with an appropriate volume of a solvent liquid to achieve the desired concentration. The dilution factor is the total number of unit volumes in which a material will be dissolved. The diluted material must then be thoroughly mixed to achieve the true dilution. For example, a 1:5 dilution (verbalize as "1 to 5" dilution) entails combining 1 unit volume of diluent (the material to be diluted) + 4 unit volumes of the solvent medium (thus, $1 + 4 = 5 =$ dilution factor). The fact that the number of moles or equivalents of solute does not change during dilution enables us to calculate the new concentration.

Simple Dilution Method

We demonstrate the simple dilution or dilution factor method in the following. Suppose we have orange juice concentration that is usually diluted with four additional cans of cold water (the dilution solvent), thus giving a dilution factor of 5 — the orange concentrate represents one unit volume to which four more cans (same unit volumes) of water have been added. Thus, the orange concentrate is now distributed through five unit volumes. This would be called a 1:5 dilution, and the orange juice is now 1/5 as concentrated at it was originally. Therefore, in a simple dilution, add one less unit volume of solvent than the desired dilution factor value.

Serial Dilution

A *serial dilution* is simply a series of simple dilutions that amplify the dilution factor quickly, beginning with a small initial quantity of material (bacterial culture, a chemical, orange juice, and so forth). The source of dilution material for each step comes from the diluted material of the previous step. In a serial dilution, the *total dilution factor* at any point is the product of the *individual dilution factor* in each step up to that point.

$$\text{Final Dilution Factor (DF)} = (DF_1)(DF_2)(DF_3) \text{ etc.} \tag{5.8}$$

To demonstrate the final dilution factor calculation, a typical lab experiment involves a three-step 1:100 serial dilution of a bacterial culture. The initial step combines 1 unit volume culture (10 μL) with 99 unit volumes of broth (990 μL) to equal a 1:100 dilution. In the next step, 1 unit volume of the 1:100 dilution is combined with 99 unit volumes of broth, now yielding a total dilution of $1:100 \times 100 = 1:10,000$ dilution. Repeated again (the third step), the total dilution would be $1:100 \times 10,000 = 1:1,000,000$ total dilution. The concentration of bacteria is now 1 million times less than in the original sample.

$C_iV_i = C_fV_f$ Method (Fixed Volumes of Specific Concentrations from Liquid Reagents)

When we are diluting solutions, the product of the concentration and volume of the initial solution must be equal to the product of the concentration and volume of the diluted solution when the same system of units is used in both solutions. Expressed as a relationship, this would be:

$$C_iV_i = C_fV_f \tag{5.9}$$

where
C_i = concentration of initial solution
V_i = volume of initial solution

C_f = concentration of final solution
V_f = volume of final solution

Example 5.11

Problem:
How much water must be added to 60 mL of 1.3-*M* HCl solution to produce a 0.5-*M* HCl solution?

Solution:

$$C_i \times V_i = C_f \times V_f$$

$$1.3 \times 60 = 0.5 \times X$$

$$X = \frac{1.3 \times 60}{0.5} = 156 \text{ ml}$$

The volume of the final solution is 156 mL. The amount of water to be added is then the difference between the volumes of the two solutions. The volume of water to be added is

$$156 - 60 = 96 \text{ ml of water}$$

Molar Solutions

Sometimes using molarity may be more efficient for calculating concentrations. A 1.0-*M* solution is equivalent to 1 formula weight (FW) (gram/mole) of chemical dissolved in 1.0 L of solvent (usually water). Formula weight is always given on the label of a chemical bottle (use molecular weight if it is not given).

Example 5.12

Problem:
Given the following data, determine how many grams of reagent to use. Chemical FW = 195 g/mol; to make 0.15 *M* solution.

Solution:

$$(195 \text{ g/mol})(0.15 \text{ mol/L}) = 29.25 \text{ g/L}$$

Example 5.13

Problem:
A chemical has a FW 190 g/mol and we need 25 mL (0.025 L) of 0.15 *M* (*M* = mol/L) solution. How many grams of the chemical must be dissolved in 25 mL of water to make this solution?

Solution:

$$\text{Number grams/desired volume (L)} = \text{desired molarity (mol/L)} \times \text{FW (g/mol)}$$

rearranging

$$\text{Number grams} = \text{desired volume (L)} \times \text{desired molarity (mol/L)} \times \text{FW (g/mol)}$$

$$\text{Number grams} = (0.025 \text{ L})(0.15 \text{ mol/L}) (190 \text{ g/mol}) = 0.7125 \text{ g/25 mL}$$

Percent Solutions

Many reagents are mixed as percent concentrations. A dry chemical is mixed as dry mass (g) per volume where number of grams per 100 milliliters = percent concentration. A 10% solution is equal to 10 g dissolved in 100 mL of solvent. In addition, if we wanted to make 3% NaCl, we would dissolve 3.0 g NaCl in 100 mL water (or the equivalent for whatever volume we need). When using liquid reagents, the percent concentration is based upon volume per volume and number of milliliters per 100 milliliters. For example, if we want to make 70% ethanol, we mix 70 mL of 100% ethanol with 30 mL water (or the equivalent for whatever volume we need).

To convert from percent solution to molarity, multiply the percent solution value by 10 to get grams per liter, then divide by the formula weight.

$$\text{Molarity} = \frac{(\% \text{ solution}) (10)}{\text{FW}} \tag{5.10}$$

Example 5.14

Problem:

Convert 6.5% solution of a chemical with FW = 351 to molarity.

Solution:

$$\frac{[6.5\text{g}/100 \text{ ml}) (10)]}{351 \text{ g/L}} = 0.1852 \text{ M}$$

To convert from molarity to percent solution, multiply the molarity by the FW and divide by 10.

$$\% \text{ Solution} = \frac{(\text{molarity})(\text{FW})}{10} \tag{5.11}$$

Example 5.15

Problem:

Convert a 0.0045-M solution of a chemical having FW 176.5 to percent solution.

Solution:

$$\frac{[0.0045 \text{ mol/L}) (176.5 \text{ g/mol})]}{10} = 0.08\%$$

5.2.2.9 Chemical Reactions

A fundamental tool in any environmental engineer's toolbox is a basic understanding of chemical reactions, reaction rates, and physical reactions. However, before discussing chemical reactivity, we review the basics of electron distribution, as well as chemical and physical changes, and discuss the important role heat plays in chemical and physical reactions

Electron Distribution

Simply stated, electron distribution around the nucleus is key to understanding chemical reactivity. Only electrons are involved in chemical change; the nuclei of atoms are not altered in any way during chemical reactions. Electrons are arranged around the nucleus in a definite pattern or series of "shells."

In general, only the "outer shell" or "valence" electrons (the ones farthest from the nucleus) are affected during chemical change. The *valence number* or *valence* of an element indicates the number of electrons involved in forming a compound — that is, the number of electrons it tends to gain or lose when combining with other elements.

> *Key point*: Positive valence indicates giving up electrons. Negative valence indicates accepting electrons.

Types of Bonding

- If these valence electrons are shared with other atoms, then a covalent bond is formed when the compound is produced.
- If these valence electrons are donated to another atom, then an ionic bond is formed when the compound is produced.
- Hydrogen bonding occurs when an atom of hydrogen is attracted by rather strong forces to two atoms instead of only one, so that it is considered to act as a bond between them. If an atom gains or loses one or more valence electrons, it becomes an ion (charged particle).
- Cations are positively charged particles.
- Anions are negatively charged particles.

Physical and Chemical Changes

In physical and chemical changes, recall that a chemical change is the change physical substances undergo when they become new or different substances. To identify a chemical change, look for observable signs, including color change; light production; smoke; bubbling or fizzing; and presence of heat.

A physical change occurs when objects undergo a change that does not change their chemical nature. This type of change involves a change in physical properties. Physical properties can be observed without changing the type of matter. Examples of physical properties include texture; size; shape; color; odor; mass; volume; density; and weight.

Heat and Chemical/Physical Reactions

An *endothermic reaction* is a chemical reaction that absorbs energy, where the energy content of the products is more than that of the reactants; heat is taken in by the system. An *exothermic reaction* is a chemical reaction that gives out energy, where the energy content of the products is less than that of the reactants; heat is given out from the system.

Types of Chemical Reactions

Chemical reactions are of fundamental importance throughout chemistry and related technologies. Although experienced chemists can sometimes predict the reactions that will occur in a new chemical system, they may overlook some alternatives. They are usually unable to make reliable predictions when the chemistry is unfamiliar to them.

> *Key point*: Few tools are available to assist in predicting chemical reactions and none at all for predicting the novel reactions that are of greatest interest. (*Note*: For more information on predicting chemical reactions, see the work by Irikura and Johnson III listed in the suggested reading at the chapter end.)

Just as classifying elements as gaseous and nongaseous is convenient, defining a way to classify chemical reactions is convenient. So many chemical reactions are possible (425 "named" reactions are listed in the Merck Index) that classifying them into four general types is helpful. The four general types of chemical reaction include:

- Combination
- Decomposition
- Replacement
- Double replacement

Combination Reactions — In a combination (or synthesis) reaction, two or more simple substances combine to form a more complex compound:

Copper (an element) + oxygen (an element) = copper oxide (a compound)

> *Key point*: Many pairs of reactants combine to give single products. The reactions happen when conditions are energetically favorable to do so.

Decomposition Reactions — Decomposition reactions occur when one compound breaks down (decomposes) into two or more substances or its elements. Basically, combination and decomposition reactions are opposites.

Hydrogen peroxide $\Rightarrow H_2O + O_2$

Replacement Reactions — Replacement reactions involve the substitution of one uncombined element for another in a compound. Two reactants yield two products.

Iron (Fe) + sulfuric acid (H_2SO_4) = hydrogen (H_2) + iron sulfate $(FeSO_4)$

Double Replacement — In a double replacement reaction, parts of two compounds exchange places to form two new compounds. Two reactants yield two products.

Sodium hydroxide + acetic acid = sodium acetate + water

> *Key point*: Note that these four classifications of chemical reactions are not based on the type of bonding, which generally can be covalent or ionic, depending on the reactants involved.
> *Interesting point*: According to the law of conservation of mass, no mass is added or removed in chemical reactions.

Specific Types of Chemical Reactions

- Hydrolysis and neutralization
- Oxidation/reduction (redox)
- Chelation
- Photolysis and polymerization
- Catalysis
- Biochemical reaction and biodegradation

- *Hydrolysis* ("hydro" means water and "lysis" means to break) is a decomposition reaction involving the splitting of water into its ions and the formation of a weak acid, base, or both.
- *Neutralization* is a double replacement reaction that unites the H^- ion and an acid with the OH^- ion of a base, forming water and a salt.

$$Acid + Acid \Rightarrow Salt + Water$$

- *Oxidation/reduction* (*redox*) reactions are combination reactions, replacement, or double replacement reactions that involve the gain and loss of electrons (changes in valence). Oxidation and reduction always occur simultaneously, in such a way that one reacting species is oxidized while the other is reduced.

  Key point: When an atom, either free or in a molecule or ion, loses electrons, it is oxidized, and its oxidation number increases. When an atom, either free or in a molecule or ion, gains electrons, it is reduced, and its oxidation number decreases.

- *Chelation* is a combination reaction in which a ligand (such as a solvent molecule or simple ion) forms more than one bond to a central ion, giving rise to complex ions or coordination compounds.

  Interesting point: The term *chelate* was first applied in 1920 by Morgan and Drew, who stated, "The adjective chelate, derived from the great claw or chela (*fr.* Greek, *chely*) of the lobster or other crustaceans, is suggested for the caliperlike groups which function as two associating units and fasten to the central atom so as to produce heterocyclic rings."

- *Free radical reaction* is any type of reaction that involves any species with an unpaired electron. Free radical reactions frequently occur in the gas phase, often proceed by chain reaction, are often initiated by light, heat, or reagents that contain unpaired bonds (such as oxides or peroxide decomposition products), and are often are very reactive.

  Interesting point: A common free radical reaction in aqueous solution is electron transfer, especially to the hydroxyl radical and to ozone:

$$R\cdot$$

- *Photolysis* ("photo" meaning light and "lysis" meaning to break) is generally a decomposition reaction in which the adsorption of light produces a photochemical reaction. Photolysis reactions form free radicals that can undergo other reactions. For example, the chlorine molecule can dissociate in the presence of high-energy light (for example, UV light):

$$Cl_2 + UV\ energy \Rightarrow Cl\cdot + Cl\cdot$$

  Important point: Photolysis is an important nonthermal technology used for treating dioxin and furan hazardous wastes.

- *Polymerization* is a combination reaction in which small organic molecules are linked together to form long chains, or complex two- and three-dimensional networks. Polymerization occurs only in molecules with double or triple bonds and usually depends on temperature, pressure, and a suitable catalyst.

  Key point: Catalysts are agents that change the speed of a chemical reaction without affecting the yield or undergoing permanent chemical change.

  Free radical chain reactions are a common mode of polymerization. The term "chain reaction" is used because each reaction produces another reactive species (another free radical) to continue the process.

  Important point: Plastics are perhaps the most common polymers, but many biopolymers are also important, including polysaccharides.

- *Biochemical reactions* are reactions that occur in living organisms.
- *Biodegradation* is a decomposition reaction that occurs in microorganisms to create smaller, less complex inorganic and organic molecules. Usually the products of biodegradation are molecular forms that tend to occur in nature.

Reaction Rates (Kinetics)

The rate of a chemical reaction is a measure of how fast the reaction proceeds (how fast reactants are consumed and products are formed):

$$A + B \text{ (reactants)} \Rightarrow C + D \text{ (products)}$$

The rate of a given reaction depends on many variables (factors), including temperature; the concentration of the reactants; catalysts; the structure of the reactants; and the pressure of gaseous reactants or products. Without considering extreme conditions, in general reaction rates increase with:

- *Increasing temperature* — two molecules will only react if they have enough energy. By heating the mixture, the energy levels of the molecules involved in the reaction are raised. Raising the temperature means the molecules move faster (kinetic theory).
- *Increasing concentration of reactants* — increasing the concentration of the reactants increases the frequency of collisions between the two reactants (collision theory).
- *Introducing catalysts* — Catalysts speed up reactions by lowering the activation energy. Only very small quantities of the catalyst are required to produce a dramatic change in the rate of the reaction because the reaction proceeds by a different pathway when the catalyst is present. Note that adding more catalyst makes absolutely no difference.
- *Increasing surface area* — the larger the surface area of a solid is, the faster the reaction is. Smaller particles have a bigger surface area than a larger particle for the same mass of solid. A simple way to visualize this is called the "bread and butter theory." If we take a loaf of bread and cut it into slices, we get extra surfaces onto which we can spread butter. The thinner we cut the slices, the more slices we get and the more butter we can spread on them. By chewing our food, we increase the surface area so that digestion goes faster.
- *Increasing the pressure on a gas* to increase the frequency of collisions between molecules — when pressure is increased, molecules are squeezed together so that the frequency of collisions between them is increased.

Reaction rates are also affected by each reaction's activation energy — the energy that the reactants must reach before they can react.

Key point: A catalyst may be recovered unaltered at the end of the reaction.

"Forward" and "backward" reactions can occur, each with a different reaction rate and associated activation energy:

$$A + B \overset{\Leftarrow}{\Rightarrow} C + D$$

For example, in a dissociation reaction that occurs readily, initially the dissociation takes place at a faster rate than recombination. Eventually, as the concentration of dissociated ions builds up, the rate of recombination catches up with the rate of dissociation. When the forward and backward reactions eventually occur at the same rate, a state of equilibrium is reached. The apparent effect is no change, even though the forward and backward reactions are still occurring.

> *Key point*: Note that the point at which equilibrium is reached is not fixed, but is also dependent on variables that include temperature, reactant concentration, pressure, reactant structure, etc.

Types of Physical Reactions

Knowledge of the physical behavior of wastes and hazardous wastes has been used to develop various unit processes for waste treatment based on physical reactions. These operations include:

- Phase separation
- Phase transitions
- Phase transfer

- *Phase separation* involves separation of components of a mixture that is already in two different phases. Types of phase separation include filtration, settling, decanting, and centrifugation.
- *Phase transition* is a physical reaction in which a material changes from one physical phase to another. Types of phase transition include distillation, evaporation, precipitation, and freeze drying (lyophilization).
- *Phase transfer* consists of the transfer of a solute in a mixture from one phase to another. Two examples of phase transfer include extraction and sorption (transfer of a substance from a solution to a solid phase).

Chemical Equations Encountered in Water/Wastewater Operations

$$Cl_2 + H_2O \leftrightarrow HCl + HOCl$$

$$NH_3 + HOCl \leftrightarrow NH_2Cl + H_2O$$

$$NH_2Cl + HOCl \leftrightarrow NHCl_2 + H_2O$$

$$NHCl_2 + HOCl \leftrightarrow NCl_3 + H_2O$$

$$Ca(OCl)_2 + Na_2CO_3 \leftrightarrow 2NaOCl + CaCO_3$$

$$Al_2(SO_4)_3 + 3CaCO_3 + 3H_2O \leftrightarrow Al_2(OH)_6 + 3CaSO_4 + 3CO_2$$

$$CO_2 + H_2O \leftrightarrow H_2CO_3$$

$$H_2CO_3 + CaCO_3 \leftrightarrow Ca(HCO_3)_2$$

$$Ca(HCO_3)_2 + Na_2CO_3 \leftrightarrow CaCO_3 + 2NaHCO_3$$

$$CaCO_3 + H_2SO_4 \leftrightarrow CaSO_4 + 2H_2CO_3$$

$$Ca(HCO_3)_2 + H_2SO_4 \leftrightarrow CaSO_4 \leftrightarrow + 2H_2CO_3$$

$$H_2S + Cl_2 \leftrightarrow 2HCl + S° \downarrow$$

$$H_2S + 4Cl_2 + 4H_2O \rightarrow H_2SO_4 + 8HCl$$

$$SO_2 + H_2O \rightarrow H_2SO_3$$

$$HOCl + H_2SO_3 \rightarrow H_2SO_4 + HCl$$

$$NH_2Cl + H_2SO_3 + H_2O \rightarrow NH_4HSO_4 + HCl$$

$$Na_2SO_4 + Cl_2 + H_2O \rightarrow Na_2SO_4 + 2HCl$$

5.2.2.10 Chemical Dosages (Water and Wastewater Treatment)

Chemicals are used extensively in water/wastewater treatment plant operations. Plant operators add chemicals to various unit processes for slime-growth control; corrosion control; odor control; grease removal; BOD reduction; pH control; sludge-bulking control; ammonia oxidation; bacterial reduction; and fluoridation; as well as for other reasons.

To apply any chemical dose correctly, the ability to make certain dosage calculations is important. One of the most frequently used calculations in water/wastewater mathematics is the conversion of milligrams per liter (mg/L) concentration to pounds per day (lb/day) or pounds (lb) dosage or loading. The general types of milligrams per liter to pounds per day or pounds calculations are for chemical dosage, BOD, COD, or SS loading/removal; pounds of solids under aeration; and WAS pumping rate. These calculations are usually made using Equation 5.12 or Equation 5.13:

$$(mg/L)(MGD\ flow)(8.34\ lb/gal) = lb/day \tag{5.12}$$

$$(mg/L)(MG\ volume)(8.34\ lb/gal) = lb \tag{5.13}$$

Note: If milligrams-per-liter concentration represents a concentration in a flow, then million-gallons-per-day (MGD) flow is used as the second factor. However, if the concentration pertains to a tank or pipeline volume, then million gallons (MG) volume is used as the second factor.

Chlorine Dosage

Chlorine is a powerful oxidizer commonly used in water treatment for purification and in wastewater treatment for disinfection, odor control, bulking control, and other applications. When chlorine is added to a unit process, we want to ensure that a measured amount is added. Two ways are used to describe the amount of chemical added or required:

- Milligrams per liter (mg/L)
- Pounds per day (lb/day)

In the conversion from milligrams per liter (or parts per million) concentration to pounds per day, we use Equation 5.14:

$$(mg/L)(MGD)(8.34) = lb/day \tag{5.14}$$

Note: In previous years, normal practice used the expression *parts per million* (ppm) as an expression of concentration because 1 mg/L = 1 ppm. However, current practice is to use milligrams per liter as the preferred expression of concentration.

Example 5.16

Problem:

Determine the chlorinator setting (pounds per day) needed to treat a flow of 8 MGD with a chlorine dose of 6 mg/L.

Solution:

$$(mg/L)(MGD)(8.34) = lb/day$$

$$(6 \text{ mg/L})(8 \text{ MGD})(8.34 \text{ lb/gal}) = lb/day$$

$$= 400 \text{ lb/day}$$

Example 5.17

Problem:

What should the chlorinator setting be (pounds per day) to treat a flow of 3 MGD if the chlorine demand is 12 mg/L and a chlorine residual of 2 mg/L is desired?

Note: The chlorine demand is the amount of chlorine used in reacting with various components of the wastewater, including harmful organisms and other organic and inorganic substances. When the chlorine demand has been satisfied, these reactions stop.

Note: To find the unknown value (pounds per day), we must first determine chlorine dose. To do this, we must use Equation 5.15.

Solution:

$$\text{Cl Dose, mg/L} = \text{Cl Demand, mg/L} + \text{Cl Residual, mg/L}$$

$$= 12 \text{ mg/L} + 2 \text{ mg/L}$$

$$= 14 \text{ mg/L} \tag{5.15}$$

Then we can make the milligrams per liter to pounds per day calculation:

$$(12 \text{ mg/L})(3 \text{ MGD})(8.34 \text{ lb/gal}) = 300 \text{ lb/day}$$

Hypochlorite Dosage

At many wastewater facilities, sodium hypochlorite or calcium hypochlorite is used instead of chlorine. The reasons for substituting hypochlorite for chlorine vary. However, with the passage of stricter hazardous chemicals regulations under OSHA and the USEPA, many facilities are deciding to substitute the hazardous chemical chlorine with nonhazardous hypochlorite. Obviously, the potential liability involved with using deadly chlorine is also a factor involved in the decision to substitute a less toxic chemical substance. For whatever reason that the wastewater treatment plant decides to substitute hypochlorite for chlorine, operators and engineers need to be aware of differences between the two chemicals.

Chlorine is a hazardous material. Chlorine gas is used in wastewater treatment applications at 100% available chlorine, an important consideration to keep in mind when making or setting chlorine feed rates. For example, if the chlorine demand and residual requires 100 lb/day chlorine, the chlorinator setting would be just that — 100 lb/24 h.

Hypochlorite is less hazardous than chlorine; similar to strong bleach, it comes in two forms: dry calcium hypochlorite (often referred to as HTH) and liquid sodium hypochlorite. Calcium hypochlorite contains about 65% available chlorine; sodium hypochlorite contains about 12 to 15% available chlorine (in industrial strengths).

Note: Because neither type of hypochlorite is 100% pure chlorine, more pounds per day must be fed into the system to obtain the same amount of chlorine for disinfection — an important economical consideration for facilities thinking about substituting hypochlorite for chlorine. Some studies indicate that such a substitution can increase overall operating costs by up to three times the cost of using chlorine.

To calculate the pounds per day hypochlorite dosage requires a two-step calculation:

Step 1. $(\text{mg/L})(\text{MGD})(8.34) = \text{lb/day}$

Step 2. $\dfrac{\text{Chlorine, lb/day}}{\text{\% Available}} \times 100 = \text{Hypochlorite, lb/day}$

Example 5.18

Problem:

A total chlorine dosage of 10 mg/L is required to treat a particular wastewater. If the flow is 1.4 MGD and the hypochlorite has 65% available chlorine, how many pounds per day of hypochlorite are required?

Solution:

Step 1. Calculate the pounds per day of chlorine required using the milligrams per liter to pounds per day equation:

$$(mg/L)(MGD)(8.34) = lb/day$$

$$(10 \text{ mg/L})(1.4 \text{ MGD})(8.34 \text{ lbs/gal}) = 117 \text{ lb/day}$$

Step 2. Calculate the pounds per day hypochlorite required. Because only 65% of the hypochlorite is chlorine, more than 117 lb/day are required:

$$\frac{117 \text{ lb/day Cl}}{65\% \text{ Available Cl}} \times 100 = 180 \text{ lb/day Hypochlorite}$$

Example 5.19

Problem:

A wastewater flow of 840,000 gpd requires a chlorine dose of 20 mg/L. If sodium hypochlorite (15% available chlorine) is used, how many pounds per day of sodium hypochlorite are required? How many gallons per day of sodium hypochlorite is this?

Solution:

Step 1. Calculate the pounds per day chlorine required:

$$(mg/L)(MGD)(8.34) = lb/day$$

$$(20 \text{ mg/L})(0.84 \text{ MGD})(8.34 \text{ lb/gal}) = 140 \text{ lb/day Cl (Chlorine)}$$

Step 2. Calculate the pounds per day sodium hypochlorite:

$$\frac{140 \text{ lb/day Cl}}{15\% \text{ Available Cl}} \times 100 = 933 \text{ lb/day Hypochlorite}$$

Step 3. Calculate the gallons per day sodium hypochlorite:

$$\frac{933 \text{ lb/day}}{8.34 \text{ lb/gal}} = 112 \text{ gal/day Sodium Hypochlorite}$$

Example 5.20

Problem:

How many pounds of chlorine gas are necessary to treat 5,000,000 gal of wastewater at a dosage of 2 mg/L?

Solution:

　　Step 1. Calculate the pounds of chlorine required:

$$V, 10^6 \text{ gal} = \text{Cl concentration (mg/L)} \times 8.34 = \text{lb Cl}$$

　　Step 2. Substitute

$$5 \times 10^6 \text{ gal} \times 2 \text{ mg/L} \times 8.34 = 83 \text{ lb Cl}$$

Additional Dosage Calculations

Example 5.21

Problem:

　　Chlorine dosage at a treatment plant averages 112.5 lb/day. Its average flow is 11.5 MGD. What is the chlorine dosage in milligrams per liter?

Solution:

$$\text{Dosage} = \frac{112.5 \text{ lb/day}}{11.5 \times 10^6 \text{ gal}}$$

$$= \frac{112.5 \text{ lb/day}}{11.5 \times 10^6 \text{ gal} \times 8.34 \text{ lb/gal}}$$

$$= \frac{1.2}{10^6}$$

$$= 1.2 \text{ ppm}$$

$$= 1.2 \text{ mg/L}$$

Example 5.22

Problem:

　　Twenty-five pounds of chlorine gas are used to treat 700,000 gal of water. The chlorine demand of the water is measured at 2.4 mg/L. What is the residual chlorine concentration in the treated water?

Solution:

$$\text{Total dosage} = 25 \text{ lb}/0.70 \text{ mil gal} = 36 \text{ lb/mil gal}$$

$$= 36 \text{ lb/mil gal} \times \frac{1 \text{ mg/L}}{8.34 \text{ lb/mil gal}}$$

$$= 4.3 \text{ mg/L}$$

$$\text{Residual Cl} = 4.3 \text{ mg/L} - 2.4 \text{ mg/L}$$

$$= 1.9 \text{ mg/L}$$

Example 5.23

Problem:

What is the daily amount of chlorine needed to treat 10 MGD of water to satisfy 2.9 mg/L chlorine demand and provide 0.6 mg/L residual chlorine?

Solution:

$$\text{Total Cl needed} = 2.9 + 0.6 = 3.5 \text{ mg/L}$$

$$\text{Daily weight} = 10 \times 10^6 \text{ gal/day} \times 8.34 \text{ lb/gal} \times 3.5 \text{ mg/L} \times 1 \text{ L/}10^6$$

$$= 292 \text{ lb/day}$$

Example 5.24

Problem:

At a 12-MGD waterworks, a pump at the rate of 0.20 gpm feeds a hydrofluosilicic acid (H_2SiF_6) with 23% by wt solution. The specific gravity of the H_2SiF_6 solution is 1.191. What is the fluoride (F) dosage?

Solution:

$$\text{Pump rate} = 0.20 \text{ gpm} \times 1440 \text{ min/day} = 288 \text{ gal/day}$$

$$\text{F applied rate} = 288 \text{ gal/day} \times 0.23 = 66.2 \text{ gal/day}$$

$$\text{Wt of F} = 66.2 \text{ gal/day} \times 8.34 \text{ lb/gal} \times 1.191$$

$$\text{Wt of water} = 12 \times 10^6 \text{ gal/day} \times 8.34 \text{ lb/gal} \times 1.0$$

$$\text{Dosage} = \text{wt of F/wt of water}$$

$$= \frac{66.2 \times 8.34 \times 1.191}{12 \times 10^6 \times 8.34 \text{ lb/gal} \times 1.0}$$

$$= \frac{6.58}{10^6}$$

$$= 6.58 \text{ mg/L}$$

Example 5.25

Problem:

A raw water flow that averages 8.6 MGD is continuously fed 10 mg/L of liquid alum with 60% strength. How much liquid alum will be used in a month (30 days)?

Solution:

$$1 \text{ mg/L} = 1 \text{ gal/mil gal}$$

$$\text{Required/day} = \frac{10}{0.60} \frac{\text{gal}}{\text{mil gal}} \times 8.6 \frac{\text{mil gal}}{\text{day}}$$

$$= 143 \text{ gal/day}$$

$$\text{Required/month} = 143 \text{ gal/day} \times 30 \text{ day/month}$$

$$= 4290 \text{ gal/month}$$

5.3 FUNDAMENTAL HYDRAULICS

Water/wastewater operators make pumpage and flow rate calculations during daily operations. In this section, we describe and perform fundamental pumping and flow rate calculations and review the foundational principles of advanced hydraulics operations for environmental engineers (introduced later in the text).

5.3.1 Principles of Water Hydraulics

Hydraulics is defined as the study of fluids at rest and in motion. Although basic principles apply to all fluids, at this time we consider only principles that apply to water/wastewater. (*Note*: Much of the basic information that follows is concerned with the hydraulics of distribution systems [piping and so forth]; however, the operators must understand [and engineers review] these basics to appreciate the function of pumps more fully.)

5.3.1.1 Weight of Air

Our study of basic water hydraulics begins with air. A blanket of air, many miles thick, surrounds the Earth. The weight of this blanket on a given square inch of the Earth's surface varies according to the thickness of the atmospheric blanket above that point. At sea level, the pressure exerted is 14.7 lb/in.^{2} (psi). On a mountaintop, air pressure decreases because the blanket is not as thick.

5.3.1.2 Weight of Water

Because water must be stored as well as moved in water supplies and because wastewater must be collected, processed in unit processes, and outfalled to its receiving body, we must consider some basic relationships in the weight of water. One cubic foot of water weighs 62.4 lb and contains 7.48 gal. One cubic inch of water weighs 0.0362 lb. Water 1 ft deep exerts a pressure of 0.43 psi on the bottom area (12 in. × 0.062 lb/in.^{3}). A column of water 2 ft high exerts 0.86 psi; one 10 ft high exerts 4.3 psi; and one 52 ft high exerts:

$$52 \text{ ft} \times 0.43 \text{ psi/ft} = 22.36 \text{ psi}$$

A column of water 2.31 ft high will exert 1.0 psi. To produce a pressure of 40 psi requires a water column:

$$40 \text{ psi} \times 2.31 \text{ ft/psi} = 92.4$$

The term *head* is used to designate water pressure in terms of the height of a column of water in feet. For example, a 10-ft column of water exerts 4.3 psi. This can be called 4.3-psi pressure or 10 ft of head. Another example: if the static pressure in a pipe leading from an elevated water storage tank is 37 psi, what is the elevation of the water above the pressure gauge? Remembering that 1 psi = 2.31 and that the pressure at the gauge is 37 psi:

$$37 \text{ psi} \times 2.31 \text{ ft/psi} = 85.5 \text{ ft (rounded)}$$

5.3.1.3 *Weight of Water Related to the Weight of Air*

The theoretical atmospheric pressure at sea level (14.7 psi) will support a column of water 34 ft high:

$$14.7 \text{ psi} \times 2.31 \text{ ft/psi} = 33.957 \text{ or } 34 \text{ ft}$$

At an elevation of 1 mi above sea level, where the atmospheric pressure is 12 psi, the column of water would be only 28 ft high (12 psi × 2.31 ft/psi = 27.72 ft or 28 ft).

If a tube is placed in a body of water at sea level (a glass, bucket, water storage reservoir, or a lake or pool, for example), water rises in the tube to the same height as the water outside the tube. The atmospheric pressure of 14.7 psi pushes down equally on the water surface inside and outside the tube. However, if the top of the tube is tightly capped and all of the air is removed from the sealed tube above the water surface — thus forming a *perfect vacuum*, the pressure on the water surface inside the tube will be 0 psi. The atmospheric pressure of 14.7 psi on the outside of the tube pushes the water up into the tube until the weight of the water exerts the same 14.7-psi pressure at a point in the tube even with the water surface outside the tube. The water will rise 14.7 psi × 2.31 ft/psi = 34 ft.

In practice, creating a perfect vacuum is impossible, so the water will rise somewhat less than 34 ft; the distance it rises depends on the amount of vacuum created.

Example 5.26

Problem:

If enough air was removed from the tube to produce an air pressure of 9.7 psi above the water in the tube, how far will the water rise in the tube?

Solution:

To maintain the 14.7 psi at the outside water surface level, the water in the tube must produce a pressure of 14.7 psi – 9.7 = 5.0 psi. The height of the column of water that will produce 5.0 psi is:

$$5.0 \text{ psi} \times 2.31 \text{ ft/psi} = 11.5 \text{ ft (rounded)}$$

5.3.1.4   Water at Rest

Stevin's law states, "The pressure at any point in a fluid at rest depends on the distance measured vertically to the free surface and the density of the fluid." Stated as a formula, this becomes

$$p = w \times h \qquad (5.16)$$

where
p = pressure in pounds per square foot (psf or lb/ft^2)
w = density in pounds per cubic foot (lb/ft^3)
h = vertical distance in feet

Example 5.27

Problem:
   What is the pressure at a point 15 ft below the surface of a reservoir?

Solution:
   To calculate this, we must know that the density of water, w, is 62.4 lb/ft^3. Thus,

$$p = w \times h$$

$$= 62.4 \text{ lb/ft}^3 \times 15 \text{ ft}$$

$$= 936 \text{ lb/ft}^2 \text{ or psf}$$

Waterworks/wastewater operators generally measure pressure in pounds per square inch rather than pounds per square foot; to convert, divide by 144 in.^{2}/ft^2 (12 in. $\times$ 12 in. = 144 in.^{2}):

$$P = \frac{936 \text{ lb/ft}^2}{144 \text{ in}^2/\text{ft}^2} = 6.5 \text{ lb/in}^2 \text{ or psi}$$

5.3.1.5   Gauge Pressure

Recall that *head* is the height that a column of water rises because of the pressure at its base. We demonstrated that a perfect vacuum plus atmospheric pressure of 14.7 psi lifts the water 34 ft. If we now open the top of the sealed tube to the atmosphere and enclose the reservoir and then increase the pressure in the reservoir, the water will again rise in the tube. Because atmospheric pressure is essentially universal, we usually ignore the first 14.7 psi of actual pressure measurements and measure only the difference between the water pressure and the atmospheric pressure; we call this *gauge pressure*.

Example 5.28

Problem:
   Water in an open reservoir is subjected to the 14.7 psi of atmospheric pressure, but subtracting this 14.7 psi leaves a gauge pressure of 0 psi. This shows that the water would rise 0 ft above the

reservoir surface. If the gauge pressure in a water main is 100 psi, how far would the water rise in a tube connected to the main?

Solution:

$$100 \text{ psi} \times 2.31 \text{ ft/psi} = 231 \text{ ft}$$

5.3.1.6 *Water in Motion*

The study of water flow is much more complicated than that of water at rest. It is necessary to understand these principles because the water/wastewater in a treatment plant or distribution/collection system is nearly always in motion (much of this motion is the result of pumping, of course).

5.3.1.7 *Discharge*

Discharge is the quantity of water passing a given point in a pipe or channel during a given period. It can be calculated by the formula:

$$Q = V \times A \tag{5.17}$$

where
Q = discharge in cubic feet per second (cfs or ft³/sec)
V = water velocity in feet per second (fps or ft/sec)
A = cross-section area of the pipe or channel in square feet (ft²)

Discharge can be converted from cubic feet per second to other units, including gallons per minute or million gallons per day by using appropriate conversion factors.

Example 5.29

Problem:
    A pipe 12 in. in diameter has water flowing through it at 10 ft/sec. What is the discharge in (a) cubic feet per second; (b) gallons per minute; and (c) million gallons per day?

Solution:
    Before we can use the basic formula, we must determine the area (A) of the pipe. The formula for the area is:

$$A = \pi \times \frac{D^2}{4} = \pi \times r^2 \tag{5.18}$$

where
π is the constant value 3.14159
D = diameter of the circle in feet
r = radius of the circle in feet

Thus, the area of the pipe is:

$$12\text{-in. pipe} = A = \pi \times \frac{D^2}{4} = 3.14159 = 0.785 \text{ ft}^2$$

$$12\text{-in. pipe} = A = \pi \times \frac{D^2}{4} = 3.14159 \times \frac{(1 \text{ ft})}{4}$$

Now, we can determine the discharge in cubic feet per second for part (a):

$$Q = V \times A = 10 \text{ ft/sec} \times 0.785 \text{ ft}^2 = 7.85 \text{ ft}^3/\text{sec or cfs}$$

For part (b), we need to know that 1 ft³/sec is 449 gal/min, so:

$$7.85 \text{ cfs} \times 449 \text{ gpm/cfs} = 3525 \text{ gpm}$$

Finally, for part (c), 1 MGD is 1.55 ft³/sec, so:

$$\frac{7.85 \text{ cfs}}{1.55 \text{ cfs/MGD}} = 5.06 \text{ MGD}$$

5.3.1.8 The Law of Continuity

The law of continuity states that the discharge at each point in a pipe or channel is the same as the discharge at any other point (provided water does not leave or enter the pipe or channel). In equation form, this becomes:

$$Q_1 = Q_2 \text{ or } A_1 V_1 = A_2 V_2 \tag{5.19}$$

Example 5.30

Problem:

A pipe 12 in. in diameter is connected to a 6-in. diameter pipe. The velocity of the water in the 12-in. pipe is 3 ft/sec. What is the velocity in the 6-in. pipe?

Solution:

Using the equation $A_1 V_1 = A_2 V_2$, we need to determine the area of each pipe:

$$12\text{-in. pipe: } A = \pi \times \frac{D^2}{4}$$

$$= 3.14159 \times \frac{(1 \text{ ft})^2}{4}$$

$$Q_1 = Q_2 \text{ or } A_1 V_1 = A_2 V_2$$

$$6\text{-in. pipe: } A = 3.14159 \times \frac{(0.5)^2}{4}$$

$$= 0.196 \text{ ft}^2$$

The continuity equation now becomes

$$(0.785 \text{ ft}^2) \times (3 \text{ ft/sec}) = (0.196 \text{ ft}^2) \times V_2$$

Solving for V_2:

$$V_2 = \frac{(0.785 \text{ ft}^2) \times (3 \text{ ft/sec})}{(0.196 \text{ ft}^2)}$$

$$= 12 \text{ ft/sec or fps}$$

5.3.1.9 Pipe Friction

The flow of water in pipes is caused by the pressure applied behind it by gravity or by hydraulic machines (pumps). The flow is retarded by the friction of the water against the inside of the pipe. The resistance of flow offered by this friction depends on the size (diameter) of the pipe, the roughness of the pipe wall, and the number and types of fittings (bends, valves, and so forth) along the pipe. It also depends on the speed of the water through the pipe — the more water one tries to pump through a pipe, the more pressure is needed to overcome the friction. Resistance can be expressed in terms of the additional pressure needed to push the water through the pipe, in either pounds per square inch or feet of head. Because it is a reduction in pressure, it is often referred to as *friction loss* or *head loss.*

Friction loss increases as:

- Flow rate increases.
- Pipe diameter decreases.
- Pipe interior becomes rougher.
- Pipe length increases.
- Pipe is constricted.
- Bends, fittings, and valves are added.

The actual calculation of friction loss is beyond the scope of this text. Many published tables give the friction loss in different types and diameters of pipe and standard fittings. Recognizing the loss of pressure or head from the friction of water flowing through a pipe is far more important.

One of the factors in friction loss is the roughness of the pipe wall. A number called the C factor indicates pipe wall roughness; the higher the C factor is, the smoother the pipe is.

> *Note*: C factor is derived from the letter C in the Hazen–Williams equation for calculating water flow through a pipe.

Some of the roughness in the pipe is the result of the piping material; cast iron pipe will be rougher than plastic, for example. Additionally, roughness increases with corrosion of the pipe material and deposited sediments in the pipe. New water pipes should have a C factor of 100 or more; older pipes can have C factors much lower than this.

In determining C factor, published tables are usually used. When friction losses for fittings are factored in, other published tables are available to make the proper determinations. Calculating the head loss from fittings by substituting the equivalent length of pipe is standard practice, and the information is available in published tables.

5.3.2 Basic Pumping Calculations

Certain computations used for determining various pumping parameters are important to the water/wastewater operator. (*Note*: The following examples are adapted from Wahren, 1997.)

5.3.2.1 Pumping Rates

Note: The rate of flow produced by a pump is expressed as the volume of water pumped during a given period.

The mathematical problems most often encountered by water/wastewater operators for determining pumping rates are often determined by using Equation 5.20 and/or Equation 5.21:

$$\text{Pumping Rate, (gpm)} = \text{gallons/minutes} \tag{5.20}$$

$$\text{Pumping Rate, (gph)} = \text{gallons/hours} \tag{5.21}$$

Example 5.31

Problem:
The meter on the discharge side of the pump reads in hundreds of gallons. If the meter shows a reading of 110 at 2:00 p.m. and 320 at 2:30 p.m., what is the pumping rate expressed in gallons per minute?

Solution:
The problem asks for pumping rate in gallons per minute, so we use Equation 5.20:

$$\text{Pumping Rate, (gpm)} = \text{gallons/minutes}$$

Step 1. To solve this problem, we must first find the total gallons pumped (determined from the meter readings):

$$
\begin{array}{r}
32,000 \text{ gallons} \\
-11,000 \text{ gallons} \\
\hline
21,000 \text{ gallons}
\end{array}
$$

Step 2. The volume was pumped between 2:00 p.m. and 2:30 p.m., for a total of 30 min. From this information, calculate the gallons-per-minute pumping rate.

$$\text{Pumping Rate, (gpm)} = \frac{21,000 \text{ gal}}{30 \text{ min}}$$

$$= 700 \text{ gpm pumping rate}$$

Example 5.32

Problem:
During a 15-min pumping test, 16,400 gal were pumped into an empty rectangular tank. What is the pumping rate in gallons per minute?

Solution:
The problem asks for the pumping rate in gallons per minute, so again we use:

$$\text{Pumping Rate, (gpm)} = \frac{\text{gallons}}{\text{minutes}}$$

$$= \frac{16{,}400 \text{ gallons}}{15 \text{ minutes}}$$

$$= 1{,}093 \text{ gpm pumping rate (rounded)}$$

Example 5.33

Problem:

A tank 50 ft in diameter is filled with water to a depth of 4 ft. To conduct a pumping test, the outlet valve to the tank is closed, and the pump is allowed to discharge into the tank. After 80 min, the water level is 5.5 ft. What is the pumping rate in gallons per minute?

Solution:

Step 1. We must first determine the volume pumped in cubic feet:

$$\text{Volume pumped} = (\text{area of circle})(\text{depth})$$

$$= (0.785)(50 \text{ ft})(50 \text{ ft})(1.5 \text{ ft})$$

$$= 2944 \text{ ft}^3 \text{ (rounded)}$$

Step 2. Convert the cubic-feet volume to gallons:

$$(2944 \text{ ft}^3)(7.48 \text{ gal/ft}^3) = 22{,}021 \text{ gallons (rounded)}$$

Step 3. The pumping test was conducted over a period of 80 min. Using Equation 5.20, calculate the pumping rate in gallons per minute.

$$\text{Pumping Rate, (gpm)} = \frac{\text{gallons}}{\text{minutes}}$$

$$= \frac{22{,}021 \text{ gallons}}{80 \text{ minutes}}$$

$$= 275.3 \text{ gpm (rounded)}$$

5.3.3 Calculating Head Loss

Note: Pump head measurements are used to determine the amount of energy a pump can or must impart to the water; they are measured in feet.

One of the principle calculations in pumping problems is used to determine head loss:

$$H_f = K(V^2/2g) \tag{5.22}$$

where
H_f = friction head
K = friction coefficient
V = velocity in pipe
G = gravity (32.17 ft/sec/sec)

5.3.4 Calculating Head

For centrifugal pumps and positive displacement pumps, several other important formulae are used to determine head. In centrifugal pump calculations, the conversion of the discharge pressure to discharge head is the norm. Positive displacement pump calculations often leave given pressures in psi.

In the following formulae, W expresses the specific weight of liquid in pounds per cubic foot. For water at 68°F, W is 62.4 lb/ft³. A water column 2.31 ft high exerts a pressure of 1 psi on 64°F water. Use the following formulae to convert discharge pressure in pounds per square inch gauge (psig) to head in feet:

- Centrifugal pumps:

$$H, ft \frac{P, psig \times 2.31}{Specific\ Gravity} \tag{5.23}$$

- Positive displacement pumps:

$$H, ft \frac{P, psig \times 144}{W} \tag{5.24}$$

To convert head into pressure:

- Centrifugal pumps:

$$P, psi \frac{H, ft \times Specific\ Gravity}{2.31} \tag{5.25}$$

- Positive displacement pumps:

$$P, psi \frac{H, ft \times W}{W} \tag{5.26}$$

5.3.5 Calculating Horsepower and Efficiency

One of the important elements of "work being done" is the "rate" at which work is done. This is called *power* and is labeled as foot-pounds per second. At some point in the past, a standard was determined: the ideal work animal (the horse, which, at the time, was the standard source of power) could move 550 lb a distance of 1 ft in 1 sec. Because large amounts of work must also be considered, this unit became known as *horsepower.*

When pushing a certain quantity of water at a given pressure, the pump performs work. 1 hp = 33,000 ft-lb/min. The two basic terms for horsepower are:

- Hydraulic horsepower (whp)
- Brake horsepower (bhp)

5.3.5.1 Hydraulic Horsepower (WHP)

One hydraulic horsepower equals:

- 550 ft-lb/sec
- 33,000 ft-lb/min
- 2545 Btu/h
- 0.746 kW
- 1014 metric hp

To calculate the hydraulic horsepower using flow in gallons per minute and head in feet, use the following formula for centrifugal pumps:

$$\text{WHP} = \frac{\text{Flow, gpm} \times \text{head, ft,} \times \text{specific gravity}}{3960} \qquad (5.27)$$

When calculating horsepower for positive displacement pumps, common practice is to use pounds per square inch for pressure. Then the hydraulic horsepower becomes:

$$\text{WHP} = \frac{\text{flow, gpm} \times \text{pressure, psi}}{3960} \qquad (5.28)$$

5.3.5.2 Pump Efficiency and Brake Horsepower (bhp)

When a motor–pump combination is used (for any purpose), neither the pump nor the motor will be 100% efficient. Simply put, not all the power supplied by the motor to the pump (called *brake horsepower,* bhp) will be used to lift the water (water or hydraulic horsepower) — some of the power is used to overcome friction within the pump. Similarly, not all of the power of the electric current driving the motor (called *motor horsepower,* mhp) will be used to drive the pump — some of the current is used to overcome friction within the motor and some is lost in the conversion of electrical energy to mechanical power.

Note: Depending on size and type, pumps are usually 50 to 85% efficient, and motors are usually 80 to 95% efficient. The efficiency of a particular motor or pump is given in the manufacturer's technical manual accompanying the unit.

A pump's brake horsepower equals its hydraulic horsepower divided by the pump's efficiency. Thus, the BHP formula becomes:

$$\text{BHP} = \frac{\text{Flow, gpm} \times \text{head, ft} \times \text{specific gravity}}{3960 \times \text{efficiency}} \qquad (5.29)$$

or

$$BHP = \frac{\text{flow, gpm} \times \text{pressure, psi}}{1714 \times \text{efficiency}} \qquad (5.30)$$

Example 5.34

Problem:

Calculate the BHP requirements for a pump handling salt water, with a flow of 600 gpm and a 40-psi differential pressure. The specific gravity of salt water at 68°F equals 1.03. The pump efficiency is 85%.

Solution:

Convert the pressure differential to total differential head, TDH = 40 × 2.31/1.03 = 90 ft (rounded).

$$BHP = \frac{600 \times 90 \times 1.03}{3960 \times 0.85}$$

$$= 16.5 \text{ hp (rounded)}$$

$$BHP = \frac{600 \times 40}{1714 \times 0.85}$$

$$= 16.5 \text{ hp (rounded)}$$

Note: Horsepower requirements vary with flow. Generally, if the flow is greater, the horsepower required to move the water will be greater.

When the motor, brake, and motor horsepower are known and the efficiency is unknown, a calculation to determine motor or pump efficiency must be done. Equation 5.31 is used to determine percent efficiency:

$$\text{Percent Efficiency} = \frac{\text{Hp output}}{\text{Hp input}} \times 100 \qquad (5.31)$$

From Equation 5.31, the specific equations to be used for motor, pump, and overall efficiency equations are:

$$\text{Percent Motor Efficiency} = \frac{bhp}{mhp} \times 100 \qquad (5.32)$$

$$\text{Percent Pump Efficiency} = \frac{whp}{bhp} \times 100 \qquad (5.33)$$

$$\text{Percent Overall Efficiency} = \frac{whp}{mhp} \qquad (5.34)$$

Example 5.35

Problem:

A pump has a water horsepower requirement of 8.5 whp. If the motor supplies the pump with 12 hp, what is the pump's efficiency?

Solution:

$$\text{Percent Pump Efficiency} = \frac{\text{whp output}}{\text{bhp supplied}} \times 100$$

$$= \frac{8.5 \text{ whp}}{12 \text{ bhp}} \times 100$$

$$= 0.71 \times 100$$

$$= 71\% \text{ (rounded)}$$

Example 5.36

Problem:

What is the pump's efficiency if electric power equivalent to 25 hp is supplied to the motor and 14 hp of work is accomplished by the pump?

Solution:

Calculate the percent of overall efficiency:

$$\text{Percent Overall Efficiency} = \frac{\text{hp output}}{\text{hp supplied}} \times 100$$

$$= \frac{14 \text{ whp}}{25 \text{ mhp}} \times 100$$

$$= 0.56 \times 100$$

$$= 56\%$$

Example 5.37

Problem:

The motor is supplied with 12 kW of power. If the brake horsepower is 14 hp, what is the efficiency of the motor?

Solution:

First, convert the kilowatt power to horsepower. Based on the fact that 1 hp = 0.746 kW, the equation becomes:

$$\frac{12 \text{ kW}}{0.746 \text{ kW/hp}} = 16.09 \text{ hp}$$

Now, calculate the percent efficiency of the motor:

$$\text{Percent Efficiency} = \frac{\text{hp output}}{\text{hp supplied}} \times 100$$

$$= \frac{14 \text{ bhp}}{16.09 \text{ mhp}} \times 100$$

$$= 87\%$$

REFERENCES

Morgan, G.T. and Drew, H.D.K. (1920). Research on residual affinity and coordination. PE II. Acetylacetones and tellurium, *J. Chem. Soc.*, 117, 1456.

Spellman, F.R. (2003). *Handbook of Water and Wastewater Treatment Plant Operations*. Boca Raton, FL: Lewis Publishers.

Types of Colloids. Accessed at http//www.ch.bris.ac.uk/webprojects2002/pdavies/types.html, 18 December 2002.

Wahren, U. (1997). *Practical Introduction to Pumping Technology*. Houston: Gulf Publishing Company.

SUGGESTED READING

Irikura, K.K. and Johnson, R.D., III (2000). Predicting unexpected chemical reactions by isopotential searching, *J. Phys. Chem. A.*, February, 2191–2194.

Missen, R.W. et al. (1999). *Introduction to Chemical Reactions Engineering and Kinetics*. New York: John Wiley & Sons.

Oxlade, C. (2002). *Materials Changes and Reactions (Chemicals in Action)*. Portsmouth, NH: Heinemann Library.

CHAPTER 6

Statistics Review

6.1 STATISTICAL CONCEPTS

Despite the protestation of Disraeli that, "there are three kinds of lies: lies, damned lies, and statistics," probably the most important step in any environmental engineering study is the *statistical analysis* of the results. The principal concept of statistics is that of variation. In conducting typical environmental studies, such as a biological sampling protocol for aquatic organisms, variation is commonly found. Variation comes from the methods employed in the sampling process or, in this example, in the distribution of organisms. Several complex statistical tests can be used to determine the accuracy of data results. In this discussion, however, only basic calculations are reviewed.

6.2 MEASURE OF CENTRAL TENDENCY

When talking statistics, we are usually estimating something on the basis of incomplete knowledge. Maybe we can only afford to test 1% of the items in which we are interested, and we want to say something about the properties of the entire lot. Perhaps we must destroy the sample by testing it. In that case, 100% sampling is not feasible because someone is supposed to get the items after we are done with them.

The questions we are usually trying to answer are "What is the central tendency of the item of interest?" and "How much dispersion about this central tendency can we expect?" Simply put, the average or averages that can be compared are measures of central tendency or central location of the data.

6.3 BASIC STATISTICAL TERMS

Basic statistical terms include the *mean* or *average*; the *median*; the *mode*; and the *range*:

- Mean — the total of the values of a set of observations divided by the number of observations
- Median — the value of the central item when the data are arrayed in size
- Mode — the observation that occurs with the greatest frequency and thus is the most "fashionable" value
- Range — the difference between the values of the highest and lowest terms

Example 6.1

Problem:

Given the following laboratory results for the measurement of dissolved oxygen (DO), find the mean, median, mode, and range. Data: 6.5 mg/L; 6.4 mg/L; 7.0 mg/L; 6.9 mg/L; 7.0 mg/L

Solution:

To find the mean:

$$Mean = \frac{(6.5 \text{ mg/L} + 6.4 \text{ mg/L} + 7.0 \text{ mg/L} + 6.0 \text{ mg/L} + 7.0 \text{ mg/L})}{5}$$

$$= 6.58 \text{ mg/L}$$

To find the mode and median, arrange in order: 6.4 mg/L; 6.5 mg/L; 6.9 mg/L; 7.0 mg/L; 7.0 mg/L.

$$Mode = 7.0 \text{ mg/L (number that appears most often)}$$

$$Median = 6.9 \text{ mg/L (central value)}$$

To find the range:

$$Range = 7.0 \text{ mg/L (highest term)} - 6.4 \text{ mg/L (lowest term)} = 0.6 \text{ mg/L}$$

The importance of using statistically valid sampling methods cannot be overemphasized; several different methodologies are available. A careful review of these methods (with the emphasis on designing appropriate sampling procedures) should be made before computing analytic results. Using appropriate sampling procedures along with careful sampling techniques provides accurate basic data.

The need for statistics in environmental engineering is driven by the discipline. As mentioned, environmental studies often deal with entities that are variable. If no variations occurred in environmental data, no need for statistical methods would occur. Over a given time interval, some variation in sampling analyses will occur. Usually, the average and the range yield the most useful information. For example, in evaluating the performance of a wastewater treatment plant, a monthly summary of flow measurements, operational data, and laboratory tests for the plant would be used.

6.4   DMR CALCULATIONS

Environmental engineers in charge of wastewater treatment facilities (typically plant or system managers) are responsible under state and federal national pollutant discharge elimination system (NPDES) permit requirements to oversee proper data recording in the daily monitoring report (DMR). In this section, we describe many of these calculations.

6.4.1   Loading Calculation

$$\text{Lb of Pollutant} = (\text{Concentration in mg/L or ppm}) \times (\text{Flow in MGD}) \times (8.34) \qquad (6.1)$$

Example 6.2

Problem:

Flow at the time of sample collection is 0.500 MGD and BOD is 10 mg/L. Determine pounds of BOD.

Solution:

$$(10) \times (0.50) \times (8.34) = 41.7 \text{ lb of BOD}$$

6.4.2 Monthly Average Loading Calculations

$$\text{Loading Average} = \frac{(L_1 + L_2 + L_3 + \dots L_N)}{N} \tag{6.2}$$

where
L = calculated loading for a sample day
N = number of samples

Example 6.3

Problem:

Given:
First sample day: flow 0.50 MGD, BOD 10 mg/L
Second sample day: flow 0.60 MGD, BOD 15 mg/L
Third sample day: flow 0.40 MGD, BOD 5 mg/L

What is the loading average?

Solution:

$$L_1 = (0.5 \text{ MGD}) (10 \text{ mg/L}) (8.34) = 41.7 \text{ lb}$$

$$L_2 = (0.6 \text{ MGD}) (15 \text{ mg/L}) (8.34) = 75.06 \text{ lb}$$

$$L_3 = (0.4 \text{ MGD}) (5 \text{ mg/L}) (8.34) = 16.68 \text{ lb}$$

$$\frac{L_3 = (41.7 \text{ lb}) 75.06 \text{ lb} + 16.68 \text{ lb}}{3} = 44.48 \text{ lb}$$

6.4.3 30-Day Average Calculation

$$\frac{(C_1 + C_2 + C_3 + \dots C_N)}{N} = C_{ave} \tag{6.3}$$

where
C = concentration of sample
N = number of samples

Example 6.4

Problem: Determine the average mg/L.

  Given:
  First sample day: BOD 10 mg/L
  Second sample day: BOD 15 mg/L
  Third sample day: BOD 5 mg/L

Solution:

$$\frac{10\ \text{mg/L} + 15\ \text{mg/L} + 5\ \text{mg/L}}{3} = 10\ \text{mg/L}$$

6.4.4 Moving Average

Conducting and establishing trend analysis for use in process control and performance evaluation is important in water and wastewater treatment plant operations. Typically, in both industries, data extending over a long period are usually available. To aid in this effort, the moving average computation is commonly used because it provides a method to develop trends for use in process control and performance evaluation.

The moving average takes all the available data into account, provides a leveling of erratic data points, and limits the length of time an individual data point will have an impact upon the computation. The moving average can be determined as an arithmetic or geometric mean and for varying periods (5, 7, or 28 days). The most common moving average is the 7-day arithmetic moving average. Because the week is the period most commonly used in water/wastewater treatment, in this section we describe the procedure for calculation of the 7-day arithmetic moving average.

  Note: A moving average can be calculated each day following completion of the initial data collection period (5, 7, or 28 days). Each day's moving average is calculated in the same way, using the most recent data period.

Procedure

  - Add all the results of tests performed during the period from Day 1 to Day 7.
  - Divide by the number of tests performed during this period.
  - This is the 7-day moving average for Day 7.
  - Repeat the procedure on Day 8 using the test results collected during the period from Day 2 to Day 8. The result of this calculation is the moving average for Day 8.
  - The same technique applies to all moving averages; only the days included in the calculation change.

$$\text{Moving Average} = \frac{\text{Test 1} + \text{Test 2} + \text{Test 3} + \text{Test 4} \ldots + \text{Test 6} + \text{Test 7}}{\text{Number of Tests Performed during the 7 Days}} = 10\ \text{mg/L} \quad (6.4)$$

Example 6.5

Problem:

The aeration tank solids concentration is determined daily. The test results for the first 10 days of the month are shown in the following chart. What is the 7-day moving average concentration on Days 7, 8, 9, and 10?

Day no.	Concentration (mg/L)
1	2330
2	3360
3	2640
4	2755
5	2860
6	2650
7	2340
8	2350
9	2888
10	2330

Solution:

$$\text{7-Day Moving Ave. for Day 7} = \frac{2330 + 3360 + 2640 + 2755 + 2860 + 2650 + 2340}{7}$$

$$= 2705 \text{ mg/L}$$

$$\text{7-Day Moving Ave. for Day 8} = \frac{3360 + 2640 + 2755 + 2860 + 2650 + 2340 + 2350}{7}$$

$$= 2708 \text{ mg/L}$$

$$\text{7-Day Moving Ave. for Day 9} = \frac{2640 + 2755 + 2860 + 2650 + 2340 + 2350 + 2888}{7}$$

$$= 2640 \text{ mg/L}$$

$$\text{7-Day Moving Ave. for Day 10} = \frac{2755 + 2860 + 2650 + 2340 + 2350 + 2888 + 2330}{7}$$

$$= 2596 \text{ mg/L}$$

6.4.5 Geometric Mean

Geometric mean (or geometric average) is a statistical calculation used for reporting bacteriological test results in water/wastewater treatment plant operations. Defined, geometric mean is a calculated mean or average appropriate for data sets containing a few values that are very high relative to the other values. To reduce the bias introduced to an arithmetic mean (average) by these very high numbers, the natural logarithms of the data are averaged. The antilog of the average is the geometric mean. Simply stated, the geometric mean is not affected by wide shifts in test results to the same extent that the arithmetic mean is affected. It can be computed using logarithms or by determining the Nth root of the product of the individual test results. Although performing each of the calculations is possible without a calculator, using one that can perform logarithm (log) functions and/or exponential (Y^x) functions is best.

6.4.5.1 Logarithm (Log) Method

To perform the calculations required to obtain a geometric mean using the log method, we must have a calculator capable of converting test results into their equivalent logarithms and converting the logarithm of the geometric mean back into its equivalent number (*antilog*):

$$\text{Geometric Mean} = \text{Antilog} \left(\frac{\log X_1 + \log X_2 + \log X_3 \ldots + \log X_n}{N, \text{ Number of Tests}} \right) \qquad (6.5)$$

Procedure

- Enter each test result into the calculator and obtain its equivalent log value. Replace any zero test result with a one and determine the log of one.
- Add all log values.
- Divide by the number of tests performed.
- Determine the antilog of this answer (the numerical equivalent of the log). The antilog is the geometric mean.

6.4.5.2 Nth Root Calculation Method

The calculated month geometric mean can also be calculated by multiplying the values of all the sample results obtained during the month (the number of samples, N) and taking the Nth root of the product:

$$\text{Geometric Mean} = \sqrt[N]{X_1 \times X_2 \times \ldots \times X_n}$$

The Nth root method requires a calculator that can multiply all the test results together and then determine the Nth root of the number.

Procedure

- Replace any zero test result with a one.
- Multiply all of the reported test values (test 1 × test 2 × test 3 × … × test N).
- Using the Nth root function (Y^x) of the calculator, determine the Nth root of the product obtained in the previous step.

Example 6.6

Problem:

    The results of the fecal coliform testing performed during the month of June are shown in the following table. What is the geometric mean of the test results computed by the log method and the Nth root method?

Test 1	20
Test 2	0
Test 3	180
Test 4	2133
Test 5	69
Test 6	96
Test 7	19
Test 8	44

Solution:

Step 1. Geometric mean by the log method:

		Log
Test 1	20	1.30103
Test 2	1	0.00000
Test 3	180	2.25527
Test 4	2133	3.32899
Test 5	69	1.83884
Test 6	96	1.98227
Test 7	19	1.27875
Test 8	44	1.64345
Geometric mean		13.62860

$$\text{Log of Geometric Mean} = \frac{13.62860}{8} = 1.703575$$

$$\text{Antilog of } 1.703575 = 50.5 \text{ or } 51$$

Step 2. Geometric mean by the Nth root method. Calculate mean by the Nth root method. Calculate the product of all the test results during the period:

$$20 \times 1 \times 180 \times 2133 \times 69 \times 96 \times 19 \times 44$$

Using the calculator, determine the eighth root of this number (eighth because there are eight test results): eighth root = 51.

6.5 STANDARD DEVIATION

In addition to simple average, moving average, geometric mean, and range calculations, it may be desirable to test the precision of test results. Standard deviation, s, is often used as an indicator of precision and is a measure of the variation (the spread in a set of observations) in the results.

Considering some of the basic theory of statistics is appropriate in order to gain better understanding of and perspective on the benefits derived from using statistical methods in environmental operations. In any set of data, the true value (mean) lies in the middle of all the measurements taken. This is true, providing the sample size is large and only random error is present in the analysis. In addition, the measurements will show a normal distribution (see Figure 6.1).

Figure 6.1 shows that 68.26% of the results fall between $M + s$ and $M - s$; 95.46% of the results lie between $M + 2s$ and $M - 2s$; and 99.74% of the results lie between $M + 3s$ and $M - 3s$. Therefore, if precise, 68.26% of all the measurements should fall between the true value estimated by the mean, plus the standard deviation and the true value minus the standard deviation. The following equation is used to calculate the sample standard deviation:

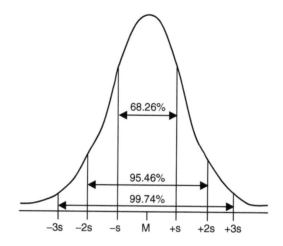

Figure 6.1   Normal distribution curve showing the frequency of a measurement.

$$s = \sqrt{\frac{\sum (X - \bar{X})2}{n-1}}$$

where:

s = standard deviation
n = number of samples
X = measurements from X to X_n
$\bar{X}$ = the mean
Σ = means to sum the values from X to X_n

Example 6.7

Problem:

Calculate the standard deviation, s, of the following dissolved oxygen values: 9.5; 10.5; 10.1; 9.9; 10.6; 9.5; 11.5; 9.5; 10.0; 9.4.

$$\bar{X} = 10.0$$

Solution:

X	X − X̄	(X − X̄)²
9.5	−0.5	0.25
10.5	0.5	0.25
10.1	0.1	0.01
9.9	−0.1	0.01
10.6	0.6	0.36
9.5	−0.5	0.25
11.5	1.5	2.25
9.5	−0.5	0.25
10.0	0	0
9.4	−0.6	0.36
		3.99

$$s = \sqrt{\frac{1}{(10-1)}} \quad (3.99)$$

$$s = \sqrt{\frac{3.99}{9}} = 0.67$$

6.6   CONCLUSION

In this chapter, we have touched only on the basics of statistics. Obviously, practicing environmental engineers need to know much more about this valuable tool. For example, engineers must understand not only elementary probability and basic statistics — with emphasis on their application in engineering and the sciences — but also the treatment of data; sampling distributions; inferences concerning means; inferences concerning variances; inferences concerning proportions; nonparametric tests; curve fitting; analysis of variance; factorial experimentation; and much more. Because these topics are beyond the scope of this text, we highly recommend Richard A. Johnson (1997) *Miller and Freund's Probability and Statistics for Engineers,* 6th ed., available from Prentice-Hall.

PART III

Math Concepts: Air Pollution Control

CHAPTER 7

Air Pollution Fundamentals

"If seven maids with seven mops

Swept it for half a year,

Do you suppose," the Walrus said,

"That they could get it clear?"

"I doubt it," said the Carpenter,

And shed a bitter tear.

Lewis Carroll

7.1 INTRODUCTION

In the last 40 years, the environmental engineering profession has expanded its societal responsibilities to include the control of air pollution from industrial sources. Though not exactly "seven maids with seven mops trying to get it clear," increasing numbers of environmental engineers are confronted with problems in this most vital area. Although the design and construction of air pollution control equipment today is accomplished with some degree of success, air pollution problems still exist. Environmental engineers of today and tomorrow must develop proficiency and improved understanding of the design and selection of air pollution control equipment in order to cope with these problems and challenges — "to get it clear."

In this spirit (contrary to point of view of the Walrus and the Carpenter), we present this chapter. In short, we simply feel that the present situation is not that grim. Why do we feel this way? Simply put, we know we can do something to control environmental air pollution. Environmental engineers who are well trained and well equipped with the proper mathematical tools and applications can make a difference when it comes time to "clear the air" that we breathe.

The USEPA focuses (as it should) enormous amounts of time on research topics related to air pollution control. In this chapter, we heavily excerpt from USEPA publications on the topic. Much of the general (basic) information provided is adapted from Spellman (1999) *The Science of Air.* The excerpted materials have been rearranged and edited to make them more concise for engineers in training and for general readers to understand the basic concepts of air pollution control mathematics and processes.

Controlling environmental air pollution begins with understanding what this pollution is. We define *environmental air pollution* as the contamination of atmospheric air in such a manner as to cause real or potential harm to human health or well being, or to damage or harm the natural surroundings without justification. Contaminants may include almost any natural or artificial composition of matter capable of being airborne (friable asbestos, for example). Contaminants may be in the form of solids particles, liquid droplets, or gases, or in combinations of these forms. Contaminants fall into two main groups: (1) those emitted directly from identifiable sources; and (2) those produced in the air by interaction between two or more primary contaminants or by reaction with normal atmospheric constituents, with or without photoactivation.

The Clean Air Act (CAA) established two types of National Ambient Air Quality Standards (NAAQS).

- Primary standards are designed to establish limits to protect public health, including the health of "sensitive" populations such as asthmatics, children, and the elderly.
- Secondary standards set limits to protect public welfare, including protection against decreased visibility and damage to animals, crops, vegetation, and buildings.

7.1.1 Six Common Air Pollutants

USEPA (2003a) has set national air quality standards for six common pollutants (also referred to as "criteria" pollutants) discharged from various sources:

- Ground-level ozone
- Nitrogen dioxide
- Particulate matter
- Sulfur dioxide
- Carbon monoxide
- Lead

7.1.1.1 Ground-Level Ozone

Ozone (O_3) is a highly reactive photochemically produced gas composed of three oxygen atoms. Although it is not usually emitted directly into the air (rather, as a secondary air pollutant), at ground level O_3 is created by a chemical reaction between oxides of nitrogen (NO_x) and volatile organic compounds (VOC) in the presence of heat and sunlight. In *The Science of Air* (1999), we characterized ozone as the Dr. Jeckel and Mr. Hyde of air pollutants. Why? Ozone has the same chemical structure, whether it occurs miles above the Earth or at ground level, and can be "good" or "bad," depending on its location in the atmosphere. "Good" (Dr. Jeckel) ozone occurs naturally in the stratosphere approximately 10 to 30 miles above the Earth's surface and forms a layer that protects life on Earth from the sun's harmful rays. In the Earth's lower atmosphere, however, ground-level ozone is considered "bad" (Mr. Hyde).

$$VOC + NO_x + Heat + Sunlight = Ozone$$

Motor vehicle exhaust and industrial emissions, gasoline vapors, and chemical solvents are some of the major sources of NO_x and VOC that help to form ozone. Sunlight and hot weather cause ground-level ozone to form in harmful concentrations in the air. As a result, ozone is known as a summertime air pollutant. Many urban areas tend to have high levels of bad ozone, but even rural areas are subject to increased ozone levels because wind carries ozone and the pollutants that combine to form it hundreds of miles away from their original sources.

7.1.1.2 Nitrogen Oxides

Nitrogen oxides (NO_x) is the generic term for a group of highly reactive gases, all of which contain nitrogen and oxygen in varying amounts. This group includes NO; NO_2; NO_3; N_2O; N_2O_3; N_2O_4; and N_2O_5; however, only two are important in the study of air pollution: nitric oxide (NO) and nitrogen dioxide (NO_2). Many of the nitrogen oxides are colorless and odorless. However, nitrogen dioxide (NO_2), a common pollutant, can often be seen, along with particles in the air, as a reddish-brown layer of air hanging over affected urban areas. Nitrogen oxides form when fuel is burned at high temperatures, as in a combustion process. The primary sources of NO_x are motor vehicles, electric utilities, and other industrial, commercial, and residential sources that burn fuels.

7.1.1.3 Particulate Matter

Particulate matter (PM) is the term for particles found in the air, including dust, dirt, soot, smoke, and liquid droplets. Particles can be suspended in the air for long periods of time. Some particles are large or dark enough to be seen as soot or smoke. Others are so small that, individually, they can only be detected with an electron microscope. Some particles are directly emitted into the air. They come from a variety of sources, including cars; trucks; buses; factories; construction sites; tilled fields; unpaved roads; stone crushing; and wood burning. Other particles may be formed in the air from the chemical change of gases. They are indirectly formed when gases from burning fuels react with sunlight and water vapor. These can result from fuel combustion in motor vehicles, at power plants, and in other industrial processes.

7.1.1.4 Sulfur Dioxide (SO_2)

Sulfur dioxide (SO_2) belongs to the family of sulfur oxide gases (SO_x). These gases dissolve easily in water. Sulfur enters the atmosphere in the form of corrosive sulfur dioxide gas, a colorless gas possessing the sharp, pungent odor of burning rubber. Also, sulfur is prevalent in raw materials, including crude oil, coal, and ore that contains common metals like aluminum, copper, zinc, lead, and iron. SO_2 gases are formed when fuels that contain sulfur (coal and oil, for example) are burned, as well as when gasoline is extracted from oil or metals are extracted from ore. SO_2 dissolves in water vapor to form acid and interacts with other gases and particles in the air to form sulfates and other products harmful to people and their environment.

Over 65% of SO_2 released to the air, or more than 13 million tons per year, come from electric utilities, especially those that burn coal. Other sources of SO_2 are industrial facilities that derive their products from raw materials like metallic ore, coal, and crude oil, or that burn coal or oil to produce process heat. Examples include petroleum refineries, and cement manufacturing and metal processing facilities. Large ships, locomotives, and some nonroad diesel equipment currently burn high sulfur fuel and release SO_2 emissions to the air in large quantities.

Two major environmental problems have developed in highly industrialized regions of the world where the atmospheric sulfur dioxide concentration has been relatively high: sulfurous smog and acid rain. Sulfurous smog is the haze that develops in the atmosphere when molecules of sulfuric acid serve as light screeners. The second problem, acid rain, is precipitation contaminated with dissolved acids, including sulfuric acid. Acid rain poses a threat to the environment by causing affected lakes to become devoid of aquatic life. Sulfur dioxide produces white to straw-colored blotches on the foliage of broad-leafed plants.

7.1.1.5 Carbon Monoxide (CO)

Carbon monoxide (CO) is a colorless, odorless, tasteless gas that is, by far, the most abundant of the primary pollutants. Formed when carbon in fuel is not burned completely, carbon monoxide is a component of motor vehicle exhaust, which contributes about 56% of all CO emissions nation-wide. Other nonroad engines and vehicles (such as construction equipment and boats) contribute about 22% of all CO emissions nationwide. Higher levels of CO generally occur in areas with heavy traffic congestion; in cities, 85 to 95% of all CO emissions come from motor vehicle exhaust. Other sources of CO emissions include industrial processes (such as metals processing and chemical manufacturing), residential wood burning, and natural sources such as forest fires. Woodstoves, gas stoves, cigarette smoke, and unvented gas and kerosene space heaters are sources of CO indoors. The highest levels of CO in the outside air typically occur during the colder months of the year when inversion conditions are more frequent. The air pollution becomes trapped near the ground beneath a layer of warm air.

7.1.1.6 Lead

Lead is a metal found naturally in the environment as well as in manufactured products. The major sources of lead emissions have historically been motor vehicles (primarily cars and trucks) and industrial sources. At present, because of the phase-out of leaded gasoline, metals processing is the major source of lead emissions to the air. The highest levels of lead in air are generally found near lead smelters; other stationary sources are waste incinerators, utilities, and lead-acid battery manufacturers. In high concentrations, lead can damage human health and the environment. Once lead enters the ecosystem, it remains there permanently. The good news? Since the 1970s, stricter emission standards have caused a dramatic reduction in lead output.

7.2 GASES

Gases are important not only because a gas can be a pollutant, but also because gases convey particulate and gaseous pollutants. For most air pollution work, expressing pollutant concentrations in volumetric terms is customary. For example, the concentration of a gaseous pollutant in parts per million (ppm) is the volume of pollutant per million parts of the air mixture. That is,

$$\text{ppm} = \frac{\text{Parts of contamination}}{\text{million parts of air}} \tag{7.1}$$

Note that calculations for gas concentrations are based on the gas laws:

- The volume of gas under constant temperature is inversely proportional to the pressure.
- The volume of a gas under constant pressure is directly proportional to the Kelvin temperature. The Kelvin temperature scale is based on absolute zero ($0°C = 273$ K).
- The pressure of a gas of a constant volume is directly proportional to the Kelvin temperature.

Thus, when measuring contaminant concentrations, we must know the atmospheric temperature and pressure under which the samples were taken. At standard temperature and pressure (STP), 1 g-mol of an ideal gas occupies 22.4 L. The STP is $0°C$ and 760 mm Hg. If the temperature is increased to $25°C$ (room temperature) and the pressure remains the same, 1 g-mol of gas occupies 24.45 L.

Sometimes converting milligrams per cubic meter (mg/m^3) — a weight-per-volume-ratio — into a volume-per-unit weight ratio is necessary. If 1 g-mol of an ideal gas at $25°C$ occupies 24.45 L is an understood fact, the following relationships can be calculated:

$$\text{ppm} = \frac{24.45}{\text{molecular wt}} \text{mg/m}^3 \qquad (7.2)$$

$$\text{mg/m}^3 = \frac{\text{molecular wt}}{24.45} \text{ppm} \qquad (7.3)$$

7.2.1 The Gas Laws

As mentioned, gases can be pollutants as well as the conveyors of pollutants. Air (which is mainly nitrogen) is usually the main gas stream. To understand the gas laws, it is imperative to have an understanding of certain terms.

- *Ideal gas* — an imaginary model of a gas that has a few very important properties:
  - The gases are assumed to be infinitely small.
  - The particles move randomly in straight lines until they collide with something (another gas molecule or the side of the container in which they are held).
  - The gas particles do not interact with each other (they do not attract or repel one another like real molecules do).
  - The energy of the particles is directly proportional to the temperature in Kelvins (in other words, the higher the temperature is, the more energy the particles have).

  These assumptions are made because they make equations a lot simpler than they would be otherwise, and because these assumptions cause negligible deviation from the ways in which actual gases behave.
- *Kelvin* — a temperature scale in which the degrees are the same size as degrees Celsius but where "0" is defined as "absolute zero," the temperature at which molecules are at their lowest energy. To convert from degrees Celsius to Kelvins (not "degrees Kelvins"), add 273.
- *Pressure* — a measure of the amount of force that a gas exerts on the container into which it is put. Units of pressure include atmospheres (1 atm is the average atmospheric pressure at sea level); torrey (equal to 1/760 of an atmosphere); millimeters of mercury (1 mm Hg is the same as 1 torr, or 1/760 atm); and kilopascals (101,325 kPa in 1 atm).
- *Standard temperature and pressure* — a set of conditions defined as 273 K and 1 atm.
- *Standard conditions (SC)* — SC is more commonly used than STP and represents typical room conditions of 20°C (70°F) and 1 atm. SC units of volume are commonly given as normal cubic meters or standard cubic feet (scf).
- *Temperature* — a measure of how much energy the particles in a gas have and defined as that property of a body that determines the flow of heat. Heat will flow from a warm body to a cold body. Several different temperature scales are in general use that depend on the freezing and boiling points of water as boundary markers for the scale. In a conventional laboratory thermometer, the boundary points are conveniently selected to relate to the known properties of water.
  - On the Celsius scale, the freezing point of water is assigned a value of 0 and the boiling point a value of 100; the distance between these two points is divided into 100 equal increments, with each increment labeled in Celsius degree (Table 7.1).
  - On the Kelvin scale, the freezing point of water is assigned a value of 273.15 K and the boiling point a value of 373.15; the distance between these two points is divided into 100 equal increments, and each increment is labeled as a Kelvin (Table 7.1).
  - On the Fahrenheit scale, the freezing point of water is assigned a value of 32 and the boiling point a value of 212; the distance between these two points is divided into 180 equal increments and each increment is labeled as a Fahrenheit degree (Table 7.1).
  - Rankine is a temperature scale with an absolute zero below which temperatures do not exist and using a degree of the same size as that used by the Fahrenheit temperature scale. Absolute zero, or 0° R, is the temperature at which molecular energy is at minimum; it corresponds to

Table 7.1 Comparison of Temperature Scales

Temperature scale	Celsius (°C)	Kelvin (K)	Fahrenheit (°F)	Rankine (°R)
Boiling point of water	+100	+373.15	+212	+671.67
	—	—	—	—
	—	—	—	—
	—	—	—	—
	↑	—	↑	—
	100 equal divisions	—	180 equal divisions	—
	↓	—	↓	—
	—	—	—	—
	—	—	—	—
Freezing point of water	0	273.15	+32	+491.67
	—	—	—	—
	—	—	—	—
	—	—	—	—
	—	—	—	—
Absolute zero	−273.15	0	−459.67	0

Note: Units of temperature important to environmental engineers include degree Celsius and Kelvins (equal to 273 plus the degree Celsius). The degree symbol (°) is not used for the Kelvin temperature scale.

a temperature of −459.67°F. Because the Rankine degree is the same size as the Fahrenheit degree, the freezing point of water (32°F) and the boiling point of water (212°F) correspond to 491.67°R and 671.67°R, respectively (Table 7.1).

We must make a distinction between the actual temperature (°C and °F) and a temperature increment (Fahrenheit degree and Celsius degree). This distinction enables the derivation of a relationship between the two temperature scales. For example, a temperature of 100°C is the same as a temperature of 212°F. A temperature difference of 100 degrees Celsius is equal to a temperature difference of 180 degrees Fahrenheit.

- *Volume* — the amount of space that an object occupies. The unit of volume can be cubic centimeters (abbreviated "cc" or "cm³"); milliliters (abbreviate "mL"; 1 mL is the same as 1 cm³); liters (abbreviated as "L" and equal to 1000 mL); or cubic meters (abbreviated "m³" – a cubic meter contains 1 million cm³).

7.2.1.1 Boyle's Law

Circa 1662, Robert Boyle stated what has come to be known as Boyle's law — the volume of any definite quantity of gas at constant temperature varies inversely as the pressure on the gas:

$$P_1V_1 = P_2V_2 \tag{7.4}$$

where
P_1 = the initial pressure of the gas
V_1 = the initial volume of the gas
P_2 = the final pressure of the gas
V_2 = the final volume of the gas

This way, if we know the initial pressure and volume of a gas and know what the final pressure will be, we can predict what the volume will be after we add pressure to it.

Example 7.1

Problem:

If we have 4 L of methane gas at a pressure of 1.0 atm, what will be the pressure of the gas if we compress it to a volume of 2.5 L?

Solution:

$$(1.0 \text{ atm})(4 \text{ L}) = (x \text{ atm})(2.5 \text{ L})$$

$$x = 1.6 \text{ atm}$$

7.2.1.2 Charles's Law

Charles observed that hydrogen (H_2), carbon dioxide (CO_2), oxygen (O_2), and air expanded by an equal amount when heated from 0 to 80°C at a constant pressure:

$$V_1/T_1 = V_2/T_2 \tag{7.5}$$

In this equation, the subscript "1" indicates the initial volume and temperature and the subscript "2" indicates the volume and temperature after the change. Temperature, incidentally, needs to be given in Kelvins and not in Celsius because if we have a temperature below 0°C, the calculation works out so that the volume of the gas is negative — a physical impossibility.

Example 7.2

Problem:

If we have 2 L of methane gas at a temperature of 40°C, what will be the volume if we heat the gas to 80°?

Solution:

The first thing to do is to convert the temperature to Kelvins (by adding 273) because Celsius cannot be used in this equation. To do this, we get that the initial temperature is 40 + 273 = 313 K and the final temperature is 80 + 273 = 353 K. We are now ready to insert these numbers into the equation:

$$2 \text{ L}/313 \text{ K} = x \text{ L}/353 \text{ K}$$

$$x = 2.26 \text{ L}$$

7.2.1.3 Gay–Lussac's Law

Gay–Lussac (1802) found that all gases increase in volume for each one degree Celsius rise in temperature and that this increase is equal to approximately 1/273.15 of the volume of the gas at 0°C:

$$P_1/T_1 = P_2/T_2 \tag{7.6}$$

If we increase the temperature of a container with fixed volume, this gas law explains how the pressure inside the container will increase.

7.2.1.4 The Combined Gas Law

This gas law combines the parameters of the preceding equations, forming

$$(P_1 V_1)/T_1 = (P_2 V_2)/T_2 \tag{7.7}$$

The advantage of the combined gas law equation is that, whenever we are changing the conditions of pressure, volume, and/or temperature for a gas, we can simply insert the numbers into this equation.

Example 7.3

Problem:
    If we have 2 L of a gas at a temperature of 420 K and decrease the temperature to 350 K, what will be the new volume of the gas?

Solution:
    To solve this problem, we use the combined gas law. Because pressure was never mentioned in this problem, we ignore it. As a result, the equation will be:

$$V_1/T_1 = V_2/T_2$$

which is the same as Charles's law. To solve, the initial volume is 2 L; the initial temperature is 420 K; and the final temperature is 350 K. The final volume, after solving the equation, should be 1.67 L.

7.2.1.5 The Ideal Gas Law

The ideal gas law combines Boyle's and Charles's laws because air cannot be compressed without its temperature changing. This gas law is an equation of state, which means that we use the basic properties of the gas to find out more about it without the need to change it in any way. Because it is an equation of state, it allows us not only to find out the pressures, volumes, and temperatures, but also to find out how much gas is present in the first place. The ideal gas law is expressed by the equation:

$$PV = nRT \tag{7.8}$$

where:
P = the pressure of the gas (in atmospheres or kilopascals)
V = the volume (in liters)
n = the number of moles
R = the ideal gas constant
T = the temperature (in Kelvins)

    The two common values for the ideal gas constant include 0.08206 L × atm/mol × K, and 8.314 × kPa/mol × K. The question is, which one do we use? The value of R used depends on the pressure given in the problem. If the pressure is given in atmospheres, use the 0.08206 value because the unit at the end of it contains "atmospheres." If the pressure is given in kilopascals, use the second value because the unit at the end contains "kPa."

The ideal gas law allows us to figure out how many grams and moles of the gas are present in a sample. After all, "moles" is the "n" term in the equation, and we already know how to convert grams to moles.

Example 7.4

Problem:
Given 4 L of a gas at a pressure of 3.4 atm and a temperature of 300 K, how many moles of gas are present?

Solution:
First, figure out what value of the ideal gas constant should be used. Because pressure is given in atmospheres, use the first one, 0.206 L × atm/mol × K. After inserting the given terms for pressure, volume, and temperature, the equation becomes:

$$(3.4 \text{ atm})(4 \text{ L}) = n (0.08206 \text{ L} \times \text{atm/mol} \times \text{K})(300 \text{ K})$$

$$n = 0.55 \text{ mol}$$

7.2.1.6 Composition of Air

The air mixture that surrounds us and that we breathe is a dynamic mixture of many components (see Table 7.2) in several respects. The moisture content of water vapor, the temperature, the pressure, and the trace gas constituents can and do vary over time and in space. The bulk of the air in the biosphere is composed of nitrogen and oxygen with various other trace gases mixed in (see Table 7.2).

Table 7.2 Approximate Composition of Dry Air (by Volume)

Component	Symbol	Concentration (%)	Concentration (ppm)
Nitrogen	N_2	78.084	780,840
Oxygen	O_2	20.9476	209,476
Argon	Ar	0.934	9,340
Carbon dioxide	CO_2	0.0314	314
Neon	Ne	0.001818	18.18
Helium	He	0.000524	5.24
Methane	CH_4	0.0002	2
Sulfur dioxide	SO_2	0–0.0001	0–1
Hydrogen	H_2	0.00005	0.5
Krypton	Kr	0.0002	2
Xenon	Xe	0.0002	2
Ozone	O_3	0.0002	2

Note: ASHRAE also reports that the molecular weight of dry air is 28.9645 g/mol based on the carbon 12 scale.

Source: *ASHRAE Handbook of Fundamentals*, 1993, Atlanta, Georgia: The American Society for Heating Refrigeration and Air Conditioning Engineers, p. 6.1 (based on the atomic weight of carbon of 12.0000).

7.3 PARTICULATE MATTER

Typically, in actual practice, the terms *particulate* (or *particle*) and *particulate matter* are used interchangeably. According to 40 CFR 51.100-90, particulate matter is defined as any airborne finely divided solid or liquid material with an aerodynamic diameter smaller than 100 μm (micro = 10^{-6}). Along with gases and water vapor, Earth's atmosphere is literally a boundless arena for particulate matter of many sizes and types. Atmospheric particulates vary in size from 0.0001 to 10,000 μm. Particulate size and shape have direct bearing on visibility. For example, a spherical particle in the 0.6-μm range can effectively scatter light in all directions, thus reducing visibility.

The types of airborne particulates in the atmosphere vary widely, with the largest sizes derived from volcanoes; tornadoes; waterspouts; burning embers from forest fires; seed parachutes; spider webs; pollen; soil particles; and living microbes. The smaller particles (the ones that scatter light) include fragments of rock; salt and spray; smoke; and particles from forested areas. The largest portion of airborne particulates is invisible. These particulates are formed by the condensation of vapors; chemical reactions; photochemical effects produced by ultraviolet radiation; and ionizing forces that come from radioactivity, cosmic rays, and thunderstorms. Airborne particulate matter is produced by mechanical weathering, breakage, and solution, or by the vapor–condensation–crystallization process (typical of particulates from the furnace of a coal-burning power plant).

We know very well that anything that goes up must eventually come down. This is typical of airborne particulates also. Fallout of particulate matter depends, obviously, mostly on size; less obvious are shape; density; weight; airflow; and injection altitude. The residence time of particulate matter also is dependent on the atmosphere's cleanup mechanisms (formation of clouds and precipitation) that work to remove them from their airborne suspended state. Some large particulates may only be airborne for a matter of seconds or minutes. Intermediate sizes may be able to stay afloat for hours or days. Finer particulates may stay airborne for a much longer duration: days, weeks, months, and even years.

Particles play an important role in atmospheric phenomena. For example, particulates provide the nuclei upon which ice particles and cloud condensation are formed, and they are essential for condensation to take place. The most important role airborne particulates play is in cloud formation. Simply put, without airborne particulate matter, no clouds would exist and, without clouds, life would be much more difficult; the cloudbursts that eventually erupted would cause an extreme devastation difficult to imagine or contemplate.

7.4 POLLUTION EMISSION MEASUREMENT PARAMETERS

Because of the gaseous and particulate emissions combustion sources can produce, they constitute a significant air quality control problem. Combustion processes can add carbon dioxide, water vapor, and heat to the atmosphere and produce a residue that must be disposed of in concentrated form. In the past, these environmental costs were tolerated in the interest of producing useful energy. However, it is becoming increasingly clear that the presence of these emissions in the atmosphere can result indirectly in a greenhouse effect and exacerbate the problem of acid rain.

Because of the environmental impact of combustion emissions, the USEPA has developed emission standards for the combustion or incineration industry. These standards usually establish the maximum allowable limit, based on volume or mass flows at specified conditions of temperature and pressure, for the discharge of specific pollutants. Emissions are measured in terms of the concentration of pollutant per volume or mass of stack (flue) gas, the pollutant mass rates, or a rate applicable to a given process. Standards fall into the following six categories:

- *Pollutant mass rate standards* — based on the fixed rate of emissions (the mass of pollutant emitted per unit time, pounds per hour or kilograms per hour)
- *Process rate standards* — establish the allowable emissions in terms of the input energy or the raw material feed of a process
- *Concentration standards* — limit the mass (weight) or volume of the pollutants in the gas leaving the stack
- *Ambient concentration standards* — pollutants, in micrograms per cubic meter, under this limit include toxic metals, organics, and hydrogen chloride
- *Reduction standards* — limits expressed as a percent reduction of the pollutants
- *Opacity standards* — the limit on the degree to which the stack emissions are visible and block the visibility of objects in the background. Stack emissions of 100% opacity totally block the view of background objects and indicate high pollutant levels. Opacity of 0% provides a clear view of the background and indicates no detectable particulate matter emissions.

7.5 STANDARD CORRECTIONS

Because combustion systems always produce stack gas of a higher temperature and pressure than those of the standards and because actual levels of pollutants emitted can be made to appear smaller if excess air is added to the stream, corrections for the differences must be made. In increases or decreases in gas temperature and pressure and the subsequent effect on gas volume, the EPA recommends answering these questions (making corrections) using the ideal gas law.

For excess air correction calculations, various federal EPA and state regulations give procedures for calculations of excess air based on dry gas (Orsat type) analyses. Based on USEPA's (1980) Method 3B — gas analysis, the percent excess air can be determined by any of the following three equivalent relationships.

$$\% \text{ Excess Air} = \frac{\text{Total Air Theoretical Air}}{\text{Theoretical Air}}(100) \tag{7.9}$$

$$\% \text{ Excess Air} = \frac{\text{Excess Air}}{\text{Theoretical Air}}(100) \tag{7.10}$$

$$\% \text{ Excess Air} = \frac{\text{Excess Air}}{\text{Total Air Excess Air}}(100) \tag{7.11}$$

Theoretical air is the amount required to convert all combustible species (mainly carbon, hydrogen, and sulfur) stoichiometrically to complete normal products of combustion (CO_2, H_2O, and SO_2). These relationships are stated as mole ratios of air, which equal volume ratios.

In its Method 3B — gas analysis for the determination of emission rate correction factor or excess air, the USEPA (2003b) provides the following explanation for determining percent excess air.

USEPA Example

Percent excess air — determine the percentage of the gas that is N_2 by subtracting the sum of the percent CO_2, percent CO, and percent O_2 from 100%. Calculate the percent excess by substituting the appropriate values of percent O_2, CO, and N_2 into Equation 7.12.

$$\% \, EA = \frac{\%O_2^- \;\; 0.5 \; \%CO}{0.264 \; \%N_2^- \;\; (\%O_2^- \;\; 0.5 \; \%CO)} (100) \qquad (7.12)$$

Equation 7.12 assumes that ambient air is used as the source of O_2 and that the fuel does not contain appreciable amounts of N_2 (as do coke oven or blast furnace gases). For cases in which appreciable amounts of N_2 are present (coal, oil, and natural gas do not contain appreciable amounts of N_2) or when oxygen enrichment is used, alternative methods are required.

REFERENCES

ASHRAE Handbook of Fundamentals (1993). Atlanta: The American Society for Heating Refrigeration and Air Conditioning Engineers.

CFR 40 51.100-90. Definition, Title 40 Parts 51.100 of Code of Federal Regulations published 1990.

Spellman, F.R. (1999). *The Science of Air*. Boca Raton, FL: CRC Press.

USEPA (2003a), Six common air pollutants. Accessed at http://www.epa.gov/air/Urbanair/6poll.html.

USEPA (2003b), Method 3B — gas analysis for the determination of emissions rate correction factor or excess air. Accessed at http://www.epa.gov/ttn/emc/promgate/m-03b.pdf.

USEPA 80/02, Combustion evaluation — Student manual, Course 427, EPA Air Pollution Training Institute (APTI), EPA450-2-80-063, February 1980.

CHAPTER 8

Gaseous Emission Control

Be it known to all within the sound of my voice,

Whoever shall be found guilty of burning coal

Shall suffer the loss of his head.

King Edward II

8.1 INTRODUCTION

Limiting gaseous emissions into the air is technically difficult as well as expensive. Although it is true that rain is "Nature's vacuum cleaner" — the only natural air-cleansing mechanism available — it is not very efficient. Good air quality depends on pollution prevention (limiting what is emitted) and sound engineering policies, procedures, and practices. The control of gaseous air emissions may be realized in a number of ways. In this chapter, we discuss many of these technologies, as well as the sources of gaseous pollutants emitted from various sources and their control points (see Figure 8.1).

The applicability of a given technique depends on the properties of the pollutant and the discharge system. In making the difficult and often complex decision of which gaseous air pollution control to employ, follow the guidelines based on experience and set forth by Buonicore and Davis (1992) in their prestigious engineering text, *Air Pollution Engineering Manual*. Table 8.1 summarizes the main techniques and technologies used to control gaseous emissions.

In the following, we discuss the air control technologies given in Table 8.1. Much of the information contained in this chapter is heavily adapted from Spellman's *The Science of Air* (1999) and USEPA-81/12 (1981), Control of Gaseous Emissions, Course 415. The excerpted materials are rearranged and edited to make the materials more accessible for the reader.

8.2 ABSORPTION

Absorption (or scrubbing) is a major chemical engineering unit operation that involves bringing contaminated effluent gas into contact with a liquid absorbent so that one or more constituents of the effluent gas are selectively dissolved into a relatively nonvolatile liquid. Key terms used when discussing the absorption process include:

- *Absorbent*: the liquid, usually water mixed with neutralizing agents, into which the contaminant is absorbed
- *Solute*: the gaseous contaminant being absorbed, such as SO_2, H_2S, and so forth

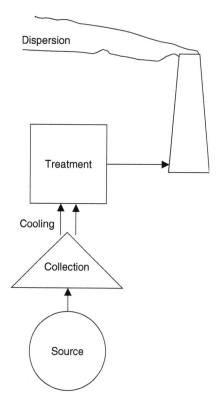

Figure 8.1 Air pollution control points. (From Spellman, F.R., 1999, *The Science of Air: Concepts and Applications*. Boca Raton, FL: CRC Press.)

Table 8.1 Comparison of Air Control Technologies

Treatment technology	Concentration and efficiency	Comments
Absorption	(<200 ppmv) 90–95% efficiency (>200 ppmv) 95%+ efficiency	Can blowdown stream be accomplished at site?
Carbon adsorption	(>200 ppmv) 90%+ efficiency (>1000 ppmv) 95%+ efficiency	Recovered organics may need additional treatment; can increase cost.
Incineration	(<100 ppmv) 90–95% efficient (>100 ppmv) 95–99% efficient	Incomplete combustion may require additional controls.
Condensation	(>2000 ppmv) 80%+ efficiency	Must have low temp. or high pressure for efficiency.

Source: Spellman, F.R., 1999, *The Science of Air*. Boca Raton, FL: CRC Press, p. 221.

- *Carrier gas*: the inert portion of the gas stream, usually flue gas, from which the contaminant is to be removed
- *Interface*: the area where the gas phase and the absorbent contact each other
- *Solubility*: the capability of a gas to be dissolved in a liquid

Absorption units are designed to transfer the pollutant from a gas phase to a liquid phase, accomplished by providing intimate contact between the gas and the liquid, which allows optimum diffusion of the gas into the solution. The actual mechanism of removal of a pollutant from the gas stream takes place in three steps: (1) diffusion of the pollutant gas to the surface of the liquid; (2) transfer across the gas–liquid interface; and (3) diffusion of the dissolved gas away from the interface into the liquid (Davis and Cornwell, 1991).

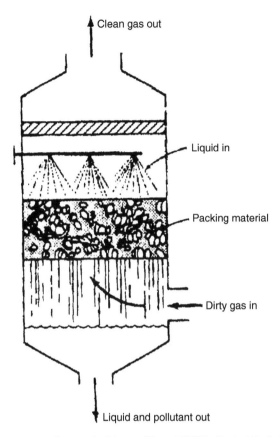

Figure 8.2   Typical countercurrent-flow packed tower. (From USEPA, Control Techniques for Gases and Particulates, 1971.)

Several types of scrubbing towers are available, including spray chambers (towers or columns); plate or tray towers; packed towers; and Venturi scrubbers. Pollutant gases commonly controlled by absorption include sulfur dioxide; hydrogen sulfide; hydrogen chloride; ammonia; and oxides of nitrogen.

The two most common absorbent units in use today are the plate and packed tower systems. Plate towers contain perforated horizontal plates or trays designed to provide large liquid–gas interfacial areas. The polluted airstream is usually introduced at one side of the bottom of the tower or column and rises up through the perforations in each plate; the rising gas prevents the liquid from draining through the openings rather than through a downpipe. During continuous operation, contact is maintained between air and liquid, allowing gaseous contaminants to be removed, with clean air emerging from the top of the tower.

The packed tower scrubbing system (see Figure 8.2) is predominantly used to control gaseous pollutants in industrial applications, where it typically demonstrates a removal efficiency of 90 to 95%. Usually vertically configured (Figure 8.2), the packed tower is literally packed with devices (see Figure 8.3) of large surface-to-volume ratio and a large void ratio that offer minimum resistance to gas flow. In addition, packing should provide even distribution of both fluid phases; be sturdy enough to support itself in the tower; and be low cost, available, and easily handled (Hesketh, 1991).

The flow through a packed tower is typically countercurrent, with gas entering at the bottom of the tower and liquid entering at the top. Liquid flows over the surface of the packing in a thin film, affording continuous contact with the gases. Though highly efficient for removal of gaseous contaminants, packed towers may create liquid disposal problems, may become easily clogged when gases with high particulate loads are introduced, and have relatively high maintenance costs.

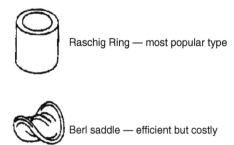 Raschig Ring — most popular type

 Berl saddle — efficient but costly

 Pall rings — good liquid distribution

Tellerette — very low unit weight

 Intalox saddle — efficient but expensive

Figure 8.3   Various packing used in packed tower scrubbers. (Adapted from Air Pollution Control Equipment, Part II, American Industrial Hygiene Association, 1968.)

8.2.1 Solubility

Solubility is a function of system temperature and, to a lesser extent, system pressure. As temperature increases, the amount of gas that can be absorbed by a liquid decreases (gases are more soluble in cold liquids than in hot liquids). Gas phase pressure can also influence solubility; by increasing the pressure of a system, the amount of gas absorbed generally increases. However, this is not a major variable in absorbers used for air pollution control because they operate at close to atmospheric pressure (USEPA-81/12, 1981).

8.2.2 Equilibrium Solubility and Henry's Law

Under certain conditions, Henry's law can express the relationship between the gas phase concentration and the liquid phase concentration of the contaminant at equilibrium. This law states that for dilute solutions in which the components do not interact, the resulting partial pressure (p) of a component A in equilibrium with other components in a solution can be expressed as

$$p = Hx_A \qquad (8.1)$$

where
p = partial pressure of contaminant in gas phase at equilibrium
H = Henry's law constant
x_A = mole fraction of contaminant or concentration of A in liquid phase

Equation 8.1 is the equation of a straight line where the slope (m) is equal to H. Henry's law can be used to predict solubility only when the equilibrium line is straight — when the solute concentrations are very dilute. In air pollution control applications this is usually the case. For example, an exhaust stream that contains a 1000-ppm SO_2 concentration corresponds to a mole fraction of SO_2 in the gas phase of only 0.001.

Another restriction on using Henry's law is that it does not hold true for gases that react or dissociate upon dissolution. If this happens, the gas no longer exists as a simple molecule. For example, scrubbing HF or HCl gases with water causes both compounds to dissociate in solution. In these cases, the equilibrium lines are curved rather than straight. Data on systems that exhibit curved equilibrium lines must be obtained from experiments.

The units of Henry's law constants are atmosphere per mole fractions. The smaller the Henry's law constant is, the more soluble the gaseous compound is in the liquid. The following example from USEPA-81/12 (1981) illustrates how to develop an equilibrium diagram from solubility data.

Example 8.1

Problem:

Given the following data for the solubility of SO_2 in pure water at 303 K (30°C) and 101.3 kPa (760 mm Hg), plot the equilibrium diagram and determine if Henry's law applies.

Solubility of SO_2 in Pure Water	
Equilibrium data	
Concentration SO_2 (g of SO_2 per 100 g/H_2O)	p (Partial pressure of SO_2)
0.5	6 kPa (42 mm Hg)
1.0	11.6 kPa (85 mm Hg)
1.5	18.3 kPa (129 mm Hg)
2.0	24.3 kPa (176 mm Hg)
2.5	30.0 kPa (224 mm Hg)
3.0	36.4 kPa (273 mm Hg)

Solution:

Step 1. The data must first be converted to mole fraction units. The mole fraction in the gas phase y is obtained by dividing the partial pressure of SO_2 by the total pressure of the system. For the first entry of the data table:

$$Y = p/P = 6 \text{ kPa}/101.3 \text{ kPa} = 0.9592 \ (0.06)$$

The mole fraction in the liquid phase x is obtained by dividing the moles of SO_2 by the total moles of liquid:

$$x = \frac{\text{moles } SO_2 \text{ in solution}}{\text{moles } SO_2 \text{ in solution } + \text{ moles } H_2O}$$

$$\text{moles of } H_2O = (100 \text{ g } H_2O = 100 \text{ g } H_2O/18 \text{ g } H_2O \text{ per mole}) = 5.55$$

For the first entry (x) of the data table:

$$x = 0.0078/(0.0078 + 5.55)$$

$$= 0.0014$$

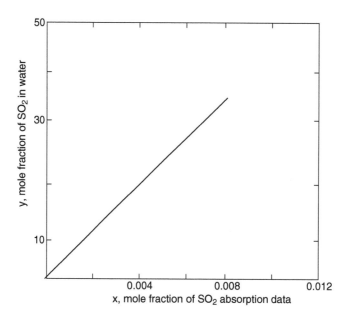

Figure 8.4 SO$_2$ absorption data for Example 8.1. (Adapted from USEPA-81/12, p. 4-8, 1981.)

Step 2. The following table is completed. The data from this table are plotted in Figure 8.4. Henry's law applies in the given concentration range, with Henry's law constant equal to 42.7 mole fraction SO$_2$ in air per mole fraction SO$_2$ in water.

Solubility data for SO$_2$			
C = (g of SO$_2$)/(100 g H$_2$O)	p (kPa)	$y = p/101.3$	$x = (C/64)/(C/64 + 5.55)$
0.5	6	0.06	0.0014
1.0	11.6	0.115	0.0028
1.5	18.3	0.18	0.0042
2.0	24.3	0.239	0.0056
2.5	30	0.298	0.007
3.0	36.4	0.359	0.0084

8.2.3 Material (Mass) Balance

The simplest way to express the fundamental engineering concept or principle of material or mass balance is to say, "Everything has to go somewhere." More precisely, the law of conservation of mass says that when chemical reactions take place, matter is neither created nor destroyed. This important concept allows us to track materials — in this case, pollutants, microorganisms, chemicals, and other materials — from one place to another.

The concept of materials balance plays an important role in environmental treatment technologies, where we assume a balance exists between the material entering and leaving the treatment process: "What comes in must equal what goes out." The concept is very helpful in evaluating process operations. In air pollution control of gas emissions using a typical countercurrent flow absorber, the solute (contaminant compound) is the material balance. Figure 8.5 illustrates a typical countercurrent flow absorber in which a material balance is drawn (USEPA-81/12, p. 4-14).

The following equation can be derived for material balance:

$$Y_1 - Y_2 = (L_m/G_m)(X_1 - X_2) \tag{8.2}$$

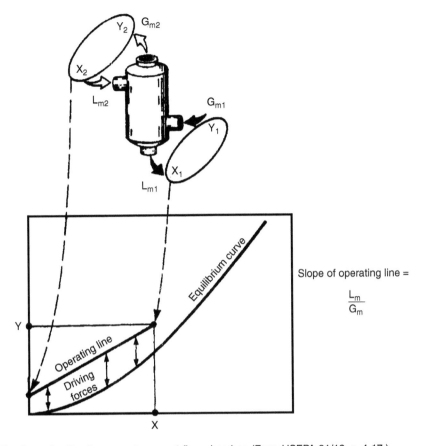

Figure 8.5 Operating line for a countercurrent flow absorber. (From USEPA-81/12, p. 4-17.)

where
Y_1 = inlet solute concentration
Y_2 = outlet solute concentration
X_1 = inlet composition of scrubbing liquid
X_2 = outlet composition of scrubbing liquid
L_m = liquid flow rate, gram-moles per hour
G_m = gas flow rate, gram-moles per hour

Equation 8.2 is the equation of a straight line. When this line is plotted on an equilibrium diagram, it is referred to as an *operating line* (see Figure 8.5). This line defines operating conditions within the absorber, that is, what is going in and what is coming out. The slope of the operating line is the liquid mass flow rate divided by the gas mass flow rate, which is the liquid-to-gas ratio or (L_m/G_m). When absorption systems are described or compared, the liquid-to-gas ratio is used extensively.

The following example (using Henry's law) illustrates how to compute the minimum liquid rate required to achieve desired removal efficiency.

Example 8.2

Problem:
Using the data and results from Example 8.1, compute the minimum liquid rate of pure water required to remove 90% of the SO_2 from a gas stream of 84.9 m³/min (3000 actual cubic feet per minute [acfm]) containing 3% SO_2 by volume. The temperature is 293 K and the pressure is 101.3 kPa (USEPA-81/12, p. 4-20).

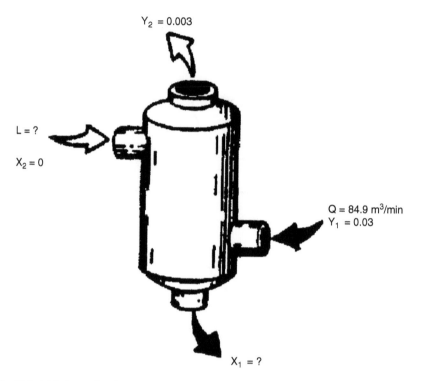

$Y_2 = 0.003$

$L = ?$

$X_2 = 0$

$Q = 84.9$ m³/min
$Y_1 = 0.03$

$X_1 = ?$

Figure 8.6   Material balance for absorber. (From USEPA-81/12, p. 4-20.)

Given:
Inlet gas solute concentration $(Y_1) = 0.03$
Minimum acceptable standards (outlet solute concentration) $(Y_2) = 0.003$
Composition of the liquid into the absorber $(X_2) = 0$
Gas flow rate $(Q) = 84.9$ m³/min
Outlet liquid concentration $(X_1) = ?$
Liquid flow rate $(L) = ?$
H = Henry's Constant

Solution:

   Step 1. Sketch and label a drawing of the system (see Figure 8.6).

   $Y_1 = 3\%$ by volume $= 0.03$

   $Y_2 = 90\%$ reduction from Y_1 or only 10% of Y_1; therefore $Y_2 = (0.10)(0.03) = 0.003$

   Step 2. At the minimum liquid rate, Y_1 and X_1 will be in equilibrium. The liquid will be saturated with
         SO_2.

$$Y_1 = H X_1$$

   From Figure 8.4,

$$H = 42.7 \text{ (mole fraction } SO_2 \text{ in air)/(mole fraction } SO_2 \text{ in water)}$$

$$0.03 = 42.7 \, X_1$$

$$X_1 = 0.000703 \text{ mole fraction}$$

Step 3. The minimum liquid-to-gas ratio is

$$Y_1 - Y_2 = (L_m / G_m)(X_1 - X_2)$$

$$(L_m / G_m) = (Y_1 - Y_2) / (X_1 - X_2)$$

$$= (0.03 - 0.003) / (0.000703 - 0)$$

$$= 38.4 \text{(g mol water)/(g mol air)}$$

Step 4. Compute the minimum required liquid flow rate. First, convert cubic meters of air to gram-moles:

At 0°C (273 K) and 101.3 kPa, there are 0.0224 m^3/g mole of an ideal gas

At 20°C (293 K): 0.0224(293 K/273 K) = 0.024 m^3/g mol

$$G_m = 84.9 \, (m^3/\text{min})(\text{g-mol air}/0.024 \, m^3)$$

$$= 3538 \text{ g-mol air/min}$$

$$L_m/G_m = 38.4 \text{ (g-mol water/g-mol of air) at minimum conditions}$$

$$L_m = 38.4(3,538)$$

$$= 136.0 \text{ kg-mol water min}$$

In mass units:

$$L = 136,000 \text{ g-mol/min (18 kg/kg-mol)}$$

$$= 2448 \text{ kg/min}$$

Step 5. Figure 8.7 illustrates the graphical solution to this problem. The slope of the minimum operating line × 1.5 = the slope of the actual operating line (line AC):

$$38.4 \times 1.5 = 57.6$$

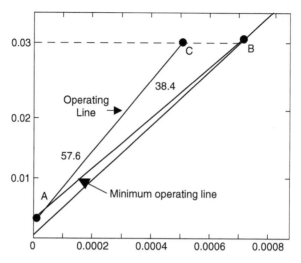

Figure 8.7   Solution to Example 8.2. (From USEPA-81/12, p. 4-21.)

8.2.4   Sizing Packed Column Diameter and Height of an Absorber

8.2.4.1   Packed Tower Absorber Diameter

The main parameter that affects the size of a packed tower is the gas velocity at which liquid droplets become entrained in the existing gas stream. Consider a packed tower operating at set gas and liquid flow rates. By decreasing the diameter of the column, the gas flow rate (meters per second or feet per second) through the column will increase. If the gas flow rate through the tower is gradually increased by using increasingly smaller diameter towers, a point is reached at which the liquid flowing down over the packing begins to be held in the void spaces between the packing. This gas-to-liquid flow ratio is termed the *loading point*. The pressure drop over the column begins to build, and the degree of mixing between the phases decreases. A further increase in gas velocity causes the liquid to fill the void spaces in the packing completely. The liquid forms a layer over the top of the packing, and no more liquid can flow down through the tower. The pressure drop increases substantially, and mixing between the phases is minimal.

This condition is referred to as *flooding,* and the gas velocity at which it occurs is the flooding velocity. Using an extremely large diameter tower eliminates this problem. However, as the diameter increases, the cost of the tower increases (USEPA-81/12, p. 4-22).

Normal practice is to size a packed column diameter to operate at a certain percent of the flooding velocity. A typical operating range for the gas velocity through the towers is 50 to 75% of the flooding velocity, assuming that by operating in this range, the gas velocity will also be below the loading point. A common and relatively simple procedure to estimate the flooding velocity (and thus minimum column diameter) is to use a generalized flooding and pressure drop correlation. One version of the flooding and pressure drop relationship in a packed tower is shown in Figure 8.8. This correlation is based on the physical properties of the gas and liquid streams and tower packing characteristics. We summarize the procedure to determine the tower diameter in the following set of calculations:

Step 1. Calculate the value of the horizontal axis (abscissa) of Figure 8.8 using Equation 8.3:

$$\text{Abscissa} = (L/G)(P_g/P_l)^{0.5} \qquad (8.3)$$

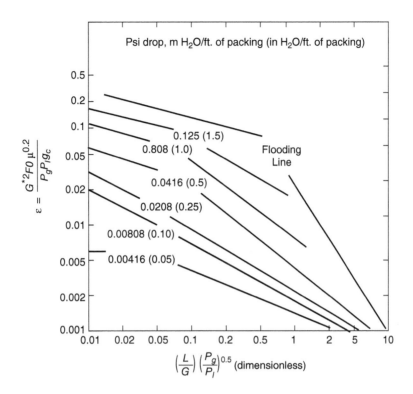

Figure 8.8 Generalized flooding and psi drop correlation. (From USEPA-81/12, p. 4-23.)

where
L = mass flow rate of liquid stream
G = mass flow rate of gas stream
P_g = gas density
P_1 = liquid density

Step 2. From the point on the abscissa (calculated in Equation 8.3), proceed up the graph to the flooding
line and read the ordinate ε.
Step 3. Using Equation 8.4, calculate the gas flow rate at flooding and solve $G*$

$$G* = \left[\frac{(\in)(P_g)(P_1)(g_c)}{F(\phi)(\mu_1)^{0.2}} \right]^{0.5}$$
(8.4)

where
$G*$ = mass flow rate of gas per unit cross-sectional area at blooding, pounds per second-square feet
ε   = ordinate value in Figure 8.8
P_g = density of gas stream, pounds per cubic foot
P_1 = density of absorbing liquid, pounds per cubic foot
g_c = gravitational constant = 9.82 m/sec² (32.2 ft/sec²)
F = packing factor, dimensionless
ϕ   = ratio of specific gravity of the scrubbing liquid to that of water, dimensionless
μ_1 = viscosity of liquid (for water = 0.8 cP = 0.0008 Pa-sec (use pascal-seconds in this equation)

Step 4. Calculate the actual gas flow rate per unit area as a fraction of the gas flow rate at flooding
(Equation 8.5).

$$G*operating = fG*flooding \qquad (8.5)$$

Step 5. Cross-sectional area of tower A is calculated from

$$Area = \frac{Total\ gas\ flow\ rate}{Gas\ flow\ rate\ per\ unit\ area} \qquad (8.6)$$

or

$$Area = G/G*_{operating} \qquad (8.7)$$

Step 6. The diameter of the tower is obtained from Equation 8.8:

$$d = (4A/\pi)^{0.5} \qquad (8.8)$$

The problem in Example 8.3 illustrates the calculation procedure for estimating the packed bed tower chamber.

Example 8.3 Tower Sizing

Problem:

For the scrubber in Example 8.2, determine the tower diameter if the operating liquid rate is 1.5 times the minimum. The gas velocity should be no greater than 75% of the flooding velocity, and the packing material is 2-in. ceramic Intalox™ saddles.

Solution:

Step 1. Calculate the value of the abscissa in Figure 8.8. From Example 8.2:

$$G_m = 3538\ g\text{-mol/min}$$

$$L_m = 2448\ kg/min$$

Convert gas molar flow to a mass flow, assuming molecular weight of the gas at 29 kg/mol:

$$G = 3538\ (kg\text{-mol/min})(29)(kg/mol) = 102.6\ kg/min$$

Adjusting the liquid flow to 1.5 times the minimum:

$$L = 1.5(2448) = 3672\ kg/min$$

The densities of water and air at 20°C are:

$$P_l = 1000\ kg/m^3$$

$$P_g = 1.17\ kg/m^3$$

Calculate the abscissa using Equation 8.3

$$\text{Abscissa} = (L / G)(P_g / P_1)^{0.5}$$

$$= (3672/102.6)(1.17/1000)^{0.5} = 1.22$$

Step 2: From Figure 8.8, determine the flooding line from 1.22. The ordinate ε is 0.019. From Equation 8.4, calculate $G*$:

$$G* = [(\varepsilon)(P_g)(P_1)(g_c)]/[F\phi(\mu_1)^{0.2}])^{0.5}$$

For water, $\phi = 1.0$, and the liquid viscosity is equal to 0.0008 Pa-sec. For 2-in. Intalox™ saddles, F = 40 ft²/ft³ or 131 m²/m³:

$$G_c = 9.82 \text{ kg/m}^3 \cdot \text{sec}$$

$$G* = \left(\frac{(0.019)(1.17)(1000)(9.82)}{(131)(1.0)(0.0008)^{0.2}} \right)^{0.5}$$

$$G* = 2.63 \text{ kg/m}^2\text{-sec at flooding}$$

Step 3. Calculate the actual gas flow rate per unit area:

$$G*o_{\text{perating}} = fG*_{\text{flooding}} = 0.75 \times 2.63 = 1.97 \text{ kg/(m}^2\text{-sec)}$$

Step 4. Calculate the tower diameter:

$$\text{Tower Area} = \text{Gas Flow Rate/G}_{\text{op}}$$

$$\frac{(102.6 \text{ kg/min})(\text{min/60 sec})}{1.97 \text{ kg/m}^2 \text{-sec}}$$

$$\text{Tower Diameter} = 1.13 \text{ A}^{0.5} = 1.05 \text{ m or at least 1 m (3.5 ft)}$$

8.2.4.2 Sizing the Packed Tower Absorber Height

The height of a packed tower refers to the depth of packing material needed to accomplish the required removal efficiency. The more difficult the separations are, the larger the packing height must be. For example, a much larger packing height is required to remove SO_2 than to remove Cl_2 from an exhaust stream using water as the absorbent because Cl_2 is more soluble in water than SO_2. Determining the proper height of packing is important because it affects the rate and efficiency of absorption (USEPA-81/12, p 4-26).

The required packing height of the tower can be expressed as

$$Z = HTU \times NTU \tag{8.9}$$

where

Z = height of packing
HTU = height of transfer unit
NTU = number of transfer units

The concept of a transfer unit comes from the operation of tray (tray/plate) tower absorbers. Discrete stages (trays or plates) of separation occur in tray/plate tower units. These stages can be visualized as a transfer unit with the number and height of each giving the total tower height. Although packed columns operate as one continuous separation process, in design terminology the process is treated as if it were broken into discrete sections (height of a transfer unit). The number and the height of a transfer unit are based on the gas or the liquid phase. Equation 8.9 can be modified to yield Equation 8.10:

$$Z = N_{OG}H_{OG} = N_{OL}H_{OL} \tag{8.10}$$

where:

Z = height of packing, meters
N_{OG} = number of transfer units based on overall gas film coefficient
H_{OG} = height of a transfer unit based on overall gas film coefficient, meters
N_{OL} = number of transfer units based on overall liquid film coefficient
H_{OL} = height of a transfer unit based on overall liquid film coefficient, meters

Values for the height of a transfer unit used in designing absorption systems are usually obtained from experimental data. To ensure the greatest accuracy, vendors of absorption equipment normally perform pilot plant studies to determine transfer unit height. When no experimental data are available, or if only a preliminary estimate of absorber efficiency is needed, generalized correlations are available to predict the height of a transfer unit. The correlations for predicting the H_{OG} or the H_{OL} are empirical in nature and a function of:

- Type of packing
- Liquid and gas flow rates
- Concentration and solubility of the contaminant
- Liquid properties
- System temperature

These correlations can be found in engineering texts. For most applications, the height of a transfer unit ranges between 0.305 and 1.22 m (1 and 4 ft). As a rough estimate, 0.6 m (2.0 ft) can be used.

The number of transfer units (NTU) can be obtained experimentally or calculated from a variety of methods. When the solute concentration is very low and the equilibrium line is straight, Equation 8.11 can be used to determine the number of transfer units (N_{OG}) based on the gas phase resistance:

$$\mathrm{Nog} = \frac{\ln\left[\left(\dfrac{(Y_1 - mX_2)}{(Y_2 - mX_2)}\right)\left(1 - \dfrac{mG_m}{Lm} + \dfrac{mG_m}{Lm}\right)\right]}{\left(1 - \dfrac{mG_m}{Lm}\right)} \tag{8.11}$$

where:

N_{OG} = transfer units
Y_1 = mole fraction of solute in entering gas

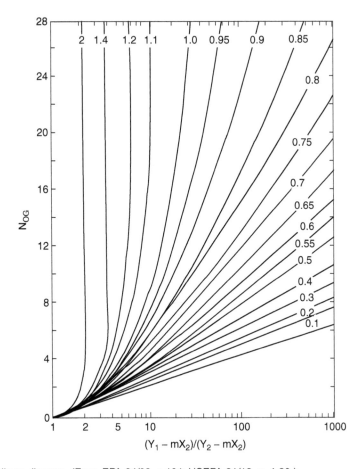

Figure 8.9   Colburn diagram. (From EPA-84/03, p.104; USEPA-81/12, p. 4-30.)

m = slope of the equilibrium line
X_2 = mole fraction of solute entering the absorber in the liquid
Y_2 = mole fraction of solute in exiting gas
G_m = molar flow rate of gas, kilogram-moles per hour
L_m = molar flow rate of liquid, kilogram-moles per hour

Equation 8.11 may be solved directly or graphically by using the Colburn diagram presented in Figure 8.9. This diagram is a plot of the N_{OG} vs. $\ln[Y_1 - mX_2)/(Y_2 - mX_2)]$, reading up the graph to the line corresponding to (mG_m/L_m), and then reading across to obtain the N_{OG}.

Equation 8.11 can be further simplified for situations in which a chemical reaction occurs or if the solute is extremely soluble. In these cases, the solute exhibits almost no partial pressure and therefore the slope of the equilibrium line approaches zero ($m = 0$). For either of these cases, Equation 8.11 reduces to Equation 8.12:

$$N_{OG} = \ln(Y_1/Y_2) \tag{8.12}$$

The number of transfer units depends only on the inlet and outlet concentration of the solute (contaminant or pollutant). For example, if the conditions in Equation 8.12 are met, 2.3 transfer units are required to achieve 90% removal of any pollutant. Equation 8.12 only applies when the equilibrium line is straight (low concentrations) and the slope approaches zero (very soluble or reactive gases). Example 8.4 illustrates a procedure to calculate the packed tower height.

Example 8.4 Tower (Column) Sizing

Problem:

From pilot plant studies of the absorption system in Example 8.2, the H_{OG} for the SO_2-water system was determined at 0.829 (2.72 ft). Calculate the total height of packing required to achieve 90% removal. The following data were taken from the previous examples.

Given:
H_{OG} = 0.829 m
m = 42.7 kg-mol H_2O per kilogram-mole of air
G_m = 3.5 kg-mol/min
L_m = (3672 kg/min)(kg/mol/18 kg) = 204 kg-mol/min
X_2 = 0 (no recirculated liquid)
Y_1 = 0.03
Y_2 = 0.003

Solution:

Step 1. Compute the N_{OG} from Equation 8.11:

$$N_{OG} = \frac{\ln\left[\left(\frac{(Y_1 - mX_2)}{(Y_2 - mX_2)}\right)\left(1 - \frac{mG_m}{Lm} + \frac{mG_m}{Lm}\right)\right]}{\left(1 - \frac{mG_m}{Lm}\right)}$$

$$N_{OG} = \frac{\ln\left[\left(\frac{0.03}{0.003}\right)\left(1 - \frac{(42.7)(3.5)}{204} + \frac{(42.7)(3.5)}{204}\right)\right]}{\left(1 - \frac{(42.7)(3.5)}{204}\right)}$$

$$N_{OG} = 4.58$$

Step 2. Calculate the total packing height:

$$Z = (H_{OG})(N_{OG})$$

$$= (0.829)(4.58)$$

$$= 3.79 \text{ m of packing height}$$

Table 8.2   Empirical Parameters for Equation 8.13

Tray	Metric ψ	Engineering ψ
Bubble cap	0.0162	0.1386
Sieve	0.0140	0.1196
Valve	0.0125	0.1069

Note: Metric ψ is expressed meters$^{0.25}$hours$^{0.5}$ per kilogram$^{0.25}$; for use with Q expressed in cubic meters per hour; and P_g expressed in kilograms per cubic meter. Engineering ψ is expressed in feet$^{0.25}$minutes$^{0.5}$ per pound$^{0.25}$; for use with Q expressed in cubic feet per minute, and P_g expressed in pounds per cubic foot.

Source: Adapted from USEPA, 1972, Wet Scrubber System Study. NTIS Report PB-213016. Research Triangle Park, NC.

8.2.4.3   Sizing the Plate (Tray) Tower

In a plate tower absorber, the scrubbing liquid enters at the top of the tower, passes over the top plate, and then down over each lower plate until it reaches the bottom. Absorption occurs as the gas, which enters at the bottom, passes up through the plate and contacts the liquid. In a plate tower, absorption occurs in a step-by-step or stage process (USEPA-81/12, p. 4-32).

The minimum diameter of a single-pass plate tower is determined by using the gas velocity through the tower. If the gas velocity is too great, liquid droplets are entrained, causing a condition known as priming. Priming occurs when the gas velocity through the tower is so great that it causes liquid on one plate to foam and rise to the plate above. Priming reduces absorber efficiency by inhibiting gas and liquid contact. For the purpose of determining tower diameter, priming in a plate tower is analogous to the flooding point in a packed tower. It determines the minimum acceptable diameter; the actual diameter should be larger. The smallest allowable diameter for a plate tower is expressed by Equation 8.13:

$$d = \Psi[Q(P_g)^{0.5}]^{0.5} \tag{8.13}$$

where:
d = plate tower diameter
ψ = empirical correlation, meters$^{0.25}$(hours)$^{0.25}$/(kilogram)$^{0.25}$
Q = volumetric gas flow, cubic meters per hour
P_g = gas density, kilograms per cubic meter

The term ψ is an empirical correlation and a function of the tray spacing and the densities of the gas and liquid streams. Values of ψ shown in Table 8.2 are for a tray spacing of 61 cm (24 in.) and a liquid specific gravity of 1.05. If the specific gravity of a liquid varies significantly from 1.05, the values for ψ in the table cannot be used.

Depending on operating conditions, trays are spaced at a minimum distance between plates to allow the gas and liquid phases to separate before reaching the plate above. Trays should be spaced to allow for easy maintenance and cleaning; they are normally spaced 45 to 70 cm (18 to 28 in.) apart. In using the information in Table 8.2 for tray spacing different from 61 cm, a correction factor must be used. Use Figure 8.10 to determine the correction factor, which is multiplied by the estimated diameter. Example 8.5 illustrates how the minimum diameter of a tray tower absorber is estimated.

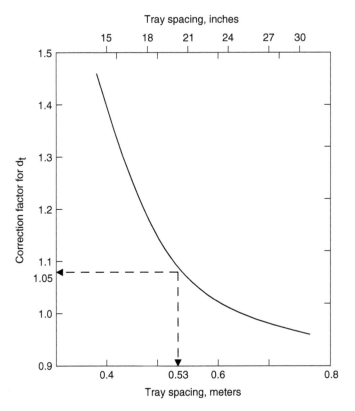

Figure 8.10   Tray spacing correction factor. (Adapted from USEPA-81/12, p. 4-33.)

Example 8.5 Plate Tower Diameter

Problem:
   For the conditions described in Example 8.2, determine the minimum acceptable diameter if the scrubber is a bubble cap tray tower absorber. The trays are spaced 0.53 m (21 in.) apart (USEPA-81/12, p. 4-34).

Solution:
   From Example 8.2 and Example 8.3:

$$\text{Gas flow rate} = Q = 84.9 \text{ m}^3/\text{min}$$

$$\text{density} = P_g = 1.17 \text{ kg/m}^3$$

From Table 8.2 for a bubble cap tray:

$$\Psi = 0.0162 \text{ m}^{0.25}(\text{h})^{0.50}/\text{kg}^{0.25}$$

Before Equation 8.13 can be used, Q must be converted to cubic meters per hour:

$$Q = 84.9(\text{m}^3/\text{min})(60 \text{ min/h}) = 5094 \text{ m}^3/\text{h}$$

Step 1. Substitute these values into Equation 8.13 for a minimum diameter d:

$$d = \Psi[Q(P_g)^{0.5}]^{0.5}$$

$$d = (0.0162)[5094]/(1.17)^{0.5}]^{0.5}$$

$$d = 1.2 \text{ m}$$

Correct this diameter for a tray spacing of 0.53 m.

Step 2. From Figure 8.10, read a correction factor of 1.05. Therefore, the minimum diameter is:

$$d = 1.2(1.05) = 1.26 \text{ m (4.13 ft)}$$

Note: This estimated diameter is a minimum acceptable diameter based on actual conditions. In practice, a larger diameter (based on maintenance and economic considerations) is usually chosen.

8.2.4.4 Theoretical Number of Absorber Plates or Trays

The several methods used to determine the number of ideal plates or trays required for a given removal efficiency can become quite complicated. One method used is a graphical technique (USEPA-81/12, p. 4-34). The number of ideal plates is obtained by drawing "steps" on an operating diagram, a procedure illustrated in Figure 8.11. This method can be rather time consuming, and inaccuracies can result at both ends of the graph.

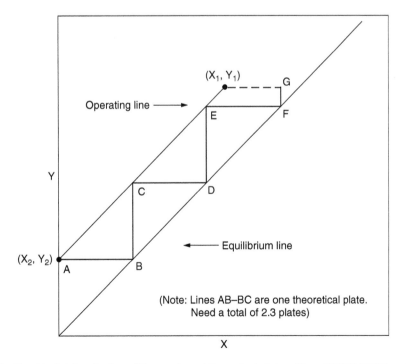

Figure 8.11 Graphical determination of the number of theoretical plates. (From USEPA-81/12, p. 4-35.)

Equation 8.14 is a simplified method of estimating the number of plates. It can only be used if the equilibrium as well as operating lines for the system are straight, a valid assumption for most air pollution control systems.

$$N_p = \ln[(Y_1 - mX_2)/(Y_2 - mX_2)(1 - mG_m/L_m) + mG_m/L_m]/\ln(L_m/mG_m) \qquad (8.14)$$

where
N_p = number of theoretical plates
Y_1 = mole fraction of solute in entering gas
X_2 = mole fraction of solute entering the tower
Y_2 = mole fraction of solute in exiting gas
m = slope of equilibrium line
G_m = molar flow rate of gas, kilogram-moles per hour
L_m = molar flow rate of liquid, kilogram-moles per hour

Equation 8.14 is used to predict the number of theoretical plates (N_p) required to achieve a given removal efficiency. The operating conditions for a theoretical plate assume that the gas and liquid stream leaving the plate are in equilibrium. This ideal condition is never achieved in practice. A larger number of actual trays is required to compensate for this decreased tray efficiency.

Three types of efficiencies are used to describe absorption efficiency for a plate tower: (1) an overall efficiency, which is concerned with the entire column; (2) Murphree efficiency, which is applicable with a single plate; and (3) local efficiency, which pertains to a specific location on a plate. The simplest of the tray efficiency concepts, the overall efficiency, is the ratio of the number of theoretical plates to the number of actual plates. Because overall tray efficiency is an oversimplification of the process, reliable values are difficult to obtain. For a rough estimate, overall tray efficiencies for absorbers operating with low-viscosity liquid normally fall in a 65 to 80% range. Example 8.6 shows a procedure to calculate the number of theoretical plates.

Example 8.6

Problem:
   Calculate the number of theoretical plates required for the scrubber in Example 8.5, using the conditions in Example 8.4. Estimate the total height of the tower if the trays are spaced at 0.53-m intervals; assume an overall tray efficiency of 70%.

Solution:
   From Example 8.5 and the previous examples, the following data are obtained:

m = slope of the equilibrium line = 42.7
G_m = 3.5 kg-mol/h
L_m = 204 kg-mol/h
X_2 = 0 (no recycle liquid)
Y_1 = 0.03
Y_2 = 0.003

   Step 1. The number of theoretical plates from Equation 8.14 is:

$$N_p = \ln[(Y_1 - mX_2)/(Y_2 - mX_2)(1 - mG_m/L_m) + mG_m/L_m]/\ln(L_m/mG_m)$$

$$= \ln[(0.03 - 0)/(0.003 - 0)[1 - 42.7(3.5)/204] + 42.7(3.5)/204]/\ln[204/(42.7)(3.5)]$$

$$= 3.94 \text{ theoretical plates}$$

Step 2. Assuming that the overall plate efficiency is 70%, the actual number of plates is:

$$\text{Actual number of plates} = 3.94/0.7$$

$$= 5.6 \text{ or } 6 \text{ plates}$$

Step 3. The height of the tower is given by:

$$Z = N_p \text{ (tray spacing)} + \text{top height}$$

The top height is the distance (freeboard) over the top plate that allows the gas–vapor mixture to separate. This distance is usually the same as the tray spacing:

$$Z = 6 \text{ plates } (0.53 \text{ m}) + 0.53 \text{ m}$$

$$= 3.71 \text{ m}$$

Note: This height is approximately the same as that predicted for the packed tower in Example 8.4. This is logical because the packed and plate towers are efficient gas absorption devices. However, due to the large number of assumptions involved, no generalization should be made.

8.3 ADSORPTION

Adsorption is a mass transfer process that involves passing a stream of effluent gas through the surface of prepared porous solids (adsorbents). The surfaces of the porous solid substance attract and hold the gas (the adsorbate) by physical or chemical adsorption. Adsorption occurs on the internal surfaces of the particles (USEPA-81/12, p. 5-1). In physical adsorption (a readily reversible process), a gas molecule adheres to the surface of the solid because of an imbalance of electron distribution. In chemical adsorption (not readily reversible), once the gas molecule adheres to the surface, it reacts chemically with it.

Several materials possess adsorptive properties. These materials include activated carbon; alumina; bone char; magnesia; silica gel; molecular sieves; strontium sulfate; and others. The most important adsorbent for air pollution control is activated charcoal. The surface area of activated charcoal will preferentially adsorb hydrocarbon vapors and odorous organic compounds from an airstream.

In an adsorption system the collected contaminant remains in the adsorption bed (in contrast to the absorption system, in which the collected contaminant is continuously removed by flowing liquid). The most common adsorption system is the fixed-bed adsorber, which can be contained in a vertical or a horizontal cylindrical shell. The adsorbent (usually activated carbon) is arranged in beds or trays in layers about 0.5 in. thick. Multiple beds may be used; in multiple-bed systems, one or more beds are adsorbing vapors while the other bed is being regenerated.

Step 1	Step 2	Step 3
Diffusion to adsorbent surface	Migration into pores of adsorbent	Monolayer buildup of adsorbate

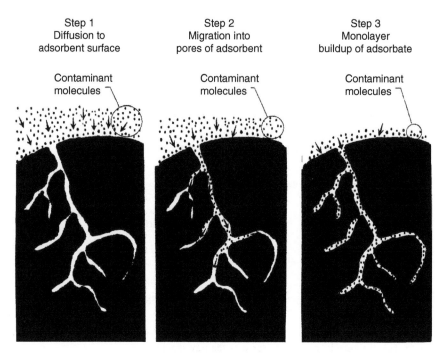

Figure 8.12 Gas collection by adsorption. (Adapted from USEPA-81/12, p. 5-2.)

The efficiency of most adsorbers is near 100% at the beginning of the operation and remains high until a breakpoint or breakthrough occurs. When the adsorbent becomes saturated with adsorbate, contaminant begins to leak out of the bed, signaling that the adsorber should be renewed or regenerated.

Although adsorption systems are high-efficiency devices that may allow recovery of product; have excellent control and response to process changes; and have the capability of being operated unattended, they also have some disadvantages. These include the need for expensive extraction schemes if product recovery is required, relatively high capital cost, and gas stream prefiltering needs (to remove any particulate capable of plugging the adsorbent bed) (Spellman, 1999).

8.3.1 Adsorption Steps

Adsorption occurs in a series of three steps (USEPA-81/12, p. 5-2). In the first step, the contaminant diffuses from the major body of the air stream to the external surface of the adsorbent particle. In the second step, the contaminant molecule migrates from the relatively small area of the external surface (a few square meters per gram) to the pores within each adsorbent particle. The bulk of adsorption occurs in these pores because the majority of available surface area is there (hundreds of square meters per gram). In the third step, the contaminant molecule adheres to the surface in the pore. Figure 8.12 illustrates this overall diffusion and adsorption process.

8.3.2 Adsorption Forces — Physical and Chemical

The adsorption process is classified as physical or chemical. The basic difference between physical and chemical adsorption is the manner in which the gas molecule is bonded to the adsorbent. In physical adsorption, the gas molecule is bonded to the solid surface by weak forces of intermolecular cohesion. The chemical nature of the adsorbed gas remains unchanged. Therefore, physical adsorption is a readily reversible process. In chemical adsorption, a much stronger bond is formed between the gas molecule and adsorbent; sharing or exchange of electrons takes place. Chemical adsorption is not easily reversible.

Physical adsorption. The forces active in physical adsorption are electrostatic in nature and are present in all states of matter: gas, liquid, and solid. They are the same forces of attraction that cause gases to condense and to deviate from ideal behavior under extreme conditions. Physical adsorption is also sometimes referred to as *van der Waals' adsorption.* The electrostatic effect that produces the van der Waal's forces depends on the polarity of the gas and the solid molecules. Molecules in any state are polar or nonpolar, depending on their chemical structure. Polar substances are those that exhibit a separation of positive and negative charges within the compound, referred to as a *permanent dipole.* Water is a prime example of a polar substance. Nonpolar substances have their positive and negative charges in one center, so they have no permanent dipole. Because of their symmetry, most organic compounds are nonpolar.

Chemical adsorption. Chemical adsorption or chemisorption results from chemical interaction between the gas and the solid. The gas is held to the surface of the adsorbent by the formation of a chemical bond. Adsorbents used in chemisorption can be pure substances or chemicals deposited on an inert carrier material. One example is using pure iron oxide chips to adsorb H_2S gases. Another is the use of activated carbon that has been impregnated with sulfur to remove mercury vapors.

All adsorption processes are exothermic, whether adsorption occurs from chemical or physical forces. In adsorption, molecules are transferred from the gas to the surface of a solid. The fast-moving gas molecules lose their kinetic energy of motion to the adsorbent in the form of heat.

8.3.3 Adsorption Equilibrium Relationships

Most available data on adsorption systems are determined at equilibrium conditions — the set of conditions at which the number of molecules arriving on the surface of the adsorbent equals the number of molecules leaving. The adsorbent bed is said to be "saturated with vapors" and can remove no more vapors from the exhaust stream. The equilibrium capacity determines the maximum amount of vapor that can be adsorbed at a given set of operating conditions. Although a number of variables affect adsorption, gas temperature and pressure are the two most important in determining adsorption capacity for a given system. Three types of equilibrium graphs are used to describe adsorption capacity:

- Isotherm at constant temperature
- Isostere at constant amount of vapors adsorbed
- Isobar at constant pressure

8.3.3.1 *Isotherm*

The most common and useful adsorption equilibrium datum is the adsorption isotherm. The isotherm is a plot of the adsorbent capacity vs. the partial pressure of the adsorbate at a constant temperature. Adsorbent capacity is usually given in weight percent expressed as gram of adsorbate per 100 g of adsorbent. Figure 8.13 is a typical example of an adsorption isotherm for carbon tetrachloride on activated carbon. Graphs of this type are used to estimate the size of adsorption systems as illustrated in Example 8.7.

Example 8.7

Problem:

A dry cleaning process exhausts a 15,000-cfm air stream containing 680-ppm carbon tetrachloride. Given Figure 8.13 and assuming that the exhaust stream is at approximately 140°F and 14.7 pounds per square inch absolute (psia), determine the saturation capacity of the carbon (USEPA-81/12, p. 5-8).

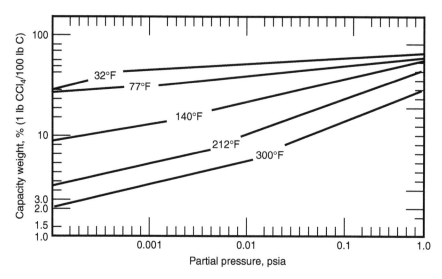

Figure 8.13 Adsorption isotherm for carbon tetrachloride on activated carbon. (Adapted from USEPA-81/12, p. 5-8.)

Solution:

Step 1. In the gas phase, the mole fraction Y is equal to the percent by volume:

$$Y = \% \text{ volume} = 680 \text{ ppm}$$

$$= 680/(10)^6$$

$$= 0.00068$$

Obtain the partial pressure:

$$p = YP = (0.00068)(14.7 \text{ psia}) = 0.01 \text{ psia}$$

Step 2. From Figure 8.13, at a partial pressure of 0.01 psia and a temperature of 140°F, the carbon capacity is read as 30%. This means that at saturation, 30 lb of vapor are removed per 100 lb of carbon in the adsorber (30 kg/100 kg).

8.3.3.2 Isostere

The isostere is a plot of the $\ln(p)$ vs. $1/T$ at a constant amount of vapor adsorbed. Adsorption isostere lines are usually straight for most adsorbate-absorbent systems. Figure 8.14 is an adsorption isostere graph for the adsorption of H_2S gas onto molecular sieves. The isostere is important in that the slope of the isostere corresponds to the heat of adsorption.

8.3.3.3 Isobar

The isobar is a plot of the amount of vapor adsorbed vs. temperature at a constant pressure. Figure 8.15 shows an isobar line for the adsorption of benzene vapors on activated carbon. Note that (as is always the case for physical adsorption) the amount adsorbed decreases with increasing temperature.

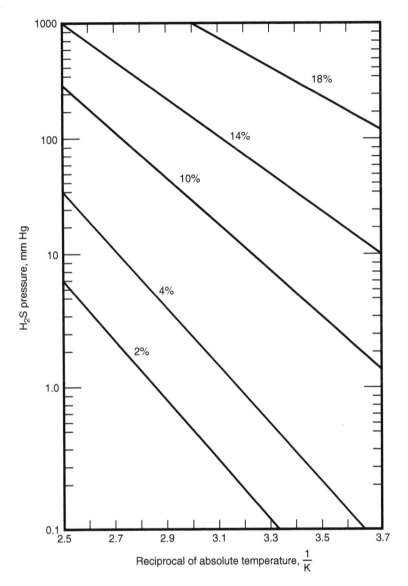

Figure 8.14 Adsorption isosteres of H_2S on 13X molecular sieve loading in percent weight. (Adapted from USEPA-81/12, p. 5-11.)

Because these three relationships (the isotherm, isostere, and isobar) were developed at equilibrium conditions, they depend on each other. By determining one of the three, the other two relationships can be determined for a given system. In the design of an air pollution control system, the adsorption isotherm is by far the most commonly used equilibrium relationship.

8.3.4 Factors Affecting Adsorption

According to USEPA-81/12, p. 5-18, a number of factors or system variables influence the performance of an adsorption system. These include:

- Temperature
- Pressure
- Gas velocity

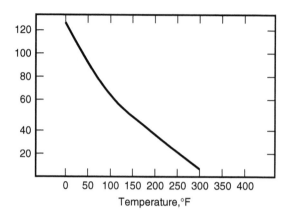

Figure 8.15   Adsorption isobar for benzene on carbon (benzene at 10.0 mm Hg). (Adapted from USEPA-81/12, p. 5-8.)

- Bed depth
- Humidity
- Contamination

We discuss these variables and their effects on the adsorption process in this subsection.

8.3.4.1   Temperature

For physical adsorption processes, the capacity of an adsorbent decreases as the temperature of the system increases. As the temperature increases, the vapor pressure of the adsorbate increases, raising the energy level of the adsorbed molecules. Adsorbed molecules now have sufficient energy to overcome the van der Waal's attraction and migrate back to the gas phase. Molecules already in the gas phase tend to stay there because of their high vapor gas phase. As a general rule, adsorber temperatures are kept below 130°F (54°C) to ensure adequate bed capacities. Temperatures above this limit can be avoided by cooling the exhaust stream before treatment.

8.3.4.2   Pressure

Adsorption capacity increases with an increase in the partial pressure of the vapor. The partial pressure of a vapor is proportional to the total pressure of the system. Any increase in pressure will increase the adsorption capacity of a system. The increase in capacity occurs because of a decrease in the mean free path of vapor at higher pressures. Simply put, the molecules are packed more tightly together. More molecules have a chance to hit the available adsorption sites, thus increasing the number of molecules adsorbed.

8.3.4.3   Gas Velocity

The gas velocity through the adsorber determines the contact or residence time between the contaminant stream and adsorbent. The residence time directly affects capture efficiency. The more slowly the contaminant stream flows through the adsorbent bed, the greater is the probability of a contaminant molecule hitting an available site. Once a molecule has been captured, it will stay on the surface until the physical conditions of the system are changed. To achieve 90%+ capture efficiency, most carbon adsorption systems are designed for a maximum gas flow velocity of 30 m/min (100 ft/min) through the adsorber. A lower limit of at least 6 m/min (20 ft/min) is maintained to avoid flow distribution problems, such as channeling.

8.3.4.4 Bed Depth

Providing a sufficient depth of adsorbent is very important in achieving efficient gas removal. If the adsorber bed depth is shorter than the required mass transfer zone, breakthrough will occur immediately, rendering the system ineffective. Computing the length of the mass transfer zone (MTZ) is difficult because it depends upon six factors: adsorbent particle size; gas velocity; adsorbate concentration; fluid properties of the gas stream; temperature; and pressure. The relationship between breakthrough capacity and MTZ is:

$$C_B = [0.5C_s(MTZ) + C_s(D - MTZ)]/D \qquad (8.15)$$

where
C_B = breakthrough capacity
C_s = saturation capacity
MTZ = mass transfer zone length
D = adsorption bed depth

Equation 8.15 is used mainly as a check to ensure that the proposed bed depth is longer than the MTZ. Actual bed depths are usually many times longer than the length of the MTZ.

The total amount of adsorbent required is usually determined from the adsorption isotherm, as illustrated in Example 8.8. Once this has been set, the bed depth can then be estimated by knowing the tower diameter and density of the adsorbent. Example 8.9 illustrates how this is done. Generally, the adsorbent bed is sized to the maximum length allowed by the pressure drop across the bed.

Example 8.8

Problem:
Assume the conditions stated in Example 8.7. Estimate the amount of carbon required if the adsorber were to operate on a 4-h cycle. The molecular weight of CCl_4 is 154 lb/lb-mol (USEPA-81/12, p. 5-18).

Solution:
From Example 8.7, we know that the carbon used will remove 30 lb of vapor for every 100 lb of carbon at saturation conditions.

Step 1. Compute the flow rate of CCl_4:

$$Q(CCl_4) = 15,000 \text{ scfm} \times 0.00068 = 10.2 \text{ scfm-}CCl_4$$

Converting to pounds per hour:

$$10.2 \text{ (ft}^3/\text{min)} \times \text{(lb-mol)}/(359 \text{ ft}^3) \times (154 \text{ lb})/\text{(lb-mol)} \times (60 \text{ min)/h} = 262.5 \text{ lb-}CCl_4$$

Step 2. The amount of carbon (at saturation) required (assuming that the working charge is twice the saturation capacity). (*Note*: This gives only a rough estimate of the amount of carbon needed.)

$$2(3500) = 7000 \text{ lb (3182 kg) carbon per 4-h cycle per adsorber}$$

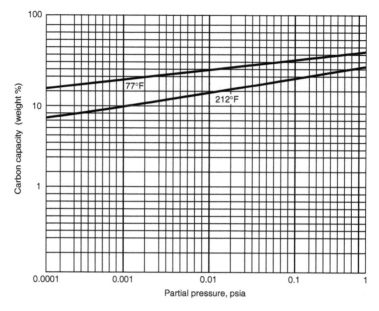

Figure 8.16 Adsorption isotherm for toluene. (Adapted from USEPA-81/05, p. 18-9.)

Example 8.9

Note: The following example is based on a number of adsorber design maximum and minimum rules of thumb and is intended as a guide to illustrate how to "red flag" any parameters that could be greatly exceeded.

Problem:

A solvent degreaser is designed to recover toluene from a 3.78 m³/sec (8000 actual cubic feet per minute [acfm]) air stream at 25°C (77°F) and atmosphere pressure. The company is planning to use a two-bed carbon adsorption system with a cycle time of 4 h. The maximum concentration of toluene is kept below 50% of the lower explosive limit (LEL) for safety purposes. Using Figure 8.16, the adsorption isotherm for toluene, and the additional operational data, estimate (USEPA-81/12, p. 5-38):

1. The amount of carbon required for a 4-h cycle
2. Square feet of surface area required based on a 0.58 m/sec (100 fpm) maximum velocity
3. Depth of the carbon bed

Given:
LEL for toluene = 1.2%
Molecular weight of toluene = 92.1 kg/kg-mol
Carbon density = 480 kg/m³ (30 lb/ft³)

Solution:

Step 1. Calculate the toluene flow rate:

$$(3.78 \text{ m}^3/\text{sec})(50\%)(1.2\%) = 0.023 \text{ m}^3/\text{sec toluene}$$

Step 2. To determine the saturation capacity of the carbon, calculate the partial pressure of toluene at the adsorption conditions:

$$p = YP$$

$$= (0.023/3.78)(14.7 \text{ psia})$$

$$= 0.089 \text{ psia}$$

From Figure 8.16, the saturation capacity of the carbon is 40% or 40 kg of toluene per 100 kg of carbon. The flow rate of toluene is:

$$(0.023 \text{ m}^3/\text{sec}) \times (\text{kg-mol}/22.4 \text{ m}^3) \times (273 \text{ K}/350 \text{ K}) \times (92.1 \text{ kg/kg-mol})$$

$$= 0.074 \text{ kg/sec of toluene}$$

Step 3. The amount of carbon at saturation for a 4-h cycle is:

$$(0.074 \text{ kg/sec toluene}) \times (100 \text{ kg carbon}/40 \text{ kg toluene}) \times (3600 \text{ sec/h}) \times (4 \text{ h})$$

$$= 2664 \text{ kg of carbon}$$

The working charge of carbon can be estimated by doubling the saturation capacity. Therefore, the working charge is:

$$\text{Working charge} = (2)(2664 \text{ kg of carbon})$$

$$= 5328 \text{ kg of carbon}$$

Step 4. The square feet of superficial surface area is the surface area set by the maximum velocity of 0.508 m/sec (100 ft/min) through the adsorber. The required surface area is:

$$A = Q/(\text{maximum velocity})$$

$$= (3.78 \text{ m}^3/\text{sec})/(0.508 \text{ m/sec})$$

$$= 7.44 \text{ m}^2$$

For a horizontal flow adsorber, this would correspond to a vessel approximately 2 m (6.6 ft) in width and 4 m (13.1 ft) in length to give 8.0 m² (87 ft²) surface area. This would supply more than the required area.

The flow rate is too high to be handled by a single vertical flow adsorber. An alternative would be to use three vessels, two adsorbing while one is being regenerated. Each vessel then must be sized to handle 1.89 m³/sec (4000 acfm). The area required for a limiting velocity of 0.508 m/sec is:

$$\text{Area to handle half flow } A = (1.89 \text{ m}^3/\text{sec})/(0.508 \text{ m/sec})$$

$$= 3.72 \text{ m}^2$$

This cross-sectional area corresponds to a vessel diameter of:

$$d = (4A/\pi)^{0.5}$$

$$= [4(3.72)/\pi]^{0.5}$$

$$= 2.18 \text{ m (7 ft)}$$

Step 5. Volume that the carbon would occupy in horizontal bed:

$$\text{Volume of carbon} = \text{weight/density}$$

$$= 5328 \text{ kg} \times (\text{m}^3/480 \text{ kg})$$

$$= 11.1 \text{ m}^3$$

Note: For the three-bed vertical system, the volume of each bed is half of this, or 5.55 m³.

Step 6. Estimate bed depth:

$$\text{Depth of carbon} = (\text{volume of carbon})/(\text{cross-sectional area of adsorber})$$

$$= (11.1 \text{ m}^3)/(7.44 \text{ m}^2)$$

$$= 1.49 \text{ m}$$

Note: The depth for the three-bed vertical system is the same because the volume and the area are halved.

8.3.4.5 *Humidity*

Activated carbon will preferentially adsorb nonpolar hydrocarbons over polar water vapor. The water vapor molecules in the exhaust stream exhibit strong attractions for each other, rather than for the adsorbent. At high relative humidity (over 50%), the number of water molecules increases until they begin to compete with the hydrocarbon molecule for active adsorption sites. This reduces the capacity and the efficiency of the adsorption system. Exhaust streams with humidity greater than 50% may require installation of additional equipment to remove some of the moisture. Coolers to remove the water are one solution; dilution air with significantly less moisture in it than the process stream has also been used. The contaminant stream may be heated to reduce the humidity, as long as the increase in temperature does not greatly affect adsorption efficiency.

8.3.4.6 *Contaminants*

Particulate matter, entrained liquid droplets, and organic compounds that have high boiling points can also reduce adsorber efficiency if present in the air stream. Micron-sized particles of dust or lint that are not filtered can cover the surface of the adsorbent. Covering active adsorption sites by an inert material is referred to as "blinding" or "deactivation." This greatly reduces the surface area of the adsorbent available to the gas molecule for adsorption. To avoid this situation, almost all industrial adsorption systems are equipped with some type of particulate matter removal device.

Entrained liquid droplets can also cause operational problems. Liquid droplets that are nonadsorbing act the same as particulate matter: the liquid covers the surface, blinding the bed. If the liquid is the same as the adsorbate, high levels of adsorption heat occur. This is especially a problem in activated carbon systems in which liquid organic droplets carried over from the process can cause bed fires from the released heat. When liquid droplets are present, some type of entrainment separator may be required.

8.4   INCINERATION

Incineration (or combustion) is a major source of air pollution; however, if properly operated, it is also a beneficial air pollution control system in which the object is to convert certain air contaminants (usually organic compounds classified as volatile organic compounds [VOCs] and/or air toxic compounds) (USEPA, 1973; Spellman, 1999). Incineration is a chemical process defined as rapid, high-temperature gas phase oxidation. The incineration equipment used to control air pollution emissions is designed to push these oxidation reactions as close to complete incineration as possible, leaving a minimum of unburned residue. Depending upon the contaminant being oxidized, equipment used to control waste gases by combustion can be divided into three categories: direct-flame incineration (or flaring); thermal incineration (afterburners); or catalytic incineration.

8.4.1   Factors Affecting Incineration for Emission Control

The operation of any incineration system used for emission control is governed by seven variables: temperature; residence time; turbulence; oxygen; combustion limit; flame combustion; and heat. For complete incineration to occur, oxygen must be available and put into contact (through turbulence) at a sufficient temperature and held at this temperature for a sufficient time. These seven variables are not independent — changing one affects the entire process.

8.4.1.1   Temperature

The rate at which a combustible compound is oxidized is greatly affected by temperature. The higher the temperature is, the faster the oxidation reaction proceeds. The chemical reactions involved in the combination of a fuel and oxygen can occur even at room temperature, but very slowly. For this reason, a pile of oily rags can be a fire hazard. Small amounts of heat are liberated by the slow oxidation of the oils; this in turn raises the temperature of the rags and increases the oxidation rate, liberating more heat. Eventually a fully engulfing fire can break out (USEPA-81/12, p. 3-2).

Most incinerators operate at higher temperature than the ignition temperature, which is a minimum temperature. Thermal destruction of most organic compounds occurs between 590 and 650°F (1100 and 1200°F). However, most incinerators are operated at 700 to 820°C (1300 to 1500°F) to convert CO to CO_2, which occurs only at these higher temperatures.

8.4.1.2   Residence Time

Much in the same manner that higher temperature and pressure affect the volume of a gas, time and temperature affect combustion. When one variable is increased, the other may be decreased with the same end result. With a higher temperature, a shorter residence time can achieve the same degree of oxidation. The reverse is also true: a higher residence time allows the use of a lower temperature. Simply stated, the residence time needed to complete the oxidation reactions in the incinerator depends partly on the rate of the reactions at the prevailing temperature and partly on the mixing of the waste stream and the hot combustion gases from the burner or burners.

The residence time of gases in the incinerator may be calculated from a simple ratio of the volume of the refractory-lined combustion chamber and the volumetric flow rate of combustion products through the chamber.

$$t = V/Q \tag{8.16}$$

where
t = residence, seconds
V = chamber volume, cubic meters
Q = gas volumetric flow rate at combustion conditions, cubic meters per second

Q is the total flow of hot gases in the combustion chamber. Adjustments to the flow rate must include any outside air added for combustion. Example 8.10 shows the determination of residence time from the volumetric flow rate of gases.

8.4.1.3 Turbulence

Proper mixing is important in combustion processes for two reasons. First, mixing of the burner fuel with air is needed to ensure complete combustion of the fuel. If not, unreacted fuel is exhausted from the stack. Second, the organic compound-containing waste gases must be thoroughly mixed with the burner combustion gases to ensure that the entire waste gas stream reaches the necessary combustion temperatures; otherwise, incomplete combustion occurs.

A number of methods are available to improve mixing of the air and combustion streams. Some of these include the use of refractory baffles, swirl fired burners, or baffle plates. Obtaining complete mixing is not easy. Unless properly designed, many of these mixing devices create dead spots and reduce operating temperatures. Inserting obstructions to increase turbulence may not be sufficient. According to USEPA (1973), in afterburner systems, the process of mixing the flame and the fume streams to obtain a uniform temperature for decomposition of pollutants is the most difficult part in the afterburner design.

8.4.1.4 Oxygen Requirement

Not only is oxygen necessary for combustion to occur, but oxygen requirements for the supplemental heat burner used in thermal incinerators must also be taken into account when sizing the burner system and the combustion chamber. To achieve complete combustion of a compound or the fuel (propane, No. 2 fuel oil, natural gas, for example), a sufficient supply of oxygen must be present in the burner flame to convert all of the carbon to CO_2. This quantity of oxygen is referred to as the stoichiometric or theoretical amount. The stoichiometric amount of oxygen is determined from a balanced chemical equation summarizing the oxidation reactions. For example, 1 mol of methane requires 2 mol of oxygen for complete combustion (USEPA-81/12, p. 3-4):

$$CH_4 + 2O_2 \rightarrow CO_2 + 2H_2O \tag{8.17}$$

If an insufficient amount of oxygen is supplied, the mixture is referred to as rich. Incomplete combustion occurs under these conditions. This reduces the peak flame temperature and creates black smoke emissions. If more than the stoichiometric amount of oxygen is supplied, the mixture is referred to as "lean." The added oxygen plays no part in the oxidation reaction and passes through the incinerator. To ensure combustion, more than the stoichiometric amount of air is used. This extra volume is referred to as excess air.

8.4.1.5 Combustion Limit

Not all mixtures of fuel and air are able to support combustion. Incinerators usually operate at organic vapor concentrations below 25% of the lower explosive limit (LEL) or lower flammability limit (LFL). Combustion does not occur readily in this range. The explosive or flammable limits for a mixture are the maximum and minimum concentration of fuel in air that will support combustion. The upper explosive limit (UEL) is defined as the concentration of fuel that produces a nonburning mixture because of a lack of oxygen. The lower explosive limit is defined as the concentration of fuel below which combustion will not be self-sustaining.

8.4.1.6 Flame Combustion

According to USEPA-81/12, p. 3-7, when fuel and air are mixed, two different mechanisms of combustion can occur. When air and fuel flowing through separate ports are ignited at the burner nozzle, a luminous yellow flame results from thermal cracking of the fuel. Cracking occurs when hydrocarbons are intensely heated before they have a chance to combine with oxygen. The cracking releases hydrogen and carbon, which diffuse into the flame to form carbon dioxide and water. The carbon particles give the flame the yellow appearance. If incomplete combustion occurs from flame temperature cooling or from insufficient oxygen, soot and black smoke form.

When the fuel and air are premixed in front of the burner nozzle, blue flame combustion occurs. The reason for the different flame color is that the fuel–air mixture is gradually heated. The hydrocarbon molecules are slowly oxidized, going from aldehydes and ketones to carbon dioxide and water. No cracking occurs, and no carbon particles are formed. Incomplete combustion results in the release of the intermediate partially oxidized compounds. Blue haze and odors are emitted from the stack.

8.4.1.7 Heat

The fuel requirement (for burners) is one of the main parameters of concern in incineration systems. The amount of fuel required to raise the temperature of the waste stream to the temperature required for complete oxidation is another area of concern. The burner fuel requirement can be estimated based on a simple heat balance of the unit and information concerning the waste gas stream. The first step in computing the heat required is to perform a heat balance around the oxidation system. From the first law of conservation of energy:

$$\text{Heat in} = \text{Heat out} + \text{Heat loss} \qquad (8.18)$$

Heat is a relative term, meaning that levels of heat must be compared at reference temperatures. To calculate the heat that exits the incinerator with the waste gas stream, the enthalpies of the inlet and outlet waste gas streams must be determined. *Enthalpy* is a thermodynamic term that includes the sensible heat and latent heat of a material. The heat content of a substance is arbitrarily taken as zero at a specified reference temperature. In the natural gas industry, the reference temperature is normally 16°C (60°F).

The enthalpy of the waste gas stream can be computed from Equation 8.19:

$$H = C_p(T - T_0) \qquad (8.19)$$

where
H = enthalpy, joules per kilogram or British thermal units per pound
C_p = specific heat at temperature T, joules per kilogram–degrees Celsius or British thermal units per pound–degrees Fahrenheit
T = temperature of the substance, degrees Celsius or degrees Fahrenheit
T_0 = reference temperature, degrees Celsius or degrees Fahrenheit

Subtracting the enthalpy of the waste stream exiting the incinerator from the waste gas stream entering the incinerator gives the heat that must be supplied by the fuel. This is referred to as a *change in enthalpy* or *heat content*. Using Equation 8.19, the enthalpy entering (T_1) is subtracted from the enthalpy leaving (T_2), giving:

$$q = m\Delta H = mC_p(T_2 - T_1) \tag{8.20}$$

where
q = heat rate, British thermal units per hour
m = mass flow rate, pounds per hour
ΔH = enthalpy change
C_p = specific heat, joules per kilogram–degrees Celsius or British thermal units per pound–degrees Fahrenheit)
T_2 = exit temperature of the substance, degrees Celsius or degrees Fahrenheit
T_1 = initial temperature, degrees Celsius or degrees Fahrenheit

8.4.2   Incineration Example Calculations

Example 8.10 Residence Time (Metric Units)

Problem:
   A thermal incinerator controls emissions from a paint-baking oven.

Given:
Cylindrical unit diameter = 1.5 m (5 ft)
Unit = 3.5 m (11.5 ft) long
Exhaust from the oven = 3.8 m³/sec (8050 scfm)

The incinerator uses 300 scfm of natural gas and operates at a temperature of 1400°F. If all the oxygen necessary for combustion is supplied from the process stream (no outside air added), what is the residence time in the combustion chamber (USEPA-81/12, p. 3-7)?

Solution:
   In solving this problem, we use the approximation that 11.5 m³ of combustion products are formed for every 1.0 m³ of natural gas burned at standard conditions (16°C and 101.3 kPa). At standard conditions, 10.33 m³ of theoretical air is required to combust 1 m³ of natural gas.

   Step 1. Determine the volume of combustion products from burning the natural gas:

$$(0.14 \text{ m}^3/\text{sec})(11.5 \text{ m}^3 \text{ of product})/(1.0 \text{ m}^3 \text{ of gas}) = 1.61 \text{ m}^3/\text{sec}$$

Step 2. Determine the air required for combustion:

$$(0.14 \text{ m}^3/\text{sec})(10.33 \text{ m}^3 \text{ of air})/(1.0 \text{ m}^3 \text{ of gas}) = 1.45 \text{ m}^3/\text{sec}$$

Step 3. Add the volumes:

$$\text{Flow from paint bake oven} = 3.8$$

$$\text{Products from combustion} = 1.61$$

$$\text{Minus the air from exhaust used on combustion} = -1.45$$

$$\text{Total volume} = 3.8 + 1.61 - 1.45 = 3.96 \text{ m}^3/\text{s}$$

Step 4. Convert the cubic meters per second calculated under the standard conditions to actual conditions:

$$(3.96 \text{ m}^3/\text{sec})(273°\text{C} + 760°\text{C})/(273°\text{C}) = 14.98 \text{ m}^3/\text{sec}$$

Step 5. Determine the chamber volume V:

$$V = \pi r^2 L$$

$$= 3.14(0.75 \text{ m})^2(3.5 \text{ m})$$

$$= 6.18 \text{ m}^3$$

Step 6. Calculate the residence time:

$$t = V/Q$$

$$= (6.18 \text{ m}^3)(14.98 \text{ m}^3/\text{sec})$$

$$= 0.41 \text{ sec}$$

Example 8.11 Incinerator Fuel Requirement

Problem:

The exhaust from a meat smokehouse contains obnoxious odors and fumes. The company plans to incinerate the 5000-acfm exhaust stream. What quantity of natural gas is required to raise the waste gas stream from a temperature 90°F to the required temperature of 1200°F? The gross heating value of natural gas is 1059 Btu/scf. Assume no heat losses (USEPA-81/12, p. 3-14).

Given:

Standard condition (state 1), $T_1 = 60°F$

Exhaust gas flow rate $V_2 = 5000$ acfm at temperature $T = 90°F$

Exhaust gas initial temperature $T_2 = 90°F$

Natural gas gross heating value = 1059 Btu/scf (standard cubic foot)

Combustion temperature $T_3 = 1200°F$

Solution:

Step 1. Correct the actual waste gas volume (V_a) to standard condition volume. The correction equation is:

$$V_1/T_1 = V_2/T_2$$

$$V_1 = V_2(T_1/T_2)$$

$$V_1 = (5000)(460 + 60)/(460 + 90)$$

$$= 4727 \text{ scfm}$$

$$= 283,620 \text{ scfh (standard cubic feet per hour)}$$

Step 2. Convert the volumetric flow rate to a mass flow rate by multiplying by the density:

$$\text{Mass flow rate} = (\text{volume rate})(\text{density})$$

Standard volume of an ideal gas at 60°F = 379.64 ft^3/(lb-mol)

Assume that the waste gas molecular weight is the same as air: 29 lb/(lb-mol):

$$\text{Density} = (\text{molecular weight})/\text{volume}$$

$$= 29/379.64$$

$$= 0.076388 \text{ lb/ft}^3$$

$$\text{Mass flow rate} = 4727(0.076388)$$

$$= 361 \text{ lb/min}$$

Step 3. Calculate the heat rate by using the ideal gas equation:

$$Q = mC_p(T_3 - T_2)$$

$$= 361(0.26)(1200 - 90)$$

$$= 104,185 \text{ Btu/min}$$

Step 4. Determine the heating value of natural gas. For natural gas, 1 scf contains 1059 Btu (given).
Step 5. Determine the natural gas quantity, W:

$$W = 104,185/1059$$

$$= 98 \text{ scfm}$$

8.5   CONDENSATION

Condensation is a process by which volatile gases are removed from the contaminant stream and changed into a liquid, a process that reduces a gas or vapor to a liquid. Condensers condense vapors to a liquid phase by increasing system pressure without a temperature change or by decreasing the system temperature to its saturation temperature without a pressure change. The common approach is to reduce the temperature of the gas stream because increasing the pressure of a gas is very expensive (USEPA-81/12, p. 6-1). Condensation is affected by the composition of the contaminant gas stream. When gases that condense under different conditions are present in the streams, condensation is hindered.

In air pollution control, a condenser can be used in two ways: for pretreatment to reduce the load problems with other air pollution control equipment or for effectively controlling contaminants in the form of gases and vapors. Two basic types of condensation equipment are available: contact scrubbers and surface condensers. In a contact scrubber (which resembles a simple spray scrubber), spraying liquid directly on the vapor stream (see Figure 8.17) cools the vapor. The cooled vapor condenses, and the water and condensate mixture are removed, treated, and disposed of.

A surface condenser is normally a shell-and-tube heat exchanger (see Figure 8.18). It uses a cooling medium of air or water in which the vapor to be condensed is separated from the cooling medium by a metal wall. Coolant flows through the tubes while the vapor is passed over the tubes, condenses on the outside of the tubes, and drains off to storage (USEPA, 1971).

In general, condensers are simple, relatively inexpensive devices that normally use water or air to cool and condense a vapor stream. Condensers are used in a wide range of industrial applications, including petroleum refining; petrochemical manufacturing; basic chemical manufacturing; dry cleaning; and degreasing.

8.5.1   Contact Condenser Calculations

In a contact condenser (see Figure 8.17) the coolant and vapor stream are physically mixed. They leave the condenser as a single exhaust stream. Simplified heat balance calculations are used to estimate the important parameters.

The first step in analyzing any heat transfer process is to set up a heat balance relationship. For a condensation system, the heat balance can be expressed as:

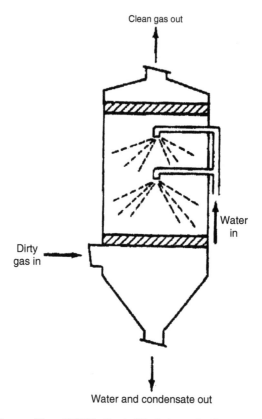

Figure 8.17   Contact condenser. (From USEPA, Control Techniques for Gases and Particulates, 1971.)

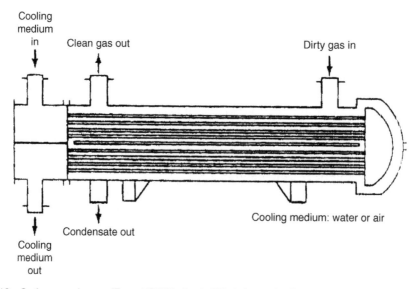

Figure 8.18   Surface condenser. (From USEPA, Control Techniques for Gases and Particulates, 1971.)

Heat in = Heat out

(Heat required to reduce vapors to the dew point) + (heat required to condense vapors)

= (heat needed to be removed by the coolant)

This heat balance is written in equation form as:

$$q = mC_p(T_{G1} - T_{dew\,point}) + mH_v = LC_p(T_{L2} - T_{L1}) \qquad (8.21)$$

where
q = heat transfer rate, British thermal units per hour
m = mass flow rate of vapor, pounds per hour
C_p = average specific heat of a gas or liquid, British thermal units per pound (degrees Fahrenheit)
T = temperature of the streams; G or gas and L for liquid coolant
H_v = heat of condensation or vaporization, British thermal units per pound
L = mass flow rate of liquid coolant, pounds per hour

In Equation 8.21, the mass flow rate m and inlet temperature T_{G1} of the vapor stream are set by the process exhaust stream. The temperature of the coolant entering the condenser T_{L1} is also set. The average specific heats C_p of both streams, the heat of condensation H_v, and the dew point temperature can be obtained from chemistry handbooks. Therefore, only the amount of coolant L and its outlet temperature must be determined. If either one of these terms is set by process restrictions (only x pounds an hour of coolant are available or the outlet temperature), then the other term can be solved for directly.

Equation 8.21 is applicable for direct contact condensers and should be used only to obtain rough estimates. The equation has a number of limitations:

- The specific heat C_p of a substance is dependent on temperature, and the temperature throughout the condenser is constantly changing.
- The dew point of a substance is dependent on its concentration in the gas phase; because the mass flow rate is constantly changing (vapors being condensed), the dew point temperature is constantly changing.
- No provision is made for cooling the vapors below the dew point. An additional term would need to be added to the left side of the equation to account for this amount of cooling.

8.5.2 Surface Condenser Calculations

According to USEPA-81/12, p. 6-3, in a surface condenser or heat exchanger (see Figure 8.18), heat is transferred from the vapor stream to the coolant through a heat exchange surface. The rate of heat transfer depends on three factors: total cooling surface available; resistance to heat transfer; and mean temperature difference between condensing vapor and coolant. This can be expressed mathematically by:

$$q = UA\Delta T_m \qquad (8.22)$$

where
q = heat transfer rate, British thermal units per hour
U = overall heat transfer coefficient, British thermal units per degree Fahrenheit – square feet per hour

Table 8.3 Typical Heat Transfer Coefficients in Tubular Heat Exchangers

Condensing vapor (shell side)	Cooling liquid	U (Btu/°F–ft²–h)
Organic solvent vapor with high percent of noncondensable gases	Water	20–60
High-boiling hydrocarbon vapor (vacuum)	Water	20–50
Low-boiling hydrocarbon vapor	Water	80–200
Hydrocarbon vapor and steam	Water	80–100
Steam	Feedwater	400–1000
Naphtha	Water	50–75
Water	Water	200–250

Source: Adapted from USEPA-81/12.

A = heat transfer surface area, square feet
ΔT_m = mean temperature difference, degrees Fahrenheit

The overall heat transfer coefficient U is a measure of the total resistance that heat experiences while being transferred from a hot body to a cold body. In a shell-and-tube condenser, cold water flows through the tubes causing vapor to condense on the outside surface of the tube wall. Heat is transferred from the vapor to the coolant. The ideal situation for heat transfer is when heat is transferred from the vapor to the coolant without any heat loss (heat resistance).

Every time heat moves through a different medium, it encounters a different and additional heat resistance. These heat resistances occur throughout the condensate; through any scale or dirt on the outside of the tube (fouling); through the tube itself; and through the film on the inside of the tube (fouling). Each of these resistances is an individual heat transfer coefficient and must be added together to obtain an overall heat transfer coefficient. An estimate of an overall heat transfer coefficient can be used for preliminary calculations. The overall heat transfer coefficients shown in Table 8.3 should be used only for preliminary estimating purposes.

In a surface heat exchanger, the temperature difference between the hot vapor and the coolant usually varies throughout the length of the exchanger. Therefore, a mean temperature difference (ΔT_m) must be used. The log mean temperature difference can be used for the special cases when the flow of both streams is completely concurrent; the flow of both streams is completely counter-current; or the temperature of the fluids remains constant (as is the case in condensing a pure liquid). The temperature profiles for these three conditions are illustrated in Figure 8.19. The log mean temperature for countercurrent flow can be expressed as shown in Equation 8.23.

$$\Delta T_m = \Delta T_{lm} = (\Delta T_1 - \Delta T_2)/\ln(\Delta T_1/\Delta T_2) \qquad (8.23)$$

where ΔT_{lm} = log mean temperature.

The value calculated from Equation 8.23 is used for single-pass heat exchangers or condensers. For multiple pass exchangers, a correction factor for the log mean temperature must be included. However, no correction factor is needed for the special case of isothermal condensation (no change in temperature) when there is a single component vapor and the gas temperature is equal to the dew point temperature.

To size a condenser, Equation 8.22 must be rearranged to solve for the surface area:

$$A = q / U\Delta T_{lm} \qquad (8.24)$$

where
A = surface area of a shell-and-tube condenser, square feet
q = heat transfer rate, British thermal units per pound

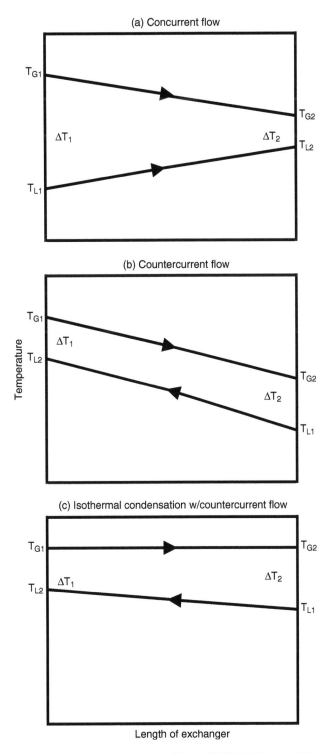

Figure 8.19 Temperature profiles in a heat exchanger. (From USEPA-81/12, p. 6-13.)

U = overall heat transfer coefficient, British thermal units per degree Fahrenheit – square feet – hour
ΔT_{lm} = log mean temperature difference, degrees Fahrenheit

Equation 8.24 is valid only for isothermal condensation of a single component. This implies that the pollutant is a pure vapor stream comprising only one specific hydrocarbon (benzene, for example) and not a mixture of hydrocarbons. Nearly all air pollution control applications involve mulitcomponent mixtures, which complicates the design procedure for a condenser.

For preliminary rough estimates of condenser size, the procedure in Example 8.12 for single-component condensation can be used for condensing multicomponents. In choosing a heat transfer coefficient, the smallest value should be used to allow as much overdesign as possible.

Example 8.12

Problem:

In a rendering plant, tallow is obtained by removing the moisture from animal matter in a cooker. Exhaust gases from the cookers contain essentially stream; however, the entrained vapors are highly odorous and must be controlled. Condensers are normally used to remove most of the moisture prior to incineration, scrubbing, or carbon adsorption (USEPA-81/12, p. 6-15).

The exhaust rate from the continuous cooker is 20,000 acfm at 250°F. The exhaust gases are 95% moisture with the remaining portion consisting of air and obnoxious organic vapors. The exhaust stream is sent first to a shell-and-tube condenser to remove the moisture, then to a carbon absorption unit. If the coolant water enters at 60°F and leaves at 120°F, estimate the required surface area of the condenser. The condenser is a horizontal, countercurrent flow system with the bottom few tubes flooded to provide subcooling.

Solution:

Step 1: Compute the pounds of steam condensed per minute:

$$20,000 \text{ acfm} \times 0.95 = 19,000 \text{ acfm steam}$$

From the ideal gas law:

$$PV = nRT$$

$$n = PV / RT$$

$$= (1 \text{ atm})(19,000 \text{ acfm})/[0.73 \text{ atm} - \text{ft}^3/\text{lb-mol} - °R$$

$$= 36.66 \text{ lb-mol/min}$$

$$m = (36.66 \text{ lb-mol/min})(18 \text{ lb/lb-mol})$$

$$= 660 \text{ lb/min of steam to be condensed}$$

Step 2. Solve the heat balance to determine q for cooling the superheated steam and condensing only:

q = (heat needed to cool steam to condensation temperature) + (heat of condensation)

$$= mC_p\Delta T + mH_v$$

The average specific heat (C_p) of steam at 250°F is roughly 0.45 Btu/lb-°F. The heat of vaporization of steam is at 212°F = 970.3 Btu/lb. Substituting into the equation:

$$q = (660 \text{ lb/min})(0.45 \text{ Btu/lb}-°\text{F})(250 - 212°\text{F}) + (660 \text{ lb/min})(970.3 \text{ Btu/lb})$$

$$q = 11,286 \text{ Btu/min} + 640,398 \text{ Btu/min} = 651,700 \text{ Btu/min}$$

Step 3. Now use Equation 8.24 to estimate the surface area for this part of the condenser:

$$A = q/U\Delta T_{lm}$$

For a countercurrent condenser, the log mean temperature is given by:

$$\Delta T_{lm} = [(T_{G1} - T_{L2}) - (T_{G2} - T_{L1})]/\ln[T_{G1} - T_{L2})/(T_{G1} - T_{L1})]$$

Remember that the superheater-condenser section is designed using the saturation temperature to calculate the log mean temperature difference:

$$\text{gas entering temperature } T_{G1} = 212°\text{F}$$

$$\text{coolant leaving temperature } T_{L2} = 120°\text{F}$$

$$\text{gas leaving temperature } T_{G2} = 212°\text{F}$$

$$\text{coolant entering temperature } T_{L1} = 60°\text{F}$$

$$\Delta T_{lm} = [(212 - 120) - (212 - 60)]/\ln[(212 - 120)/(212 - 60)] = 119.5°\text{F}$$

The overall heat transfer coefficient U is assumed to be 100 Btu/°F-ft²-h. Substituting the appropriate values into Equation 8.24:

$$A = [(651,700 \text{ Btu/min})(60 \text{ min/h})]/[(100 \text{ Btu/°F-ft}^2\text{-h})(119.5°\text{F})]$$

$$= 3272 \text{ ft}^2$$

Step 4. Estimate the total size of the condenser. Allow for subcooling of the water (212°F–160°F). 160°F is an assumed safe margin. Refiguring the heat balance for cooling the water:

$$q = UA\Delta T_m$$

where

m = 660 lb/min (assuming all the steam is condensed)

q = (660 lb/min)(1Btu/lb – °F)(212 – 160)°F = 34,320 Btu/min

ΔT_{lm} = [(212 – 120) – (160 – 60)]/ln[(212 – 120)/(160 – 60)] = 96°F

A = $q/U\Delta T_{lm}$

For cooling water with a water coolant, U is assumed to be 200 Btu/°F-ft^2-h.

$$A = (34{,}320 \text{ Btu/min})/[(200 \text{ Btu/°F-ft}^2\text{-h})(96°F)$$

$$= 1.79 \text{ ft}^2 \text{ or } 2 \text{ ft}^2$$

Step 5. The total area needed is

$$A = 3272 + 2$$

$$= 3274 \text{ ft}^2$$

As illustrated by this example, the area for subcooling is very small compared to the area required for condensing.

REFERENCES

American Industrial Hygiene Association (1968). *Air Pollution Manual: Control Equipment, Part II*. Detroit: AIHA.

Buonicore, A.J. and Davis, W.T. (1992) (Eds.). *Air Pollution Engineering Manual*. New York: Van Nostrand Reinhold.

Davis, M.L. and Cornwell, D.A. (1991). *Introduction to Environmental Engineering*. New York: McGraw-Hill.

Hesketh, H.E. (1991). *Air Pollution Control: Traditional and Hazardous Pollutants*. Lancaster, PA: Technomic Publishing Co., Inc.

Spellman, F.R. (1999). *The Science of Air: Concepts and Applications*. Boca Raton, FL: CRC Press.

USEPA (1971). Control Techniques for Gases and Particulates.

USEPA (1972). Wet Scrubber System Study. NTIS Report PB-213016. Research Triangle Park, NC.

USEPA (1973). *Air Pollution Engineering Manual*, 2nd ed. AP-40. Research Triangle Park, NC.

USEPA-81/12 (1981). Control of gaseous emissions, Course 415, USEPA Air Pollution Training Institute (APTI), EPA450-2-81-005.

USEPA-81/05 (1981). Control of gaseous emissions, Course 415, USEPA Air Pollution Training Institute (APTI), EPA450-2-81-006.

USEPA-84/03 (1984). Wet scrubber plan review, Course 51412C, USEPA Air Pollution Training Institute (APTI), EPA450-2-82-020.

CHAPTER 9

Particulate Emission Control

Fresh air is good if you do not take too much of it; most of the achievements and pleasures of life are in bad air.

Oliver Wendell Holmes

9.1 PARTICULATE EMISSION CONTROL BASICS

Particle or particulate matter is defined as tiny particles or liquid droplets suspended in the air; they can contain a variety of chemical components. Larger particles are visible as smoke or dust and settle out relatively rapidly. The tiniest particles can be suspended in the air for long periods of time and are the most harmful to human health because they can penetrate deep into the lungs. Some particles are directly emitted into the air from pollution sources.

Constituting a major class of air pollutants, particulates have a variety of shapes and sizes; as either liquid droplet or dry dust, they have a wide range of physical and chemical characteristics. Dry particulates are emitted from a variety of different sources in industry, mining, construction activities, and incinerators, as well as from internal combustion engines — from cars, trucks, buses, factories, construction sites, tilled fields, unpaved roads, stone crushing, and wood burning. Dry particulates also come from natural sources such as volcanoes, forest fires, pollen, and windstorms. Other particles are formed in the atmosphere by chemical reactions.

When a flowing fluid (engineering and science applications consider liquid and gaseous states as fluid) approaches a stationary object (a metal plate, fabric thread, or large water droplet, for example), the fluid flow will diverge around that object. Particles in the fluid (because of inertia) will not follow stream flow exactly, but tend to continue in their original directions. If the particles have enough inertia and are located close enough to the stationary object, they collide with the object and can be collected by it. This important phenomenon is depicted in Figure 9.1.

9.1.1 Interaction of Particles with Gas

To understand the interaction of particles with the surrounding gas, knowledge of certain aspects of the kinetic theory of gases is necessary. This kinetic theory explains temperature, pressure, mean free path, viscosity, and diffusion in the motion of gas molecules (Hinds, 1986). The theory assumes gases — along with molecules as rigid spheres that travel in straight lines — contain a large number of molecules that are small enough so that the relevant distances between them are discontinuous.

Air molecules travel at an average of 1519 ft/sec (463 m/sec) at standard conditions. Speed decreases with increased molecule weight. As the square root of absolute temperature increases, molecular velocity increases. Thus, temperature is an indication of the kinetic energy of gas

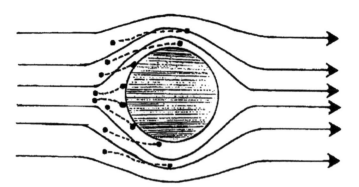

Figure 9.1   Particle collection of a stationary object. (Adapted from USEPA-84/03, p. 1-5.)

molecules. When molecular impact on a surface occurs, pressure develops and is directly related to concentration. Gas viscosity represents the transfer of momentum by randomly moving molecules from a faster moving layer of gas to an adjacent slower moving layer of gas. Viscosity of a gas is independent of pressure but will increase as temperature increases. Finally, diffusion is the transfer of molecular mass without any fluid flow (Hinds, 1986). Diffusion transfer of gas molecules is from a higher to a lower concentration. Movement of gas molecules by diffusion is directly proportional to the concentration gradient, inversely proportional to concentration, and proportional to the square root of absolute temperature.

The *mean free path,* kinetic theory's most critical quantity, is the average distance a molecule travels in a gas between collisions with other molecules. The mean free path increases with increasing temperature and decreases with increasing pressure (Hinds, 1986).

The Reynolds number characterizes gas flow, a dimensionless index that describes the flow regime. The Reynolds number for gas is determined by the following equation:

$$\mathrm{Re} = \frac{pU_G D}{\eta} \tag{9.1}$$

where
Re = Reynolds number
p = gas density, pounds per cubic foot (kilograms per cubic meter)
U_g = gas velocity, feet per second (meters per second)
D = characteristic length, feet (meters)
η = gas viscosity, lbm/ft·sec (kg/m·sec)

The Reynolds number helps to determine the flow regime, the application of certain equations, and geometric similarity (Baron and Willeke, 1993). Flow is laminar at low Reynolds numbers and viscous forces predominate. Inertial forces dominate the flow at high Reynolds numbers, when mixing causes the streamlines to disappear.

9.1.2   Particulate Collection

Particles are collected by gravity, centrifugal force, and electrostatic force, as well as by impaction, interception, and diffusion. Impaction occurs when the center of mass of a particle diverging from the fluid strikes a stationary object. Interception occurs when the particle's center of mass closely misses the object but, because of its size, the particle strikes the object. Diffusion occurs when small particulates happen to "diffuse" toward the object while passing near it. Particles that strike the object by any of these means are collected if short-range forces (chemical, electrostatic, and so forth) are strong enough to hold them to the surface (Copper and Alley, 1990).

Different classes of particulate control equipment include gravity settlers, cyclones, electrostatic precipitators, wet (Venturi) scrubbers, and baghouses (fabric filters). In the following sections we discuss many of the calculations used in particulate emission control operations. Many of the calculations presented are excerpted from USEPA-81/10.

9.2 PARTICULATE SIZE CHARACTERISTICS AND GENERAL CHARACTERISTICS

As we have said, particulate air pollution consists of solid and/or liquid matter in air or gas. Airborne particles come in a range of sizes. From near molecular size, the size of particulate matter ranges upward and is expressed in micrometers (μm — one millionth of a meter). For control purposes, the lower practical limit is about 0.01 μm. Because of the increased difficulty in controlling their emission, particles of 3 μm or smaller are defined as *fine particles*. Unless otherwise specified, concentrations of particulate matter are by mass.

Liquid particulate matter and particulates formed from liquids (very small particles) are likely to be spherical in shape. To express the size of a nonspherical (irregular) particle as a diameter, several relationships are important. These include:

- Aerodynamic diameter
- Equivalent diameter
- Sedimentation diameter
- Cut diameter
- Dynamic shape factor

9.2.1 Aerodynamic Diameter

Aerodynamic diameter, d_a, is the diameter of a unit density sphere (density = 1.00 g/cm^3) that would have the same settling velocity as the particle or aerosol in question. Note that because USEPA is interested in how deeply a particle penetrates into the lung, the agency is more interested in nominal aerodynamic diameter than in the other methods of assessing size of nonspherical particles. Nevertheless, a particle's nominal aerodynamic diameter is generally similar to its conventional, nominal physical diameter.

9.2.2 Equivalent Diameter

Equivalent diameter, d_e, is the diameter of a sphere that has the same value of a physical property as that of the nonspherical particle and is given by

$$d_e = \left(\frac{6V}{\pi} \right)^{1/3} \tag{9.2}$$

where V is volume of the particle.

9.2.3 Sedimentation Diameter

Sedimentation diameter, or Stokes diameter, d_s, is the diameter of a sphere that has the same terminal settling velocity and density as the particle. In density particles, it is called the reduced sedimentation diameter, making it the same as aerodynamic diameter. The dynamic shape factor accounts for a nonspherical particle settling more slowly than a sphere of the same volume.

9.2.4   Cut Diameter

Cut diameter, d_c, is the diameter of particles collected with 50% efficiency, i.e., individual efficiency $\varepsilon_I = 0.5$, and half penetrate through the collector — penetration $Pt = 0.5$.

$$Pt_I = 1 - \varepsilon_I \tag{9.3}$$

9.2.5   Dynamic Shape Factor

Dynamic shape factor, χ, is a dimensionless proportionality constant relating the equivalent and sedimentation diameters:

$$\chi = \left(\frac{d_e}{d_s}\right)^2 \tag{9.4}$$

The d_e equals d_s for spherical particles, so χ for spheres is 1.0.

9.3   FLOW REGIME OF PARTICLE MOTION

Air pollution control devices collect solid or liquid particles via the movement of a particle in the gas (fluid) stream. For a particle to be captured, the particle must be subjected to external forces large enough to separate it from the gas stream. According to USEPA-81/10, p. 3-1, forces acting on a particle include three major forces as well as other forces:

- Gravitational force
- Buoyant force
- Drag force
- Other forces (magnetic, inertial, electrostatic, and thermal force, for example)

The consequence of acting forces on a particle results in the settling velocity — the speed at which a particle settles. The settling velocity (also known as the terminal velocity) is a constant value of velocity reached when all forces (gravity, drag, buoyancy, etc.) acting on a body are balanced — that is, when the sum of all the forces is equal to zero (no acceleration). To solve for an unknown particle settling velocity, we must determine the flow regime of particle motion. Once the flow regime has been determined, we can calculate the settling velocity of a particle.

The flow regime can be calculated using the following equation (USEPA-81/10, p. 3-10):

$$K = d_p(gp_p p_a/\mu^2)^{0.33} \tag{9.5}$$

where
K = a dimensionless constant that determines the range of the fluid-particle dynamic laws
d_p = particle diameter, centimeters or feet
g = gravity force, cm/sec^2 or ft/sec^2
p_p = particle density, grams per cubic centimeter or pounds per cubic foot
p_a = fluid (gas) density, grams per cubic centimeter or pounds per cubic foot
μ = fluid (gas) viscosity, grams per centimeter-second or pounds per foot-second

USEPA-81/10, p. 3-10, lists the K values corresponding to different flow regimes as:

- Laminar regime (also known as Stokes' law range): $K < 3.3$
- Transition regime (also known as intermediate law range): $3.3 < K, 43.6$
- Turbulent regime (also known as Newton's law range): $K > 43.6$

According to USEPA-81/10, p. 3-10, the K value determines the appropriate range of the fluid-particle dynamic laws that apply.

- For a laminar regime (Stokes' law range), the terminal velocity is

$$v = gp_p(d_p)^2 / (18\mu) \tag{9.6}$$

- For a transition regime (intermediate law range), the terminal velocity is:

$$v = 0.153g^{0.71}(d_p)^{1.14}(p_p)^{0.71} / [\mu^{0.43}(p_a)^{0.29}] \tag{9.7}$$

- For a turbulent regime (Newton's law range), the terminal velocity is:

$$v = 1.74(gd_p p_p / p_a)^{0.5} \tag{9.8}$$

When particles approach sizes comparable with the mean free path of fluid molecules (also known as the Knudsen number, Kn), the medium can no longer be regarded as continuous because particles can fall between the molecules at a faster rate than that predicted by aerodynamic theory. Cunningham's correction factor, which includes thermal and momentum accommodation factors based on the Millikan oil-drop studies and which is empirically adjusted to fit a wide range of Kn values, is introduced into Stoke's law to allow for this slip rate (Hesketh, 1991; USEPA-84/09, p. 58):

$$v = gp_p(d_p)^2 C_f / (18\,\mu) \tag{9.9}$$

where
C_f = Cunningham correction factor = $1 + (2A\lambda/d_p)$
A = $1.257 + 0.40e^{-1.10}d_p/2\lambda$
λ = free path of the fluid molecules (6.53×10^{-6} cm for ambient air)

Example 9.1

Problem:
Calculate the settling velocity of a particle moving in a gas stream. Assume the following information (USEPA-81/10, p. 3-11):

Given:
d_p = particle diameter = 45 μm (45 microns)
g = gravity forces = 980 cm/sec^2
p_p = particle density = 0.899 g/cm^3
p_a = fluid (gas) density = 0.012 g/cm^3
μ = fluid (gas) viscosity = 1.82×10^{-4} g/cm-sec
C_f = 1.0 (if applicable)

Solution:

Step 1. Use Equation 9.5 to calculate the K parameter to determine the proper flow regime:

$$K = d_p(gp_p p_a / \mu^2)0^{.33}$$

$$= 45 \times 10^{-4}[980(0.899)(0.012)/1.82 \times 10^{-4})^2]^{0.33}$$

$$= 3.07$$

The result demonstrates that the flow regime is laminar.

Step 2. Use Equation 9.9 to determine the settling velocity:

$$v = gp_p(d_p)^2 C_f / (18\mu)$$

$$= 980(0.899)(45 \times 10^{-4})^2(1)/[18(1.82 \times 10^{-4})]$$

$$= 5.38 \text{ cm/sec}$$

Example 9.2

Problem:

Three differently sized fly ash particles settle through the air. Calculate the particle terminal velocity (assume the particles are spherical) and determine how far each will fall in 30 sec.

Given:
Fly ash particle diameters = 0.4, 40, 400 µm
Air temperature and pressure = 238°F, 1 atm
Specific gravity of fly ash = 2.31

Because the Cunningham correction factor is usually applied to particles equal to or smaller than 1 µm, check how it affects the terminal settling velocity for the 0.4-µm particle.

Solution:

Step 1. Determine the value for K for each fly ash particle size settling in air. Calculate the particle density using the specific gravity given:

$$P_p = \text{particle density} = (\text{specific gravity of fly ash})(\text{density of water})$$

$$= 2.31(62.4)$$

$$= 144.14 \text{ lb/ft}^3$$

Calculate the density of air:

$p =$ air density $=$ PM/RT

$\qquad = (1)(29)/(0.7302)(238 + 460) = 0.0569$ lb/ft^3

$\mu =$ air viscosity $= 0.021$ cp $= 1.41 \times 10^{-5}$ lb/ft-sec (USEPA-84/09, p. 167)

Determine the flow regime (K):

$$K = d_p (g p_p p_a / \mu^2)^{0.33}$$

For $d_p = 0.4$ μm:

$K = [(0.4)/(25{,}400)(12)][32.2(144.14)(0.0569)/(1.41 \times 10^{-5})^2]^{0.33} = 0.0144$

where 1 ft $= 25{,}400(12)$ μm (USEPA-84/09, p. 183)

For $d_p = 40$ μm:

$K = [(40)/(25{,}400)(12)][32.3(144.14)(0.0569)/(1.41 \times 10^{-5})^2]^{0.33} = 1.44$

For $d_p = 400$ μm:

$K = [(400)/(25{,}400)(12)][32.2(144.14)(0.0569)/(1.41 \times 10^{-5})^2]^{0.33} = 14.4$

Step 2. Select the appropriate law, determined by the numerical value of K:
$\quad K < 3.3$; Stokes' law range
$\quad 3.3 < K < 43.6$; intermediate law range
$\quad 43.6 < K < 2360$; Newton's law range
$\quad$ For $d_p = 0.4$ μm, the flow regime is laminar (USEPA-81/10, p. 3-10)
$\quad$ For $d_p = 40$ μm, the flow regime is also laminar
$\quad$ For $d_p = 400$ μm, the flow regime is the transition regime

For $d_p = 0.4$ μm:

$$v = g p_p (d_p)^2 / (18\mu)$$

$$= (32.2)[(0.4)/25{,}400)(12)]^2 (144.14)/(18)(1.41 \times 10^{-5})$$

$$= 3.15 \times 10^{-5} \text{ ft/sec}$$

For $d_p = 40$ μm:

$$v = g p_p (d_p)^2 / (18\mu)$$

$$= (32.2)[(40)/25{,}400)(12)]^2 (144.14)(18)(1.41 \times 10^{-5})$$

$$= 0.315 \text{ ft/sec}$$

For d_p = 400 μm (use transition regime equation):

$$v = 0.153g^{0.71} \, (d_p)^{1.14}(p_p)^{0.71}/(\mu^{0.43}p^{0.29})$$

$$= 0.153(32.2)^{0.71}[(400)/(25,400)(12)]^{1.14}(144.14)^{0.71}/[(1.41 \times 10^{-5})^{0.43}(0.0569)^{0.29}]$$

$$= 8.90 \text{ ft/sec}$$

Step 4. Calculate distance.
    For d_p = 40 μm, distance = (time)(velocity):

$$\text{Distance} = 30(0.315) = 9.45 \text{ ft}$$

For d_p = 400 μm, distance = (time)(velocity):

$$\text{Distance} = 30(8.90) = 267 \text{ ft}$$

For d_p = 0.4 μm, without Cunningham correction factor, distance = (time)(velocity):

$$\text{Distance} = 30(3.15 \times 10^{-5}) = 94.5 \times 10^{-5} \text{ft}$$

For d_p = 0.4 μm with Cunningham correction factor, the velocity term must be corrected. For our purposes, assume particle diameter = 0.5 μm and temperature = 212°F to find the C_f value. Thus, C_f is approximately equal to 1.446.

$$\text{The corrected velocity} = vC_f = 3.15 \times 10^{-5}(1.446) = 4.55 \times 10^{-5} \text{ ft/sec}$$

$$\text{Distance} = 30(4.55 \times 10^{-5}) = 1.365 \times 10^{-3} \text{ ft}$$

Example 9.3

Problem:

Determine the minimum distance downstream from a cement dust-emitting source that will be free of cement deposit. The source is equipped with a cyclone (USEPA-84/09, p. 59).

Given:
Particle size range of cement dust = 2.5 to 50.0 μm
Specific gravity of the cement dust = 1.96
Wind speed = 3.0 mi/h

The cyclone is located 150 ft above ground level. Assume ambient conditions are at 60°F and 1 atm. Disregard meteorological aspects.

μ = air viscosity at 60°F = 1.22×10^{-5} lb/ft-sec (USEPA-84/09, p. 167)
μm (1 μm = 10^{-6}) = 3.048×10^{5} ft (USEPA-84/09, p. 183)

Solution:

Step 1. A particle diameter of 2.5 μm is used to calculate the minimum distance downstream free of dust because the smallest particle will travel the greatest horizontal distance.

Step 2. Determine the value of K for the appropriate size of the dust. Calculate the particle density (p_p) using the specific gravity given:

$$p_p = \text{(specific gravity of fly ash)(density of water)}$$

$$= 1.96(62.4)$$

$$= 122.3 \text{ lb/ft}^3$$

Calculate the air density (p). Use modified ideal gas equation, $PV = nR_uT = (m/M)R_uT$

$$P = \text{(mass)(volume)}$$

$$= PM / R_u T$$

$$= (1)(29)/[0.73(60 + 460)] = 0.0764 \text{ lb/ft}^3$$

Determine the flow regime (K):

$$K = d_p (gp_p p_a / \mu^2)^{0.33}$$

For $d_p = 2.5$ μm:

$$K = [(2.5)/(25,400)(12)][32.2)(122.3)(0.0764)/(1.22 \times 10^{-5})^2]^{0.33} = 0.104$$

where 1 ft = 25,400(12) μm = 304,800 μm (USEPA-84/09, p. 183)

Step 3. Determine which fluid-particle dynamic law applies for the preceding value of K. Compare the K value of 0.104 with the following range:
$K < 3.3$; Stokes' law range
$3.3 \leq K < 43.6$; intermediate law range
$43.6 < K < 2360$; Newton's law range
The flow is in the Stokes' law range; thus it is laminar.

Step 4. Calculate the terminal settling velocity in feet per second. For Stokes' law range, the velocity is

$$v = gp_p (d_p)^2 / (18\ \mu)$$

$$= (32.2)[2.5/(25,400)(12)]^2 (122.3)/(18)(1.22 \times 10^{-5})$$

$$= 1.21 \times 10^{-3} \text{ ft/sec}$$

Step 5. Calculate the time for settling:

$$t = \text{(outlet height)/(terminal velocity)}$$

$$= 150/1.21 \times 10^{-3}$$

$$= 1.24 \times 10^5 \text{ sec} = 34.4 \text{ h}$$

Step 6. Calculate the horizontal distance traveled:

$$\text{Distance} = \text{(time for descent)(wind speed)}$$

$$= (1.24 \times 10^5)(3.0/3600)$$

$$= 103.3 \text{ miles}$$

9.4 PARTICULATE EMISSION CONTROL EQUIPMENT CALCULATIONS

Different classes of particulate control equipment include gravity settlers; cyclones; electrostatic precipitators; wet (Venturi) scrubbers (discussed in Chapter 10); and baghouses (fabric filters). In the following section, we discuss calculations used for each of the major types of particulate control equipment.

9.4.1 Gravity Settlers

Gravity settlers have long been used by industry for removing solid and liquid waste materials from gaseous streams. Simply constructed (see Figure 9.2 and Figure 9.3), a gravity settler is actually nothing more than a large chamber in which the horizontal gas velocity is slowed, allowing particles to settle out by gravity. Gravity settlers have the advantage of having low initial cost and are relatively inexpensive to operate because not much can go wrong. Although simple in design,

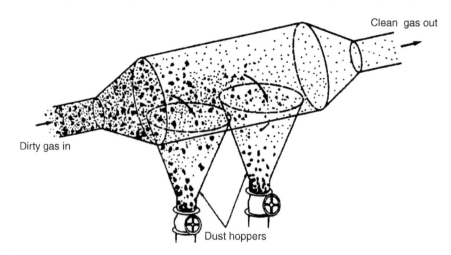

Figure 9.2   Gravitational settling chamber. (From USEPA, Control Techniques for Gases and Particulates, 1971.)

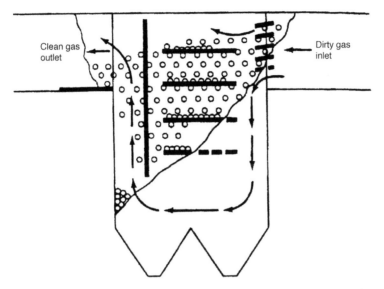

Figure 9.3 Baffled gravitational settling chamber. (From USEPA, Control Techniques for Gases and Particulates, 1971.)

gravity settlers require a large space for installation and have relatively low efficiency, especially for removal of small particles (<50 μm).

9.4.2 Gravity Settling Chamber Theoretical Collection Efficiency

The theoretical collection efficiency of the gravity-settling chamber (USEPA-81/10, p. 5-4) is given by the expression:

$$\eta = (v_y L)(v_x H) \tag{9.10}$$

where
η = fractional efficiency of particle size d_p (one size)
v_y = vertical settling velocity
v_x = horizontal gas velocity
L = chamber length
H = chamber height (greatest distance a particle must fall to be collected)

As mentioned, the settling velocity can be calculated from Stokes' law. As a rule of thumb, Stokes' law applies when the particle size d_p is less than 100 μm in size. The settling velocity is

$$v_t = [g(d_p)^2(p_p - p_a)]/(18 \mu) \tag{9.11}$$

where
v_t = settling velocity in Stokes' law range, meters per second (feet per second)
g = acceleration due to gravity, 9.8 m/sec² (32.1 ft/sec²)
d_p = particle diameter, microns
p_p = particle density, kilograms per cubic meter (pounds per cubic foot)
p_a = gas density, kilograms per cubic meter (pounds per cubic foot)

μ = gas viscosity, pascal-seconds (pound-feet-second)
$p_a = N/m^2$
N = kilogram-meters per sec^2

Equation 9.11 can be rearranged to determine the minimum particle size that can be collected in the unit with 100% efficiency. The minimum particle size $(d_p)*$ in microns is given as:

$$(d_p)* = \{v_t(18\mu) / [g(p_p - p_a)]\}^{0.5} \tag{9.12}$$

Because the density of the particle p_p is usually much greater than the density of gas p_a, the quantity $p_p - p_a$ reduces to p_p. The velocity can be written as:

$$V = Q / BL \tag{9.13}$$

where
Q = volumetric flow
B = chamber width
L = chamber length

Equation 9.12 is reduced to:

$$(d_p)* = (18\mu Q) / [gp_pBL]\}^{0.5} \tag{9.14}$$

The efficiency equation can also be expressed as:

$$\eta = [(gp_pBLN_c) / (18\mu Q)](d_p)^2 \tag{9.15}$$

where N_c = number of parallel chambers: 1, for simple settling chamber and N trays +1, for a Howard settling chamber.
Equation 9.15 can be written as:

$$\eta = 0.5[(gp_pBLN_c) / (18\mu Q)](d_p)^2 \tag{9.16}$$

and the overall efficiency can be calculated using

$$\eta_{TOT} = \Sigma\eta_i w_i \tag{9.17}$$

where
η_{TOT} = overall collection efficiency
η_i = fractional efficiency of specific size particle
w_i = weight fraction of specific size particle

When flow is turbulent, Equation 9.18 is used.

$$\eta = \exp[-(Lv_y / Hv_x)] \tag{9.18}$$

In using Equation 9.10 through Equation 9.16, note that Stokes' law does not work for particles greater than 100 μm.

9.4.3 Minimum Particle Size

Most gravity settlers are precleaners that remove the relatively large particles (>60 μm) before the gas stream enters a more efficient particulate control device such as a cyclone, baghouse, electrostatic precipitator (ESP), or scrubber.

Example 9.4

Problem:
A hydrochloric acid mist in air at 25°C is collected in a gravity settler. Calculate the smallest mist droplet (spherical in shape) collected by the settler. Stokes' law applies; assume the acid concentration is uniform through the inlet cross-section of the unit (USEPA-84/09, p. 61).

Given:
Dimensions of gravity settler = 30 ft wide, 20 ft high, 50 ft long
Actual volumetric flow rate of acid gas in air = 50 ft³/sec
Specific gravity of acid = 1.6
Viscosity of air = 0.0185 cp = 1.243 × 10^{-5} lb/ft-sec
Density of air = 0.076 lb/ft³

Solution:

Step 1. Calculate the density of the acid mist using the specific gravity given:

$$p_p = \text{particle density} = (\text{specific gravity of fly ash})(\text{density of water})$$

$$1.6(62.4) = 99.84 \text{ lb/ft}^3$$

Step 2. Calculate the minimum particle diameter in feet and microns, assuming that Stokes' law applies. For Stokes' law range:

$$\text{Minimum } d_p = (18\mu Q/gp_pBL)^{0.5}$$

$$\text{Minimum } d_p = [(18)(1.243 \times 10^{-5})(50)/(32.2)(99.84)(30)(50)]^{0.5}$$

$$= 4.82 \times 10^{-5} \text{ ft}$$

$$= (4.82 \times 10^{-5} \text{ ft})(3.048 \times 10^5 \text{ μm/ft})$$

$$= 14.7 \text{ μm}$$

Table 9.1 Particle Size Distribution Data

Particle size range, μm	Average particle diameter, μm	Inlet Grains/scf	Inlet wt%
0–20	10	0.0062	2.7
20–30	25	0.0159	6.9
30–40	35	0.0216	9.4
40–50	45	0.0242	10.5
50–60	55	0.0242	10.5
60–70	65	0.0218	9.5
70–80	75	0.0161	7
80–94	85	0.0218	9.5
94	94	0.0782	34
Total		0.23	100

Example 9.5

Problem:

A settling chamber that uses a traveling grate stoker is installed in a small heat plant. Determine the overall collection efficiency of the settling chamber, given the operating conditions, chamber dimensions, and particle size distribution data (USEPA-84/09, p. 62).

Given:
Chamber width = 10.8 ft
Chamber height = 2.46 ft
Chamber length = 15.0 ft
Volumetric flow rate of contaminated air stream = 70.6 scfs
Flue gas temperature = 446°F
Flue gas pressure = 1 atm
Particle concentration = 0.23 gr/scf
Particle specific gravity = 2.65
Standard conditions = 32°F, 1 atm

Particle size distribution data of the inlet dust from the traveling grate stoker are shown in Table 9.1. Assume that the actual terminal settling velocity is one-half of the Stokes' law velocity.

Solution:

Step 1. Plot the size efficiency curve for the settling chamber. The size efficiency curve is needed to calculate the outlet concentration for each particle size (range). These outlet concentrations are then used to calculate the overall collection efficiency of the settling chamber. The collection efficiency for a settling chamber can be expressed in terms of the terminal velocity, volumetric flow rate of contaminated stream, and chamber dimensions:

$$\eta = vBL/Q = [gp_p(d_p)^2/(18\mu)](BL/Q)$$

where
η = fractional collection efficiency
v = terminal settling velocity
B = chamber width
L = chamber length
Q = volumetric flow rate of the stream

Step 2. Express the collection efficiency in terms of the particle diameter d_p. Replace the terminal settling velocity in the preceding equation with Stokes' law. Because the actual terminal settling velocity is assumed to be one half of the Stokes' law velocity (according to the given problem statement), the velocity equation becomes:

$$v = g(d_p)^2 p_p / 36\mu$$

$$\eta = [gp_p(d_p)^2/36\mu)(BL/Q)$$

Determine the viscosity of the air in pounds per foot-second:

$$\text{Viscosity of air at } 446°F = 1.75 \times 10^{-5} \text{ lb/ft-sec}$$

Determine the particle density in pounds per cubic foot:

$$P_p = 2.65(62.4) = 165.4 \text{ lb/ft}^3$$

Determine the actual flow rate in actual cubic feet per second. To calculate the collection efficiency of the system at the operating conditions, the standard volumetric flow rate of contaminated air of 70.6 scfs is converted to actual volumetric flow of 130 acfs:

$$Q_a = Q_s(T_a / T_s)$$

$$= 70.6(446 + 460)/(32 + 460)$$

$$= 130 \text{ acfs}$$

Express the collection efficiency in terms of d_p, with d_p in feet. Also express the collection efficiency in terms of d_p, with d_p in microns.

Use the following equation; substitute values for p_p, g, B, L, μ, and Q in consistent units. Use the conversion factor for feet to microns. To convert d_p from square feet to square microns, d_p is divided by $(304,800)^2$.

$$\eta = [gp_p(d_p)^2/36\mu)(BL/Q)$$

$$= (32.2)(165.4)(10.8)(15)(d_p)^2/[(36)(1.75 \times 10^{-5})(130)(304,800)^2]$$

$$= 1.134 \times 10^{-4}(d_p)^2$$

where d_p is in microns.

Calculate the collection efficiency for each particle size. For a particle diameter of 10 μm:

$$\eta = (1.134 \times 10^{-4})(d_p)^2 = (1.134 \times 10^{-4})(10)^2 = 1.1 \times 10^{-2} = 1.1\%$$

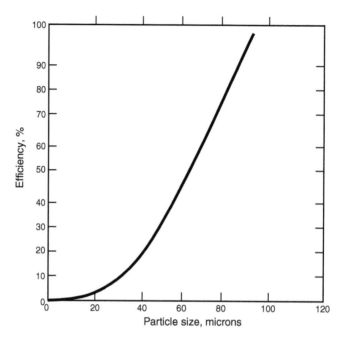

Figure 9.4   Size efficiency curve for settling chamber. (Adapted from USEPA-84/09, 136.)

Table 9.2   Collection Efficiency
for Each Particle Size

d_p, μm	η, %
94	100
90	92
80	73
60	41
40	18.2
20	4.6
10	1.11

Table 9.2 provides the collection efficiency for each particle size. The size efficiency curve for the settling chamber is shown in Figure 9.4; read off the collection efficiency of each particle size from this figure.

Calculate the overall collection efficiency (Table 9.3).

$$\eta = \Sigma w_i \eta_i$$

$$= (0.027)(1.1) + (0.069)(7.1) + (0.094)(14.0) + (0.105)(23.0) + (0.105)$$

$$(34.0) + (0.095)(48.0) + ((0.070)(64.0) + (0.095)(83.0) + (0.340)(100.0)$$

$$= 59.0\%$$

Table 9.3   Data for Calculation of Overall Collection Efficiency

d_p, μm	Weight fraction, w_i	η_i
10	0.027	1.1
25	0.069	7.1
35	0.094	14
45	0.105	23
55	0.105	34
65	0.095	48
75	0.07	64
85	0.095	83
94	0.34	100
Total	1	

9.4.4   Cyclones

Cyclones — the most common dust removal devices used within industry (Strauss, 1975) — remove particles by causing the entire gas stream to flow in a spiral pattern inside a tube. They are the collectors of choice for removing particles greater than 10 μm in diameter. By centrifugal force, the larger particles move outward and collide with the wall of the tube. The particles slide down the wall and fall to the bottom of the cone, where they are removed. The cleaned gas flows out the top of the cyclone (see Figure 9.5).

Cyclones have low construction costs and relatively small space requirements for installation; they are much more efficient in particulate removal than settling chambers. However, note that the cyclone's overall particulate collection efficiency is low, especially on particles below 10 μm in size, and they do not handle sticky materials well. The most serious problems encountered with cyclones are with airflow equalization and with their tendency to plug (Spellman, 1999). They are often installed as precleaners before more efficient devices such as electrostatic precipitators and baghouses (USEPA-81/10, p. 6-1) are used. Cyclones have been used successfully at feed and grain mills; cement plants; fertilizer plants; petroleum refineries; and other applications involving large quantities of gas containing relatively large particles.

9.4.4.1   Factors Affecting Cyclone Performance

The factors that affect cyclone performance include centrifugal force, cut diameter, pressure drop, collection efficiency, and summary of performance characteristics. Of these parameters, the cut diameter is the most convenient way of defining efficiency for a control device because it gives an idea of the effectiveness for a particle size range. As mentioned earlier, the cut diameter is defined as the size (diameter) of particles collected with 50% efficiency. Note that the cut diameter, $[d_p]_{cut}$, is a characteristic of the control device and should not be confused with the geometric mean particle diameter of the size distribution. A frequently used expression for cut diameter where collection efficiency is a function of the ratio of particle diameter to cut diameter is known as the Lapple cut diameter equation (method) (Copper and Alley, 1990):

$$[d_p]_{cut} = \{9\mu B/[2\pi n_t (p_p - p_g)]\} \tag{9.19}$$

where

$[d_p]_{cut}$ = cut diameter, feet (microns)
μ   = viscosity, pounds per foot-second (pascal-seconds) or (kilograms per meter-second)
B = inlet width, feet (meters)

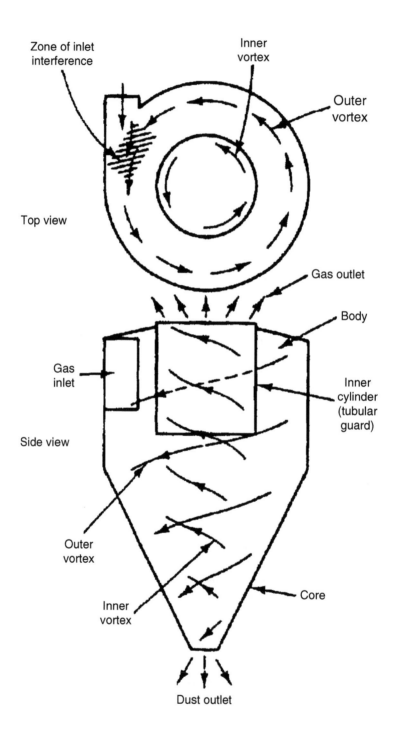

Figure 9.5 Convection reverse-flow cyclone. (From USEPA, Control Techniques for Gases and Particulates, 1971.)

n_t = effective number of turns (5 to 10 for common cyclones)
v_i = inlet gas velocity, feet per second (meters per second)
p_p = particle density, pounds per cubic foot (kilograms per cubic meter)
p_g = gas density, pounds per cubic foot (kilograms per cubic meter)

Example 9.6

Problem:
   Determine the cut size diameter and overall collection efficiency of a cyclone, given the particle size distribution of a dust from a cement kiln (USEPA-84/09, p. 66).

Given:
Gas viscosity μ = 0.02 centipoises (cP) = 0.02(6.72 × 10⁻⁴) lb/ft-sec
Specific gravity of the particle = 2.9
Inlet gas velocity to cyclone = 50 ft/sec
Effective number of turns within cyclone = 5
Cyclone diameter = 10 ft
Cyclone inlet width = 2.5 ft
Particle size distribution data are shown in Table 9.4.

Table 9.4 Particle Size Distribution Data

Average particle size in range d_p, μm	% wt
1	3
5	20
10	15
20	20
30	16
40	10
50	6
60	3
>60	7

Solution:

Step 1. Calculate the cut diameter $[d_p]_{cut}$, which is the particle collected at 50% efficiency. For cyclones:

$$[d_p]_{cut} = \{9\mu B_c/[2\pi n_{ti}(p_p - p_g)]\}\,0.5$$

where
μ = gas viscosity, pounds per foot-second
B_c = cyclone inlet width, feet
n_t = number of turns
v_i = inlet gas velocity, feet per second
p_p = particle density, pounds per cubic foot
p = gas density, pounds per cubic foot

   Determine the value of $p_p - p$. Because the particle density is much greater than the gas density, $p_p - p$ can be assumed to be p_p:

$$P_p - p = p_p = 2.9(62.4) = 180.96 \text{ lb/ft}^3$$

Calculate the cut diameter:

$$[d_p]c_{ut} = [(9)(0.02)(6.72 \times 10^{-4})(2.5)/(2\pi)5(50)(180.96)]^{0.5}$$

$$= 3.26 \times 10^{-5} \text{ ft}$$

$$= 9.94 \text{ }\mu\text{m}$$

Step 2. Complete the size efficiency table (Table 9.5) using Lapple's method (Lapple, 1951). As mentioned, this method provides the collection efficiency as a function of the ratio of particle diameter to cut diameter. Use the equation

$$\eta = 1 \quad (1.0)/[1.0 + (d_p/[d_p]_{cut})^2]$$

Table 9.5   Size Efficiency Table

d_p, μm	w_i	$d_p/[d_p]_{cut}$	η_i,%	$w_i\eta_i$,%
1	0.03	0.1	0	0
5	0.2	0.5	20	4
10	0.15	1	50	7.5
20	0.2	2	80	16
30	0.16	3	90	14.4
40	0.1	4	93	9.3
50	0.06	5	95	5.7
60	0.03	6	98	2.94
>60	0.07	—	100	7

Step 3. Determine overall collection efficiency:

$$\Sigma w_i n_i(\%) = 0 + 4 + 7.5 + 16 + 14.4 + 9.3 + 5.7 + 2.94 + 7$$

$$= 66.84\%$$

Example 9.7

Problem:

An air pollution control officer has been asked to evaluate a permit application to operate a cyclone as the only device on the ABC Stoneworks plant's gravel drier (USEPA-84/09, p. 68).

Given (design and operating data from permit application):
Average particle diameter = 7.5 μm
Total inlet loading to cyclone = 0.5 gr/ft^3 (grains per cubic foot)
Cyclone diameter = 2.0 ft
Inlet velocity = 50 ft/sec
Specific gravity of the particle = 2.75
Number of turns = 4.5 turns

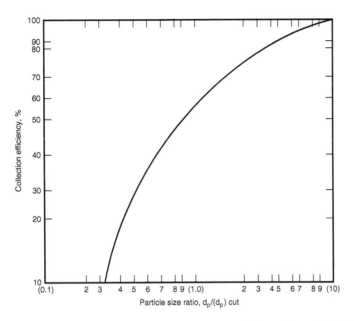

Figure 9.6   Lapple's curve: cyclone efficiency vs. particle size ratio. (Adapted from USEPA-84/09, p 70.)

Operating temperature = 70°F
Viscosity of air at operating temperature = 1.21×10^{-5} lb/ft-sec
The cyclone is a conventional one.

Air pollution control agency criteria:
Maximum total outlet loading = 0.1 gr/ft³
Cyclone efficiency as a function of particle size ratio is provided in Figure 9.6 (Lapple's curve).

Solution:

Step 1. Determine the collection efficiency of the cyclone. Use Lapple's method, which provides collection efficiency and values from a graph relating efficiency to the ratio of average particle diameter to the cut diameter. Again, the cut diameter is the particle diameter collected at 50% efficiency (see Figure 9.6). Calculate the cut diameter using the Lapple method (Equation 9.19):

$$[d_p]_{cut} = \{9\mu B_c / [2\pi n_{tvi}(p_p - p_g)]\}\,0.5$$

Determine the inlet width of the cyclone, B_c. The permit application has established this cyclone as conventional. The inlet width of a conventional cyclone is one fourth of the cyclone diameter:

$$B_c = \text{cyclone diameter}/4 = 2.0/4 = 0.5 \text{ ft}$$

Determine the value of $p_p - p$. Because the particle density is much greater than the gas density, $p_p - p$ can be assumed to be p_p:

$$p_p - p = p_p = 2.75(62.4) = 171.6 \text{ lb/ft}^3$$

Calculate the cut diameter:

$$[d_p]_{cut} = \{9\mu B_c/[2\pi n_{tvi}(p_p - p_g)]\}$$

$$= [(9)(1.21 \times 10^{-5})(0.5)/(2\pi)4.5(50)(171.6)]^{0.5}$$

$$= 4.57 \text{ microns}$$

Calculate the ratio of average particle diameter to the cut diameter:

$$d_p/[d_p]_{cut} = 7.5/4.57 = 1.64$$

Determine the collection efficiency utilizing Lapple's curve (see Figure 9.6):

$$\eta = 72\%$$

Step 2. Calculate the required collection efficiency for the approval of the permit:

$$\mu = [(\text{inlet loading} - \text{outlet loading})/(\text{inlet loading})](100)$$

$$= [(0.5 - 0.1)/(0.5)](100)$$

$$= 80\%$$

Step 3. Should the permit be approved? Because the collection efficiency of the cyclone is lower than the collection efficiency required by the agency, the permit should not be approved.

9.4.6 Electrostatic Precipitator (ESP)

Electrostatic precipitators (ESPs) have been used as effective particulate-control devices for many years. Usually used to remove small particles from moving gas streams at high collection efficiencies, ESPs are extensively used when dust emissions are less than 10 to 20 μm in size, with a predominant portion in the submicron range (USEPA-81/10, p. 7-1). Widely used in power plants for removing fly ash from the gases prior to discharge, electrostatic precipitators apply electrical force to separate particles from the gas stream. A high voltage drop is established between electrodes, and particles passing through the resulting electrical field acquire a charge. The charged particles are attracted to and collected on an oppositely charged plate, and the cleaned gas flows through the device. Periodically, the plates are cleaned by rapping to shake off the layer of dust that accumulates; the dust is collected in hoppers at the bottom of the device (see Figure 9.7). ESPs have the advantages of low operating costs; capability for operation in high-temperature applications (to 1300°F); low-pressure drop; and extremely high particulate (coarse and fine) collection efficiencies; however, they have the disadvantages of high capital costs and space requirements.

9.4.6.1 Collection Efficiency

According to USEPA-81/10, p. 7-9, ESP collection efficiency can be expressed by the following two equations:

- Migration velocity equation
- Deutsch–Anderson equation

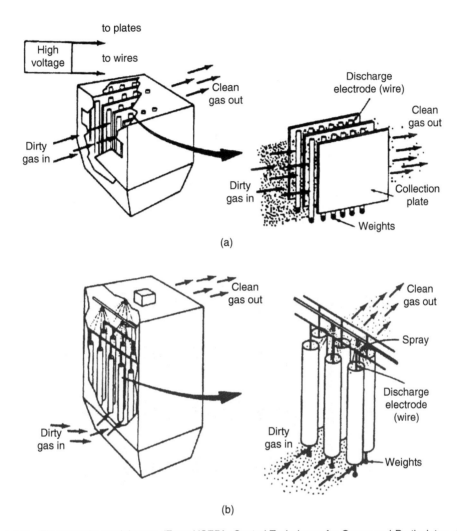

Figure 9.7   Electrostatic precipitators. (From USEPA, Control Techniques for Gases and Particulates, 1971.)

Particle migration velocity. The migration velocity *w* (sometimes referred to as the drift velocity) represents the parameter at which a group of dust particles in a specific process can be collected in a precipitator; it is usually based on empirical data. The migration velocity can be expressed in terms of:

$$w = d_p E_o E_p / (4\pi\mu) \tag{9.20}$$

where
w = migration velocity
d_p = diameter of the particle, microns
E_o = strength of field in which particles are charged, volts per meter (represented by peak voltage)
E_p = strength of field in which particles are collected, volts per meter (normally, the field close to the collecting plates)
μ = viscosity of gas, pascal-seconds

Migration velocity is quite sensitive to the voltage, as the presence of the electric field twice in Equation 9.20 demonstrates. Therefore, the precipitator must be designed using the maximum electric field for maximum collection efficiency. The migration velocity is also dependent on particle size; larger particles are collected more easily than smaller ones.

Particle migration velocity can also be determined by the following equation:

$$w = qE_p / (4\pi\mu r) \tag{9.21}$$

where
w = migration velocity
q = particle charge (charges)
E_p = strength of field in which particles are collected, volts per meter (normally, the field close to the collecting plates)
μ = viscosity of gas, pascal-seconds
r = radius of the particle, microns

Deutsch–Anderson equation. This equation or derivatives thereof are used extensively throughout the precipitator industry. Specifically, it has been used to determine the collection efficiency of the precipitator under ideal conditions. Though scientifically valid, a number of operating parameters can cause the results to be in error by a factor of two of more. Therefore, this equation should be used only for making preliminary estimates of precipitation collection efficiency. The simplest form of the equation is:

$$\eta = 1 - \exp(-wA/Q) \tag{9.22}$$

where
η = fractional collection efficiency
A = collection surface area of the plates
Q = volumetric flow rate
w = drift velocity

9.4.6.2 Precipitator Example Calculations

Example 9.8

Problem:
A horizontal parallel-plate electrostatic precipitator consists of a single duct, 24 ft high and 20 ft deep, with an 11-in. plate-to-plate spacing. Given collection efficiency at a gas flow rate of 4200 actual cubic feet per minute (acfm), determine the bulk velocity of the gas, outlet loading, and drift velocity of this electrostatic precipitator. Also calculate revised collection efficiency if the flow rate and the plate spacing are changed (USEPA-84/09, p. 71).

Given:
Inlet loading = 2.82 gr/ft^3 (grains per cubic foot)
Collection efficiency at 4200 acfm = 88.2%
Increased (new) flow rate = 5400 acfm
New plate spacing = 9 in.

Solution:

Step 1. Calculate the bulk flow (throughout) velocity *v*. The equation for calculating throughput velocity is

$$V = Q / S$$

where
Q = volumetric flow rate
S = cross-sectional area through which the gas passes

$$V = Q / S$$

$$= (4200)/[(11/12)(24)]$$

$$= 191 \text{ ft/min}$$

$$= 3.2 \text{ ft/sec}$$

Step 2. Calculate outlet loading. Remember that

$$\eta = (\text{fractional}) = (\text{inlet loading} - \text{outlet loading})/(\text{inlet loading})$$

Therefore,

$$\text{outlet loading} = (\text{inlet loading})(1 - \eta)$$

$$= (2.82)(1 - 0.882)$$

$$= 0.333 \text{ gr/ft}^3$$

Step 3. Calculate the drift velocity, which is the velocity at which the particle migrates toward the collection electrode with the electrostatic precipitator. Recall that Equation 9.22, the Deutsch–Anderson equation, describes the collection efficiency of an electrostatic precipitator:

$$\eta = 1 \ \exp(-wA/Q)$$

where
η = fractional collection efficiency
A = collection surface area of the plates
Q = volumetric flow rate
w = drift velocity

Calculate the collection surface area *A*. Remember that the particles will be collected on both sides of the plate.

$$A = (2)(24)(20) = 960 \text{ ft}^2$$

Calculate the drift velocity w. Because the collection efficiency, gas flow rate, and collection surface area are now known, the drift velocity can easily be found from the Deutsch–Anderson equation:

$$\eta = 1 - \exp(-wA/Q)$$

$$0.882 = 1 - \exp[-(960)(w)/(4200)]$$

Solving for w:

$$w = 9.36 \text{ ft/min}$$

Step 4. Calculate the revised collection efficiency when the gas volumetric flow rate is increased to 5400 cfm. Assume the drift velocity remains the same:

$$\eta = 1 - \exp(-wA/Q)$$

$$= 1 - \exp(-(960)(9.36)/(5400)]$$

$$= 0.812$$

$$= 81.2\%$$

Step 5. Does the collection efficiency change with changed plate spacing? No. Note that the Deutsch–Anderson equation does not contain a plate-spacing term.

Example 9.9

Problem:

Calculate the collection efficiency of an electrostatic precipitator containing three ducts with plates of a given size, assuming a uniform distribution of particles. Also, determine the collection efficiency if one duct is fed 50% of the gas and the other passages are fed 25% each (USEPA-84/09, p. 73).

Given:
Volumetric flow rate of contaminated gas = 4000 acfm
Operating temperature and pressure = 200°C and 1 atm
Drift velocity = 0.40 ft/sec
Size of the plate = 12 ft long and 12 ft high
Plate-to-plate spacing = 8 in.

Solution:

Step 1. What is the collection efficiency of the electrostatic precipitator with a uniform volumetric flow rate to each duct? Use the Deutsch–Anderson equation to determine the collection efficiency of the electrostatic precipitator.

$$\eta = 1 - \exp(-wA/Q)$$

Calculate the collection efficiency of the electrostatic precipitator using the Deutsch–Anderson equation. The volumetric flow rate (Q) through a passage is one third of the total volumetric flow rate:

$$Q = (4000)/(3)(60)$$

$$= 22.22 \text{ acfs}$$

$$\eta = 1 - \exp(-wA/Q)$$

$$= 1 - \exp[-(288)(0.4)/(22.22)]$$

$$= 0.9944$$

$$= 99.44\%$$

Step 2. What is the collection efficiency of the electrostatic precipitator, if one duct is fed 50% of gas and the others 25% each? The collection surface area per duct remains the same. What is the collection efficiency of the duct with 50% of gas, η_1? Calculate the volumetric flow rate of gas through the duct in actual cubic feet per second:

$$Q = (4000)/(2)(60) = 33.33 \text{ acfs}$$

Calculate the collection efficiency of the duct with 50% of gas:

$$\eta_1 = 1 - \exp[-(288)(0.4)/(33.33)]$$

$$= 0.9684$$

$$= 96.84\%$$

What is the collection efficiency (η_2) of the duct with 25% of gas flow in each? Calculate the volumetric flow rate of gas through the duct in actual cubic feet per second:

$$Q = (4000)/(4)(60) = 16.67 \text{ acfs}$$

Calculate the collection efficiency (η_2) of the duct with 25% of gas:

$$\eta_2 = 1 - \exp[-(288)(0.4)/(16.67)]$$

$$= 0.9990$$

$$= 99.90\%$$

Calculate the new overall collection efficiency. The equation becomes:

Table 9.6   Particle Size Distribution

Weight range	Average particle size d_p, µm
0–20%	3.5
20–40%	8
40–60%	13
60–80%	19
80–100%	45

$$\eta_t = (0.5)(\eta_1) + (2)(0.25)(\eta_2)$$

$$= (0.5)(96.84) + (2)(0.25)(99.90)$$

$$= 98.37\%$$

Example 9.10

Problem:

A vendor has compiled fractional efficiency curves describing the performance of a specific model of an electrostatic precipitator. Although these curves are not available, the cut diameter is known. The vendor claims that this particular model will perform with a given efficiency under particular operating conditions. Verify this claim and make certain the effluent loading does not exceed the standard set by USEPA (USEPA-84/09, p. 75).

Given:
Plate-to-plate spacing = 10 in.
Cut diameter = 0.9 µm
Collection efficiency claimed by the vendor = 98%
Inlet loading = 14 gr/ft³
USEPA standard for the outlet loading = 0.2 gr/ft³ (maximum)

The particle size distribution is given in Table 9.6.

A Deutsch–Anderson type of equation describing the collection efficiency of an electrostatic precipitator is:

$$\eta = 1 - \exp(-Kd_p)$$

where
η = fractional collection efficiency
K = empirical constant
d_p = particle diameter

Solution:

Step 1. Is the overall efficiency of the electrostatic precipitator equal to or greater than 98%? Because the weight fractions are given, collection efficiencies of each particle size are needed to calculate the overall collection efficiency.

Determine the value of K by using the given cut diameter. Because the cut diameter is known, we can solve the Deutsch–Anderson type equation directly for K.

Table 9.7 Collection Efficiency for Each Particle Size

Weight fraction w_i	Average particle size d_p, μm	η_i
0.2	3.5	0.9325
0.2	8	0.9979
0.2	13	0.9999
0.2	19	0.9999
0.2	45	0.9999

$$\eta = 1 - \exp(-kd_p)$$

$$0.5 = 1 - \exp[-K(0.9)]$$

Solving for K,

$$K = 0.77$$

Calculate the collection efficiency using the Deutsch–Anderson equation where $d_p = 3.5$:

$$\eta = 1 - \exp[(-0.77)(3.5)]$$

$$= 0.9325$$

Table 9.7 shows the collection efficiency for each particle size. Calculate the overall collection efficiency.

$$\eta = \Sigma w_i \eta_i$$

$$= (0.2)(0.9325) + (0.2)(0.9979) + (0.2)(0.9999) + (0.2)(0.9999) + (0.2)(0.9999)$$

$$= 0.9861$$

$$= 98.61\%$$

where
η = overall collection efficiency
w_i = weight fraction of the ith particle size
ηi = collection efficiency of the ith particle size

Is the overall collection efficiency greater than 98%? Yes

Step 2. Does the outlet loading meet USEPA's standard? Calculate the outlet loading in grains per cubic foot:

$$\text{Outlet loading} = (1.0 - \eta)(\text{inlet loading})$$

where η is the fractional efficiency for the preceding equation.

$$\text{Outlet loading} = [(1.0 - 0.9861)(14)]$$

$$= 0.195 \text{ gr/ft}^3$$

Is the outlet loading less than 0.2 gr/ft³? Yes.

Step 3. Is the vendor's claim verified? Yes.

9.4.7 Baghouse (Fabric) Filters

"The term *baghouse* encompasses an entire family of collectors with several types of filter bag shapes, cleaning mechanisms, and body shapes" (Heumann, 1997). Baghouse filters (fabric filters or, more properly, tube filters) are the most commonly used air pollution control filtration systems. In much the same manner as the common vacuum cleaner, fabric filter material capable of removing most particles as small as 0.5 μm and substantial quantities of particles as small as 0.1 μm is formed into cylindrical or envelope bags and suspended in the baghouse (see Figure 9.8). The particulate-laden gas stream is forced through the fabric filter, and as the air passes through the fabric, particulates accumulate on the cloth, providing a cleaned airstream. As particulates build up on the inside surfaces of the bags, the pressure drop increases. Before the pressure drop becomes

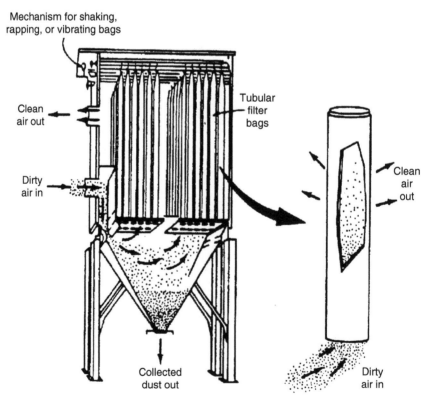

Figure 9.8 Typical simple fabric filter baghouse design. (From USEPA, Control Techniques for Gases and Particulates, 1971.)

too severe, the bags must be relieved of some of the particulate layer. The particulates are period-ically removed from the cloth by shaking or by reversing the airflow.

Fabric filters are relatively simple to operate, provide high overall collection efficiencies (up to 99%+), and are very effective in controlling submicrometer particles, but they do have limitations. These include relatively high capital costs; high maintenance requirements (bag replacement, etc.); high space requirements; and flammability hazards for some dusts.

9.4.7.1 Air-to-Filter (Media) Ratio

The air-to-filter (cloth) ratio is a measurement of the velocity (filtration velocity) of the air passing through the filter media. The ratio definition is the volume of air (expressed in cubic feet per minute or cubic meters per hour) divided by the filter media area (expressed in square feet or square meters). Generally, the smaller a particle diameter is, the more difficult it is to filter, thereby requiring a lower A/C value. The formula used to express air-to-filter (cloth) ratio (A/C ratio), filtration velocity, or face velocity (terms can be used interchangeably) is:

$$v_f = Q / A_c \tag{9.23}$$

where
v_f = filtration velocity, feet per minute (centimeters per second)
Q = volumetric air flow rate, cubic feet per minute (cubic centimeters per second)
A_c = area of cloth filter, square feet (square centimeters)

9.4.7.2 Baghouse Example Calculations

Example 9.11

Problem:
    The manufacturer does baghouse sizing. A simple check or estimate of the amount of baghouse cloth needed for a given process flow rate can be computed by using the A/C ratio equation (Equation 9.23) (USEPA-81/10, p. 8-34):

$$v_f = Q / A_c \text{ or } A_c = Q / v_f$$

For example, if the process gas exhaust rate is given as $4.72 \times 10^6 \text{cm}^3/\text{sec}$ (10,000 ft^3/min) and the filtration velocity is 4 cm/sec (A/C is 4:1 (cm^3/sec)/cm^2), the cloth area would be

$$A_c = 4.72 \times 10^6 / 4$$

$$= 118 \text{ m}^2 \text{ (cloth required)}$$

Solution:
    To determine the number of bags required in the baghouse, use the formula:

$$A_b = \pi dh$$

where

Table 9.8   Filter Bag Properties

Property	Filter bag A	Filter bag B	Filter bag C	Filter bag D
Tensile strength	Excellent	Above average	Fair	Excellent
Recommended maximum temp, °F	260	275	260	220
Resistance factor	0.9	1	0.5	0.9
Cost per bag, $	26.00	38.00	10.00	20.00
Standard size	8 in. × 16 ft	10 in. × 16 ft	1 ft × 16 ft	1 ft × 20 ft

A_b = area of bag, meters (feet)
d = bag diameter, meters (feet)
h = bag height, meters (feet)

If the bag diameter is 0.203 m (8 in.) and the bag height is 3.66 m (12 ft), the area of each bag is

$$A_b = 3.14(0.203)(3.66)$$

$$= 2.33 \text{ m}^2$$

The calculated number of bags in the baghouse is:

$$\text{Number of bags} = 118/2.33$$

$$= 51 \text{ bags}$$

Example 9.12

Problem:
    A proposal to install a pulse jet fabric filter system for cleaning an air stream containing particulate matter must be evaluated. Select the most appropriate filter bag, considering performance and cost (USEPA-84/09, p. 84).

Given:
Volumetric flow rate of polluted air stream = 10,000 scfm (60°F, 1 atm)
Operating temperature = 250°F
Concentration of pollutants = 4 gr/ft^3
Average air-to-cloth ratio (A/C ratio) = 2.5 cfm/ft^2 cloth
Collection efficiency requirement = 99%

Table 9.8 lists information given by filter bag manufacturers. Assume no bag has an advantage from the standpoint of durability under the operating conditions for which the bag is to be designed.

Solution:

    Step 1. Eliminate from consideration bags that, on the basis of given characteristics, are unsatisfactory.
        Considering the operating temperature and the bag tensile strength required for a pulse jet system:
        • Bag D is eliminated because its recommended maximum temperature (220°F) is below the operating temperature of 250°F.
        • Bag C is also eliminated because a pulse jet fabric filter system requires the tensile strength of the bag to be at least above average.

Step 2. Determine comparative costs of the remaining bags. Total cost for each bag type is the number of bags times the cost per bag. No single individual bag type is more durable than the other.

Establish the cost per bag. From the information given in Table 9.8, the cost per bag is $26.00 for Bag A and $38.00 for Bag B.

Determine number of bags, N, for each type. The number of bags required, N, is the total filtering area required, divided by the filtering area per bag.

Calculate the total filter area A_t. Calculate given flow rate to actual cubic feet per minute, Q_a.

$$Q_a = (10,000)(250 + 460)/(60 + 460) = 13,654 \text{ acfm}$$

Establish filtering capacity v_f. This is given. The A/C ratio, expressed in cubic feet per minute per square foot, is the same as the filtering velocity, which is given previously as 2.5 cfm/ft² cloth. From the information given in Table 9.8, the filtering velocity is:

$$v_f = 2.5 \text{ cfm/ft}^2$$

$$= 2.5 \text{ ft/min}$$

Calculate the total filtering cloth area, A_c, from the actual cubic feet per minute and filtering velocity determined before:

$$A_c = Q_a/v_f = 13,654/2.5 = 5461.6 \text{ ft}^2$$

Calculate the filtering area per bag. Bags are assumed to be cylindrical; the bag area is $A = \pi Dh$, where D = bag diameter and h = bag length:

$$\text{For bag A: } A = \pi Dh = \pi(8/12)(16) = 33.5 \text{ ft}^2$$

$$\text{For bag B: } A = \pi Dh = \pi(10/12)(16) = 41.9 \text{ ft}^2$$

Determine the number of bags required, N. N = (filtering cloth area of each bag A_c)/(bag area A):

$$\text{For bag A: } N = A_c / A = 5461.6/33.5 = 163$$

$$\text{For bag B: } N = 5461.6/41.9 = 130$$

Determine the total cost for each bag:

$$\text{For bag A: total cost} = (N)(\text{cost per bag}) = (163)(26.00) = \$4238$$

$$\text{For bag B: total cost} = (130)(38.00) = \$4940$$

Step 3. Select the most appropriate filter bag, considering the performance and cost. Because the total cost for bag A is less than for bag B, select bag A.

Example 9.13

Problem:

Determine the number of filtering bags required and cleaning frequency for a plant equipped with a fabric filter system. Operating and design data are given below (USEPA-84/09, p. 86).

Given:
Volumetric flow rate of the gas stream = 50,000 acfm
Dust concentration = 5.0 gr/ft^3
Efficiency of the fabric filter system = 98.0%
Filtration velocity = 10 ft/min
Diameter of filtering bag = 1.0 ft
Length of filtering bag = 15 ft

The system is designed to begin cleaning when the pressure drop reaches 8.9 in. of water. The pressure drop is given by:

$$\Delta p = 0.2v_f + 5c(v_f)^2 t$$

where
Δp = pressure drop, inches H_2O
v_f = filtration velocity, feet per minute
c = dust concentration, pounds per cubic foot
t = time since the bags were cleaned, minutes

Solution:

Step 1. What is the number of bags N needed? To calculate N, we need the total required surface area of the bags and the surface area of each bag.

Calculate the total required surface area of the bags A_c in square feet

$$A_c = q/v_f$$

where
A_c = total surface area of the bags
q = volumetric flow rate
v_f = filtering velocity

$$A_c = q/v_f$$

$$= 50,000/10$$

$$= 5000 \text{ ft}^2$$

Calculate the surface area of each bag A, in square feet:

$$A = \pi Dh$$

where
A = surface area of a bag
D = diameter of the bag
h = length of the bag

$$A = \pi Dh$$

$$= \pi(1.0)(15)$$

$$= 47.12 \text{ ft}^2$$

Calculate the required number of bags N:

$$N = A_c / A$$

$$= 5000/47.12$$

$$= 106$$

Step 2. Calculate the required cleaning frequency:

$$\Delta p = 0.2 v_f + 5c(vf)^2 t$$

Because Δp is given as 8.0 in. H_2O, the time since the bags were cleaned is calculated by solving the preceding equation:

$$5.0 \text{ gr/ft}^3 = 0.0007143 \text{ lb/ft}^3 \text{ and } \Delta p = 0.2 v_f + 5c(v_t)^2 t$$

$$8.0 = (0.2)(10) + (5)(0.0007143)(10)^2 t$$

Solving for t,

$$t = 16.8 \text{ min}$$

Example 9.14

Problem:
    An installed baghouse is presently treating a contaminated gas stream. Suddenly, some of the bags break. We must now estimate this baghouse system's new outlet loading (USEPA-84/09, p. 88).

Given:
Operation conditions of the system = 60°F, 1 atm
Inlet loading = 4.0 gr/acf
Outlet loading before bag failure = 0.02 gr/acf
Volumetric flow rate of contaminated gas = 50,000 acfm

Number of compartments = 6
Number of bags per compartment = 100
Bag diameter = 6 in.
Pressure drop across the system = 6 in. H$_2$O
Number of broken bags = 2 bags

Assume that all the contaminated gas emitted through the broken bags is the same as that passing through the tube sheet thimble.

Solution:

Step 1. Calculate the collection efficiency and penetration before the bag failures. Collection efficiency is a measure of a control device's degree of performance; it specifically refers to degree of removal of pollutants. *Loading* refers to the concentration of pollutants, usually in grains of pollutants per cubic foot of contaminated gas streams. Mathematically, the collection efficiency is defined as:

$$\eta = [(\text{inlet loading} - \text{outlet loading})/(\text{inlet loading})](100)$$

From the preceding equation, the collected amount of pollutants by a control unit is the product of collection efficiency η and inlet loading. The inlet loading minus the amount collected gives the amount discharged to the atmosphere.

Another term used to describe the performance or collection efficiency of control devices is penetration P_t. USEPA-84/09, p. 3.3 gives it as:

$$P_t = 1 - \eta / 100 \quad \text{(fractional basis)}$$

$$P_t = 100 - \eta \quad \text{(percent basis)}$$

The following equation describes the effect of bag failure on baghouse efficiency:

$$P_{t1} = P_{t2} + P_{tc}$$

$$Ptc = 0.582(\Delta p)^{0.5}/\phi$$

$$\phi = Q/(LD^2(Y + 460)^{0.5})$$

where
P_{t1} = penetration after bag failure
P_{t2} = penetration before bag failure
P_{tc} = penetration correction term, contribution of broken bags to P_{t1}
Δp = pressure drop, inches H$_2$O
ϕ = dimensionless parameter
Q = volumetric flow rate of contaminated gas, actual cubic feet per minute
L = number of broken bags
D = bag diameter, inches
T = temperature, degrees Fahrenheit

Collection efficiency η is:

$$\eta = (\text{inlet loading} - \text{outlet loading})/(\text{inlet loading})$$

$$= (4.0 - 0.02)/(4.0)$$

$$= 0.005$$

$$= 99.5\%$$

Penetration is:

$$P_t = 1.0 - \eta$$

$$= 0.005$$

Step 2. Calculate the bag failure parameter ϕ, a dimensionless number:

$$\phi = Q/(LD^2(T + 460)^{0.5})$$

$$= 50,000/(2)(6)^2(60 + 460)^{0.5}$$

$$= 30.45$$

Step 3. Calculate the penetration correction P_{tc}; this determines penetration from bag failure:

$$P_{tc} = 0.582(\Delta p)^{0.5}/\phi$$

$$= (0.582)(6)^{0.5}/30.45$$

$$= 0.0468$$

Step 4. Calculate the penetration and efficiency after the two bags failed. Use the results of steps 1 and 3 to calculate P_{t1}:

$$P_{t1} = P_{t2} + P_{tc}$$

$$= 0.005 + 0.0468$$

$$= 0.0518$$

$$\eta^* = 1 - 0.0518$$

$$= 0.948$$

Step 5. Calculate the new outlet loading after the bag failures. Relate inlet loading and new outlet loading to the revised efficiency or penetration:

$$\text{New outlet loading} = (\text{inlet loading})P_{t1}$$

$$= (4.0)(0.0518)$$

$$= 0.207 \text{ gr/acf}$$

Example 9.15

Problem:

A plant emits 50,000 acfm of gas containing a dust loading of 2.0 gr/ft³. A particulate control device is employed for particle capture; the dust captured from the unit is worth $0.01/lb. Determine the collection efficiency for which the cost of power equals the value of the recovered material. Also determine the pressure drop in inches of H_2O at this condition (USEPA-84/09, p. 122).

Given:
Overall fan efficiency = 55%
Electric power cost = $0.06/kWh

For this control device, assume that the collection efficiency is related to the system pressure drop Δp through the equation:

$$\eta = \Delta p/(\Delta p + 5.0)$$

where
Δp = pressure drop, pounds per square foot
η = fractional collection efficiency

Solution:

Step 1. Express the value of the dust collected in terms of collection efficiency η:

$$\text{Amount of dust collected} = (Q)(\text{inlet loading})(\eta)$$

Note that the collected dust contains 7000 grains per pound.

$$\text{The value of dust collected} = 50,000(\text{ft}^3/\text{min})2(\text{gr/ft}^3)(1/7000)(\text{lb/gr}) \times 0.01(\$/\text{lb})\eta$$

$$= 0.143 \, \eta \, \$/\text{min}$$

Step 2. Express the value of the dust collected in terms of pressure drop Δp. Recall that $\eta = \Delta p/(\Delta p + 5.0)$.

$$\text{The value of dust collected} = 0.143[\Delta p/(\Delta p + 5.0)]\$/\text{min}$$

Step 3. Express the cost of power in terms of pressure drop Δp:

$$\text{Bhp} = Q\Delta p/\eta' = \text{brake horsepower}$$

where
η' = fan efficiency
Δp = pressure drop, pounds$_f$ per square foot
Q = volumetric flow rate

$$\text{Cost of power} = \Delta p(\text{lb}_f/\text{ft}^2)(50,000)[(\text{ft}^3/\text{min})(1/44,200)(\text{kW-min/ft-lb}_f)(1/0.55)$$

$$\times (0.06)(\$/\text{kWh})(1/60)(\text{h/min})]$$

$$= 0.002\Delta p \ \$/\text{min}$$

Step 4. Set the cost of power equal to the value of dust collected and solve for Δp in pounds$_f$ per square foot. This represents breakeven operation. Then, convert this pressure drop to inches of H_2O. To convert from pounds$_f$ per square foot to inches of H_2O, divide by 5.2.

$$(0.143)\Delta p/(\Delta p + 5) = 0.002\Delta p$$

Solving for Δp:

$$\Delta p = 66.5 \ \text{lb}_f/\text{ft}^2 = 12.8 \ \text{in. } H_2O$$

Step 5. Calculate the collection efficiency using the calculated value of Δp.

$$\eta = 66.5/(66.5 + 5) = 0.93$$

$$= 93.0\%$$

Example 9.16

Problem:
Determine capital, operating, and maintenance costs on an annualized basis for a textile dye and finishing plant (with two coal-fired stoker boilers), where a baghouse is employed for particulate control. Use the given operating, design, and economic factors (USEPA-84/09, p. 123).

Given:
Exhaust volumetric flow from two boilers = 70,000 acfm
Overall fan efficiency = 60%
Operating time = 6240 h/yr
Surface area of each bag = 12.0 ft^2
Bag type = Teflon® felt
Air-to-cloth ratio = 5.81 acfm/ft^2
Total pressure drop across the system = 17.16 lb$_f$/ft^2
Cost of each bag = $75.00
Installed capital costs = $2.536/acfm

Cost of electrical energy = $0.03/kWh
Yearly maintenance cost = $5000 plus yearly cost to replace 25% of the bags
Salvage value = 0
Interest rate (i) = 8%
Lifetime of baghouse (m) = 15 yr
Annual installed capital cost (AICC) = (installed capital cost) $\{i(1 + i)^m/[(1 + i)^m - 1]\}$

Solution:

Step 1. What is the annual maintenance cost? Calculate the number of bags N:

$$N = Q/(\text{air-to-cloth ratio})(A)$$

where
Q = total exhaust volumetric flow rate
A = surface area of a bag

$$N = Q/(\text{air-to-cloth ratio})(A)$$

$$= (70,000)/(5.81)(12)$$

$$= 1004 \text{ bags}$$

Calculate the annual maintenance cost in dollars per year.

Annual maintenance cost = $5000/year + cost of replacing 25% of the bags each year

$$= \$5000 + (0.25)(1004)(75.00)$$

$$= \$23,825/\text{year}$$

Step 2. What is the annualized installed cost (AICC)? Calculate the installed capital cost in dollars:

Installed capital cost = $(Q)(\$2.536/\text{acfm})$

$$= (70,000)(2.536)$$

$$= \$177,520$$

Calculate the AICC using the equation given previously:

AICC = (installed capital cost) $\{i(1 + i)^m/[(1 + i)^m - 1]\}$

$$= (177,520)\{0.08(1 + 0.08)^{15}/[(1 + 0.08)^{15} - 1]\}$$

$$= \$20,740/\text{yr}$$

Step 3. Calculate the operating cost in dollars per year:

$$\text{Operating cost} = Q\Delta p(\text{operating time})(0.03/\text{kWh}/E)$$

Because 1 ft-lb/sec = 0.0013558 kW,

$$\text{Operating cost} = (70,000/60)(17.16)(6240)(0.03)(0.0012558)/0.6$$

$$= \$8470/\text{yr}$$

Step 4. Calculate the total annualized cost in dollars per year:

$$\text{Total annualized cost} = (\text{maintenance cost}) + \text{AICC} + (\text{operating cost})$$

$$= 23,825 + 20,740 + 8470$$

$$= \$53,035/\text{yr}$$

REFERENCES

Baron, P.A. and Willeke, K. (1993). Gas and particle motion, in K. Willeke and P.A. Baron (Eds.), *Aerosol Measurement: Principles, Techniques and Applications*. New York: Van Nostrand Reinhold.

Cooper, C.D. and Alley, F.C. (1990). *Air Pollution Control*. Philadelphia: Waveland Press.

Hesketh, H.E. (1991). *Air Pollution Control: Traditional and Hazardous Pollutants*. Lancaster, PA: Technomic Publishing Company, Inc.

Heumann, W.L. (1997). *Industrial Air Pollution Control Systems*. New York: McGraw-Hill.

Hinds, W.C. (1986). Aerosol Technology: Properties, Behavior, and Measurement of Airborne Particulates. New York: John Wiley & Sons.

Lapple, C.E. (1951). *Fluid and Particle Mechanics*. Newark, DE: University of Delaware.

Spellman, F.R. (1999). *The Science of Air: Concepts & Applications*. Lancaster, PA: Technomic Publishing Company, Inc.

Strauss, W. (1975). *Industrial Gas Cleaning*, 2nd ed. Oxford: Pergamon Press.

USEPA (1969). Control Techniques for Particulate Air Pollutants._

USEPA (1971). Control Techniques for Gases and Particulates.

USEPA-81/10 (1984). Control of particulate emissions, course 413, USEPA Air Pollution Training Institute, EPA450-2-80-066.

USEPA-84/09 (1984). Control of gaseous and particulate emissions, Course SI:412D, USEPA Air Pollution Training Institute (APTI), EPA450-2-84-007.

USEPA-84/03 (1984). Wet scrubber plan review, Course SI:412C, USEPA Air Pollution Training Institute (APTI), EPA-450-2-82-020.

CHAPTER 10

Wet Scrubbers for Emission Control

I durst not laugh for fear of opening my lips and receiving the bad air.

William Shakespeare

10.1 INTRODUCTION

How do scrubbers work? To answer this question, we need only look to a critical part of Earth's natural pollution control system: how rain cleans the lower atmosphere. Obviously, this is most evidenced by the freshness of the air following a rainstorm. The simplicity of spraying water into a gas stream to remove a relatively high percentage of contaminants has contributed to scrubbers' extensive use within industry since the early 1900s. Heumann and Subramania (1997) point out that "most pollution control problems are solved by the selection of equipment based upon two simple questions … (1) will the equipment meet the pollution control requirements? and (2) which selection will cost the least?"

How are scrubbing systems capabilities evaluated? They are evaluated based on empirical relationships, theoretical models, and pilot scale test data. Two important parameters in the design and operation of wet scrubbing systems as a function of the process being controlled are dust properties and exhaust gas characteristics. Particle size distribution is the most critical parameter in choosing the most effective scrubber design and determining the overall collection efficiency.

In operation, scrubbers are considered universal control devices because they can control particulate and/or gaseous contaminants. Numerous types of scrubbers are available, including wet scrubbers, wet–dry scrubbers, and dry–dry scrubbers. Scrubbers use chemicals to accomplish contaminant removal, whereby the gaseous contaminants are absorbed or converted to particles, then wasted or removed from the stream.

In this chapter, our focus is on the calculations used for wet scrubber systems. Much of the information is excerpted from Spellman (1999); USEPA-84/03; USEPA-84/09; and USEPA-81/10.

10.1.1 Wet Scrubbers

Wet scrubbers (or collectors) have found widespread use in cleaning contaminated gas streams (acid mists, foundry dust emissions, and furnace fumes, for example) because of their ability to remove particulate and gaseous contaminants effectively. These types of scrubbers vary in complexity from simple spray chambers used to remove coarse particles to high-efficiency systems (Venturi types) that remove fine particles.

Whichever system is used, operation employs the same basic principles of inertial impingement or impaction, and interception of dust particles by droplets of water. The larger, heavier water

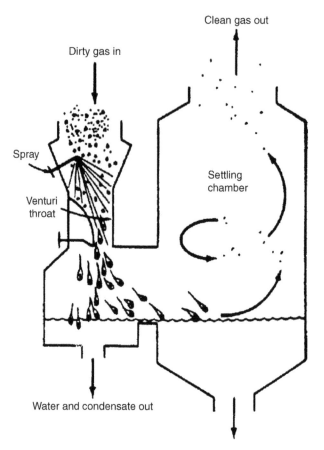

Figure 10.1 Venturi wet scrubber. (From USEPA, Control Techniques for Gases and Particulates, 1971).

droplets are easily separated from the gas by gravity. The solid particles can then be independently separated from the water, or the water can be otherwise treated before reuse or discharge. Increasing the gas velocity or the liquid droplet velocity in a scrubber increases the efficiency because of the greater number of collisions per unit time. For the ultimate in wet scrubbing, where high collection efficiency is desired, Venturi scrubbers are used. These scrubbers operate at extremely high gas and liquid velocities with a very high-pressure drop across the Venturi throat. Venturi scrubbers (see Figure 10.1) are most efficient for removing particulate matter in the size range of 0.5 to 5 μm, which makes them especially effective for the removal of submicron particulates associated with smoke and fumes.

Although wet scrubbers require relatively small space for installation, have low capital costs, and can handle high-temperature, high-humidity gas streams, their power and maintenance costs are relatively high. They may also create water disposal problems, their corrosion problems are more severe than dry systems, and their final product is collected wet (Spellman, 1999).

10.2 WET SCRUBBER COLLECTION MECHANISMS AND EFFICIENCY (PARTICULATES)

As mentioned, wet scrubbers capture relatively small dust particles with large liquid droplets. This is accomplished by generating easily collected large particles by combining liquid droplets with relatively small dust particles. In this process, the dust particles are grown into larger particles by several methods. These include impaction; diffusion; direct interception; electrostatic attraction; condensation; centrifugal force; and gravity.

10.2.1 Collection Efficiency

Collection efficiency is commonly expressed in terms of penetration, which is defined as the fraction of particles in the exhaust system that passes through the scrubber uncollected. Simply put, penetration is the opposite of the fraction of particles collected (USEPA-84/03, p. 9-3). It is expressed as:

$$P_t = 1 - \eta \qquad (10.1)$$

where
P_t = penetration
η = collection efficiency (expressed as a fraction)

Wet scrubbers usually have an efficiency curve that fits the relationship of:

$$\eta = 1 - e^{\{-[f(system)]\}} \qquad (10.2)$$

where
η = collection efficiency (expressed as a fraction)
e = exponential function
f(system) = some function of the scrubbing system variables

Substituting for efficiency, penetration can be expressed as:

$$P_t = 1 - \eta$$

$$= 1 - [1 - e^{\{-[f(system)]\}}]$$

$$= e^{\{-[f(system)]\}}$$

10.2.2 Impaction

In a wet scrubbing system, particulate matter tends to follow the streamlines of the exhaust system. When liquid droplets are introduced into the exhaust stream, however, particulate matter cannot always follow these streamlines as they diverge around the droplet. Instead, because of the particle's mass, the particles break away from the streamlines and impact on the droplet. When gas stream velocity exceeds 0.3 m/sec (1 ft/sec), impaction is the predominant collection mechanism. Most scrubbers do operate with gas stream velocities well above 0.3 m/sec, allowing particles with diameters greater than 1.0 μm to be collected by this mechanism (USEPA-84/03, p. 1-4). Impaction increases as the velocity of the particles in the exhaust stream increases relative to the liquid droplet's velocity. In addition, as the size of the liquid droplet decreases, impaction also increases.

In impaction, after the design, the key parameter is known as the impaction parameter ψ (USEPA-81/10, p. 9-5) and is expressed by:

$$\psi = C_f p_p v(d_p)^2 / 18 d_d \mu \qquad (10.3)$$

where

p_p = particle density
v = gas velocity at Venturi throat, feet per second
d_p = particle diameter, feet
d_d = droplet diameter, feet
μ = gas viscosity, pounds per foot-second
C_f = Cunningham correction factor

The collection efficiency associated with this impaction effect (USEPA-81/10, p. 9-7) is expressed as:

$$\eta_{impaction} = f(\psi)$$

10.2.3 Interception

If a small particle is moving around an obstruction (a water droplet, for example) in the flow stream, it may come in contact with (be intercepted by) that object because of the particle's physical size. This interception of particles on the collector usually occurs on the sides before reaching the top or bottom. Because the center of a particle follows the streamlines around the droplet, collusion occurs if the distance between the particle and droplet is less than the radius of the particle. Collection of particles by interception results in an increase in overall collection efficiency.

This effect is characterized by the separation number, which is the ratio of the particle diameter to the droplet diameter (USEPA-81/10, p. 9-8) and is expressed as:

$$d_p/d_d$$

where

d_p = particle diameter, feet
d_d = droplet diameter, feet

The collection efficiency associated with this interception effect as a function is d_p and d_d or:

$$\eta_{interception} = f(d_p/d_d)$$

10.2.4 Diffusion

Very small particles with aerodynamic particle diameters of less than 0.1 μm are primarily collected by Brownian diffusion because they have little inertial impaction (bumping) due to their small mass. Interception is limited by their reduced physical size. This bumping causes them to move first one way and then another randomly (diffused) through the gas. The Brownian diffusion process leading to particle capture is most often described by a dimensionless parameter called the Peclet number, Pe (USEPA-81/10, p. 9-8):

$$Pe = 3\pi\mu v d_p d_d/(C_f k_B T) \tag{10.4}$$

where

Pe = Peclet number
μ = gas viscosity

v = gas stream velocity
d_p = particle diameter
d_d = droplet diameter
C_f = Cunningham correction factor
k_B = Boltzmann's constant
T = temperature of gas stream

Equation 10.4 shows that, as temperature increases, Pe decreases. As temperature increases, gas molecules move around faster than they do at lower temperatures. This action leads to increased bumping of the small particles, increased random motion, and increased collection efficiency by this mechanism. Collection efficiency by the diffusion process is generally expressed as:

$$\eta_{\text{diffusion}} = f(1/Pe) \tag{10.5}$$

Equation 10.5 shows that as the Peclet number decreases, collection efficiency by diffusion increases.

10.2.5 Calculation of Venturi Scrubber Efficiency

Several models are available for the calculation of Venturi particle collection efficiency. USEPA-84/03, p. 9-1, and USEPA-81/10, p. 9-1, list the following models:

- Johnstone equation
- Infinite throat model
- Cut power method
- Contact power theory
- Pressure drop

10.2.5.1 Johnstone Equation

The collection efficiency for liquid Venturi scrubbers (considering only the predominant mechanism of inertial impaction) is often determined with an equation from Johnstone, given as:

$$\eta = 1 - \exp[-k(Q_L/Q_G)\sqrt{\psi}]$$

where
η = fractional collection efficiency
k = correlation coefficient whose value depends on the system geometry and operating conditions, typically 0.1 to 0.2, 1000 acf/gal
Q_L/Q_G = liquid-to-gas ratio, gal/1000 acf
ψ = $C_f p_p v(d_p)^2/18 d_d \mu$ = inertial impaction parameter
p_p = particle density
v = gas velocity at Venturi throat, feet per second
d_p = particle diameter, feet
d_d = droplet diameter, feet
μ = gas viscosity, pounds per foot-second
C_f = Cunningham correction factor

10.2.5.2 Infinite Throat Model

Another method for predicting particle collection efficiency in a Venturi scrubber is the infinite throat model (Yung et al., 1977). This model is a refined version of the Calvert correlation given in the Wet Scrubber System Study (Calvert et al., 1972). The equations presented in the infinite throat model assume that the water in the throat section of the Venturi captures all particles. Two studies found that this method correlated very well with actual Venturi scrubber operating data (Yung et al., 1977; Calvert et al., 1972).

The equations listed in the model can be used to predict the penetration (P_t) for one particle size (diameter) or for the overall penetration (P_t^*), which is obtained by integrating the entire particle-size distribution. The equations are provided below (USEPA-84/03, p. 9-4):

$$\ln P_t(d_p) = -B\{[4K_{po} + 4.2 - 5.02(K_{po})^{0.5}(1 + 0.7/K_{po})\tan^{-1}(K_{po}/0.7)^{0.5}]/(Kpo + 0.7)\} \quad (10.6)$$

where
$P_t(d_p)$ = penetration for one particle size
B = parameter characterizing the liquid-to-gas ratio, dimensionless
K_{po} = inertial parameter at throat entrance, dimensionless

Note: Equation 10.6 was developed assuming that the Venturi has an infinite-sized throat length (l). This is valid only when l in the following equation is greater than 2.0.

$$l = (3l_t C_D p_g)/(2d_d p_t)$$

where
l = throat length parameter, dimensionless
l_t = Venturi throat length, centimeters
C_D = drag coefficient for the liquid at the throat entrance, dimensionless
p_g = gas density, grams per cubic centimeter
p_t = liquid density, grams per cubic centimeter
d_d = droplet diameter, centimeters

The following equation is known as the Nukiyama and Tanasawa (1983) equation.

$$d_d = 50/v_{gt} + 91.8(L/G)^{1.5} \quad (10.7)$$

where
d_d = droplet diameter, centimeters
v_{gt} = gas velocity in the throat, centimeters per second
L/G = liquid-to-gas ratio, dimensionless

The parameter characterizing the liquid-to-gas ratio (B) can be calculated using Equation 10.8.

$$B = (L/G)p_l/(p_g C_D) \quad (10.8)$$

where
B = parameter characterizing the liquid-to-gas ratio, dimensionless
L/G = liquid-to-gas ratio, dimensionless

p_g = gas density, grams per cubic centimeter
p_l = liquid density, grams per cubic centimeter
C_D = drag coefficient for the liquid at the throat entrance, dimensionless

The inertial parameter at the throat entrance is calculated in Equation 10.9:

$$K_{po} = (d_p)^2 v_{gt}/(9\mu_g d_d) \tag{10.9}$$

where
K_{po} = inertial parameter at throat entrance, dimensionless
d_p = particle aerodynamic resistance diameter, centimeters
v_{gt} = gas velocity in the throat, centimeters per second
μ_g = gas viscosity, grams per second-centimeter
d_d = droplet diameter, centimeters

The value for K_{pg} is calculated using Equation 10.10:

$$K_{pg} = (d_{pg})^2 v_{gt}/(9\mu_g d_d) \tag{10.10}$$

where
K_{pg} = inertial parameter for mass-median diameter, dimensionless
d_{pg} = particle aerodynamic geometric mean diameter, centimeters
v_{gt} = gas velocity in the throat, centimeters per second
μ_g = gas velocity, grams per second-centimeter
d_d = droplet diameter, centimeters

The value for C_D is calculated using Equation 10.11:

$$C_D = 0.22 + (24/N_{Reo})[1 + 0.15(N_{Reo})^{0.6}] \tag{10.11}$$

where
C_D = drag coefficient for the liquid at the throat entrance, dimensionless
N_{Reo} = Reynolds number for the liquid droplet at the throat inlet, dimensionless

The Reynolds number is determined in Equation 10.12:

$$N_{Reo} = v_{gt} d_d / v_g \tag{10.12}$$

where
N_{Reo} = Reynolds number for the liquid droplet at the throat inlet, dimensionless
v_{gt} = gas velocity in the throat, centimeters per second
d_d = droplet diameter, centimeters
v_g = gas kinematic viscosity, square centimeters per second

The value for d_{pg} is calculated by using Equation 10.13:

$$d_{pg} = d_{ps}(C_f \times p_p)^{0.5} \tag{10.13}$$

where

d_{pg} = particle aerodynamic geometric mean diameter, micrometers A (where A$[=](g/cm^3)^{0.5}$, where
　　　[=] means "has units of")
d_{ps} = particle physical, or Stokes diameter, micrometers
C_f = Cunningham slip correction factor, dimensionless
p_p = particle density, grams per cubic centimeter

The Cunningham slip correction factor (C_f) can be found by solving Equation 10.14:

$$C_f = 1 + [(6.21 \times 10^{-4})T]/d_{ps} \qquad (10.14)$$

where

C_f = Cunningham slip correction factor, dimensionless
T　= absolute temperature, Kelvin
d_{ps} = particle physical, or Stokes diameter, micrometers

Example 10.1 illustrates how to use the infinite throat model to predict the performance of a Venturi scrubber. When using the equations given in the model, make sure that the units for each equation are consistent.

Example 10.1

Problem:

Cheeps Disposal Inc. plans to install a hazardous-waste incinerator to burn liquid and solid waste materials. The exhaust gas from the incinerator will pass through a quench spray, then into a Venturi scrubber, and finally though a packed bed scrubber. Caustic added to the scrubbing liquor will remove any HCl from the flue gas and will control the pH of the scrubbing liquor. The uncontrolled particulate emissions leaving the incinerator are estimated to be 1100 kg/h (maximum average). Local air pollution regulations state that particulate emissions must not exceed 10 kg/h. Using the following data, estimate the particulate collection efficiency of the Venturi scrubber (USEPA-84/03, p. 9-8).

Given:
Mass-median particle size (physical) d_{ps} = 9.0 μm
Geometric standard deviation σ_{gm} = 2.5
Particle density p_p = 1.9 g/cm^3
Gas viscosity μ_g = 2.0 × 10^{-4} g/cm-sec
Gas kinematic viscosity v_g = 0.2 cm^2/sec
Gas density p_g = 1.0 kg/m^3
Gas flow rate Q_G = 15 m^3/sec
Gas velocity in Venturi throat v_{gt} = 9000 cm/sec
Gas temperature (in Venturi) T_g = 80°C
Water temperature T_l = 30°C
Liquid density p_l = 1000 kg/m^3
Liquid flow rate Q_L = 0.014 m^3/sec
Liquid-to-gas ratio L/G = 0.0009 L/m^3

Solution:

Step 1. Calculate the Cunningham slip correction factor. The mass-median particle size (physical) d_{ps} is 9.0 μm. Because the particle aerodynamic geometric mean diameter d_{pg} is not known, we must use Equation 10.13 to calculate d_{pg}, and Equation 10.14 to calculate the Cunningham slip correction factor C_f. From Equation 10.14:

$$C_f = 1 + [(6.21 \times 10^{-4})T]/d_{ps}$$

$$= 1 + [(6.21 \times 10^{-4})(273 + 80]/9$$

$$= 1.024$$

From Equation 10.13

$$d_{pg} = d_{ps}(C_f \times p_p)^{0.5}$$

$$= 9 \, \mu m(1.024 \times 1.9 \, g/cm^3)^{0.5}$$

$$= 12.6 \, \mu mA$$

$$= 12.6 \times 10^{-4} \, cmA$$

where $A[=](g/cm^3)^{0.5}$

Note: If the particle diameter is given as the aerodynamic geometric mean diameter d_{pg} and expressed in units of μmA, this step is not required.

Step 2. Calculate the droplet diameter d_d from Equation (10.7) (Nukiyama and Tanasawa equation):

$$d_d = 50/v_{gr} + 91.8(L/G)^{1.5}$$

where
d_d = droplet diameter, centimeters
v_{gr} = gas velocity in the throat, centimeters per second
L/G = liquid-to-gas ratio, dimensionless

$$d_d = 50/(9000 \, cm/sec) + 91.8(0.0009)^{1.5}$$

$$= 0.00080 \, cm$$

Step 3. Calculate the inertial parameter for the mass-median diameter K_{pg}, using Equation 10.10:

$$K_{pg} = (d_{pg})^2 v_{gt}/(9\mu_g d_d)$$

where

K_{pg} = inertial parameter for mass-median diameter, dimensionless
d_{pg} = particle aerodynamic geometric mean diameter, centimeters
v_{gt} = gas velocity in the throat, centimeters per second
μ_g = gas velocity, grams per second-centimeter
d_d = droplet diameter, centimeters

$$K_{pg} = (12.6 \times 10^{-4}\text{cm})^2(9000 \text{ cm/sec})/\{[9(2.0 \times 10^{-4} \text{ (g/cm-sec)}(0.008 \text{ cm})]\}$$

$$= 992$$

Step 4. Calculate the Reynolds number N_{Reo}, using Equation 10.12:

$$N_{Reo} = v_{gt}d_d/v_g$$

where

N_{REO} = Reynolds number for the liquid droplet at the throat inlet, dimensionless
v_{gt} = gas velocity in the throat, centimeters per second
d_d = droplet diameter, centimeters
v_g = gas kinematic viscosity, square centimeters per second

$$N_{Reo} = v_{gt}d_d/v_g$$

$$= (9000 \text{ cm/sec})(0.008 \text{ cm})(0.2 \text{ cm}^2/\text{sec})$$

$$= 360$$

Step 5. Calculate the drag coefficient for the liquid at the throat entrance C_D, using Equation 10.11.

$$C_D = 0.22 + (24/N_{Reo})[1 + 0.15(N_{Reo})^{0.6}]$$

where

C_D = drag coefficient for the liquid at the throat entrance, dimensionless
N_{Reo} = Reynolds number for the liquid droplet at the throat inlet, dimensionless

$$C_D = 0.22 + (24/N_{Reo})[1 + 0.15(N_{Reo})^{0.6}]$$

$$= 0.22 + (24/360)[(1 + 0.15(360)^{0.6}]$$

$$= 0.628$$

Step 6. Now, calculate the parameter characterizing the liquid-to-gas ratio B, using Equation 10.8:

$$B = (L/G)p_l/(p_gC_D)$$

where

B	=	parameter characterizing the liquid-to-gas ratio, dimensionless
L/G	=	liquid-to-gas ratio, dimensionless
p_g	=	gas density, grams per cubic centimeter
p_l	=	liquid density, grams per cubic centimeter
C_D	=	drag coefficient for the liquid at the throat entrance, dimensionless

$$B = (L/G)p_l/(p_g C_D)$$

$$= (0.0009)(1000 \text{ kg/m}^3)/(1.0 \text{ kg/m}^3)(0.628)$$

$$= 1.43$$

Step 7. The geometric standard deviation σ_{gm} is 2.5. The overall penetration P_t* is 0.008.

Step 8. The collection efficiency can be calculated using the equation:

$$\eta = 1 - P_t*$$

$$= 1 - 0.008$$

$$= 0.992$$

$$= 99.2\%$$

Step 9. Determine whether the local regulations for particulate emissions are being met. The local regulations state that the particulate emissions cannot exceed 10 kg/h. The required collection efficiency is calculated by using the equation:

$$\eta_{required} = (\text{dust}_{in} - \text{dust}_{out})/\text{dust}_{in}$$

where

dust$_{in}$ = dust concentration leading into the Venturi
dust$_{out}$ = dust concentration leaving the Venturi

$$\eta_{required} = (1100 \text{ kg/h} - 10 \text{ kg/h})/1100 \text{ kg/h}$$

$$= 0.991$$

$$= 99.1\%$$

The estimated efficiency of the Venturi scrubber is slightly higher than the required efficiency.

10.2.5.3 Cut Power Method

The cut power method is an empirical correlation used to predict the collection efficiency of a scrubber. In this method, penetration is a function of the cut diameter of the particles to be collected by the scrubber. Recall that the cut diameter is the diameter of the particles collected by the scrubber with at least 50% efficiency. Because scrubbers have limits to the size of particles they can collect, knowledge of the cut diameter is useful in evaluating the scrubbing system (USEPA-84/03, p. 9-11).

In the cut power method, penetration is a function of particle diameter and is given as:

$$P_t \ = \ \exp[-A_{cut}d_p(B_{cut})] \tag{10.15}$$

where
P_t = penetration
A_{cut} = parameter characterizing the particle size distribution
B_{cut} = empirically determined constant, depending on the scrubber design
d_p = aerodynamic diameter of the particle

Penetration, calculated by Equation 10.15, is given for only one particle size (d_p). To obtain the overall penetration, the equation can be integrated over the log-normal particle size distribution. By mathematically integrating P_t over a log-normal distribution of particles and by varying the geometric standard deviation σ_{gm} and the geometric mean particle diameter d_{pg}, the overall penetration P_t^* can be obtained.

Example 10.2

Problem:
Given conditions similar to those used in the infinite throat section example, estimate the cut diameter for a Venturi scrubber. The following data are approximate (USEPA-84/03, p. 9-12).

Given:
Geometric standard deviation σ_{gm} = 2.5
Particle aerodynamic geometric mean diameter d_{pg} = 12.6 μmA
Required efficiency = 99.1% or 0.991

Solution:

Step 1. For an efficiency of 99.1%, the overall penetration can be calculated from

$$P_t^* \ = \ 1 - \eta$$

$$= \ 1 - 0.991$$

$$= \ 0.009$$

Step 2. The overall penetration is 0.009 and the geometric standard deviation is 2.5. The figures in USEPA-84/03, p. 9-12, provide us with the following information:

$$P_t^* \ = \ 0.009$$

$$\sigma_{gm} = 2.5$$

$$(d_p)_{cut}/d_{pg} = 0.09$$

Step 3. The cut diameter $(d_p)_{cut}$ is calculated from

$$(d_p)_{cut}/d_{pg} = 0.09(12.6 \, \mu mA) = 1.134 \, \mu mA$$

Example 10.3

Problem:

A particle size analysis indicated that (USEPA-81/10, p. 9-14):

d_{gm} (geometric mean particle diameter) = 12 μm
σ_{gm} (standard deviation of the distribution) = 3.0
η (wet collector efficiency) = 99%

If a collection efficiency of 99% were required to meet emission standards, what would the cut diameter of the scrubber need to be?

Solution:

Step 1. Write the penetration (P_t) equation:

$$P_t^* = 1 - \eta$$

$$= 1 - 0.99$$

$$= 0.01$$

From the figures in USEPA-84/03, p. 9-12, we read $(d_p)_{cut}/d_{gm}$, for $P_t^* = 0.01$, and $\sigma_{gm} = 3.0$; $[d_p]_{cut}/d_{gm}$ equals 0.063. Because $d_{gm} = 12$ μm, the scrubber must be able to collect particles of size $0.063 \times 12 = 0.76$ μm (with at least 50% efficiency) to achieve an overall scrubber efficiency of 99%.

10.2.5.4 Contact Power Theory

A more general theory for estimating collection efficiency is the contact power theory based on a series of experimental observations made by Lapple and Kamack (1955). The fundamental assumption of the theory is: "When compared at the same power consumption, all scrubbers give substantially the same degree of collection of a given dispersed dust, regardless of the mechanism involved and regardless of whether the pressure drop is obtained by high gas flow rates or high water flow rates" (Lapple and Kamack, 1955).

In other words, collection efficiency is a function of how much power the scrubber uses, not of how the scrubber is designed — an assumption with a number of implications in the evaluation and selection of wet collectors. Once we know the amount of power needed to attain certain collection efficiency, the claims about specially located nozzles, baffles, etc. can be evaluated more objectively. For example, the choice between two different scrubbers with the same power requirements may depend primarily on ease of maintenance (USEPA-81/10, p. 9-16; USEPA-84/03, p. 9-13).

Semrau (1960, 1963) developed the contact power theory from the work of Lapple and Kamack (1955). This theory is empirical in approach and relates the total pressure loss (P_T) of the system to the collection efficiency. The total pressure loss is expressed in terms of the power expended in injecting the liquid into the scrubber, plus the power needed to move the process gas through the system (USEPA-84/03, p. 9-13):

$$P_T = P_G + P_L$$

$$P_G = 0.157\Delta p \qquad\qquad\qquad (10.16)$$

$$P_L = 0.583 p_t (Q_L/Q_G))$$

where
P_T = total contacting power (total pressure loss), kilowatt hours per 100 m³ (horsepower per 1000 acfm)
P_G = power input from gas stream, kilowatt hours per 100 m³ (horsepower per 1000 acfm)
P_L = contacting power from liquid injection, kilowatt hours per 100 m³ (horsepower per 1000 acfm)

Note: The total pressure loss P_T should not be confused with penetration P_t.

The power expended in moving the gas through the system P_G is expressed in terms of the scrubber pressure drop:

$$P_G = 2.724 \times 10^{-4} \Delta P, \text{ kWh/1000 m}^3 \text{ (metric units)} \qquad (10.17)$$

or

$$P_G\ 0.1575 \Delta p, \text{ hp/1000 acfm (British/US customary units)}$$

where Δp = pressure drop, kilopascals (inches H_2O)
The power expended in the liquid stream P_L is expressed as

$$P_L = 0.28 p_L (Q_L/Q_G), \text{ kWh/1000 m}^3 \text{ (metric units)} \qquad (10.18)$$

or

$$P_G = 0.583 p_L (Q_L/Q_G), \text{ hp/1000 acfm (British/US customary units)}$$

where
p_L = liquid inlet pressure, 100 kPa (pounds per square inch)
Q_L = liquid feed rate, cubic meters per hour (gallons per minute)
Q_G = gas flow rate, cubic meters per hour (cubic feet per minute)

The constants given in the expressions for P_G and P_L incorporate conversion factors to put the terms on a consistent basis. The total power can therefore be expressed as:

$$P_T = P_G + P_L$$

$$= 2.724 \times 10^{-4}\Delta p + 0.28 p_L (Q_L/Q_G), \text{ kWh/1000 m}^3 \text{ (metric units)} \quad (10.19)$$

or

$$P_T = 0.1575\Delta p + 0.583 p_L (Q_L/Q_G), \text{ hp/1000 acfm (British/US customary units)}$$

Correlate this with scrubber efficiency by using the following equations:

$$\eta = 1 - \exp[-f(\text{system})] \quad (10.20)$$

where $f(\text{system})$ is defined as:

$$f(\text{system}) = N_t = \alpha(P_T)^\beta \quad (10.21)$$

and where
N_t = number of transfer units
P_T = total contacting power
α and β = empirical constants determined from experiment and dependent on characteristics of the particles

The efficiency then becomes:

$$\eta = 1 - \exp[-\alpha(P_T)^\beta] \quad (10.22)$$

Note: The values of α and β (used in metric or British units) can be found in USEPA-84/03, p. 9-15.

Scrubber efficiency is also expressed as the number of transfer units (USEPA-81/10, p. 9-17):

$$N_t = \alpha(P_T)^\beta = \ln[1/(1 - \eta)] \quad (10.23)$$

where
N_t = number of transfer units
η = fractional collection efficiency
α and β = characteristic parameters for the type of particulates collected

Unlike the cut power and Johnstone theories, the contact power theory cannot predict efficiency from a given particle size distribution. The contact power theory gives a relationship that is independent of the size of the scrubber. With this observation, a small pilot scrubber could first be used to determine the pressure drop needed for the required collection efficiency. The full-scale scrubber design could then be scaled up from the pilot information. Consider the following example:

Example 10.4

Problem:
Stack test results for a wet scrubber used to control particulate emissions from a foundry cupola reveal that the particulate emissions must be reduced by 85% to meet emission standards.

If a 100-acfm pilot unit is operated with a water flow rate of 0.5 gal/min at a water pressure of 80 psi, what pressure drop (Δp) would be needed across a 10,000-acfm scrubber unit (USEPA-84/03, p. 9-15; USEPA-81/10, p. 9-18)?

Solution:

Step 1. From the table in USEPA-84/03, p. 9-15, read the α and β parameters for foundry cupola dust:

$$\alpha = 1.35$$

$$\beta = 0.621$$

Step 2. Calculate the number of transfer units N_t using Equation 10.20:

$$\eta = 1 - \exp(-N_t)$$

$$N_t = \ln[1/(1 - \eta)]$$

$$= \ln[1/(1 - 0.85)]$$

$$= 1.896$$

Step 3. Calculate the total contacting power P_T using Equation 10.21:

$$N_t = \alpha(P_T)^\beta$$

$$1.896 = 1.35 \, (P_T)^{0.621}$$

$$1.404 = (P_T)^{0.621}$$

$$\ln 1.404 = 0.621(\ln P_T)$$

$$0.3393 = 0.621(\ln P_T)$$

$$0.5464 = \ln P_T$$

$$P_T = 1.73 \text{ hp/1000 acfm}$$

Step 4. Calculate the pressure drop Δp using Equation 10.19:

$$P_T = 0.1575\Delta p + 0.583 p_L (Q_L/Q_G)$$

$$1.73 = 0.1575\Delta p + 0.583(80)(0.5/100)$$

$$\Delta p = 9.5 \text{ in. } H_2O$$

10.2.5.5 Pressure Drop

A number of factors affect particle capture in a scrubber. One of the most important for many scrubber types is pressure drop, which is the difference in pressure between the scrubbing process inlet and outlet. Static pressure drop of a system is dependent on the system's mechanical design, as well as on the required collection efficiency: the sum of the energy required to accelerate and move the gas stream and the frictional losses as the gases move through the scrubbing system (UESPA-84/03, p. 9-17).

The following factors affect the pressure drop in a scrubber:

- Scrubber design and geometry
- Gas velocity
- Liquid-to-gas ratio

As with calculating collection efficiency, no single equation can predict pressure drop for all scrubbing systems.

Many theoretical and empirical relationships are available for estimating the pressure drop across a scrubber. Generally, the most accurate are those developed by scrubber manufacturers for *their* particular scrubbing systems. Because of the lack of validated models, users should consult the vendor's literature to estimate pressure drop for the particular scrubbing device of concern.

One widely accepted expression was developed for Venturis. The correlation proposed by Calvert (Yung et al., 1977) is:

$$\Delta p = 8.24 \times 10^{-4}(v_{gt})^2(L/G) \quad \text{(metric units)} \tag{10.24}$$

or

$$\Delta p = 4.0 \times 10^{-5}(v_{gt})^2(L/G) \quad \text{(English units)} \tag{10.25}$$

where
Δp = pressure drop, centimeters H_2O (inches H_2O)
v_{gt} = gas velocity in the Venturi throat, centimeters per second (feet per second)
L/G = liquid-to-gas ratio, dimensionless (actually in liters per cubic meter [gallons per 1000 ft³])

Using Equation 10.24 for the conditions given in the example in the infinite throat model section, we obtain:

$$v_{gt} = 9000 \text{ cm/sec}$$

$$L/G = 0.0009 \text{ L/m}^3$$

$$\Delta p = 8.24 \times 10^{-4}(9000)^2(0.0009)$$

$$= 60 \text{ cm } H_2O$$

10.3 WET SCRUBBER COLLECTION MECHANISMS AND EFFICIENCY (GASEOUS EMISSIONS)

Although Venturi scrubbers are used predominantly for control of particulate air pollutants, these devices can simultaneously function as absorbers. Consequently, absorption devices used to remove gaseous contaminants are referred to as *absorbers* or *wet scrubbers* (USEPA-84/03, p. 1-7). To remove a gaseous pollutant by absorption, the exhaust stream must be passed through (brought into contact with) a liquid. The process involves three steps:

1. The gaseous pollutant diffuses from the bulk area of the gas phase to the gas–liquid interface.
2. The gas moves (transfers) across the interface to the liquid phase. This step occurs extremely rapidly once the gas molecules (pollutants) arrive at the interface area.
3. The gas diffuses into the bulk area of the liquid, thus making room for additional gas molecules to be absorbed. The rate of absorption (mass transfer of the pollutant from the gas phase to the liquid phase) depends on the diffusion rates of the pollutant in the gas phase (first step) and in the liquid phase (third step).

To enhance gas diffusion, and therefore absorption, steps include:

• Providing a large interfacial contact area between the gas and liquid phases
• Providing good mixing of the gas and liquid phases (turbulence)
• Allowing sufficient "residence" or "contact" time between the phases for adsorption to occur

10.4 ASSORTED VENTURI SCRUBBER EXAMPLE CALCULATIONS

In this section, we provide several example calculations environmental engineers would be expected to perform for problems dealing with a Venturi scrubber design; scrubber overall collection efficiency; scrubber plan review; spray tower; packed tower review; and tower height and diameter.

10.4.1 Scrubber Design of a Venturi Scrubber

Example 10.5

Problem:
    Calculate the throat area of a Venturi scrubber to operate at specified collection efficiency (USEPA-84/09, p. 77).

Given:
Volumetric flow rate of process gas stream = 11,040 acfm (at 68°F)
Density of dust = 187 lb/ft³
Liquid-to-gas ratio = 2 gal/1000 ft³
Average particle size = 3.2 μm (1.05×10^{-5} ft)
Water droplet size = 48 μm (1.575×10^{-4} ft)
Scrubber coefficient $k = 0.14$

Required collection efficiency = 98%
Viscosity of gas = 1.23×10^{-5} lb/ft-sec
Cunningham correction factor = 1.0

Solution:

Step 1. Calculate the inertial impaction parameter, ψ, from Johnstone's equation. (Johnstone's equation describes the collection efficiency of a Venturi scrubber [USEPA-81/10, p. 9-11]).

$$\eta = 1 - \exp[-k(Q_L/Q_G)\sqrt{\psi}]$$

where
η = fractional collection efficiency
k = correlation coefficient whose value depends on the system geometry and operating conditions, typically 0.1 to 0.2, 1000 acf/gal
Q_L/Q_G = liquid-to-gas ratio, gallons per 1000 acf
ψ = $C_f p_p v(d_p)^2/18d_d\mu$ = inertial impaction parameter
p_p = particle density
v = gas velocity at Venturi throat, feet per second
d_p = particle diameter, feet
d_d = droplet diameter, feet
μ = gas viscosity, pounds per foot-second
C_f = Cunningham correction factor

Step 2. From the calculated value of the preceding ψ (internal impaction parameter), back calculate the gas velocity at the Venturi throat v.
Calculate ψ:

$$\eta = 1 - \exp[-k(Q_L/Q_G)\sqrt{\psi}]$$

$$0.98 = 1 - \exp[-0.14(2)\sqrt{\psi}]$$

Solving for ψ:

$$\psi = 195.2$$

Calculate v:

$$\psi = C_f p_p v(d_p)^2/18d_d\mu$$

$$v = 18\psi d_d\mu/C_f p_p(d_p)^2$$

$$= (18)(195.2)(1.575 \times 10^{-4})(1.23 \times 10^{-5})/(1)(187)(1.05 \times 10^{-5})^2$$

$$= 330.2 \text{ ft/sec}$$

Table 10.1   Particle Size Distribution Data

d_p, Microns	Weight, percent
<0.1	0.01
0.1–0.5	0.21
0.5–1.0	0.78
1.0–5.0	13
5.0–10.0	16
10.0–15.0	12
15.0–20.0	8
>20.0	50

Step 3. Calculate the throat area S, using gas velocity at the Venturi throat v:

$$S = \text{(volumetric flow rate)/(velocity)}$$

$$= 11,040/(60)(330.2)$$

$$= 0.557 \text{ ft}^2$$

Example 10.6

Problem:

Calculate the overall collection efficiency of a Venturi scrubber that cleans a fly ash-laden gas stream, given the liquid-to-gas ratio, throat velocity, and particle size distribution (USEPA-84/09, p. 79).

Given
Liquid-to-gas ratio = 8.5 gal/1000 ft³
Throat velocity = 227 ft/sec
Particle density of fly ash = 43.7 lb/ft³
Gas viscosity = 1.5×10^{-5} lb/ft-sec

The particle size distribution data are given in Table 10.1. Use Johnstone's equation with a k value of 0.2 to calculate the collection efficiency. Ignore the Cunningham correction factor effect.

Solution:

Step 1. What are the parameters used in Johnstone's equation?

$$\text{Johnstone's equation} \quad \eta = 1 - \exp[-k(Q_L/Q_G)\sqrt{\psi}]$$

where
η = fractional collection efficiency
k = correlation coefficient whose value depends on the system geometry and operating conditions, typically 0.1 to 0.2, 1000 acf/gal
Q_L/Q_G = liquid-to-gas ratio, gallons per 1000 acf
ψ = $C_f p_p v(d_p)^2/18d_d \mu$ = inertial impaction parameter
p_p = particle density
v = gas velocity at Venturi throat, feet per second
d_p = particle diameter, feet

d_d = droplet diameter, feet
μ = gas viscosity, pounds per foot-second
C_f = Cunningham correction factor

Calculate the average droplet diameter in feet. The average droplet diameter may be calculated using the equation:

$$d_d = (16,400/V) + 1.45(Q_L/Q_G)^{1.5}$$

where d_d = droplet diameter, microns.

$$d_d = (16,400/272) + 1.45(8.5)^{1.5}$$

$$= 96.23 \text{ microns}$$

$$= 3.156 \times 10^{-4} \text{ ft}$$

Express the inertial impaction parameter in terms of d_p (feet):

$$\psi = C_f p_p v(dp)^2/18d_d\mu$$

$$= (1)(43.7)(272)(d_p)^2/18(3.156 \times 10^{-4})(1.5 \times 10^{-5})$$

$$= 1.3945 \times 10^{11}(d_p)^2$$

Express the fractional collection efficiency η_i, in terms of d_{pi} (d_p in feet):

$$\eta = 1 - \exp[-k(Q_L/Q_G)\sqrt{\psi}]$$

$$\eta_i = 1 - \exp[-(0.2)(8.5)(1.3945 \times 10^{11}(d_p)^2)^{0.5}]$$

$$= 1 - \exp[-6.348 \times 10^5 d_{pi}]$$

Step 2. Calculate the collection efficiency for each particle size appearing in Table 10.1:

For $d_p = 0.05$ micron (1.64×10^{-7} ft), for example:

$$\eta_i = 1 - \exp[-6.348 \times 10^5 \, d_{pi}]$$

$$= 1 - \exp[-6.348 \times 10^5(1.64 \times 10^{-7})]$$

$$= 0.0989$$

$$w_i\eta_i = 0.01(0.0989) = 9.89 \times 1^{-4}$$

Table 10.2 Particle Size Data

d_p, Feet	w_i, Percent	η_i	$w_i\eta_i$, Percent
1.64×10^{-7}	0.01	0.0989	9.89×10^{-4}
9.84×10^{-7}	0.21	0.4645	0.0975
2.62×10^{-6}	0.78	0.8109	0.6325
9.84×10^{-6}	13	0.9981	12.98
2.62×10^{-5}	16	1	16
4.27×10^{-5}	12	1	12
5.91×10^{-5}	8	1	8
6.56×10^{-5}	50	1	50

Table 10.2 shows the results (rounded) of the preceding calculations for each particle size.

Step 3. Calculate the overall collection efficiency:

$$\eta = \Sigma w_i \eta_i$$

$$= 9.89 \times 10^{-4} + 0.0975 + 0.6325 + 12.980 + 16.00 + 12.00 + 8.00 + 50.00$$

$$= 99.71\%$$

Example 10.7

Problem:

A vendor proposes to use a spray tower on a lime kiln operation to reduce the discharge of solids to the atmosphere. The inlet loading must be reduced to meet state regulations. The vendor's design calls for a certain water pressure drop and gas pressure drop across the tower. Determine whether this spray tower will meet state regulations. If the spray tower does not meet state regulations, propose a set of operating conditions that will meet the regulations (USEPA-84/09, p. 81).

Given:
Gas flow rate = 10,000 acfm
Water rate = 50 gal/min
Inlet loading = 5.0 gr/ft³
Maximum gas pressure drop across the unit = 15 in. H_2O
Maximum water pressure drop across the unit = 100 psi
Water pressure drop = 80 psi
Gas pressure drop across the tower = 5.0 in. H_2O
State regulations require a maximum outlet loading of 0.05 grains per cubic foot. Assume that the contact power theory applies (USEPA-81/10, p. 9-15).

Solution:

Step 1. Calculate the collection efficiency based on the design data given by the vendor. The contact power theory is an empirical approach that relates particulate collection efficiency and pressure drop in wet scrubber systems. It assumes that particulate collection efficiency is a sole function of the total pressure loss for the unit:

$$P_T = P_G + P_L$$

$$P_G = 0.157\Delta p$$

$$P_L = 0.583 p_L (Q_L / Q_G)$$

where
P_T = total pressure loss, horsepower per 1000 acfm
P_G = contacting power based on gas stream input, horsepower per 1000 acfm
Δp = pressure drop across the scrubber, inches H_2O
P_L = contacting power based on liquid stream energy input, horsepower per 1000 acfm
p_L = liquid inlet pressure, pounds per square inch
Q_L = liquid feed rate, gallons per minute
Q_G = gas flow rate, cubic feet per minute

The scrubber collection efficiency is also expressed as the number of transfer units:

$$N_t = \alpha(P_T)^\beta = \ln[1/(1-\eta)]$$

where
N_t = number of transfer units
η = fractional collection efficiency
α and β = characteristic parameters for the type of particulates collected

Calculate the total pressure loss P_T. To calculate the total pressure loss, we need the contacting power for the gas stream energy input and liquid stream energy input.

Calculate the contacting power based on the gas stream energy input P_G in hp/1000 acfm. Because the vendor gives the pressure drop across the scrubber, we can calculate P_G:

$$P_G = 0.157 \, \Delta p$$

$$= (0.157)(5.0)$$

$$= 0.785 \text{ hp}/1000 \text{ acfm}$$

Calculate the contacting power based on the liquid stream energy input P_L, in horsepower per 1000 acfm. Because the liquid inlet pressure and liquid-to-gas ratio are given, we can calculate P_L:

$$P_L = 0.583 p_L (Q_L / Q_G)$$

$$= 0.583(80)(50/10,000)$$

$$= 0.233 \text{ hp}/1000 \text{ acfm}$$

Calculate the total pressure loss P_T, in horsepower per 1000 acfm:

$$P_T = P_G + P_L$$

$$= 0.785 + 0.233$$

$$= 1.018 \text{ hp/1000 acfm}$$

Calculate the number of transfer units N_t:

$$N_t = \alpha(P_T)^\beta$$

The values of α and β for a lime kiln operation are 1.47 and 1.05, respectively. These coefficients have been previously obtained from field test data. Therefore,

$$N_t = \alpha(P_T)^\beta$$

$$= (1.47)(1.018)^{1.05}$$

$$= 1.50$$

Calculate the collection efficiency based on the design data given by the vendor:

$$N_t = \ln[1/(1 - \eta)]$$

$$1.50 = \ln[1/(1 - \eta)]$$

Solving for η:

$$\eta = 0.777$$

$$= 77.7\%$$

Step 2. Calculate collection efficiency required by state regulations. Because the inlet loading is known and the outlet loading is set by the regulations, we can readily calculate the collection efficiency:

$$\text{Collection efficiency} = [(\text{inlet loading} - \text{outlet loading})(\text{inlet loading})](100)$$

$$= [(5.0 - 0.05)/(5.0)](100)$$

$$= 99.0\%$$

Step 3. Does the spray tower meet the regulations? No. The collection efficiency, based on the design data given by the vendor, should be higher than the collection efficiency required by the state regulations.

Step 4. Assuming the spray tower does not meet the regulations, propose a set of operating conditions that will meet the regulations.

Note that the calculation procedure is now reversed.

Calculate the total pressure loss P_T, using the collection efficiency required by the regulations in horsepower per 1000 acfm.

Calculate the number of transfer units for the efficiency required by the regulations:

$$N_t = \ln[1/(1 - \eta)]$$

$$\ln[1/(1 - 0.99)]$$

$$= 4.605$$

Calculate the total pressure loss P_T, in horsepower per 1000 acfm:

$$N_t = \alpha(P_T)^\beta$$

$$4,605 = 1.47(P_T)^{1.05}$$

Solving for P_T:

$$P_T = 2.96 \text{ hp/1000 acfm}$$

Calculate the contacting power based on the gas stream energy input P_G, using a Δp of 15 in. H_2O. A pressure drop Δp of 15 in. H_2O is the maximum value allowed by the design

$$P_G = 0.157 \, \Delta p$$

$$= (0.157)(15)$$

$$= 2.355 \text{ hp/1000 acfm}$$

Calculate the contacting power based on the liquid stream energy input P_L:

$$P_L = P_T + P_G$$

$$= 2.96 - 2.355$$

$$= 0.605 \text{ hp/1000 acfm}$$

Calculate Q_L/Q_G in gallons per actual cubic feet, using a p_L of 100 psi:

$$P_L = 0.583 p_L (Q_L/Q_G)$$

$$Q_L/Q_G = P_L/0.583p_L$$

$$= 0.605/(0.583)(100)$$

$$= 0.0104$$

Determine the new water flow rate Q_L', in gallons per minute:

$$(Q_L) = (Q_L/Q_G)(10,000 \text{ acfm})$$

$$= 0.0104(10,000 \text{ acfm})$$

$$= 104 \text{ gal/min}$$

What is the new set of operating conditions that will meet the regulations?

$$Q_L' = 104 \text{ gal/min}$$

$$P_T = 2.96 \text{ hp/1000 acfm}$$

10.4.2 Spray Tower

Example 10.8

Problem:

A steel pickling operation emits 300 ppm HCl (hydrochloric acid), with peak values of 500 ppm, 15% of the time. The airflow is a constant 25,000 acfm at 75°F and 1 atm. Only sketchy information was submitted with the scrubber permit application for a spray tower. We are requested to determine if the spray unit is satisfactory (USEPA-84/09, p. 100).

Given
Emission limit = 25 ppm HCl
Maximum gas velocity allowed through the water = 3 ft/sec
Number of sprays = 6
Diameter of the tower = 14 ft

The plans show a countercurrent water spray tower. For a very soluble gas (Henry's law constant approximately zero), the number of transfer units (N_{OG}) can be determined by the following equation:

$$N_{OG} = \ln(y_1/y_2)$$

where
y_1 = concentration of inlet gas
y_2 = concentration of outlet gas

In a spray tower, the number of transfer units N_{OG} for the first (or top) spray is about 0.7. Each lower spray has only about 60% of the N_{OG} of the spray above it. The final spray, if placed in the inlet duct, has an N_{OG} of 0.5.

The spray sections of a tower are normally spaced at 3-ft intervals. The inlet duct spray adds no height to the column.

Solution:

Step 1. Calculate the gas velocity through the tower:

$$V = Q/S$$

$$= Q/(\pi D^2/4)$$

where
V = velocity
Q = actual volumetric gas flow
S = cross-sectional area
D = diameter of the tower

$$V = Q/(\pi D^2/4)$$

$$= 25,000/[\pi(14)^2/4]$$

$$= 162.4 \text{ ft/min}$$

$$= 2.7 \text{ ft/sec}$$

Step 2. Does the gas velocity meet the requirement? Yes, because the gas velocity is less than 3 ft/sec.

Step 3. Calculate the number of overall gas transfer units N_{OG} required to meet the regulation. Recall that

$$N_{OG} = \ln(y_1/y_2)$$

where
y_1 = concentration of inlet gas
y_2 = concentration of outlet gas

Use the peak value for inlet gas concentration:

$$N_{OG} = \ln(y_1/y_2)$$

$$= \ln(500/25)$$

$$= 3.0$$

Step 4. Determine the total number of transfer units provided by a tower with six spray sections. Remember that each lower spray has only 60% of the efficiency of the section above it (because of back mixing of liquids and gases from adjacent sections). Spray section N_{OG} values are derived accordingly:

$$\text{Top spray } N_{OG} = 0.7 \text{ (given)}$$

$$\text{2nd spray } N_{OG} = 0.7(0.6) = 0.42$$

$$\text{3rd spray } N_{OG} = 0.42(0.6) = 0.252$$

$$\text{4th spray } N_{OG} = 0.252(0.6) = 0.1512$$

$$\text{5th spray } N_{OG} = 0.1512(0.6) = 0.0907$$

$$\text{Inlet } N_{OG} = 0.5 \text{ (given)}$$

$$\text{Total } N_{OG} = 0.7 + 0.42 + 0.252 + 0.1512 + 0.0907 + 0.5 = 2.114$$

This value is below the required value of 3.0.

Step 5. Calculate the outlet concentration of gas.

$$N_{OG} = \ln(y_1/y_2)$$

$$y_1/y_2 = \exp(NOG)$$

$$= \exp(2.114)$$

$$= 8.28$$

$$y_2 = 500/8.28$$

$$= 60.4 \text{ ppm}$$

Step 6. Does the spray tower meet the HCl regulation? Because y_2 is greater than the required emission limit of 25 ppm, the spray unit is not satisfactory.

10.4.3 Packed Tower

Example 10.9

Pollution Unlimited, Inc. has submitted plans for a packed ammonia scrubber on an air stream containing NH_3. The operating and design data are given by Pollution Unlimited. We remember approving plans for a nearly identical scrubber for Pollution Unlimited in 1978. After consulting our old files, we find all the conditions were identical except for the gas flow rate. What is our recommendation (USEPA-84/09, p. 102)?

Given
Tower diameter = 3.57 ft
Packed height of column = 8 ft
Gas and liquid temperature = 75°F
Operating pressure = 1.0 atm
Ammonia-free liquid flow rate (inlet) = 1000 lb/ft²-h
Gas flow rate = 1575 acfm
Gas flow rate in the 1978 plan = 1121 acfm
Inlet NH_3 gas composition = 2.0 mol%
Outlet NH_3 gas composition = 0.1 mol%
Air density = 0.0743 lb/ft³
Molecular weight of air = 29
Henry's law constant m = 0.972
Molecular weight of water = 18
Emission regulation = 0.1% NH_3
Colburn chart

Solution:

Step 1. What is the number of overall gas transfer units N_{OG}? The number of overall gas transfer units N_{OG} is used when calculating packing height requirements. It is a function of the extent of the desired separation and the magnitude of the driving force through the column (the displacement of the operating line from the equilibrium line).
　Calculate the gas molar flow rate G_m and liquid molar flow rate L_m, in pound-moles per square foot-hour. The values of G_m and L_m are found on the Colburn chart (EPA-84/03, p. 104; EPA-81/12, p. 4-30). As mentioned earlier, this chart graphically predicts the value of N_{OG}.
　Calculate the cross-sectional area of the tower S, in square feet:

$$S = \pi D^2 / 4$$

where
S = cross-sectional area of the tower
D = diameter of the tower

$$S = \pi D^2 / 4$$

$$= (\pi)(3.57)^2 / (4)$$

$$= 10.0 \ ft^2$$

Calculate the gas molar flow rate G_m, in pound-moles per square foot-hour:

$$G_m = Q_p / SM$$

where

G_m = gas molar flow rate, pound-moles per square foot-hour
Q = volumetric flow rate of gas stream
p = density of air
S = cross-sectional area of the tower
M = molecular weight of air

$$G_m = Q_p/SM$$

$$= (1575)(0.0743)/(10.0)(29)$$

$$= 0.404 \text{ lb-mol/ft}^2\text{-min}$$

$$= 24.2 \text{ lb-mol/ft}^2\text{-h}$$

Calculate the liquid molar flow rate L_m, in pound-moles per square foot-hour:

$$L_m = L/M_L$$

where
L_m = liquid molar flow rate, pound-moles per square foot-hour
L = liquid mass velocity, pounds per square foot-hour
M_L = liquid molecular weight

$$L_m = L/M_L$$

$$= (1000)/(18)$$

$$= 55.6 \text{ lb mol/ft}^2\text{-h}$$

Calculate the value of mG_m/L_m:

where
m = Henry's law constant
G_m = gas molar flow rate, pound-moles per square foot-hour
L_m = liquid molar flow rate, pound-moles per square foot-hour

$$mG_m/L_m = (0.972)(24.2/55.6)$$

$$= 0.423$$

Calculate the value of $(y_1 - mx_2)/(y_2 - mx_2)$, the abscissa of the Colburn chart:

where
y_1 = inlet gas mole fraction
y_2 = outlet gas mole fraction
x_2 = inlet liquid mole fraction

m = Henry's law constant

$$(y_1 - mx_2)/(y_2 - mx_2) = [0.02 - (0.972)(0)]/[0.001 - (0.972)(0)]$$

$$= 20.0$$

Determine the value of N_{OG} from the Colburn chart. From the Colburn chart, use the values of $(y_1 - mx_2)/(y_2 - mx_2)$ and mG_m/L_m to find the value of N_{OG}:

$$N_{OG} = 4.3$$

Step 2. What is the height of an overall gas transfer unit H_{OG}? The height of an overall gas transfer unit H_{OG} is also used to calculate packing height requirements. H_{OG} values in air pollution are almost always based on experience. H_{OG} is a strong function of solvent viscosity and difficulty of separation, increasing with increasing values of both.

Calculate the gas mass velocity G, in pounds per square foot-hour:

$$G = pQ/S$$

where
G = gas mass velocity, pounds per square foot-hour
S = cross-sectional area of the tower
p = density of air

$$G = pQ/S$$

$$= (1575)(0.0743)/10.0$$

$$= 11.7 \text{ lb/ft}^2\text{-min}$$

$$= 702 \text{ lb/ft}^2\text{-h}$$

H_{OG} value is 2.2 ft (determined from USEPA chart).

Step 3. What is the required packed column height Z, in feet?

$$Z = (N_{OG})(H_{OG})$$

where
Z = height of packing
H_{OG} = height of an overall gas transfer unit
N_{OG} = number of transfer units

$$Z = (N_{OG})(H_{OG})$$

$$= (4.3)(2.2)$$

$$= 9.46 \text{ ft}$$

Step 4. Compare the packed column height of 8 ft specified by Pollution Unlimited, Inc. to the height calculated previously. What is the recommendation? The submission is disapproved because the calculated height (9.46 ft) is higher than that (8 ft) proposed by the company.

10.4.4 Packed Column Height and Diameter

Example 10.10

Problem:

A packed column is designed to absorb ammonia from a gas stream. Given the operating conditions and type of packing (see below), calculate the height of packing and column diameter (USEPA-84/09, p. 106).

Given
Gas mass flow rate = 5000 lb/h
NH_3 concentration in inlet gas stream = 2.0 mol%
Scrubbing liquid = pure water
Packing type = 1-in. Raschig rings
Packing factor, $F = 160$
H_{OG} of the column = 2.5 ft
Henry's law constant $m = 1.20$
Density of gas (air) = 0.075 lb/ft³
Density of water = 62.4 lb/ft³
Viscosity of water = 1.8 cp
Generalized flooding and pressure drop correction graph (USEPA-84/09, p. 107)
Figure 10.2: graphical representation of the packed column

The unit operates at 60% of the flooding gas mass velocity; the actual liquid flow rate is 25% more than the minimum and 90% of the ammonia must be collected to meet state regulations.

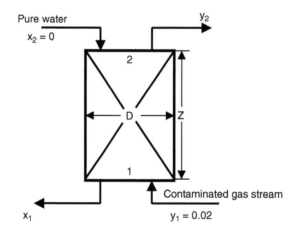

Z = packed column height
D = column diameter
y_1 = inlet gas composition
y_2 = outlet gas composition
x_1 = outlet liquid composition
x_2 = inlet liquid composition

Figure 10.2 Graphical representation of the packed column. (Adapted from USEPA-84/09, p. 107).

Solution:

Step 1. What is the number of overall gas transfer units N_{OG}? Remember that the height of packing Z is given by:

$$Z = (N_{OG})(H_{OG})$$

where
Z = height of packing
H_{OG} = height of an overall gas transfer unit
N_{OG} = number of transfer units

Because H_{OG} is given, we only need N_{OG} to calculate Z. N_{OG} is a function of the liquid and gas flow rates; however, it is usually available for most air pollution applications.

What is the equilibrium outlet liquid composition x_1 and the outlet gas composition y_2 for 90% removal? Recall that we need the inlet and outlet concentrations (mole fractions) of both streams to use the Colburn chart.

Calculate the equilibrium outlet concentration x_1* at y_1 = 0.02. According to Henry's law, x_1* at y_1/m, the equilibrium outlet liquid composition is needed to calculate the minimum L_m/G_m:

where
x_1* = outlet concentration
y_1 = inlet gas mole fraction
m = Henry's law constant
L_m = liquid molar flow rate, pound-moles per square foot-hour
G_m = gas molar flow rate, pound-moles per square foot-hour

$$x_1{}^* = y_1/m$$

$$= (0.02)/(1.20)$$

$$= 0.0167$$

Calculate y_2 for 90% removal. Because state regulations require the removal of 90% of NH_3, by material balance, 10% NH_3 will remain in the outlet gas stream:

$$y_2 = (0.1y_1)/[(1 - y_1) + (0.1)y_1]$$

where
y_1 = inlet gas mole fraction
y_2 = outlet gas mole fraction

$$y_2 = (0.1y_1)/[(1 - y_1) + (0.1)y_1]$$

$$(0.1)(0.02)/[(1 - 0.02) + (0.1)(0.02)]$$

$$= 0.00204$$

Determine the minimum ratio of molar liquid flow rate to molar gas flow rate $(L_m/G_m)_{min}$ by a material balance. Material balance around the packed column:

$$G_m(y_1 - y_2) = L_m(x_1{}^* - x_2)$$

$$(L_m/G_m)_{min} = (y_1 - y_2)/(x_1{}^* - x_2)$$

where
y_1 = inlet gas mole fraction
y_2 = outlet gas mole fraction
$x_1{}^*$ = outlet concentration
x_2 = inlet liquid mole fraction
L_m = liquid molar flow rate, pound-moles per square foot-hour
G_m = gas molar flow rate, pound-moles per square foot-hour

$$(L_m/G_m)_{min} = (y_1 - y_2)/(x_1{}^* - x_2)$$

$$= (0.02 - 0.00204)/(0.0167 - 0)$$

$$= 1.08$$

Calculate the actual ratio of molar liquid flow rate to molar gas flow rate (L_m/G_m). Remember that the actual liquid flow rate is 25% more than the minimum based on the given operating conditions:

$$(L_m/G_m) = 1.25 \, (L_m/G_m)_{min}$$

$$= (1.25)(1.08)$$

$$= 1.35$$

Calculate the value of $(y_1 - mx_2)/(y_2 - mx_2)$, the abscissa of the Colburn chart:

where
y_1 = inlet gas mole fraction
y_2 = outlet gas mole fraction
x_1 = outlet liquid mole fraction
x_2 = inlet liquid mole fraction
m = Henry's law constant

$$(y_1 - mx_2)/(y_2 - mx_2) = [(0.02) - (1.2)(0)]/[(0.00204) - (1.2)(0)]$$

$$= 9.80$$

Calculate the value of mG_m/L_m:

where

m = Henry's law constant
G_m = gas molar flow rate, pound-moles per square foot-hour
L_m = liquid molar flow rate, pound-moles per square foot-hour

Even though the individual values of G_m and L_m are not known, the ratio of the two has been previously calculated:

$$mG_m/L_m = (1.2)/(1.35)$$

$$= 0.889$$

Determine number of overall gas transfer units N_{OG} from the Colburn chart using the values calculated previously (9.80 and 0.899). From the Colburn chart $N_{OG} = 6.2$.

Step 2. Calculate the height of packing Z:

$$Z = (N_{OG})(H_{OG})$$

where

Z = height of packing
H_{OG} = height of an overall gas transfer unit
N_{OG} = number of transfer units

$$Z = (N_{OG})(H_{OG})$$

$$= (6.2)(2.5)$$

$$= 15.5 \text{ ft}$$

Step 3. What is the diameter of the packed column? The actual gas mass velocity must be determined. To calculate the diameter of the column, we need the flooding gas mass velocity. USEPA's generalized flooding and pressure drop correction graph (USEPA-84/09, p. 107) is used to determine the flooding gas mass velocity. The mass velocity is obtained by dividing the mass flow rate by the cross-sectional area.
Calculate the flooding gas mass velocity G_f.
Calculate the abscissa of USEPA's generalized flooding and pressure drop correction graph, $(L/G)(p/p_L)^{0.5}$:

$$(L/G)(p/p_L)^{0.5} = (L_m/G_m)(18/29)(p/p_L)^{0.5}$$

where

18/29 = ratio of molecular weight of water to air
L = liquid mass velocity, pounds per second-square foot
G = gas mass velocity, pounds per second-square foot
p = gas density
p_L = liquid density
G_m = gas molar flow rate, pound-moles per square foot-hour
L_m = liquid molar flow rate, pound-moles per square foot-hour

Note that the L and G terms (in USEPA's generalized flooding and pressure drop correction graph) are based on mass and not moles:

$$(L/G)(p/p_L)^{0.5} = (1.35)(18/29)(0.075/62.4)^{0.5}$$

$$= 0.0291$$

Determine the value of the ordinate at the flooding line using the calculated value of the abscissa:
Ordinate $= G^2 F\psi(\mu_L)^{0.2}/P_L p g_c$
F = packing factor = 160 for 1-in. Raschig rings
ψ = ratio, density of water/density of liquid
g_c = 32.2 lb-ft/lb$_f$-sec^2
μ_L = viscosity of liquid, centipoise
G = gas mass velocity, pounds per square foot-second
From USEPA's generalized flooding and pressure drop correction graph,

$$G^2 F\psi(\mu_L)^{0.2}/p_L p g_c = 0.19$$

Solve the abscissa for the flooding gas mass velocity G_f, in pounds per square foot-second. The G value becomes G_f for this case. Thus,

$$G_f = [0.19(p_L p g_c)/(F\psi(\mu_L)^{0.2})]^{0.5}$$

$$= [(0.19)(62.4)(0.075)(32.2)/(160)(1)(1.8)^{0.2}]^{0.5}$$

$$= 0.400 \text{ lb/ft}^2\text{-sec}$$

Calculate the actual gas mass velocity G_{act}, in pounds per square foot-second:

$$G_{act} = 0.6 G_f$$

$$= (0.6)(0.400)$$

$$= 0.240 \text{ lb/ft}^2\text{-sec}$$

$$= 864 \text{ lb/ft}^2\text{-h}$$

Calculate the diameter of the column in feet:

$$S = (\text{mass flow rate of gas stream})/G_{act}$$

$$= 5000/G_{act}$$

$$S = \pi D^2/4$$

$$\pi D^2/4 = 5000/G_{act}$$

$$D = [(4(5000))/(\pi G_{act})]^{0.5}$$

$$= 2.71 \text{ ft}$$

10.5   SUMMARY OF KEY POINTS

Use the following approaches to evaluate the capabilities of scrubbing systems: empirical relationships, theoretical models, and pilot scale tests.

- Two important parameters in the design and operation of wet scrubbing systems that are a function of the process being controlled are dust properties and exhaust gas characteristics.
- Particle size distribution is the most critical parameter in choosing the most effective scrubber design and determining the overall collection efficiency.
- Static pressure drop of a system is dependent on the mechanical design of the system and collection efficiency required.
- The scrubber used most often to remove particulate matter from exhaust systems is a Venturi scrubber.
- The term *penetration* is defined as the fraction of particles that passes through a scrubber uncollected.
- No one simple equation can be used to estimate scrubber collection efficiency for all scrubber types.
- Efficient particle removal requires high gas-to-liquid (relative) velocities.
- The infinite throat model is used to estimate particle collection in Venturi scrubbers.
- The contact power theory is dependent on pilot test data to determine required collection efficiency.
- The total pressure loss, or contacting power, of the scrubbing system is represented by $P_T = P_G + P_L$, the symbol P_T.
- Efficient particle removal requires high gas-to-liquid (relative) velocities.
- According to the contact power theory, the higher the pressure drop is across the scrubbing system, the higher the collection efficiency will be.
- The following factors affect the pressure drop of a scrubbing system: scrubber design, gas velocity, and liquid-to-gas ratio.

REFERENCES

Calvert, S.J., Goldschmid, D., Leith, D., and Metha, D. (1972). Wet Scrubber System Study. Vol 1, *Scrubber Handbook*. USEPA-R2-72-118a. US Environmental Protection Agency.

Heumann, W.L. and Subramania, V. (1997). Particle scrubbing. In *Industrial Air Pollution Control Systems*. William L. Heumann (Ed.). New York: McGraw-Hill.

Lapple, C.E. and Kamack, H.J. (1955). Performance of wet dust scrubbers. *Chem. Eng. Prog.* 51:110–121.

Nukiyama, S. and Tanasawa, Y. (1983). An experiment on atomization of liquid by means of air stream (in Japanese). *Trans. Soc. Mechanical Eng.*, Japan. 4:86.

Semrau, K.T. (1960). Correlation of dust scrubber efficiency. *J. Air Pollution Control Assoc.*, 10:200–207.

Semrau, K.T. (1963). Dust scrubber design — a critique on the state of the art. *J. Air Control Assoc.* 13:587–593.

Spellman, F.R. (1999). *The Science of Air: Concepts & Applications*. Lancaster, PA: Technomic Publishing Co. Inc.

USEPA (1971). Control Techniques for Gases and Particulates.

USEPA-81/10 (1981). Control of particulate emissions, Course 413, USEPA Air Pollution Training Institute (APTI), USEPA450-2-80-066.

USEPA-84/02 (1984). Wet scrubber plan review, Course SI:412C, USEPA Air Pollution Training Institute (APTI), EPA450-2-82-020.

USEPA-84/03, Web scrubber plan review, Course SI: 412C, USEPA Air Pollution Training Institute (APTI), EPA450-2-82-020, March, 1984.

USEPA-84/09 (1984). Control of gaseous and particulate emission, Course SI:412D. USEPA Air Pollution Training Institute (APTI), USEPA450-2-84-007.

Yung, S., Calvert, S. and Barbarika, J.F. (1977). Venturi Scrubber Performance Model. USEPA 600/2-77-172. U.S. Environmental Protection Agency, Cincinnati, OH.

PART IV

Math Concepts: Water Quality

CHAPTER 11

Running Waters

In terms of practical usefulness the waste assimilation capacity of streams as a water resource has its basis in the complex phenomenon termed stream self-purification. This is a dynamic phenomenon reflecting hydrologic and biologic variations, and the interrelations are not yet fully understood in precise terms. However, this does not preclude applying what is known. Sufficient knowledge is available to permit quantitative definition of resultant stream conditions under expected ranges of variation to serve as practical guides in decisions dealing with water resource use, development, and management.

C.J. Velz (1970)

11.1 BALANCING THE "AQUARIUM"

An outdoor excursion to the local stream can be a relaxing and enjoyable undertaking. However, when the wayfarer arrives at the local stream, spreads a blanket on its bank, and then looks out upon its flowing mass only to discover a parade of waste and discarded rubble bobbing along and cluttering the adjacent shoreline and downstream areas, he quickly loses any feeling of relaxation or enjoyment. The sickening sensation the observer feels increases as he closely scrutinizes the putrid flow. He recognizes the rainbow-colored shimmer of an oil slick, interrupted here and there by dead fish and floating refuse, and the prevailing slimy fungal growth. At the same time, the observer's sense of smell is alerted to the noxious conditions. Along with the fouled water and the stench of rot-filled air, the observer notices the ultimate insult and tragedy: signs warning "DANGER — NO SWIMMING or FISHING." The observer soon realizes that the stream before him is not a stream at all; it has become little more than an unsightly drainage ditch. He has discovered what ecologists have known and warned about for years: contrary to popular belief, rivers and streams do not have an infinite capacity for pollution (Spellman, 1996).

Before the early 1970s, occurrences such as the one just described were common along the rivers and streams near main metropolitan areas throughout most of the United States. Many aquatic habitats were fouled through uncontrolled industrialization waste and resource management practices. Obviously, our streams and rivers were not always in such deplorable condition. Before the Industrial Revolution of the 1800s, metropolitan areas were small and sparsely populated. Thus, river and stream systems within or next to early communities received insignificant quantities of discarded waste. Early on, these systems were able to compensate for the small amount of wastes they received; when wounded (polluted), Nature had a way of fighting back by providing rivers' and streams' flowing waters with the ability to restore themselves through their self-purification process. Only when humans gathered in great numbers to form great cities were the stream systems not able to recover from receiving great quantities of refuse and other wastes.

What exactly is it that human populations do to rivers and streams? Halsam (1990) pointed out that man's actions are determined by his expediency. Add to this that most people do not realize that we have the same amount of water as we did millions of years ago. Through the water cycle, we continually reuse that same water — we are using water that was used by the ancient Romans and Greeks. Increased demand puts enormous stress on our water supply. Human populations—people—are the cause of this stress. What people do to rivers and streams upsets the delicate balance between levels of pollution and the purification process. Anyone who has kept fish knows what happens when the aquarium or pond water becomes too fouled. In a sense, we tend to unbalance the aquarium for our own water supplies.

With the advent of industrialization, local rivers and streams became deplorable cesspools that worsened with time. For example, during the Industrial Revolution, the removal of horse manure and garbage from city streets became a pressing concern. Moran and colleagues point out that "none too frequently, garbage collectors cleaned the streets and dumped the refuse into the nearest river" (Moran et al., 1986). Halsam reports that as late as 1887, river keepers gained full employment by removing a constant flow of dead animals from a river in London. The prevailing attitude of that day was "I don't want it anymore; throw it into the river" (Halsam, 1990).

Once we understood the dangers of unclean waters, any threat to the quality of water destined for use for drinking and recreation quickly angered those affected. Fortunately, since the 1970s we have moved to correct the stream pollution problem. Through scientific study and incorporation of wastewater treatment technology, we have started to restore streams to their natural condition.

Fortunately, through the phenomenon of self-purification, the stream aids us in this effort to restore a stream's natural water quality.

A balance of biological organisms is normal for all streams. Clean, healthy streams have certain characteristics in common. For example, as mentioned earlier, one property of streams is their ability to dispose of small amounts of pollution. However, if streams receive unusually large amounts of waste, the stream life will change and attempt to stabilize such pollutants; that is, the biota will attempt to balance the aquarium. However, if the stream biota are not capable of self-purifying, then the stream may become a lifeless body. This self-purification process relates to the purification of organic matter only. In this chapter, we discuss only organic stream pollution and self-purification.

11.1.1 Sources of Stream Pollution

Sources of stream pollution are normally classified as point or nonpoint sources. A point source (PS) discharges effluent, such as wastewater from sewage treatment and industrial plants. Simply put, a point source is usually easily identified as "end of the pipe" pollution: it emanates from a concentrated source or sources. In addition to organic pollution received from the effluents of sewage treatment plants, other sources of organic pollution include runoffs and dissolution of minerals throughout an area that are not from one or more concentrated sources.

Nonconcentrated sources are known as nonpoint sources (NPSs). NPS pollution, unlike pollution from industrial and sewage treatment plants, comes from many diffuse sources. Rainfall or snowmelt moving over and through the ground carries NPS pollution. As the runoff moves, it picks up and carries away natural and man-made pollutants, finally depositing them into streams, lakes, wetlands, rivers, coastal waters, and even our underground sources of drinking water. These pollutants include:

- Excess fertilizers, herbicides, and insecticides from agricultural lands and residential areas
- Oil, grease, and toxic chemicals from urban runoff and energy production
- Sediment from improperly managed construction sites, crop and forest lands, and eroding stream banks

- Salt from irrigation practices and acid drainage from abandoned mines
- Bacteria and nutrients from livestock, pet wastes, and faulty septic systems

Atmospheric deposition and hydromodification are also sources of nonpoint source pollution (USEPA, 1994).

Of particular interest to environmentalists in recent years have been agricultural effluents. As a case in point, take, for example, farm silage effluent, estimated as more than 200 times as potent (in terms of BOD) as treated sewage (Mason, 1990).

Nutrients are organic and inorganic substances that provide food for microorganisms, including bacteria, fungi, and algae. Nutrients are supplemented by sewage discharge. The bacteria, fungi, and algae are consumed by the higher trophic levels in the community. Because of a limited amount of dissolved oxygen (DO), each stream has a limited capacity for aerobic decomposition of organic matter without becoming anaerobic. If the organic load received is above that capacity, the stream becomes unfit for normal aquatic life and cannot support organisms sensitive to oxygen depletion (Smith, 1974).

Effluent from a sewage treatment plant is most commonly disposed of in a nearby waterway. At the point of entry of the discharge, a sharp decline in the concentration of DO in the stream occurs. This phenomenon is known as the oxygen sag. Unfortunately (for the organisms that normally occupy a clean, healthy stream), when the DO is decreased, a concurrent massive increase in BOD occurs as microorganisms use the DO to help break down the organic matter. When the organic matter is depleted, the microbial population and BOD decline, while the DO concentration increases, assisted by stream flow (in the form of turbulence) and by the photosynthesis of aquatic plants. This self-purification process is very efficient, and the stream suffers no permanent damage as long as the quantity of waste is not too high. Obviously, understanding this self-purification process is important in preventing stream ecosystem overload.

As urban and industrial centers continue to grow, waste disposal problems also grow. Because wastes have increased in volume and are much more concentrated than in earlier years, natural waterways must have help in the purification process. Wastewater treatment plants, which function to reduce the organic loading that raw sewage would impose on discharge into streams, provide this help. Wastewater treatment plants use three stages of treatment: primary, secondary, and tertiary. In breaking down the wastes, a secondary wastewater treatment plant uses the same type of self-purification process found in any stream ecosystem. Small bacteria and protozoans (one-celled organisms) begin breaking down the organic material. Aquatic insects and rotifers are then able to continue the purification process. Eventually, the stream recovers and shows little or no effects of the sewage discharge. This phenomenon is known as *natural stream purification* (Spellman and Whiting, 1999).

11.2 IS DILUTION THE SOLUTION?

In the early 1900s, wastewater disposal practices were based on the premise that "the solution is dilution." Dilution was considered the most economical means of wastewater disposal into running waters (primarily rivers) and thus good engineering practice (Clark et al., 1977; Velz, 1970). Early practices in the field evolved around mixing-zone concepts based on the lateral, vertical, and longitudinal dispersion characteristics of the receiving waters (Peavy et al., 1975). Various formulae predicting space and dispersion characteristics for diluting certain pollutants to preselected concentrations were developed. Highly polluted discharged water was viewed as acceptable because the theory prevailed that the stream or river would eventually purify itself. Actually, a more accurate perception was that discharged wastewater "out of sight" was wastewater "out of mind" as well — leave it to the running water body to dilute the wastestream, and let the folks down river worry about their water supply.

Although dilution is a powerful factor in self-cleansing mechanisms of surface waters, its success has limitations. Dilution is a viable tool of running waters' ability to self-purify only if discharging is limited to relatively small quantities of waste into relatively large bodies of water. One factor that impedes this ability to self-purify is the increasing number of "dumpers" into the water body. "Growth in population and industrial activity, with attendant increases in water demand and wastewater quantities, precludes the use of many streams for dilution of raw or poorly treated wastewaters" (Peavy et al., 1975).

11.2.1 Dilution Capacity of Running Waters

Within limits, dilution is an effective means of dealing with a discharged wastestream. Immediately beyond the point of discharge, the process of mixing and dilution begins. However, complete mixing does not take place at the outfall. Instead, a waste plume is formed that gradually widens. The length and width (dispersion) of the plume depend upon running water geometry, flow velocity, and flow depth (Gupta, 1997). Beyond the mixing zone, the dilution capacity of running waters can be calculated using the principles of mass balance relation, using worst case scenario conditions — a 7-day, 10-year low flow for stream flow condition. A simple dilution equation can be written as

$$C_d = \frac{Q_s C_s + Q_w C_w}{Q_s + Q_w} \tag{11.1}$$

where
C_d = completely mixed constituent concentration downstream of the effluent, milligrams per liter
Q_s = stream flow upstream of the effluent, cubic feet per second
C_s = constituent concentration of upstream flow, milligrams per liter
Q_w = flow of the effluent, cubic feet per second
C_w = constituent concentration of the effluent, milligrams per liter

Example 11.1

Problem:
    A power plant pumps 25 ft³/sec from a stream with a flow of 180 ft³/sec. The discharge of the plant's ash pond is 22 ft³/sec. The boron concentrations for upstream water and effluent are 0.053 and 8.7 mg/L, respectively. Compute the boron concentration in the stream after complete mixing.

Solution:

$$C_d = \frac{Q_s C_s + Q_w C_w}{Q_s + Q_w}$$

$$\frac{(180 - 25)(0.053) + 22 \times 8.7}{(180 - 25) + 22} = 1.13 \text{ mg/L}$$

11.3 DISCHARGE MEASUREMENT

The total discharge for a running water body can be estimated by float method with wind and other surface effects, by dye study, or by actual subsection flow measurement, depending on cost, time, personnel, local conditions, etc. The discharge in a stream cross-section is typically measured from a subsection using the following equation:

$$Q = \text{Sum (mean depth} \times \text{width} \times \text{mean velocity)}$$

$$Q = \sum_{n=1}^{n} = \frac{1}{2}(h_n + h_{n-1})(w_n - w_{n-1}) \times \frac{1}{2}(v_n + v_{n-1}) \qquad (11.2)$$

where
Q = discharge, cubic feet per second
w_n = nth distance from initial point 0, feet
h_n = nth water depth, feet
v_n = nth velocity, feet per second

A velocity meter measures velocity (v).

11.4 TIME OF TRAVEL

Dye study or computation is used to determine the time of travel of running water. Running water time of travel and river or stream geometry characteristics can be computed using a volume displacement model. The time of travel is determined at any specific reach as the channel volume of the reach divided by the flow:

$$t = V/Q \times 1/86,400 \qquad (11.3)$$

where
t = time of travel at a stream reach, days
V = stream reach volume, cubic feet or cubic meter
Q = average stream flow in reach, cubic feet per second or cubic meters per meter

Example 11.2

Problem:
The cross-section areas at river miles 63.5, 64.0, 64.5, 65.0, and 65.7 are, respectively, 270, 264, 263, 258, 257, and 260 ft^2 at a surface water elevation. The average flow is 32.3 ft^3/sec. Find the time of travel for a reach between river miles 63.5 and 65.7.

Solution:

Step 1. Find the area in the reach:

$$\text{Average area} = 1/6(270 + 264 + 263 + 258 + 257 + 260) = 262 \text{ ft}^2$$

Step 2. Find volume:

$$\text{Distance of the reach} = (65.7 - 63.5) \text{ mi}$$

$$= 2.2 \text{ miles} \times 5280 \text{ ft/mi}$$

$$= 11,616 \text{ ft}$$

$$V = 262 \text{ ft}^2 \times 11{,}616 \text{ ft}$$

$$= 3{,}043{,}392 \text{ ft}^3$$

Step 3. Find t:

$$t = V/Q \times 1/86{,}400$$

$$\frac{3{,}043{,}392 \text{ ft}}{32.3 \times 86{,}400}$$

$$= 1.1 \text{ days}$$

11.5   DISSOLVED OXYGEN (DO)

A running water system produces and consumes oxygen. It gains oxygen from the atmosphere and from plants as a result of photosynthesis. Because of its churning, running water dissolves more oxygen than still water, such as that in a lake, reservoir, or pond. Respiration by aquatic animals, decomposition, and various chemical reactions consume oxygen.

Dissolved oxygen (DO) in running waters is as critical to the good health of stream organisms as is gaseous oxygen to humans. Simply put, DO is essential to the respiration of aquatic organisms, and its concentration in streams is a major determinant of the species composition of biota in the water and underlying sediments. DO in streams has a profound effect on the biochemical reactions that occur in water and sediments, which in turn affect numerous aspects of water quality, including the solubility of many lotic elements, as well as aesthetic qualities of odor and taste. For these reasons, DO historically has been one of the most frequently measured indicators of water quality (Hem, 1985).

In the absence of substances that cause its depletion, the DO concentration in running waters approximates the saturation level for oxygen in water in contact with the atmosphere, and decreases with increasing water temperature from about 14 mg/L at freezing to about 7 mg/L at 86°F (30°C). For this reason, in ecologically healthy streams, the dissolved-oxygen concentration depends primarily on temperature, which varies with season and climate.

Criteria for defining desirable DO concentration often are differentiated as applicable to cold-water biota, such as trout and their insect prey, and the more low oxygen-tolerant species of warm-water ecosystems. Because of the critical respiratory function of DO in aquatic animals, criteria often are expressed in terms of the short-term duration and frequency of occurrence of minimum concentration rather than long-term average concentrations. Studies cited by the USEPA of the dependence of freshwater biota on DO suggest that streams in which the concentration is less than 6.5 mg/L for more than about 20% of the time generally are not capable of supporting trout or other cold-water fish, and such concentrations could impair population growth among some warm-water game fish, such as largemouth bass (USEPA, 1986). Streams in which the DO-deficit concentration is greater than 4 mg/L for more than 20% of the time generally cannot support cold- or warm-water game fish. DO deficit refers to the difference between the saturation and measured concentrations of DO in a water sample; it is a direct measure of the effects of oxygen-demanding substances on DO in streams.

Major sources of substances that cause depletion of DO in streams are discharges from municipal and industrial wastewater treatment plants; leaks and overflows from sewage lines and septic tanks;

stormwater runoff from agricultural and urban land; and decaying vegetation, including aquatic plants from the stream and detrital terrestrial vegetation. DO is added to stream water by the process of aeration (waterfalls, riffles) and the photosynthesis of plants.

DO saturation (DO_{sat}) values for various water temperatures can be computed using the American Society of Civil Engineer's equation (ASCE, 1960). This equation represents saturation values for distilled water (β 1.0), at sea-level pressure. Even though water impurities can increase or decrease the saturation level, for most cases β is assumed to be unity (1).

$$DO_{sat} = 14.652 - 0.41022T + 0.0079910T^2 - 0.000077774T^3 \qquad (11.4)$$

where
DO_{sat} = dissolved oxygen saturation concentration, milligrams per liter
T = water temperature, degrees Celsius

Example 11.3

Problem:
Calculate DO saturation concentration for water temperature at 0, 10, 20, and 30°C, assuming $\beta = 1.0$.

Solution:

   a. at $T = 0$°C

DO_{sat} = 14.652 − 0 + 0 − 0 = 14.652 mg/L

   b. at $T = 10$°C

DO_{sat} = 14.652 − 0.41022 × 10 + 0.0079910 × 10^2 − 0.000077774 × 10^3 = 11.27 mg/L

   c. at $T = 20$°C

DO_{sat} = 14.652 − 0.41022 × 20 + 0.0079910 × 20^2 − 0.000077774 × 20^3 = 9.02 mg/L

   d. at $T = 30$°C

DO_{sat} = 14.652 − 0.41022 × 30 + 0.0079910 × 30^2 − 0.000077774 × 30^3 = 7.44 mg/L

11.5.1 DO Correction Factor

Because of differences in air pressure caused by air temperature changes, and for elevation above the mean sea level (MSL), the DO saturation concentrations generated by the formula must be corrected. Calculate the correction, using Equation 11.5.

$$f = \frac{2116.8 - (0.08 - 0.000115A)E}{2116.8} \qquad (11.5)$$

where

f = correction factor for above MSL
A = air temperature, degrees Celsius
E = elevation of the site, feet above MSL

Example 11.4

Problem:

Find the correction factor of DO_{sat} value for water at 640 ft above the MSL and air temperature of 25°C. What is DO_{sat} at a water temperature of 20°C?

Solution:

Step 1.

$$f = \frac{2116.8 - (0.08 - 0.000115A)E}{2116.8}$$

$$= \frac{2116.8 - (0.08 - 0.000115 \times 25)640}{2116.8}$$

$$f = \frac{2116.8 - 49.4}{2116.8} = 0.977$$

Step 2. Compute DO_{sat}. $T = 20°C$.

$$DO_{sat} = 9.02 \text{ mg/L}$$

With an elevation correction factor of 0.977

$$DO_{sat} = 9.02 \text{ mg/L} \times 0.977 = 8.81 \text{ mg/L}$$

11.6 BIOCHEMICAL OXYGEN DEMAND

Biochemical oxygen demand (BOD) measures the amount of oxygen consumed by microorganisms in decomposing organic matter in stream water. BOD also measures the chemical oxidation of inorganic matter (the extraction of oxygen from water via chemical reaction). A test is used to measure the amount of oxygen consumed by these organisms during a specified period of time (usually 5 days at 20°C). The rate of oxygen consumption in a stream is affected by a number of variables: temperature; pH; presence of certain kinds of microorganisms; and type of organic and inorganic material in the water.

BOD directly affects the amount of dissolved oxygen in running waters. The greater the BOD is, the more rapidly oxygen is depleted in the system. This means less oxygen is available to higher forms of aquatic life. The consequences of high BOD are the same as those for low dissolved oxygen: aquatic organisms become stressed, suffocate, and die. Sources of BOD include leaves and wood debris; dead plants and animals; animal manure; effluents from pulp and paper mills, wastewater treatment plants, feedlots, and food-processing plants; failing septic systems; and urban stormwater runoff.

11.6.1 BOD Test Procedure

Standard BOD test procedures can be found in *Standard Methods for the Examination of Water and Wastewater* (APHA, AWWA, and WEF, 1995). When the dilution waste is seeded, oxygen uptake (consumed) is assumed to be the same as the uptake in the seeded blank. The difference between the sample BOD and the blank BOD, corrected for the amount of seed used in the sample, is the true BOD. Formulae for calculation of BOD are (APHA, AWWA, and WEF, 1995):

When dilution water is not seeded:

$$\text{BOD, mg/L} = \frac{D_1 - D_2}{P} \qquad (11.6)$$

When dilution water is seeded:

$$\text{BOD, mg/L} = \frac{(D_i - D_e) - (B_i - B_e) f}{P} \qquad (11.7)$$

where
D_1, D_i = DO of diluted sample immediately after preparation, milligrams per liter
D_2, D_e = DO of diluted sample after incubation at 20°C, milligrams per liter
P = decimal volumetric fraction of sample used; milliliters of sample/300 mL
B_i = DO of seed control before incubation, milligrams per liter
B_e = DO of seed control after incubation, milligrams per liter
f = ratio of seed in diluted sample to seed in seed control
P = percent seed in diluted sample/percent seed in seed control

If seed material is added directly to the sample and to control bottles, f = volume of seed in diluted sample/volume of seed in seed control

11.6.2 Practical BOD Calculation Procedure

The following BOD calculation procedures for unseeded and seeded samples are commonly used.

11.6.2.1 Unseeded BOD Procedure

1. Select the dilutions that meet the test criteria
2. Calculate the BOD for each selected dilution, using the following formula:

$$\text{BOD, mg/L} = \frac{(DO_{Start}, \text{mg/L} - DO_{Final}, \text{mg/L}) \times 300 \text{ mL}}{\text{Sample Volume, mL}} \qquad (11.8)$$

Example 11.5

Problem:
Determine BOD, milligrams per liter, given the following data:

- Initial DO = 8.2 mg/L
- Final DO = 4.4 mg/L
- Sample size = 5 mL

Solution:

$$BOD, mg/L = \frac{(8.2 - 4.4)}{5} = 228 \ mg/L$$

11.6.2.2 Seeded BOD Procedure

1. Select dilutions that meet the test criteria.
2. Calculate the BOD for each selected dilution, using the following formula:

$$BOD, mg/L = \frac{[(DO_{Start}, mg/L - DO_{Final}, mg/L) - Seed \ Correction] \times 300 \ mL}{Sample \ Volume, mL} \quad (11.9)$$

Note: Seed correction is calculated using:

$$Seed \ Correction, mg/L = \frac{BOD_{Start}}{300 \ mL} \times mL \ Seed \ in \ the \ Sample \ Dilution \quad (11.10)$$

Example 11.6

Problem:

A series of seed dilutions were prepared in 300-mL BOD bottles using seed material (settled raw wastewater) and unseeded dilution water. The average BOD for the seed material was 204 mg/L. One milliliter of the seed material was also added to each bottle of a series of sample dilutions. Given the data for two samples in the following table, calculate the seed correction factor (SC) and BOD of the sample.

Bottle #	mL Sample	mL Seed/bottle	DO Initial	mg/L Final	Depletion, mg/L
12	50	1	8.0	4.6	3.4
13	75	1	7.7	3.9	2.8

Solution:

Step 1. Calculate the BOD of each milliliter of seed material.

$$BOD/mL \ of \ Seed = \frac{204 \ mg/L}{300 \ mL} = 0.68 \ mg/L \ BOD/mL \ seed$$

Step 2. Calculate the SC factor:

$$SC = 0.68 \ mg/L \ BOD/mL \ seed \times 1 \ mL \ seed/bottle = 0.68 \ mg/L$$

Step 3. Calculate the BOD of each sample dilution:

$$BOD, mg/L, Bottle \ \#12 = \frac{3.4 - 0.68}{50 \ mL} \times 300 = 16.3 \ mg/L$$

$$\text{BOD, mg/L, Bottle \#13} = \frac{3.8 - 0.68}{75 \text{ mL}} \times 300 = 12.5 \text{ mg/L}$$

Step 4. Calculate reported BOD:

$$\text{Reported BOD} = \frac{16.3 + 12.5}{2} = 14.4 \text{ mg/L}$$

11.7 OXYGEN SAG (DEOXYGENATION)

Biochemical oxygen demand is the amount of oxygen required to decay or break down a certain amount of organic matter. Measuring the BOD of a stream is one way to determine pollution levels. When too much organic waste, such as raw sewage, is added to the stream, all of the available oxygen will be used up. The high BOD reduces the DO because they are interrelated. A typical DO vs. time or distance curve is somewhat spoon shaped because of the reaeration process. This spoon-shaped curve, commonly called the oxygen sag curve, is obtained using the Streeter–Phelps equation (to be discussed later).

Simply stated, an oxygen sag curve is a graph of the measured concentration of DO in water samples collected (1) upstream from a significant point source (PS) of readily degradable organic material (pollution); (2) from the area of the discharge; and (3) from some distance downstream from the discharge, plotted by sample location. The amount of DO is typically high upstream, diminishes at and just downstream from the discharge location (causing a sag in the line graph), and returns to the upstream levels at some distance downstream from the source of pollution or discharge. The oxygen-sag curve is illustrated in Figure 11.1.

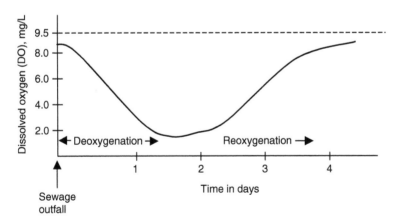

Figure 11.1   Oxygen-sag curve. (From Spellman, F.R., 1996, *Stream Ecology & Self-Purification*. Lancaster, PA: Technomic Publishing Company.)

From the figure, we see that the percentage of DO vs. time or distance shows a characteristic sag, which occurs because the organisms breaking down the wastes use up the DO in the decomposition process. When the wastes are decomposed, recovery takes place and the DO rate rises again.

Several factors determine the extent of recovery. First, the minimum level of dissolved oxygen found below a sewage outfall depends on the BOD strength and quantity of the waste, as well as other factors. These other factors include velocity of the stream, stream length, biotic content, and the initial DO content (Porteous, 1992).

The rates of reaeration and deoxygenation determine the amount of DO in the stream. If no reaeration occurs, the DO will reach zero in a short period of time after the initial discharge of sewage into the stream. Because of reaeration (the rate of which is influenced directly by the rate of deoxygenation), enough compensation for aerobic decomposition of organic matter occurs. When the stream velocity is too low and the stream too deep, the DO level may reach zero.

Depletion of oxygen causes an oxygen deficit, which in turn causes absorption of atmospheric oxygen at the air–liquid interface. Thorough mixing from turbulence brings about effective reaeration. A shallow, rapid stream has a higher rate of reaeration (constantly saturated with oxygen) and purifies itself faster than a deep sluggish one will (Smith, 1974).

> *Important concept*: Reoxygenation of a stream is effected through aeration, absorption, and photosynthesis. Riffles and other natural turbulence in streams enhance aeration and oxygen absorption. Aquatic plants add oxygen to the water through transpiration. Oxygen production from photosynthesis of aquatic plants, primarily blue-green algae, slows down or ceases at night, creating a diurnal or daily fluctuation in DO levels in streams. The amount of DO a stream can retain increases as water temperatures cool and concentration of dissolved solids diminishes.

11.8 STREAM PURIFICATION: A QUANTITATIVE ANALYSIS

> *Note*: The important concepts presented in this section are heavily excerpted from Spellman's *Stream Ecology & Self-Purification* (1996).

Before sewage is dumped into a stream, determining the maximum BOD loading for the stream is important to avoid rendering it septic. The most common method of ultimate wastewater disposal is discharge into a selected body of water. The receiving water, stream, lake, or river is given the final job of purification. The degree of purification that takes place depends upon the flow or volume, oxygen content, and reoxygenation ability of the receiving water. Self-purification is a dynamic variable, changing from day to day, and closely following the hydrological variation characteristic of each stream. Additional variables include stream runoff, water temperature, reaeration, and the time of passage down the stream.

The purification process is carried out by several different aquatic organisms. Mathematical expressions help in determining the oxygen response of the receiving stream. Keep in mind, however, that because the biota and conditions in various parts of the stream change (that is, decomposition of organic matter in a stream is a function of degradation by microorganisms, and oxygenation by reaeration — competing processes working simultaneously), quantifying variables and results is difficult.

Streeter and Phelps first described the most common and well-known mathematical equation for oxygen sag for streams and rivers in 1925. The Streeter–Phelps equation is:

$$D = \frac{K_1 L_A}{K_2 - K_1}[e^{-K_1 t} - e^{-K_2 t}] + D_A e^{-K_2 t} \qquad (11.11)$$

where

D = dissolved oxygen deficit (parts per million)

t = time of flow (days)

L_A = ultimate BOD of the stream after the waste enters

D_A = initial oxygen deficit (before discharge) (ppm)

K_1 = BOD rate coefficient (per day)

K_2 = reaeration constant (per day)

Table 11.1   Typical Reaeration Constants (K_2) for Water Bodies

Water body	Ranges of K_2 at 20°C
Backwaters	0.10–0.23
Sluggish streams	0.23–0.35
Large streams (low velocity)	0.35–0.46
Large streams (normal velocity)	0.46–0.69
Swift streams	0.69–1.15
Rapids	>1.15

Source: From Spellman, F.R., 1996, *Stream Ecology & Self-Purification*. Lancaster, PA: Technomic Publishing Co. Inc.

Note: K_1 (or deoxygenation constant) is the rate at which microbes consume oxygen for aerobic decomposition of organic matter. Use the following equation to calculate K_1:

$$y = L(1 - 10^{-Kt}) \quad \text{or} \quad K_1 = \frac{-\log(1\ y/L)}{5t} \tag{11.12}$$

where:
y = BOD$_5$ (5 days BOD)
L = ultimate or BOD$_{21}$
K_1 = deoxygenation constant
t = time in days (5 days)

K_2 (reaeration constant) is the reaction characteristic of the stream and varies from stretch to stretch depending on the velocity of the water; the depth; the surface area exposed to the atmosphere; and the amount of biodegradable organic matter in the stream. K_2 is given in Table 11.1.

The reaeration constant for a fast moving, shallow stream is higher than for a sluggish stream or a lake. For shallow streams where vertical gradient and sheer stress exist, the reaeration constant is commonly found using the following formulation:

$$K_{2\,(20°C)} = \frac{48.6S^{1/4}}{H^{5/4}} \tag{11.13}$$

The reaeration constant for turbulence typical in deep streams can be found by using the following equation:

$$K_{2\,(20°C)} = \frac{13.0V^{1/2}}{H^{3/2}} \tag{11.14}$$

where:
K_2 = reaeration constant
V = velocity of stream (feet per second)
H = stream depth (feet)
S = slope of stream bed (feet per foot)

The Streeter–Phelps equation should be used with caution. Remember that this equation assumes that variable conditions (flow, BOD removal and oxygen demand rate, depth and temperature throughout the stream) are constant. In other words, the equation assumes that all conditions are the same (or constant) for every stream. However, this is seldom true; streams, of course, change from reach to reach and each stream is different. Additionally, because rivers and streams are longer than they are wide, organic pollution mixes rapidly in their surface waters. Because some rivers and streams are wider than others, the mixing of organic pollutants with river or stream water does not occur at the same rate in different rivers and streams.

Example 11.7

Problem:

Calculate the oxygen deficit in a stream after pollution. Use the following equation and parameters for a stream to calculate the oxygen deficit D in the stream after pollution.

$$D = \frac{K_1 L_A}{K_2 - K_1} \left[e^{-K1t} - e^{-K2t} \right] + D_A e^{-K2t}$$

Parameters:
Pollution enters stream at point X
t = 2.13
L_A = 22 mg/L (of pollution and stream at point X)
D_A = 2 mg/L
K_1 = 0.280/day (base e)
K_2 = 0.550/day (base e)

Note: To convert log base e to base 10, divide by 2.31.

Solution:

$$D = \frac{0.280 \times 22}{0.550 - 0.280} [e^{-0.280 \times 2.13} - e^{-0.550 \times 2.13}] + 2e^{-0.550 \times 2.13}$$

$$\frac{6.16}{0.270} [10^{-0.258} - 10^{-0.510}] + 2 \times 10^{-0.510}$$

$$= 22.81 [0.5520 - 0.3090] + 2 \times 0.3090$$

$$= 22.81 \times 0.243 + 0.6180 \, \text{mg/L}$$

$$= 6.16 \, \text{mg/L}$$

Example 11.8

Problem:

Calculate deoxygenation constant K_1 for a domestic sewage with BOD_5, 135 mg/L and BOD_{21}, 400 mg/L.

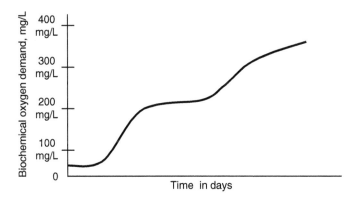

Figure 11.2   Sketch of the carbonaceous and nitrogenous BOD in a waste sample. (From Spellman, F.R., 1996, *Stream Ecology & Self-Purification*. Lancaster, PA: Technomic Publishing Company.)

Solution:

$$K_1 = \frac{-\log\left(1 - \dfrac{BOD_1}{BOD_{21}}\right)}{t}$$

$$\frac{-\log\left(1 - \dfrac{135}{400}\right)}{5}$$

$$\frac{-\log 0.66}{5}$$

$$\frac{0.1804}{5}$$

$$= 0.361/day$$

BOD occurs in two different phases (see Figure 11.2). The first is called the carbonaceous BOD (CBOD) because mainly organic or carbonaceous material is broken down. The second phase is called the nitrogenous phase. Here, nitrogen compounds are decomposed, requiring oxygen. This is of particular concern when conducting tests for discharge permit compliance, especially if nitrification is known to occur.

The Streeter–Phelps equation provides a rough estimate of the ecological conditions in a stream. Small variations in the stream may cause the DO to be higher or lower than the equation indicates. However, this equation may be used in several different ways. As we have said, the quantity of the waste is important in determining the extent of environmental damage. The Streeter–Phelps equation may be used to determine whether a certain stream has the capability of handling the estimated flow of wastes. If the stream receives wastes from other sites, then the equation may be helpful in determining whether the stream has recovered fully before it receives the next batch of wastes and, thus, whether it will be able to recover from the succeeding waste effluents.

REFERENCES

ACSE (1960). American Society of Civil Engineering Committee on Sanitary Engineering Research. Solubility of atmospheric oxygen in water. *J. Sanitary Eng. Div.* 86: 41–53.

APHA, AWWA, and WEF. (1995). *Standard Methods for the Examination of Water and Wastewater.* 19th ed. Washington, D.C.: American Public Health Association.

Clark, J.W., Viessman, W., Jr., and Hammer, M.J. (1977). *Water Supply and Pollution Control*, 3rd ed. New York: Harper & Row.

Gupta, R.S. (1997). *Environmental Engineering and Science: An Introduction.* Rockville, MD: Government Institutes.

Halsam, S.M. (1990). *River Pollution: An ecological perspective.* New York: Bellhaven Press.

Hem, J.D. (1985). Study and interpretation of the chemical characteristics of natural water, 3rd ed. Washington, D.C.: U.S. Geological Survey Water-Supply Paper.

Mason, C.F. (1990). Biological aspects of freshwater pollution. In *Pollution: Causes, Effects, and Control.* R.M. Harrison, Ed. Cambridge, Great Britain: The Royal Society of Chemistry.

Moran, J.M., Morgan, M.D., and Wiersma, J.H. (1986). *Introduction to Environmental Science.* New York: W.H. Freeman & Company.

Peavy, H.S., Rowe, D.R., and Tchobanoglous, G. (1975). *Environmental Engineering.* New York: McGraw-Hill, Inc.

Porteous, A. (1992). *Dictionary of Environmental Science and Technology* (revised ed.). New York: John Wiley & Sons.

Smith, R.I. (1974). *Ecology and Field Biology.* New York: Harper & Row.

Spellman, F.R. (1996). *Stream Ecology & Self-Purification.* Lancaster, PA: Technomic Publishing Company.

Spellman, F.R. and Whiting, N.E. (1999). *Water Pollution Control Technology: Concepts and Applications.* Rockville, MD: Government Institutes.

Streeter, J.P. and Phelps, E.B. (1925). A study of the pollution and natural purification of the Ohio River. Cincinnati: U.S. Public Health Service, Bulletin No. 146.

USEPA (1986). *Quality Criteria for Water 1986.* Washington, D.C.: United States Environmental Protection Agency.

USEPA (1994). *What Is Nonpoint Source Pollution?* Washington, D.C.: United States Environmental Protection Agency, EPA-F-94-005.

Velz, C.J. (1970). *Applied Stream Sanitation.* New York: Wiley-Interscience.

CHAPTER 12

Still Waters

12.1 INTRODUCTION

Consider a river pool, isolated by fluvial processes and time from the main stream flow. We are immediately struck by one overwhelming impression: it appears so still ... so very still ... still enough to soothe us. The river pool provides a kind of poetic solemnity, if only at the pool's surface. No words of peace, no description of silence or motionlessness can convey the perfection of this place, in this moment stolen out of time. We ask ourselves, "The water is still, but does the term still correctly describe what we are viewing? Is there any other term we can use besides that? Is there any other kind of *still*?"

Yes, of course, we know many ways to characterize still. For sound or noise, *still* can mean inaudible, noiseless, quiet, or silent. With movement (or lack of movement), *still* can mean immobile, inert, motionless, or stationary. At least this is how the pool appears at its surface to the casual visitor. The visitor sees no more than water and rocks.

The rest of the pool? We know very well that a river pool is more than just a surface. How does the rest of the pool (the subsurface, for example) fit the descriptors we tried to use to characterize its surface? Maybe they fit; maybe they do not. In time, we will go beneath the surface, through the liquid mass, to the very bottom of the pool to find out. For now, remember that images retained from first glances are almost always incorrectly perceived, incorrectly discerned, and never fully understood.

On second look, we see that the fundamental characterization of this particular pool's surface is correct enough. Wedged in a lonely riparian corridor — formed by river bank on one side and sand bar on the other — it is between a youthful, vigorous river system on its lower end and a glacier- and artesian-fed lake on its headwater end. The surface of the large pool, almost entirely overhung by mossy old Sitka spruce (at least at this particular location), is indeed still. In the proverbial sense, the pool's surface is as still and as flat as a flawless sheet of glass.

The glass image is a good one because, like perfect glass, the pool's surface is clear, crystalline, unclouded, definitely transparent, and yet deceptive. The water's clarity, accentuated by its bone-chilling coldness, is apparent at close range. Further back, we see only the world reflected in the water —the depths are hidden and unknown. Quiet and reflective, the polished surface of the water perfectly reflects in mirror-image reversal the spring greens of the forest at the pond's edge, without the slightest ripple. Up close, looking straight into the bowels of the pool, we are struck by the water's transparency. In the motionless depths, we do not see a deep, slow-moving reach with muddy bottom typical of a river or stream pool; instead, we clearly see the warm variegated tapestry of blues, greens, and blacks stitched together with threads of fine, warm-colored sand that carpets the bottom at least 12 feet below. Still waters can run deep.

No sounds emanate from the pool. The motionless, silent water does not, as we might expect, lap against its bank or bubble or gurgle over the gravel at its edge. Here, the river pool, held in temporary bondage, is patient, quiet, waiting, withholding all signs of life from its surface visitor.

Then the reality check occurs: the present stillness, like all feelings of calm and serenity, could be fleeting, you think. Of course, you would be correct because there is nothing still about a healthy river pool. At this exact moment, true clarity is present; it just needs to be perceived.

We toss a small stone into the river pool and watch the concentric circles ripple outward as the stone drops through the clear depths to the pool bottom. For a brief instant, we are struck by the obvious: the stone sinks to the bottom, following the laws of gravity, just as the river flows, according to those same inexorable laws, downhill in its search for the sea. As we watch, the ripples die away, leaving as little mark as the usual human lifespan creates in the waters of the world, and disappear as if they had never been. Now the river water is as before, still. At the pool's edge, we look down through the depth to the very bottom — the substrate.

We determine that the pool bottom is not flat or smooth, but instead is pitted and mounded occasionally with discontinuities. Gravel mounds alongside small corresponding indentations — small, shallow pits — make it apparent to us that gravel was removed from the indentations and piled into slightly higher mounds. From our topside position, as we look down through the cool liquid, the exact height of the mounds and the depth of the indentations are difficult for us to judge; our vision is distorted through several feet of water. However, we can detect movement near and through the low gravel mounds (where female salmon buried their eggs and where their young grow until they are old enough to fend for themselves): water flow — an upwelling of groundwater. This water movement explains our ability to see the variegated color of pebbles. The mud and silt that would normally cover these pebbles has been washed away by the water's subtle movement. Obviously, in the depths, our still water is not as still as it first appeared.

The slow, inexorable flow of water in and out of the pool, along with the up-flowing of groundwater through the pool's substrate and through the salmon redds (nests) is only a small part of the activities occurring within the pool, which include the air above it, the vegetation surrounding it, and the damp bank and sandbar forming its sides.

If we could look at a cross-sectional slice of the pool, at the water column, the surface of the pool may carry animals that can literally walk on water. The body of the pool may carry rotifers and protozoa and bacteria — tiny microscopic animals — as well as many fish. Fish inhabit hidden areas beneath large rocks and ledges, to escape predators. Going down further in the water column, we come to the pool bed. This is called the benthic zone, and certainly the greatest number of creatures lives here, including larvae and nymphs of all sorts; worms; leeches; flatworms; clams; crayfish; dace; brook lampreys; sculpins; suckers; and water mites.

We need to go down even further into the pool bed to see the whole story. How far this goes and what lives here, beneath the water, depend on whether it is a gravelly bed or a silty or muddy one. Gravel allows water, with its oxygen and food, to reach organisms that live underneath the pool. Many of the organisms found in the benthic zone may also be found underneath in the hyporheal zone.

To see the rest of the story, we need to look at the pool's outlet and where its flow enters the main river. In the riffles, the water runs fast and is disturbed by flowing over rocks. Only organisms that cling very well, such as net-winged midges; caddisflies; stoneflies; some mayflies; dace; and sculpins can spend much time here, and the plant life is restricted to diatoms and small algae. Riffles are good places for mayflies, stoneflies, and caddisflies to live because they offer plenty of gravel in which to hide.

At first, we struggled to find the proper words to describe the river pool. Eventually, we settled on still waters. We did this because of our initial impression and because of our lack of understanding and lack of knowledge. Even knowing what we know now, we might still describe the river pool as still waters. However, in reality, we must call the pool what it really is — a dynamic habitat. Because each river pool has its own biological community, all members are interwoven with each other in complex fashion, all depending on each other. Thus, our river pool habitat is part of a complex, dynamic ecosystem. On reflection, we realize that anything dynamic certainly cannot be accurately characterized as still — including our river pool (Spellman and Drinan, 2001).

12.2 STILL WATER SYSTEMS

Freshwater systems may be conveniently considered in two classes: running water and still (standing) water. No sharp distinction can be made between the two classes. Lakes are defined as basins filled with water with no immediate means of flowing to the sea, containing relatively still waters. Ponds are small lakes in which rooted plants on the top layer reach to the bottom. Reservoirs are usually man-made impoundments of potable water. Lakes, ponds, and reservoirs are sensitive to pollution inputs because they flush out their contents relatively slowly. Lakes undergo eutrophication, an aging process caused by the inputs of organic matters and siltation. Simply put, lakes, ponds, and reservoirs (that is, all still waters) are temporary holding basins.

12.3 STILL WATER SYSTEM CALCULATIONS

Environmental engineers involved with still water system management are generally concerned with determining and measuring lake, pond, or reservoir morphometric data, which is commonly recorded on preimpoundment topographic maps. Determining and maintaining water quality in still water systems is also a major area of concern for environmental engineers. Water quality involves the physical, chemical, and biological integrity of water resources. USEPA and other regulatory agencies promulgate water quality goals for protection of water resources in watershed management. Again, most still water data are directly related to the morphological features of the water basin.

Mapping the water basin should be the centerpiece of any comprehensive study of a still water body. Calculations made from the map allow the investigator to accumulate and relate a lot of data concerning the still water body system. In determining and measuring a still water body's water quality, several different models are used. The purpose of modeling is to help the environmental engineer organize an extended project. Modeling is a direct measurement method intended for a smaller body of water (a lake, pond, or reservoir). For example, water budget models and energy budget (lake evaporation) models can be used.

12.3.1 Still Water Body Morphometry Calculations

Still water body volume (V), shoreline development (SDI), and mean depth ($\overline{D}$) can be calculated using the formulae provided by Wetzel (1975) and Cole (1994).

12.3.1.1 Volume

The volume of a still water body can be calculated when the area circumscribed by each isobath (i.e., each subsurface contour line) is known. Wetzel's formula for water body volume is:

$$V = \sum_{i=0}^{n} h/3 \, (A_i + A_{i+1} + \sqrt{A_i \times A_{i+1}}) \qquad (12.1)$$

where
V = volume, cubic feet, acre-feet, or cubic meters
h = depth of the stratum, feet or meters
i = number of depth stratum
A_i = area at depth i, square feet, acres, or square meters

Cole's formula for the volume of water between the shoreline contour (z_0) and the first subsurface contour (z_1) is as follows:

$$V_{z1} - z_0 = 1/3(Az_0 + Az_1 + \sqrt{Az_0 + Az_1})(z_1 - z_0) \qquad (12.2)$$

where

z_0 = shoreline contour
z_1 = first subsurface contour
Az_0 = total area of the water body
Az_1 = area limited by the z_1 line

12.3.1.2 Shoreline Development Index (D_L)

The development of the shoreline is a comparative figure relating the shoreline length to the circumference of a circle with the same area as the still water body. The smallest possible index would be 1.0. For the following formula, L and A must be in consistent units (meters and square meters) for this comparison:

$$D_L = \frac{L}{2\sqrt{\pi A}} \qquad (12.3)$$

where

L = length of shoreline, miles or meters
A = surface area of lake, acres, square feet, or square meters

12.3.1.3 Mean Depth

The still water body volume divided by its surface area will yield the mean depth. Remember to keep units the same. If volume is in cubic meters, then area must be in square meters. The equation would be:

$$\overline{D} = \frac{V}{A} \qquad (12.4)$$

where

$\overline{D}$ = mean depth, feet or meters
V = volume of lake, cubic feet, acre-feet, or cubic meters
A = surface area, square feet, acres, or square meters

Example 12.1

Problem:

A pond has a shoreline length of 8.60 miles; the surface area is 510 acres, and its maximum depth is 8.0 ft. The areas for each foot depth are 460, 420, 332, 274, 201, 140, 110, 75, 30, and 1. Calculate the volume of the lake, shoreline development index, and mean depth of the pond.

Solution:

Step 1. Compute volume of the pond:

$$V = \sum_{i=0}^{n} h/3(A_i + A_{i+1} + \sqrt{A_i \times A_{i+1}})$$

$$= 1/3[(510 + 460 + \sqrt{510 \times 460}) + (460 + 420 + \sqrt{460 \times 420})$$

$$+ (420 + 332 + \sqrt{420 \times 332}) + (332 + 274 + \sqrt{332 \times 274})$$

$$+ (274 + 201 + \sqrt{274 \times 201}) + (201 + 140 + \sqrt{201 \times 140})$$

$$+ (140 + 110 + \sqrt{140 \times 110}) + (110 + 75 + \sqrt{110 \times 75})$$

$$+ (75 + 30 + \sqrt{75 \times 30}) + (30 + 1 + \sqrt{30 \times 0})]$$

$$= 1/3[6823]$$

$$= 2274 \text{ acre-ft}$$

Step 2. Compute shoreline development index:

$$D_L = \frac{L}{2\sqrt{\pi A}}$$

A = 510 acres = 510 acres × 1 sq. mi/640 acres = 0.7969 sq. mi

$$D_L = \frac{8.60 \text{ miles}}{2\sqrt{3.14 \times 0.7969 \text{ sq. mi}}}$$

$$\frac{8.60 \text{ mi}}{3.16}$$

$$= 2.72$$

Step 3. Compute mean depth:

$$\overline{D} = \frac{V}{A}$$

$$\frac{2274 \text{ acre-ft}}{510 \text{ acres}}$$

$$= 4.46 \text{ ft}$$

Other morphometric information can be calculated by the following formulae.

Bottom Slope

$$S = \frac{\overline{D}}{D_m} \tag{12.5}$$

where
S = bottom slope
$\overline{D}$ = mean depth, feet or meters
D_m = maximum depth, feet or meters

Volume Development (D$_v$)

According to Cole (1994), another morphometric parameter is volume development, D_v. This compares the shape of the (still water) basin to an inverted cone with a height equal to D_m and a base equal to the (still water body's) surface area:

$$D_v = 3 \frac{\overline{D}}{D_m} \tag{12.6}$$

Water Retention Time

$$RT = \frac{\text{Storage capacity, acre-ft or m}^3}{\text{Annual runoff, acre-ft/yr or m}^3\text{/yr}} \tag{12.7}$$

where RT = retention time, years.

Ratio of Drainage Area to Still Water Body Capacity R

$$R = \frac{\text{Drainage area, acre or m}^2}{\text{Storage capacity, acre-ft or m}^3} \tag{12.8}$$

Example 12.2

Problem:
   Assume annual rainfall is 38.8 in. and watershed drainage is 10,220 acres. Using the data provided in Example 12.1, calculate the bottom slope, volume development, water retention time, and ratio of drainage area to lake capacity.

Solution:

Step 1. Bottom slope:

$$S = \frac{\overline{D}}{D_m}$$

$$= \frac{4.46 \text{ ft}}{8.0}$$

$$= 0.56$$

Step 2. Volume development:

$$D_v = 3 \frac{\overline{D}}{D_m}$$

$$= 3 \times 0.56$$

$$= 1.68$$

Step 3. Water retention time:

Storage capacity V = 2274 acre-ft

Annual runoff = 38.8 in./yr × 10,220 acres = 38.8 in./yr × feet per 12 in. × 10,220 acres = 33,045 acre-ft/yr

$$RT = \frac{\text{Storage capacity}}{\text{Annual runoff}}$$

$$= \frac{2274 \text{ acre-ft}}{33,045 \text{ acre-ft/yr}}$$

$$= 0.069 \text{ yr}$$

Step 4. Ratio of drainage area to lake capacity:

$$R = \frac{\text{Drainage area}}{\text{Storage capacity}}$$

$$= \frac{10,220}{2274}$$

$$= \frac{4.49}{1}$$

12.4 STILL WATER SURFACE EVAPORATION

In lake and reservoir management, knowledge of evaporative processes is important to the environmental engineer in understanding how to determine water losses through evaporation. Evaporation increases the storage requirement and decreases the yield of lakes and reservoirs. Several models and empirical methods are used for calculating lake and reservoir evaporative processes. In the following, we present applications used for the water budget and energy budget models, along with four empirical methods: the Priestly–Taylor; Penman; DeBruin–Keijman; and Papadakis equations.

12.4.1 Water Budget Model

The water budget model for lake evaporation is used to make estimations of lake evaporation in some areas. It depends on an accurate measurement of the inflow and outflow of the lake and is expressed as:

$$\Delta S = P + R + GI - GO - E - T - O \tag{12.9}$$

where
ΔS = change in lake storage, millimeters
P = precipitation, millimeters
R = surface runoff or inflow, millimeters
GI = groundwater inflow, millimeters
GO = groundwater outflow, millimeters
E = evaporation, millimeters
T = transpiration, millimeters
O = surface water release, millimeters

If a lake has little vegetation and negligible groundwater inflow and outflow, lake evaporation can be estimated by:

$$E = P + R - O \pm \Delta S \tag{12.10}$$

Note: Much of the following information is adapted from USGS (2003) and Mosner and Aulenbach (2003).

12.4.2 Energy Budget Model

According to Rosenberry et al. (1993), the energy budget (Lee and Swancar, 1996; Equation 12.12) is recognized as the most accurate method for determining lake evaporation. Mosner and Aulenbach (2003) point out that it is also the most costly and time-consuming method. The evaporation rate, E_{EB}, is given by (Lee and Swancar, 1996):

$$E_{EB}, \text{cm/day} = \frac{Q_s - Q_r + Q_a + Q_{ar} - Q_{bs} + Q_v - Q_x}{L(1 + BR) + T_0} \tag{12.11}$$

where
E_{EB} = evaporation, in centimeters per day
Q_s = incident shortwave radiation, in calories per square centimeter per day

Q_r = reflected shortwave radiation, in calories per square centimeter per day
Q_a = incident longwave radiation from atmosphere, in calories per square centimeter per day
Q_{ar} = reflected longwave radiation, in calories per square centimeter per day
Q_{bs} = longwave radiation emitted by lake, in calories per square centimeter per day
Q_v = net energy advected by streamflow, ground water, and precipitation, in calories per square centimeter per day
Q_x = change in heat stored in water body, in calories per square centimeter per day
L = latent heat of vaporization, in calories per gallon
BR = Bowen ratio, dimensionless
T_0 = water-surface temperature (degrees Celsius)

12.4.3 Priestly–Taylor Equation

Winter and colleagues (1995) point out that the Priestly-Taylor equation, used to calculate potential evapotranspiration (PET) or evaporation as a function of latent heat of vaporization and heat flux in a water body, is defined by the equation:

$$\text{PET, cm/day} = \alpha(s/s + \gamma))[(Q_n - Q_x)/L] \qquad (12.12)$$

where
PET = potential evapotranspiration, centimeters per day
α = 1.26; Priestly–Taylor empirically derived constant, dimensionless
$(s/s + \gamma)$ = parameters derived from slope of saturated vapor pressure-temperature curve at the mean air temperature; γ is the psychrometric constant; and s is the slope of the saturated vapor pressure gradient, dimensionless
Q_n = net radiation, calories per square centimeter per day
Q_x = change in heat stored in water body, calories per square centimeter per day
L = latent heat of vaporization, calories per gallon

12.4.4 Penman Equation

Winter et al. (1995) point out that the Penman equation for estimating potential evapotranspiration, E_0, can be written as:

$$E_0 = \frac{(\Delta/\gamma)\, H_e + E_a}{(\Delta/\gamma) + 1} \qquad (12.13)$$

where
Δ = slope of the saturation absolute humidity curve at the air temperature
γ = the psychrometric constant
H_e = evaporation equivalent of the net radiation
E_a = aerodynamic expression for evaporation

12.4.5 DeBruin–Keijman Equation

The DeBruin–Keijman equation (Winter et al., 1995) determines evaporation rates as a function of the moisture content of the air above the water body, the heat stored in the still water body, and the psychrometric constant, which is a function of atmospheric pressure and latent heat of vaporization.

$$PET, cm/day = [SVP/0.95SVP + 0.63\gamma(Q_n - Q_x)] \qquad (12.14)$$

where SVP = saturated vapor pressure at mean air temperature, millibars per Kelvin. All other terms have been defined previously.

12.4.6 Papadakis Equation

The Papadakis equation (Winger et al., 1995) does not account for the heat flux that occurs in the still water body to determine evaporation. Instead, the equation depends on the difference in the saturated vapor pressure above the water body at maximum and minimum air temperatures; evaporation is defined by the equation:

$$PET, cm/day = 0.5625[E_0 max - (E_0 min-2)] \qquad (12.15)$$

where all terms have been defined previously.

REFERENCES

Cole, G.A. (1994). *Textbook of Limnology*, 4th ed. Prospect Heights, IL: Waveland Press.

Lee, T.M. and Swancar, A. (1996). Influence of evaporation, ground water, and uncertainty in the hydrologic budget of Lake Lucerne, a seepage lake in Polk County, Florida. U.S. Geologic Survey Water-Supply Paper 2439.

Mosner, M.S. and Aulenbach, B.T. (2003). Comparison of methods used to estimate lake evaporation for a water budget of Lake Seminole, Southwestern Georgia and Northwestern Florida. U.S. Geological Survey, Atlanta, Georgia.

Rosenberry, D.O., Sturrock, A.M., and Winter, T.C. (1993). Evaluation of the energy budget method of determining evaporation at Williams Lake, Minnesota, using alternative instrumentation and study approaches. *Water Resources Res.*, 29(8): 2473–2483.

Spellman, F.R. and Drinan, J. (2001). *Stream Ecology & Self-Purification*, 2nd ed. Lancaster, PA: Technomic Publishing Company.

Wetzel, R.G. (1975). *Limnology*. Philadelphia, PA: W.B. Saunders Company.

Winter, T.C., Rosenberry, D.O., and Sturrock, A.M. (1995). Evaluation of eleven equations for determining evaporation for a small lake in the north central United States, *Water Resources Res.*, 31(4), 983–993.

CHAPTER 13

Groundwater

Once polluted, groundwater is difficult, if not impossible, to clean up, since it contains few decomposing microbes and is not exposed to sunlight, strong water flow, or any of the other natural purification processes that cleanse surface water.

Eugene P. Odum; *Ecology and Our Endangered Life Support Systems*

13.1 GROUNDWATER AND AQUIFERS

Part of the precipitation that falls on land may infiltrate the surface, percolate downward through the soil under the force of gravity, and become what is known as *groundwater*. Like surface water, groundwater is an extremely important part of the hydrologic cycle. Almost half of the people in the U.S. obtain their public water supply from groundwater. Overall, the U.S. has more groundwater than surface water, including the water in the Great Lakes. Unfortunately, pumping it to the surface for use is sometimes uneconomical and, in recent years, the pollution of groundwater supplies from improper disposal has become a significant problem (Spellman, 1996).

Groundwater is found in saturated layers called *aquifers* that lie under the Earth's surface. Aquifers are made up of a combination of solid material such as rock and gravel, and open spaces called *pores*. Regardless of the type of aquifer, the groundwater in the aquifer is always in motion. Aquifers that lie just under the Earth's surface are in the zone of saturation and are called *unconfined aquifers* (see Figure 13.1). The top of the zone of saturation is the water table. An unconfined aquifer is only contained on the bottom and is dependent on local precipitation for recharge. This type of aquifer is often referred to as a *water table aquifer*.

The actual amount of water in an aquifer is dependent upon the amount of space available between the various grains of material that make up the aquifer. The amount of space available is called *porosity*. The ease of movement through an aquifer depends upon how well the pores are connected. The ability of an aquifer to pass water is called *permeability*. Types of aquifers include:

- Unconfined aquifers: a primary source of shallow well water (see Figure 13.1). Because these wells are shallow, they are subject to local contamination from hazardous and toxic materials that provide increased levels of nitrates and microorganisms, including fuel and oil; agricultural runoff containing nitrates and microorganisms; and septic tanks. (Note that this type of well may be classified as groundwater under the direct influence of surface water [GUDISW] and therefore require treatment for control of microorganisms [disinfection]).
- Confined aquifers: aquifers sandwiched between two impermeable layers that block the flow of water. The water in a confined aquifer is under hydrostatic pressure. It does not have a free water table (see Figure 13.2).

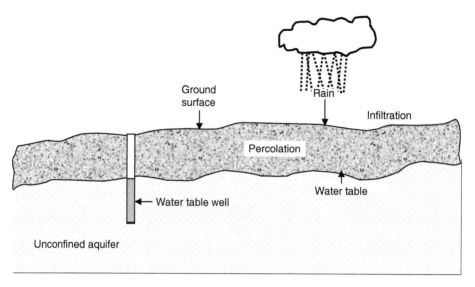

Figure 13.1 Unconfined aquifer. (From Spellman, F.R., 1996, *Stream Ecology & Self-Purification*. Lancaster, PA: Technomic Publishing Company.)

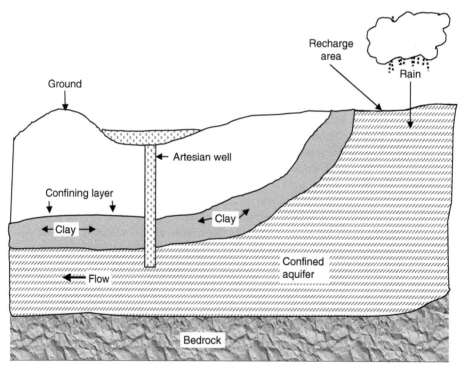

Figure 13.2 Confined aquifer. (From Spellman, F.R., 1996, *Stream Ecology & Self-Purification*. Lancaster, PA: Technomic Publishing Company.)

Confined aquifers are artesian aquifers. A well drilled in an artesian aquifer is called an *artesian well* and commonly yields large quantities of high-quality water. A well in a confined aquifer is normally referred to as a *deep* well and is not generally affected by local hydrological events. A confined aquifer is recharged by rain or snow in the mountains, where it is close to the surface of the Earth. Because the recharge area is some distance from areas of possible contamination of the confined aquifer, the possibility of contamination is usually very low. However, once contaminated, such aquifers may take centuries to recover.

When groundwater exits the Earth's crust, it is called a *spring*. The water in a spring can originate from a water table aquifer or from a confined aquifer. Only water from a confined aquifer spring is considered desirable for a public water system.

13.1.1 Groundwater Quality

Generally, groundwater possesses high chemical, bacteriological, and physical quality. When pumped from an aquifer composed of a mixture of sand and gravel, groundwater is often used without filtration (if not directly influenced by surface water). It can also be used without disinfection if it has a low coliform count. However (as pointed out earlier), groundwater can become contaminated. For example, when septic systems fail; saltwater intrudes; improper disposal of wastes occurs;, chemicals are improperly stock-piled; underground storage tanks leak; hazardous materials are spilled; fertilizers and pesticides are misapplied; and mines are improperly abandoned, groundwater can become contaminated.

When groundwater is removed from its underground water-bearing stratum via a well, water flows toward the center of the well. In a water table aquifer, this movement causes the water table to sag toward the well. This sag is called the *cone of depression*. The shape and size of the cone is dependent on the relationship between the pumping rate and the rate at which water can move toward the well. If the movement rate is high, the cone is shallow and its growth stable. The area included in the cone of depression is called the *zone of influence*; any contamination in this zone will be drawn into the well.

13.1.2 GUDISW

Groundwater under the direct influence of surface water (GUDISW) is not classified as a groundwater supply. When a supply is designated as GUDISW, the state's surface water rules apply to the source rather than the groundwater rules. The Surface Water Treatment Rule of the Safe Drinking Water Act requires each site to determine which groundwater supplies are influenced by surface water (when surface water can infiltrate a groundwater supply and could contaminate it with *Giardia*, viruses, turbidity, or organic material from the surface water source). To determine whether a groundwater supply is under the direct influence of surface water, USEPA has developed procedures that focus on significant and relatively rapid shifts in water quality characteristics, including turbidity, temperature, and pH. When these shifts can be closely correlated with rainfall or other surface water conditions or when certain indicator organisms associated with surface water are found, the source is said to be under the direct influence of surface water.

13.2 AQUIFER PARAMETERS

Certain aquifer parameters are relevant to determining the available volume of water and the ease of its withdrawal. We identify and define these relevant parameters in this section.

13.2.1 Aquifer Porosity

Aquifer porosity is defined as the ratio of the volume of voids (open spaces) in the soil to the total volume. Simply stated, porosity is the volume of open space and is often determined using the equation:

$$\varphi = \frac{V_{void}}{V_{total}}$$ (13.1)

Two basic types of porosity are common: *primary*, formed at the time the rock was deposited, and *secondary*, formed later (by dissolution of carbonate in caverns). Well-sorted materials tend to have higher porosities than poorly sorted ones. Fine-grained sediments tend to have higher porosities than coarse-grained sediments, although they are often poorly connected. Some typical values of porosity are clay, 55%; fine sand, 45%; sand and gravel, 20%; sandstone, 15%; and limestone, 15%. The interconnected or effective porosity (ϕ_e) is the most important in hydrology, and $\leq\phi$.

13.2.2 Specific Yield (Storage Coefficient)

Specific yield is the percentage of water that is free to drain from the aquifer under the influence of gravity. It is not equal to porosity because the molecular and surface tension forces in the pore spaces keep some of the water in the voids. Specific yield reflects the amount of water available for development (Davis and Cornwell, 1991). Specific yield and storage coefficient may be used interchangeably for unconfined aquifers.

13.2.3 Permeability (*K*)

Permeability describes the measure of an aquifer's ability to transmit water under a sloping piezometric surface. It is defined as the discharge that occurs through a unit cross-section of aquifer.

13.2.4 Transmissivity (*T*)

Transmissivity describes the capacity of an aquifer to transmit water. It is the product of hydraulic conductivity (permeability) and the aquifer's saturated thickness:

$$T = Kb$$ (13.2)

where
T = transmissivity of an aquifer, gallons per day per foot or cubic meters per day-meter
K = permeability, gallons per day per square foot (per day-square meter)
b = thickness of aquifer, feet or meters

A rough estimation of T is found by multiplying specific capacity by 2000 (USEPA, 1994).

Example 13.1

Problem:
    If an aquifer's thickness is 60 ft, estimate the permeability of the aquifer with transmissibility of 30,000 gpm/ft.

Solution:
    Rearranging Equation 13.2:

$$K = T/b = (30,000 \text{ gpm/ft})/60 \text{ ft}$$

$$= 500 \text{ gpm/ft}^2$$

13.2.5 Hydraulic Gradient and Head

The height of the potentiometric surface at any point in an aquifer is the hydraulic gradient. Stated differently, the hydraulic gradient is the slope of the piezometric surface. The difference in elevation from one point to another along the hydraulic gradient is a measure of pressure. This elevation difference is called *pressure head*.

We usually express the amount of mechanical energy that groundwater contains in terms of the hydraulic head (h), the total mechanical energy per unit weight. The hydraulic head is conveniently measured as the elevation to which water will rise in an open pipe, relative to some reference level. Therefore, the hydraulic head has units of length — for example, the elevation to which water will rise.

Two main components contribute to the mechanical energy or the hydraulic head of groundwater: potential energy due to gravity and pressure exerted on the water. Kinetic energy, a third energy, is caused by movement of water and is very small compared to the other two energies because groundwater flows very slowly and the energy can therefore be neglected. In terms of hydraulic head, the potential energy is expressed as the elevation head (z) or simply the elevation of the point of interest relative to some reference level. The energy of fluid pressure is expressed as the pressure head (h_p). The pressure head is equivalent to the height of the water column overlying the point of interest. The total hydraulic head is then given by:

$$h = z + h_p \qquad (13.3)$$

According to Baron (2003), groundwater will move from areas of high mechanical energy to areas with low mechanical energy. The hydraulic gradient of a flow system of interest is defined as the difference in hydraulic head between two points of interest (dh) and the flow distance between these two points (dl) or, in mathematical terms:

$$\text{gradient } h = dh/dl \qquad (13.4)$$

13.2.6 Flow Lines and Flow Nets

A flow line is an imaginary line that follows the path that a parcel of groundwater would follow as it flowed through an aquifer. These lines are useful tools for visualizing the flow of groundwater. Flow lines can be constructed from equipotential lines or lines of equal hydraulic head. The combination of equipotential lines and flow lines results in a flow net — basically, a graphical solution of the two-dimensional Laplace equation (Fetter, 1994).

13.3 GROUNDWATER FLOW

Groundwater flow for a steady-state condition in which the water table or piezometric head does not change within a specified time is expressed by the following equations (Gupta, 1997):

$$\text{Pore velocity or advection, } v = \frac{K\,(h_1 - h_2)}{nL} \qquad (13.5)$$

where pore area of flow, $A_v = nA$.

Because $Q = Av$, combining the two gives us Darcy's law:

$$\text{Rate of groundwater flow, } Q = \frac{K\,(h_1 - h_2)A}{L} \qquad (13.6)$$

where
Q = rate of groundwater flow
v = pore velocity or advection
K = hydraulic conductivity
A = aquifer cross-section area through which flow takes place
h_1 = water head at upstream end
h_2 = water head at downstream end
L = distance between h_1 and h_2
n = porosity

Note: The term $(h_1 - h_2)/L$ is the hydraulic gradient.

Example 13.2

Problem:

An irrigation ditch runs parallel to a pond; they are 2200 ft apart. A pervious formation of 40-ft average thickness connects them. Hydraulic conductivity and porosity of the pervious formation are 12 ft/day and 0.55, respectively. The water level in the ditch is at an elevation of 120 ft and 110 ft in the pond. Determine the rate of seepage from the channel to the pond.

Solution:

$$\text{Hydraulic gradient, I} = \frac{h_1 - h_2}{L} = \frac{120 - 110}{2200} = 0.0045$$

For each 1 ft width:

$$A = 1 \times 40 = 40 \text{ ft}^2$$

From Equation 13.6:

$$Q = (12 \text{ ft/day})(0.0045)(40 \text{ ft}^2) = 2.16 \text{ ft}^3/\text{day/ft width}$$

From Equation 13.5:

$$\text{Seepage velocity, v} = \frac{K(h_1 - h_2)}{nL} = \frac{(12)(0.0045)}{0.55} = 0.098 \text{ ft/day}$$

13.4 GENERAL EQUATIONS OF GROUNDWATER FLOW

The combination of Darcy's law and a statement of mass conservation results in general equations describing the flow of groundwater through a porous medium. These general equations are partial differential equations in which the spatial coordinates in all three dimensions (x, y, and z) and the time are all independent variables.

To derive the general equations, Darcy's law and the law of mass conservation are applied to a small volume of aquifer, the *control volume*. The law of mass conservation is basically an account of all the water that goes into and out of the control volume. That is, all the water that goes into the control volume must come out, or a change must occur in water storage in the control volume.

Applying these two laws to a confined aquifer results in Laplace's equation, a famous partial differential equation that can also be used to describe many other physical phenomena (for example, the flow of heat through a solid) (Baron, 2003):

$$d^2 \, h/dx^2 \, + \, d^2h/dy^2 \, + \, d^2/dz^2 \, = \, 0$$

Applying Darcy's law and the law of mass conservation to two-dimensional flow in an unconfined aquifer results in the Boussinesq equation:

$$d/dx \, (h \, dh/ds) \, + \, d/dx \, (h \, dh/dy) \, = \, S_y/K \, dh/dt$$

where S_y is the specific yield of the aquifer.

If the drawdown in the aquifer is very small compared with the saturated thickness, the variable thickness, h, can be replaced with an average thickness that is assumed to be constant over the aquifer. The Boussinesq equation can then be simplified to:

$$d^2h/dx^2 \, + \, d^2h/dy^2 \, = \, S_y/(Kb) \, dh/dt$$

Describing groundwater flow in confined and unconfined aquifers by using these general partial differential equations is difficult to solve directly. However, these differential equations can be simplified to algebraic equations for the solution of simple cases (for example, one-dimensional flow in a homogenous porous medium). Another approach is to use a flow net (described earlier) to solve Laplace's equation graphically for relatively simple cases. More complex cases, however, must be solved mathematically, most commonly with computerized groundwater modeling programs. The most popular of these programs is MODFLOW-2000, published by the United States Geological Society (USGS).

13.4.1 Steady Flow in a Confined Aquifer

If steady movement of groundwater occurs in a confined aquifer and the hydraulic heads do not change over time, we can use another derivation of Darcy's law directly to determine how much water is flowing through a unit width of aquifer, using the following equation:

$$q' \, = \, - \, Kb \, dh/dl \qquad\qquad (13.7)$$

where
q' = flow per unit width (L^2/T)
K = hydraulic conductivity (L/T)
b = aquifer thickness (L)
dh/dl = hydraulic gradient (dimensionless)

13.4.2 Steady Flow in an Unconfined Aquifer

Steady flow of water through an unconfined aquifer can be described by Dupuit's equation:

$$q' = -\tfrac{1}{2} K ((h_1^2 - h_2^2)/L) \qquad (13.8)$$

where
h_1 and h_2 = the water level at two points of interest
L = the distance between these two points

Based on Dupuit's assumptions, this equation states that the hydraulic gradient is equal to the slope of the water table; the streamlines are horizontal; and equipotentials are vertical. This equation is useful, particularly in field evaluations of the hydraulic characteristics of aquifer materials.

REFERENCES

Baron, D. (2003). Water: California's Precious Resource. Accessed at http://www.cs.Csubak.edu/Geology/Faculty/Baron/SuppGWNotes-2.htm.

Davis, M.L. and Cornwell, D.A. (1985). *Introduction to Environmental Engineering*, 2nd ed. New York: McGraw-Hill, Inc.

Fetter, C.W. (1994). *Applied Hydrology*, 3rd ed. New York: Prentice Hall.

Gupta, R.S. (1997). *Environmental Engineering and Science: An Introduction*. Rockville, MD: Government Institutes.

Odum, E.P. (1997). *Ecology and Our Endangered Life-Support Systems*. New York: Sinauer Associates.

Spellman, F.R. (1996). *Stream Ecology & Self-Purification*. Lancaster, PA: Technomic Publishing Company.

USEPA. (1994). *Handbook: Ground Water and Wellhead Protection*, EPA/625/R-94/001. Washington, D.C.: United States Environmental Protection Agency.

CHAPTER 14

Basic Hydraulics

14.1 INTRODUCTION

The word "hydraulic" is derived from the Greek words "hydro" (meaning water) and "aulis" (meaning pipe). Originally, "hydraulics" referred only to the study of water at rest and in motion (flow of water in pipes or channels). Today it is taken to mean the flow of any "liquid" in a system. *Hydraulics* — the study of fluids at rest and in motion — is essential for an understanding of how water/wastewater systems work, especially water distribution and wastewater collection systems.

14.2 BASIC CONCEPTS

Air pressure (at sea level) = 14.7 lb/in.^{2} (pounds per square inch or psi)

The relationship shown above is important because our study of hydraulics begins with air — a blanket of air, many miles thick, surrounding the Earth. The weight of this blanket on any given square inch of the Earth's surface varies according to the thickness of the atmospheric blanket above that point. As shown, at sea level, the pressure exerted is 14.7 lb/in.^{2} (psi). On a mountaintop, air pressure decreases because the blanket is not as thick.

$$1 \text{ ft}^3 \text{ H}_2\text{O} = 62.4 \text{ lb}$$

The relationship shown above is also important to us: cubic feet and pounds are used to describe a volume of water. A defined relationship exists between these two methods of measurement. The specific weight of water is defined relative to a cubic foot: 1 ft^3 of water weighs 62.4 lb. This relationship is true only at a temperature of 4°C and at a pressure of 1 atm (known as standard temperature and pressure [STP] 14.7 lb/in.^{2} at sea level containing 7.48 gal). The weight varies so little that, for practical purposes, we use this weight from a temperature of 0 to 100°C. One cubic inch of water weighs 0.0362 lb. Water 1 ft deep exerts a pressure of 0.43 lb/in^2 on the bottom area (12 in. × 0.0362 lb/in.^{3}). A column of water 2 ft high exerts 0.86 psi, one 10 ft high exerts 4.3 psi, and one 55 ft high exerts:

$$55 \text{ ft} \times 0.43 \text{ psi/ft} = 23.65 \text{ psi}$$

A column of water 2.31 ft high exerts 1.0 psi. To produce a pressure of 50 psi requires a water column:

$$50 \text{ psi} \times 2.31 \text{ ft/psi} = 115.5 \text{ ft}$$

A second relationship is also important:

$$1 \text{ gal H}_2\text{O} = 8.34 \text{ lb}$$

At standard temperature and pressure, 1 ft³ of water contains 7.48 gal. With these two relationships, we can determine the weight of 1 gal of water:

$$\text{wt. of gallon of water} = 62.4 \text{ lb} \div 7.48 \text{ gal} = 8.34 \text{ lb/gal}$$

Thus,

$$1 \text{ gal H}_2\text{O} = 8.34 \text{ lb}$$

Example 14.1

Problem:
Find the number of gallons in a reservoir that has a volume of 855.5 ft³.

Solution:

$$855.5 \text{ ft}^3 \times 7.48 \text{ gal/ft}^3 = 6{,}399 \text{ gallons (rounded)}$$

The term *head* is used to designate water pressure in terms of the height of a column of water in feet. For example, a 10-ft column of water exerts 4.3 psi. This can be called 4.3-psi pressure or 10 ft of head. Another example: if the static pressure in a pipe leading from an elevated water storage tank is 45 psi, what is the elevation of the water above the pressure gauge? Remembering that 1 psi = 2.31 ft and that the pressure at the gauge is 45 psi:

$$45 \text{ psi} \times 2.31 \text{ ft/psi} = 104 \text{ ft}$$

In demonstrating the relationship of the weight of water to the weight of air, we can say theoretically that the atmospheric pressure at sea level (14.7 psi) will support a column of water 34 ft high:

$$14.7 \text{ psi} \times 2.31 \text{ ft/psi} = 34 \text{ ft}$$

At an elevation of 1 mile above sea level, where the atmospheric pressure is 12 psi, the column of water would be only 28 ft high (12 psi × 2.31 ft/psi = 28 ft [rounded]).

If a glass or clear plastic tube is placed in a body of water at sea level, the water will rise in the tube to the same height as the water outside the tube. The atmospheric pressure of 14.7 psi will push down equally on the water surface inside and outside the tube. However, if the top of the tube is tightly capped and all of the air is removed from the sealed tube above the water surface, thus forming a perfect vacuum, the pressure on the water surface inside the tube will be 0 psi. The atmospheric pressure of 14.7 psi on the outside of the tube will push the water up into the tube until the weight of the water exerts the same 14.7 psi pressure at a point in the tube even with the water surface outside the tube. The water will rise: 14.7×2.31 ft/psi = 34 ft.

In practice, creating a perfect vacuum on Earth is impossible, so the water will rise somewhat less than 34 ft; the distance depends on the amount of vacuum created. If, for example, enough air were removed from the tube to produce an air pressure of 9.7 psi above the water in the tube, how far would the water rise in the tube? To maintain the 14.7 psi at the outside water surface level, the water in the tube must produce a pressure of 14.7 psi – 9.7 psi = 5.0 psi. The height of the column of water that will produce 5.0 psi is: 5.0 psi $\times 2.31$ ft/psi = 11.5 ft.

14.2.1 Stevin's Law

Stevin's law deals with water at rest. Specifically, it states: "The pressure at any point in a fluid at rest depends on the distance measured vertically to the free surface and the density of the fluid." Stated as a formula, this becomes:

$$p = w \times h \qquad\qquad\qquad (14.1)$$

where
p = pressure in pounds per square foot
w = density in pounds per cubic foot
h = vertical distance in feet

Example 14.2

Problem:
   What is the pressure at a point 18 ft below the surface of a reservoir?

Solution:
   Note: To calculate this, we must know that the density of the water, w, is 62.4 lb/ft³.

$$p = w \times h$$

$$= 62.4 \text{ lb/ft}^3 \times 18 \text{ ft}$$

$$= 1123 \text{ lb/ft}^2 \text{ or } 1123 \text{ psf}$$

Water and wastewater professionals generally measure pressure in pounds per square *inch* rather than pounds per square *foot*; to convert, divide by 144 in.²/ft² (12 in. $\times$ 12 in. = 144 in.²):

$$P = \frac{1123 \text{ psf}}{144 \text{ in.}^2/\text{ft}^2} = 7.8 \text{ lb/in.}^2 \text{ or psi (rounded)}$$

14.2.2 Density and Specific Gravity

When we say that iron is heavier than aluminum, we say that iron has greater density than aluminum. In practice, what we are really saying is that a given volume of iron is heavier than the same volume of aluminum. *Density* is the mass per unit volume of a substance.

Consider a tub of lard and a large box of cold cereal, each with a mass of 600 g. The density of the cereal would be much less than the density of the lard; the cereal occupies a much larger volume than the lard occupies.

The density of an object can be calculated by using the formula:

$$\text{Density} = \frac{\text{Mass}}{\text{Volume}} \tag{14.2}$$

In general use, perhaps the most common measures of density are pounds per cubic foot (lb/ft^3) and pounds per gallon (lb/gal):

- 1 ft^3 of water weighs 62.4 lb; density = 62.4 lb/ft^3
- 1 gal of water weighs 8.34 lb; density = 8.34 lb/gal

The density of a dry material (including cereal, lime, soda, and sand) is usually expressed in pounds per cubic foot. The density of a liquid, (including liquid alum, liquid chlorine, or water) can be expressed as pounds per cubic foot or as pounds per gallon. The density of a gas (including chlorine gas, methane, carbon dioxide, or air) is usually expressed in pounds per cubic foot.

The density of a substance like water changes slightly as the temperature of the substance changes. This occurs because substances usually increase in volume (or size; they expand) as they become warmer. Because of this expansion with warming, the same weight is spread over a larger volume, so the density is lower when a substance is warm than when it is cold.

Specific gravity is the weight (or density) of a substance compared to the weight (or density) of an equal volume of water. (*Note*: The specific gravity of water is 1.) This relationship is easily seen when a cubic foot of water (weight 62.4 lb) is compared to a cubic foot of aluminum (weight 178 lb). Aluminum is 2.7 times as heavy as water.

Finding the specific gravity of a piece of metal is not that difficult. All we have to do is to weigh the metal in air, then weigh it under water. Its loss of weight is the weight of an equal volume of water. To find the specific gravity, divide the weight of the metal by its loss of weight in water:

$$\text{Specific Gravity} = \frac{\text{Weight of Substance}}{\text{Weight of Equal Volume of Water}} \tag{14.3}$$

Example 14.3

Problem:

Suppose a piece of metal weighs 150 lb in air and 85 lb under water. What is the specific gravity?

Solution:

Step 1. 150 lb – 85 lb = 65-lb loss of weight in water
Step 2.:

$$\text{Specific Gravity} = \frac{150}{65} = 2.3$$

Note: In a calculation of specific gravity, it is essential that the densities be expressed in the same units.

The specific gravity of water is 1, which is the standard, the reference against which all other liquid or solid substances are compared. Specifically, any object that has a specific gravity greater than 1 will sink in water (rocks, steel, iron, grit, floc, sludge). Substances with a specific gravity of less than 1 will float (wood, scum, gasoline). With the total weight and volume of a ship taken into consideration, its specific gravity is less than 1; therefore, it can float.

The most common use of specific gravity in water plant operations is in gallons-to-pounds conversions. In many cases, the liquids being handled have a specific gravity of 1.00 or very nearly 1.00 (between 0.98 and 1.02), so 1.00 may be used in the calculations without introducing significant error. However, in calculations involving a liquid with a specific gravity of less than 0.98 or greater than 1.02, the conversions from gallons to pounds must consider specific gravity. The technique is illustrated in the following example.

Example 14.4

Problem:
A basin holds 1455 gal of a certain liquid. If the specific gravity of the liquid is 0.94, how many pounds of liquid are in the basin?

Solution:
Normally, for a conversion from gallons to pounds, we would use the factor 8.34 lb/gal (the density of water), if the substance's specific gravity were between 0.98 and 1.02. However, in this instance, the substance has a specific gravity outside this range, so the 8.34 factor must be adjusted.

Multiply 8.34 lb/gal by the specific gravity to obtain the adjusted factor:

Step 1. (8.34 lb/gal) (0.94) = 7.84 lb/gal (rounded)
Step 2. Then, convert 1455 gal to pounds using the corrected factor:

$$(1455 \text{ gal}) (7.84 \text{ lb/gal}) = 11,407 \text{ lb (rounded)}$$

14.2.3 Force and Pressure

Water exerts forces against the walls of its container, whether stored in a tank or flowing in a pipeline. We can also say that it exerts *pressure*. However, force and pressure are different, although they are closely related.

- *Force* is the push or pull influence that causes motion. In the English system, force and weight are often used in the same way. The weight of a cubic foot of water is 62.4 lb. The force exerted on the bottom of a 1-ft cube is 62.4 lb (see Figure 14.1). If we stack two cubes on top of one another, the force on the bottom is 124.8 lb.

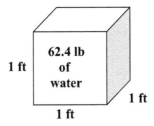

Figure 14.1 One cubic foot of water weighs 62.4 lb. (From Spellman, F.R. and Drinan, J., 2001, *Water Hydraulics*. Lancaster, PA, Technomic Publishing Company.)

- *Pressure* is a force per unit of area. In equation form, this can be expressed as:

$$P = \frac{F}{A} \qquad (14.4)$$

where
P = pressure
F = force
A = area over which the force is distributed

Pounds per square inch or pounds per square foot are common expressions of pressure. The pressure on the bottom of the cube is 62.4 lb/ft^2 (see Figure 14.1). Expressing pressure in pounds per square inch (psi) is normal. This is easily accomplished by determining the weight of 1 in.^{2} of a cube 1 ft high. If we have a cube that is 12 in. on each side, the number of square inches on the bottom surface of the cube is $12 \times 12 = 144$ in.^{2}. By dividing the weight by the number of square inches, we can determine the weight on each square inch:

$$\text{psi} = \frac{62.4 \text{ lb/ft}}{144 \text{ in.}^2} = 0.433 \text{ psi/ft}$$

This is the weight of a column of water 1-in. square and 1 ft tall. If the column of water were 2 ft tall, the pressure would be 2 ft $\times$ 0.433 psi/ft = 0.866.

With this information, we can convert feet of head to pounds per square inch by multiplying the feet of head times 0.433 psi/ft.

Example 14.5

Problem:
   A tank is mounted at a height of 90 ft. Find the pressure at the bottom of the tank.

Solution:

$$90 \text{ ft} \times 0.433 \text{ psi/ft} = 39 \text{ psi (rounded)}$$

If we wanted to make the conversion of pounds per square inch to feet, we would divide the pounds per square inch by 0.433 psi/ft.

Example 14.6

Problem:
   Find the height of water in a tank if the pressure at the bottom of the tank is 22 psi.

Solution:

$$\text{height in feet} = \frac{22 \text{ psi}}{0.433 \text{ psi/ft}} = 51 \text{ ft (rounded)}$$

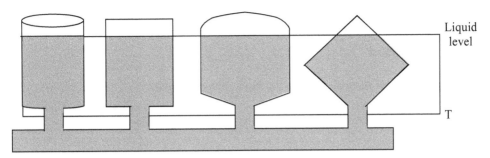

Figure 14.2 Hydrostatic pressure. (From Spellman, F.R. and Drinan, J., 2001, *Water Hydraulics*. Lancaster, PA, Technomic Publishing Company.)

14.2.4 Hydrostatic Pressure

Figure 14.2 shows a number of connected and differently shaped open containers of water. Note that the water level is the same in each container, regardless of its shape or size. This occurs because pressure is developed within water (or any other liquid) by the weight of the water above. If the water level in any one container were to be momentarily higher than that in any of the other containers, the higher pressure at the bottom of this container would cause some water to flow into the container with the lower liquid level. Also, the pressure of the water at any level (such as line T) is the same in each of the containers. Pressure increases because of the weight of the water. The farther down from the surface, the more pressure is created. This illustrates that the weight, not the volume, of water contained in a vessel determines the pressure at the bottom of the vessel.

Some very important principles always apply for hydrostatic pressure (Nathanson, 1997):

- The pressure depends only on the depth of water above the point in question (not on the water surface area).
- The pressure increases in direct proportion to the depth.
- The pressure in a continuous volume of water is the same at all points that are at the same depth.
- The pressure at any point in the water acts in all directions at the same depth.

14.2.5 Head

Head is defined as the vertical distance water/wastewater must be lifted from the supply tank to the discharge, or as the height a column of water would rise due to the pressure at its base. In a perfect vacuum, atmospheric pressure of 14.7 psi lifts water 34 ft. If we open the top of the sealed tube to the atmosphere and enclose the reservoir and then increase the pressure in the reservoir, the water will rise in the tube. Because atmospheric pressure is essentially universal, we usually ignore the first 14.7 psi of actual pressure measurements and measure only the difference between the water pressure and the atmospheric pressure; we call this *gauge pressure*. For example, water in an open reservoir is subjected to the 14.7 psi of atmospheric pressure, but subtracting this amount leaves a gauge pressure of 0 psi. This shows that the water would rise 0 ft above the reservoir surface. If the gauge pressure in a water main were 120 psi, the water would rise in a tube connected to the main:

$$120 \text{ psi} \times 2.31 \text{ ft/psi} = 277 \text{ ft}$$

The *total head* includes the vertical distance the liquid must be lifted (static head); the loss to friction (friction head); and the energy required to maintain the desired velocity (velocity head):

$$\text{Total Head } = \text{ Static Head } + \text{ Friction Head } + \text{ Velocity Head} \tag{14.5}$$

14.2.5.1 Static Head

Static head is the actual vertical distance the liquid must be lifted:

$$\text{Static Head } = \text{ Discharge Elevation } - \text{ Supply Elevation} \tag{14.6}$$

Example 14.7

Problem:

The supply tank is located at elevation 118 ft. The discharge point is at elevation 215 ft. What is the static head in feet?

Solution:

$$\text{Static Head, ft } = \text{ 215 ft } - \text{ 118 ft } = \text{ 97 ft}$$

14.2.5.2 Friction Head

Friction head is the equivalent distance of the energy that must be supplied to overcome friction. Engineering references include tables showing the equivalent vertical distance for various sizes and types of pipes, fittings, and valves. The total friction head is the sum of the equivalent vertical distances for each component:

$$\text{Friction Head, ft } = \text{ Energy Losses Due to Friction} \tag{14.7}$$

14.2.5.3 Velocity Head

Velocity head is the equivalent distance of the energy consumed in achieving and maintaining the desired velocity in the system (the height of a fluid column):

$$\text{Velocity Head, ft } = \text{ Energy Losses to Maintain Velocity} \tag{14.8}$$

or

$$V_h = \frac{V^2}{2g}$$

14.2.5.4 Total Dynamic Head (Total System Head)

$$\text{Total Head } = \text{ Static Head } + \text{ Friction Head } + \text{ Velocity Head} \tag{14.9}$$

14.2.5.5 Pressure/Head

The pressure exerted by water/wastewater is directly proportional to its depth or head in the pipe, tank or channel. If the pressure is known, the equivalent head can be calculated:

$$\text{Head, ft } = \text{ Pressure, psi } \times 2.31 \text{ ft/psi} \qquad (14.10)$$

or

$$PE = pAV \text{ (for water flow in pipe)}$$

where
p = pressure at a cross section
A = pipe cross-sectional area, square centimeters or square inches
V = mean velocity

Example 14.8

Problem:
    The pressure gauge on the discharge line from the influent pump reads 72.3 psi. What is the equivalent head in feet?

Solution:

$$\text{Head, ft } = 72.3 \times 2.31 \text{ ft/psi } = 167 \text{ ft}$$

14.2.5.6 Head/Pressure

If the head is known, the equivalent pressure can be calculated by:

$$\text{Pressure, psi } = \frac{\text{Head, ft}}{2.31 \text{ ft/psi}} \qquad (14.11)$$

Example 14.9

Problem:
    The tank is 22 ft deep. What is the pressure in pounds per square inch at the bottom of the tank when it is filled with water?

Solution:

$$\text{Pressure, psi } = \frac{22 \text{ ft}}{2.31 \text{ ft/psi}} = 9.52 \text{ psi}$$

14.3 FLOW/DISCHARGE RATE: WATER IN MOTION

The study of fluid flow is much more complicated than that of fluids at rest, but understanding these principles is important because water in waterworks and distribution systems and in wastewater treatment plants and collection systems is nearly always in motion. *Discharge* (or flow) is the quantity of water passing a given point in a pipe or channel during a given period of time. Stated another way for open channels: the flow rate through an open channel is directly related to the velocity of the liquid and the cross-sectional area of the liquid in the channel.

$$\text{Flow (Q), cfs} = \text{Area, ft}^2 \times \text{v, fps} \qquad (14.12)$$

where
Q = Flow/discharge in cubic feet per second
A = Cross-sectional area of the pipe or channel, square feet
V = water velocity in feet per second

Example 14.10

Problem:
The channel is 6 ft wide and the water depth is 3 ft. The velocity in the channel is 4 ft/sec. What is the discharge or flow rate in cubic feet per second?

Solution:

$$\text{Flow, cfs} = 6 \text{ ft} \times 3 \text{ ft} \times 4 \text{ ft/second} = 72 \text{ cfs}$$

Discharge or flow can be recorded as gallons per day (gpd), gallons per minute (gpm), or cubic feet per second (cfs). Flows treated by many waterworks or wastewater treatment plants are large and often referred to in million gallons per day (MGD). The discharge or flow rate can be converted from cubic feet per second to other units such as gallons per minute or million gallons per day by using appropriate conversion factors.

Example 14.11

Problem:
A pipe 12 in. in diameter has water flowing through it at 10 ft/sec. What is the discharge in (a) cubic feet per second; (b) gallons per minute; and (c) million gallons per day? Before we can use the basic formula from Chapter 2 (Equation 2.12), we must determine the area A of the pipe. The formula for the area of a circle is:

$$A = \pi \times \frac{D^2}{4} = \pi \times r^2 \qquad (14.13)$$

where
D = diameter of the circle in feet
r = radius of the circle in feet

Solution:

Step 1. Area of the pipe is:

$$A = \pi \times \frac{D^2}{4} = 3.14 \times \frac{(1 \text{ ft})^2}{4} = 0.785 \text{ ft}^2$$

Now we can determine the discharge in cubic feet per second (part a):

$$Q = V \times A = 10 \text{ ft/sec} \times 0.785 \text{ ft}^2 = 7.85 \text{ ft}^3/\text{sec or cfs}$$

Step 2. For part (b), 1 cfs is 449 gpm, so 7.85 cfs × 449 gpm/cfs = 3520 gpm.
Step 3. Finally, for part (c), 1 MGD is 1.55 cfs, so:

$$\frac{7.85 \text{ cfs}}{1.55 \dfrac{\text{cfs}}{\text{MGD}}} = 5.06 \text{ MGD}$$

14.3.1 Area/Velocity

The *law of continuity* states that the discharge at each point in a pipe or channel is the same as the discharge at any other point (if water does not leave or enter the pipe or channel). That is, under the assumption of steady state flow, the flow that enters the pipe or channel is the same flow that exits the pipe or channel. In equation form, this becomes:

$$Q_1 = Q_2 \text{ or } A_1 V_1 = A_2 V_2 \tag{14.14}$$

Example 14.12

Problem:
A pipe 12 in. in diameter is connected to a 6-in. diameter pipe. The velocity of the water in the 12-in. pipe is 3 fps. What is the velocity in the 6-in. pipe?

Solution:
Using the equation $A_1 V_1 = A_2 V_2$, we need to determine the area of each pipe:

$$12 \text{ in.: } A = \pi \times \frac{D^2}{4}$$

$$= 3.14 \times \frac{(1 \text{ft})^2}{4}$$

$$= 0.785 \text{ ft}^2$$

$$6 \text{ in.: } A = 3.14 \times \frac{(0.5)^2}{4}$$

$$= 0.196 \text{ ft}^2$$

The continuity equation now becomes

$$(0.785 \text{ ft}^2) \times \left(3 \, \frac{\text{ft}}{\text{sec}} = (0.196 \text{ ft}^2) \times V_2 \right)$$

Solving for V_2

$$V_2 = \frac{(0.785 \text{ ft}^2) \times (3 \text{ ft/sec})}{(0.196 \text{ ft}^2)} = (0.196 \text{ ft}^2)$$

$$= 12 \text{ ft/sec or fps}$$

14.3.2 Pressure/Velocity

In a closed pipe flowing full (under pressure), the pressure is indirectly related to the velocity of the liquid. This principle, when combined with the principle discussed in the previous section, forms the basis for several flow measurement devices (Venturi meters and rotameters), as well as the injector used for dissolving chlorine into water, and chlorine, sulfur dioxide and/or other chemicals into wastewater.

$$\text{Velocity}_1 \times \text{Pressure}_1 = \text{Velocity}_2 \times \text{Pressure}_2 \qquad (14.15)$$

or

$$V_1 P_1 = V_2 P_2$$

14.4 BERNOULLI'S THEOREM

In the 1700s, the Swiss physicist and mathematician Samuel Bernoulli developed the calculation for the total energy relationship from point to point in a steady state fluid system. Before discussing Bernoulli's energy equation, we explain the basic principle behind Bernoulli's equation.

Earlier we pointed out that water (and any other hydraulic fluid) in a hydraulic system possesses two types of energy: kinetic and potential. *Kinetic energy* is present when the water is in motion; the faster the water moves, the more kinetic energy is used. *Potential energy* is a result of the water pressure. The total energy of the water is the sum of the kinetic and potential energy. Bernoulli's principle states that the total energy of the water (fluid) always remains constant. Therefore, when the water flow in a system increases, the pressure must decrease. We stated earlier that when water starts to flow in a hydraulic system, the pressure drops. When the flow stops, the pressure rises again. The pressure gauges shown in Figure 14.3 indicate this balance clearly.

14.4.1 Bernoulli's Equation

In a hydraulic system, total energy head is equal to the sum of three individual energy heads. This can be expressed as:

$$\text{Total Head} = \text{Elevation Head} + \text{Pressure Head} + \text{Velocity Head}$$

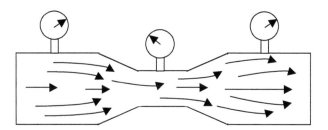

Figure 14.3   Demonstrates Bernoulli's principle. (From Spellman, F.R. and Drinan, J., 2001, *Water Hydraulics*. Lancaster, PA, Technomic Publishing Company.)

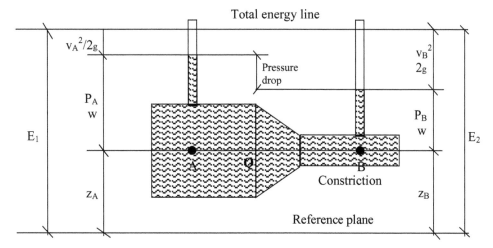

Figure 14.4   Shows the result of the law of conservation. Since the velocity and kinetic energy of the water flowing in the constricted section must increase, the potential energy may decrease. This is observed as a pressure drop in the constriction. (Adapted from Nathanson, J.A., 1997, *Basic Environmental Technology: Water Supply, Waste Management, and Pollution Control*, 2nd ed. Upper Saddle River, NJ: Prentice Hall.)

This can be expressed mathematically as:

$$E = z + \frac{p}{w} + \frac{v^2}{2g} \tag{14.16}$$

where
E = total energy head
z = height of the water above a reference plane, feet
p = pressure, pounds per square inch
w = unit weight of water, 62.4 lb/ft^3
v = flow velocity, feet per second
g = acceleration due to gravity, 32.2 ft/sec^2

Consider the constriction in section of pipe shown in Figure 14.4. We know, based on the law of energy conservation, that the total energy head at section A, E_1, must equal the total energy head at section B, E_2, and using Equation 14.16, we get Bernoulli's equation:

$$z_A = \frac{P_A}{w} + \frac{v_A^2}{2g} = z_B + \frac{P_B}{w} + \frac{V_B^2}{2g} \qquad (14.17)$$

The pipeline system shown in Figure 14.4 is horizontal. Therefore, we can simplify Bernoulli's equation because $z_A = z_B$. Because they are equal, the elevation heads cancel out from both sides, leaving:

$$\frac{P_A}{w} + \frac{v_A^2}{2g} + \frac{P_B}{w} + \frac{V_B^2}{2g} \qquad (14.18)$$

In Figure 14.4, as water passes through the constricted section of the pipe (section B), we know from continuity of flow that the velocity at section B must be greater than the velocity at section A because of the smaller flow area at section B; the velocity head in the system increases as the water flows into the constricted section. However, the total energy must remain constant. For this to occur, the pressure head, and therefore the pressure, must drop. In effect, pressure energy is converted into kinetic energy in the constriction.

The fact that the pressure in the narrower pipe section (constriction) is less than the pressure in the bigger section seems to defy common sense, but it does follow logically from continuity of flow and conservation of energy. The fact that a pressure difference exists allows measurement of flow rate in the closed pipe.

Example 14.13

Problem:

In Figure 14.4, the diameter at section A is 8 in. and at section B is 4 in. The flow rate through the pipe is 3.0 cfs, and the pressure at section A is 100 psi. What is the pressure in the constriction at section B?

Solution:

Step 1. Compute the flow area at each section, as follows:

$$A_A = \frac{\pi (0.666 \text{ ft})^2}{4} = 0.349 \text{ ft}^2 \text{ (rounded)}$$

Step 2. From $Q = A \times V$ or $V = Q/A$, we get:

$$V_A = \frac{3.0 \text{ ft}^3/\text{sec}}{0.349 \text{ ft}^2} = 8.6 \text{ ft/sec (rounded)}$$

$$V_B = \frac{3.0 \text{ ft}^3/\text{sec}}{0.087 \text{ ft}^2} = 34.5 \text{ ft/sec (rounded)}$$

Step 3. Applying Equation 14.18, we get:

$$\frac{100 \times 144}{62.4} + \frac{8.6^2}{2 \times 32.2} + \frac{P_B \times 144}{62.4} + \frac{34.5^2}{2 \times 32.2}$$

$$231 + 1.15 = 2.3p_B + 18.5$$

and

$$p_B = \frac{232.2 - 18.5}{2.3} = \frac{213.7}{2.3} = 93 \text{ psi}$$

14.5 CALCULATING MAJOR HEAD LOSS

The first practical equation used to determine pipe friction was developed in about 1850 by Darcy, Weisbach, and others. The equation or formula now known as the Darcy–Weisbach equation for circular pipes is:

$$h_f = f \frac{LV^2}{D2g} \tag{14.19}$$

In terms of the flow rate Q, the equation becomes:

$$h_f = \frac{8fLQ^2}{\pi^2 g D^5} \tag{14.20}$$

where
h_f = head loss, feet
f = coefficient of friction
L = length of pipe, feet
V = mean velocity, feet per second
D = diameter of pipe, feet
g = acceleration due to gravity, 32.2 ft/sec^2
Q = flow rate, cubic feet per second

The Darcy–Weisbach formula as such was meant to apply to the flow of any fluid; into this, the friction factor was incorporated, as well as the degree of roughness and an element called the *Reynold's number,* based on the viscosity of the fluid and the degree of turbulence of flow. This formula is used primarily for determining head loss calculations in pipes. For making this determination in open channels, the Manning equation was developed during the later part of the 19th century. Later, this equation was used for open channels as well as closed conduits.

In the early 1900s, a more practical equation, the Hazen–Williams equation, was developed for use in making calculations related to water pipes and wastewater force mains:

$$Q = 0.435 \times CD^{2.63} \times S^{0.54} \tag{14.21}$$

where
Q = flow rate, cubic feet per second
C = coefficient of roughness (C decreases with roughness)
D = hydraulic radius R, feet
S = slope of energy grade line, feet per foot

14.5.1 *C* Factor

C factor, as used in the Hazen–Williams formula, designates the coefficient of roughness. *C* does not vary appreciably with velocity, and by comparing pipe types and ages, it includes only the concept of roughness, ignoring fluid viscosity and Reynold's number. Based on experience and experimentation, accepted tables of *C* factors have been established for pipes and are given in engineering tables. Generally, *C* factor decreases by one with each year of pipe age. Flow for a newly designed system is often calculated with a *C* factor of 100, based on averaging it over the life of the pipe system.

> *Note*: A high *C* factor means a smooth pipe; a low *C* factor means a rough pipe.

14.6 CHARACTERISTICS OF OPEN-CHANNEL FLOW

McGhee (1991) points out that basic hydraulic principles apply in open channel flow (with water depth constant) although no pressure exists to act as the driving force. Velocity head is the only natural energy this water possesses, and at normal water velocities, this is a small value ($V^2/2g$). Several parameters can be (and often are) used to describe open-channel flow. However, we begin our discussion with a few characteristics, including laminar or turbulent; uniform or varied; and subcritical, critical, or supercritical.

14.6.1 Laminar and Turbulent Flow

Laminar and turbulent flow in open channels is analogous to that in closed pressurized conduits (for example, pipes). Understand, however, that flow in open channels is usually turbulent. In addition, there is no important circumstance in which laminar flow occurs in open channels in water or wastewater unit processes or structures.

14.6.2 Uniform and Varied Flow

Flow can be a function of time and location. If the flow quantity is invariant, it is said to be steady. *Uniform* flow is flow in which the depth, width, and velocity remain constant along a channel. That is, if the flow cross section does not depend on the location along the channel, the flow is said to be uniform. *Varied* or *nonuniform* flow involves a change in these, with a change in one producing a change in the others. Most circumstances of open-channel flow in water/wastewater systems involve varied flow. The concept of uniform flow is valuable, however, in that it defines a limit to which the varied flow may be considered to be approaching in many cases.

14.6.3 Critical Flow

Critical flow (flow at the critical depth and velocity) defines a state of flow between two flow regimes. This flow coincides with minimum specific energy for a given discharge and maximum discharge for a given specific energy. Critical flow occurs in flow measurement devices at or near free discharges and establishes controls in open-channel flow. Critical flow occurs frequently in water/wastewater systems and is very important in their operation and design.

14.6.4 Parameters Used in Open Channel Flow

The three primary parameters used in open channel flow are: hydraulic radius, hydraulic depth, and slope, S.

14.6.4.1 Hydraulic Radius

The hydraulic radius is the ratio of area in flow to wetted perimeter:

$$r_H = \frac{A}{P} \tag{14.22}$$

where
r_H = hydraulic radius
A = the cross-sectional area of the water
P = wetted perimeter

Consider, for example, that in open channels, maintaining the proper velocity is of primary importance because, if velocity is not maintained, flow stops (theoretically). To maintain velocity at a constant level, the channel slope must be adequate to overcome friction losses. As with other flows, calculation of head loss at a given flow is necessary, and the Hazen–Williams equation is useful ($Q = .435 \times C \times d^{2.63} \times S^{.54}$). Keep in mind that the concept of slope has not changed. Instead, we are now measuring, or calculating for, the physical slope of a channel (feet per foot), equivalent to head loss.

The preceding seems logical, but presents a problem with the diameter. In conduits that are not circular (grit chambers, contact basins, streams, and rivers), or in pipes only partially full (drains, wastewater gravity mains, sewers, etc.), in which the cross-sectional area of the water is not circular, no diameter exists. With no diameter in a situation in which the cross-sectional area of the water is not circular, we must use another parameter to designate the size of the cross section, as well as the amount of the cross section that contacts the sides of the conduit. This is where the *hydraulic radius* (r_H) comes in. The hydraulic radius is a measure of the efficiency with which the conduit can transmit water. Its value depends on pipe size and amount of fullness. Simply put, we use the hydraulic radius to measure how much of the water is in contact with the sides of the channel, or how much of the water is not in contact with the sides (see Figure 14.5).

14.6.4.2 Hydraulic Depth

The hydraulic depth is the ratio of area in flow to the width of the channel at the fluid surface. (Note that another name for hydraulic depth is the *hydraulic mean depth* or *hydraulic radius*.)

$$d_H = \frac{A}{w} \tag{14.23}$$

where
d_H = hydraulic depth
A = area in flow
w = width of the channel at the fluid surface

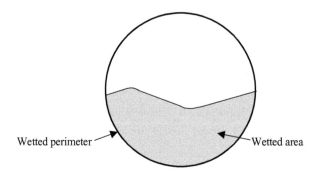

Figure 14.5 Hydraulic radius. (From Spellman, F.R. and Drinan, J., 2001, *Water Hydraulics*. Lancaster, PA, Technomic Publishing Company.)

14.6.4.3 Slope, S

The slope, *S*, in open channel equations is the slope of the energy line. If the flow is uniform, the slope of the energy line will parallel the water surface and channel bottom. In general, the slope can be calculated from the Bernoulli equation as the energy loss per unit length of channel:

$$S = \frac{Dh}{Dl} \tag{14.24}$$

14.7 OPEN-CHANNEL FLOW CALCULATIONS

We stated earlier that the calculation for head loss at a given flow is typically accomplished by using the Hazen–Williams equation. We also stated in open-channel flow problems that, although the concept of slope has not changed, the problem arises with the diameter. Again, in pipes only partially full where the cross-sectional area of the water is not circular, there is no diameter. Thus, the hydraulic radius is used for these noncircular areas.

In the original version of the Hazen–Williams equation, the hydraulic radius was incorporated. Similar versions developed by Chezy (pronounced "Shay-zee") and Manning, and others, incorporated the hydraulic radius as well. For open channels, Manning's formula is the most commonly used:

$$Q = \frac{1.5}{n} \times A \times R^{.66} \times s^{.5} \tag{14.25}$$

where
Q = channel discharge capacity, cubic feet per second
1.5 = constant
n = channel roughness coefficient
A = cross-sectional flow area, square feet
R = hydraulic radius of the channel, feet
S = slope of the channel bottom, dimensionless

Recall that we defined the hydraulic radius of a channel as the ratio of the flow area to the wetted perimeter *P*. In formula form, $R = A/P$. The new component is *n* (the roughness coefficient)

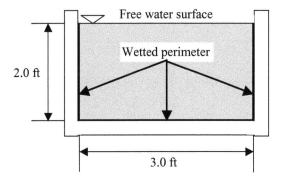

Figure 14.6 For Example 14.15.

and depends on the material and age for a pipe or lined channel and on topographic features for a natural streambed. It approximates roughness in open channels and can range from a value of 0.01 for a smooth clay pipe to 0.1 for a small natural stream. The value of n commonly assumed for concrete pipes or lined channels is 0.013; n values decrease as the channels get smoother.

The following example illustrates the application of Manning's formula for a channel with a rectangular cross section.

Example 14.14

Problem:

A rectangular drainage channel is 3 ft wide and is lined with concrete, as illustrated in Figure 14.6. The bottom of the channel drops in elevation at a rate of 0.5 per 100 ft. What is the discharge in the channel when the depth of water is 2 ft?

Solution:

Assume $n = 0.013$. Referring to Figure 14.6, we see that the cross-sectional flow area $A = 3$ ft $\times$ 2 ft $= 6$ ft^2, and the wetted perimeter $P = 2$ ft $+ 3$ ft $+ 2$ ft $= 7$ ft. The hydraulic radius $R = A/P = 6$ ft^2/7 ft $= 0.86$ ft. The slope $S = 0.5/100 = 0.005$. Assume the constant is 2.0.

Applying Manning's formula,

$$Q = \frac{2.0}{0.013} \times 6 \times 0.86^{.66} \times 0.005^{.5}$$

$$Q = 59 \text{ ft}^3$$

REFERENCES

McGhee, T.J. (1991). *Water Supply and Sewerage*, 6th ed. New York: McGraw-Hill, Inc.

Nathanson, J.A. (1997). *Basic Environmental Technology: Water Supply, Waste Management, and Pollution Control*. Upper Saddle River, NJ: Prentice Hall.

CHAPTER 15

Water Treatment Process Calculations

15.1 INTRODUCTION

Gupta (1997) points out that because of huge volume and flow conditions, the quality of natural water cannot be modified significantly within a body of water. Consequently, the quality control approach is directed to water withdrawn from a source for a specific use; the drawn water is treated prior to use. Typically, the overall treatment of water (for potable use) consists of physical and chemical methods of treatment — unlike wastewater treatment, where physical, chemical, and/or biological unit processes are used, depending on the desired quality of the effluent and operational limitations.

The physical unit operations used in water treatment include:

- *Screening* — used to remove large-sized floating and suspended debris.
- *Mixing* — coagulant chemicals (alum, for example) are mixed with the water to make tiny particles stick together.
- *Flocculation* — water mixed with coagulants is given low-level motion to allow particles to meet and floc together.
- *Sedimentation* (settling) — water is detained for a time sufficient to allow flocculated particles to settle by gravity.
- *Filtration* — fine particles that remain in the water after settling and some microorganisms present are filtered out by sending the water through a bed of sand and coal.

Chemical unit processes used in treating raw water, depending on regulatory requirements and the need for additional chemical treatment, include disinfection, precipitation, adsorption, ion exchange, and gas transfer. A flow diagram of a conventional water treatment system is shown in Figure 15.1.

> *Note*: In the following sections (for water and wastewater treatment processes presented in Chapter 16), we present basic, often used daily operational calculations (operator's math) along with engineering calculations (engineer's math) used for solving more complex computations. This presentation method is in contrast to normal presentation methods used in many engineering texts. We deviate from the norm based on our practical real-world experience; we have found that environmental engineers tasked with managing water or wastewater treatment plants are responsible not only for computation of many complex math operations (engineering calculations) but also for overseeing proper plant operation (including math operations at the operator level). Obviously, engineers are well versed in basic math operations; however, they often need to refer to example plant operation calculations in a variety of texts. In this text, the format used, although unconventional, is designed to provide basic operations math as well as more complex engineering math in one ready reference.

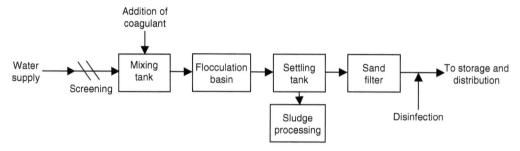

Figure 15.1 Conventional water treatment model.

15.2   WATER SOURCE AND STORAGE CALCULATIONS

Approximately 40 million cubic miles of water cover or reside within the Earth. The oceans contain about 97% of all water on Earth; the other 3% is fresh water. Snow and ice on the surface of Earth contain about 2.25% of the water; usable ground water is approximately 0.3%, and surface fresh-water is less than 0.5%. In the U.S., for example, average rainfall is approximately 2.6 ft (a volume of 5900 km³). Of this amount, approximately 71% evaporates (about 4200 km³), and 29% goes to stream flow (about 1700 km³).

Uses of freshwater include manufacturing; food production; domestic and public needs; recreation; hydroelectric power production; and flood control. Stream flow withdrawn annually is about 7.5% (440 km³). Irrigation and industry use almost half of this amount (3.4% or 200 km³ per year). Municipalities use only about 0.6% (35 km³ per year) of this amount.

Historically, in the U.S., water usage is increasing (as might be expected). For example, in 1900, 40 billion gal of fresh water were used. In 1975, the total increased to 455 billion gal. Projected use for 2000 (the latest published data) was about 720 billion gal.

The primary sources of fresh water include:

- Water captured and stored rainfall in cisterns and water jars
- Groundwater from springs, artesian wells, and drilled or dug wells
- Surface water from lakes, rivers, and streams
- Desalinized seawater or brackish groundwater
- Reclaimed wastewater

15.2.1   Water Source Calculations

Water source calculations covered in this subsection apply to wells and pond or lake storage capacity. Specific well calculations discussed include well drawdown; well yield; specific yield; well-casing disinfection; and deep-well turbine pump capacity.

15.2.1.1   Well Drawdown

Drawdown is the drop in the level of water in a well when water is being pumped (see Figure 15.2). Drawdown is usually measured in feet or meters. One of the most important reasons for measuring drawdown is to make sure that the source water is adequate and not being depleted. The data collected to calculate drawdown can indicate if the water supply is slowly declining. Early detection can give the system time to explore alternative sources, establish conservation measures, or obtain any special funding that may be needed to get a new water source.

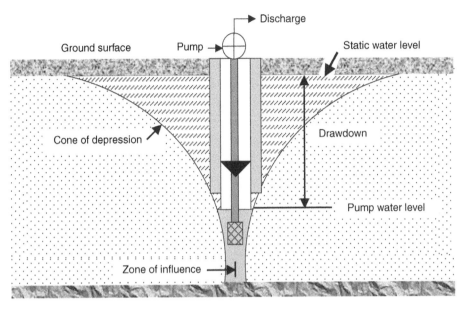

Figure 15.2 Hydraulic characteristics of a well.

Well drawdown is the difference between the pumping water level and the static water level:

$$\text{Drawdown, ft} = \text{Pumping Water Level, ft} - \text{Static Water Level, ft} \qquad (15.1)$$

Example 15.1

Problem:
The static water level for a well is 70 ft. If the pumping water level is 90 ft, what is the drawdown?

Solution:

$$\text{Drawdown, ft,} = \text{Pumping Water Level, ft,} - \text{Static Water Level, ft}$$

$$= 90 \, \text{ft} - 70 \, \text{ft}$$

$$= 20 \, \text{ft}$$

Example 15.2

Problem:
The static water level of a well is 122 ft. The pumping water level is determined using the sounding line. The air pressure applied to the sounding line is 4.0 psi, and the length of the sounding line is 180 ft. What is the drawdown?

Solution:
First, calculate the water depth in the sounding line and the pumping water level:

$$\text{Water Depth in Sounding Line} = (4.0 \text{ psi}) (2.31 \text{ ft/psi})$$

$$= 9.2 \text{ ft}$$

$$\text{Pumping Water Level} = 180 \text{ ft} - 9.2 \text{ ft} = 170.8 \text{ ft}$$

Then, calculate drawdown as usual:

$$\text{Drawdown, ft} = \text{Pumping Water Level, ft} - \text{Static Water Level, ft}$$

$$= 170.8 \text{ ft} - 122 \text{ ft}$$

$$= 48.8 \text{ ft}$$

15.2.1.2   Well Yield

Well yield is the volume of water per unit of time produced from well pumping. Usually, well yield is measured in terms of gallons per minute (gpm) or gallons per hour (gph). Sometimes, large flows are measured in cubic feet per second (cfs). Well yield is determined using the following equation.

$$\text{Well Yield, gpm} = \frac{\text{Gallons Produced}}{\text{Duration of Test, min}} \qquad (15.2)$$

Example 15.3

Problem:

Once the drawdown level of a well stabilized, operators determined that the well produced 400 gal during the 5-min test. What is the well yield in gpm^2?

Solution:

$$\text{Well Yield, gpm} = \frac{\text{Gallons Produced}}{\text{Duration of Test, min}}$$

$$= \frac{400 \text{ gallons}}{5 \text{ minutes}}$$

$$= 80 \text{ gpm}$$

Example 15.4

Problem:

During a 5-min test for well yield, a total of 780 gal are removed from the well. What is the well yield in gallons per minute? in gallons per hour?

Solution:

$$\text{Well Yield, gpm} = \frac{\text{Gallons Removed}}{\text{Duration of Test, min}}$$

$$= \frac{780 \text{ gallons}}{5 \text{ minutes}}$$

$$= 156 \text{ gpm}$$

Then convert gallons per minute flow to gallons per hour flow:

$$(156 \text{ gal/min}) (60/\text{hr}) = 9360 \text{ gph}$$

15.2.1.3 Specific Yield

Specific yield is the discharge capacity of the well per foot of drawdown. The specific yield may range from 1-gpm/ft drawdown to more than 100-gpm/ft drawdown for a properly developed well. Specific yield is calculated using the equation:

$$\text{Specific Yield, gpm/ft} = \frac{\text{Well Yield, gpm}}{\text{Drawdown, ft}} \qquad (15.3)$$

Example 15.5

Problem:
   A well produces 260 gpm. If the drawdown for the well is 22 ft, what is the specific yield in gallons per minute per foot, and what is the specific yield in gallons per minute per foot of drawdown?

Solution:

$$\text{Specific Yield, gpm/ft} = \frac{\text{Well Yield, gpm}}{\text{Drawdown, ft}}$$

$$= \frac{260 \text{ gpm}}{22 \text{ ft}}$$

$$= 11.8 \text{ gpm/ft}$$

Example 15.6

Problem:
   The yield for a particular well is 310 gpm. If the drawdown for this well is 30 ft, what is the specific yield in gallons per minute per foot of drawdown?

Solution:

$$\text{Specific Yield, gpm/ft} = \frac{\text{Well Yield, gpm}}{\text{Drawdown, ft}}$$

$$= \frac{310 \text{ gpm}}{30 \text{ ft}}$$

$$= 10.3 \text{ gpm/ft}$$

15.2.1.4 Well Casing Disinfection

A new, cleaned, or repaired well normally contains contamination that may remain for weeks unless the well is thoroughly disinfected. This may be accomplished by ordinary bleach in a concentration of 100 ppm (parts per million) of chlorine. The amount of disinfectant required is determined by the amount of water in the well. The following equation is used to calculate the pounds of chlorine required for disinfection:

$$\text{Chlorine, lb} = (\text{Chlorine, mg/L}) (\text{Casing Vol., MG}) (8.34 \text{ lb/gal}) \qquad (15.4)$$

Example 15.7

Problem:
    A new well is to be disinfected with chlorine at a dosage of 50 mg/L. If the well casing diameter is 8 in. and the length of the water-filled casing is 110 ft, how many pounds of chlorine will be required?

Solution:
    First, calculate the volume of the water-filled casing:

$$(0.785) (.67) (67) (110 \text{ ft}) (7.48 \text{ gal/ft}^3) = 290 \text{ gallons}$$

Then, determine the pounds of chlorine required, using the milligrams-per-liter to pounds equation:

$$\text{Chlorine, lb} = (\text{chlorine, mg/L})(\text{Volume, MG}) (8.34 \text{ lb/gal})$$

$$(50 \text{ mg/L}) (0.000290 \text{ MG}) (8.34 \text{ lb/gal}) = 0.12 \text{ lb Chlorine}$$

15.2.1.5 Deep-Well Turbine Pump Calculations

Deep well turbine pumps are used for high-capacity deep wells. Usually consisting of more than one stage of centrifugal pumps, the pump is fastened to a pipe called the pump column; the pump is located in the water. The pump is driven from the surface through a shaft running inside the pump column, and the water is discharged from the pump up through the pump column to the surface. The pump may be driven by a vertical shaft, electric motor at the top of the well, or some other power source, usually through a right angle gear drive located at the top of the well. A modern version of the deep-well turbine pump is the submersible type pump in which the pump, along with a close-coupled electric motor built as a single unit, is located below water level in the well. The motor is built to operate submerged in water.

15.2.3 Vertical Turbine Pump Calculations

The calculations pertaining to well pumps include head, horsepower, and efficiency calculations.

Discharge head is measured to the pressure gauge located close to the pump discharge flange. The pressure (psi) can be converted to feet of head using the equation:

$$\text{Discharge Head, ft} = (\text{press, psi}) (2.31 \text{ ft/psi}) \tag{15.5}$$

Total pumping head (field head) is a measure of the lift *below* the discharge head pumping water level (discharge head). Total pumping head is calculated as:

$$\text{Pumping Head, ft} = \text{Pumping Water Level, ft} + \text{Discharge Head, ft} \tag{15.6}$$

Example 15.8

Problem:

The pressure gauge reading at a pump discharge head is 4.1 psi. What is this discharge head expressed in feet?

Solution:

$$(4.1 \text{ psi}) (2.31 \text{ ft/psi}) = 9.5 \text{ ft}$$

Example 15.9

Problem:

The static water level of a pump is 100 ft. The well drawdown is 26 ft. If the gauge reading at the pump discharge head is 3.7 psi, what is the total pumping head?

Solution:

$$\text{Total Pumping Head, ft} = \text{Pumping Water Level, ft} + \text{Discharge Head, ft} \tag{15.7}$$

$$= (100 \text{ ft} + 26 \text{ ft}) + (3.7 \text{ psi}) (2.31 \text{ ft/psi})$$

$$= 126 \text{ ft} + 8.5 \text{ ft}$$

$$= 134.5 \text{ ft}$$

Five types of horsepower calculations are used for vertical turbine pumps; a general understanding of these five horsepower types is important:

- *Motor horsepower* refers to the horsepower supplied to the motor. The following equation is used to calculate motor horsepower:

$$\text{Motor hp (input hp)} = \frac{\text{Field bhp}}{\dfrac{\text{Motor Efficiency}}{100}} \tag{15.8}$$

- *Total brake horsepower* (*bhp*) refers to the horsepower output of the motor. The following equation is used to calculate total bhp:

$$\text{Total bhp} = \text{Field bhp} + \text{Thrust Bearing Loss, hp} \tag{15.9}$$

- *Field horsepower* refers to the horsepower required at the top of the pump shaft. The following equation is used to calculate field horsepower:

$$\text{Field bhp} = \text{Bowl bhp} + \text{Shaft Loss, hp} \tag{15.10}$$

- *Bowl* or *laboratory horsepower* refers to the horsepower at the entry to the pump bowls. The following equation is used to calculate bowl horsepower:

$$\text{Bowl bhp (Lab bhp)} = \frac{(\text{Bowl Head, ft}) (\text{Capacity, gpm})}{\dfrac{(3960) (\text{Bowl Efficiency})}{100}} \tag{15.11}$$

- *Water horsepower* refers to the horsepower at the pump discharge. The following equation is used to calculate water hp:

$$\text{Water hp} = \frac{(\text{Field Head, ft}) (\text{Capacity, gpm})}{3960} \tag{15.12}$$

or the equivalent equation:

$$\text{Water hp} = \frac{(\text{Field Head, ft}) (\text{Capacity, gpm})}{33,000 \text{ ft-lb/min}}$$

Example 15.10

Problem:

The pumping water level for a well pump is 150 ft, and the discharge pressure measured at the pump discharge centerline is 3.5 psi. If the flow rate from the pump is 700 gpm, what is the water horsepower? (Use Equation 15.12.)

Solution:

First, calculate the field head. The discharge head must be converted from psi to ft:

$$(3.5 \text{ psi}) (2.31 \text{ ft/psi}) = 8.1 \text{ ft}$$

The water horsepower is therefore:

$$150 \text{ ft} + 8.1 \text{ ft} = 158.1 \text{ ft}$$

The water horsepower can now be determined:

$$= \frac{150 \text{ ft} + 8.1 \text{ ft} = 158.1 \text{ ft}}{33,000 \text{ ft-lb/min}}$$

$$= 28 \text{ whp}$$

Example 15.11

Problem:
   The pumping water level for a pump is 170 ft. The discharge pressure measured at the pump discharge head is 4.2 psi. If the pump flow rate is 800 gpm, what is the water horsepower? (Use Equation 15.12.)

Solution:
   First, determine the field head by converting the discharge head from psi to ft:

$$(4.2 \text{ psi}) (2.31 \text{ ft/psi}) = 9.7 \text{ ft}$$

Now, calculate the field head:

$$170 \text{ ft} + 9.7 \text{ ft} = 179.7 \text{ ft}$$

and then calculate the water horsepower:

$$whp = \frac{(179.7 \text{ ft}) (800 \text{ gpm})}{3960}$$

$$= 36 \text{ whp}$$

Example 15.12

Problem:
   A deep-well vertical turbine pump delivers 600 gpm. If the lab head is 185 ft and the bowl efficiency is 84%, what is the bowl horsepower? (Use Equation 15.11.)

Solution:

$$\text{Bowl bhp} = \frac{\text{(Bowl Head, ft) (Capacity, gpm)}}{\dfrac{\text{(3960) (Bowl Efficiency)}}{100}}$$

$$= \frac{\text{(185 ft) (600 gpm)}}{\dfrac{\text{(3960) (84.0)}}{100}}$$

$$= \frac{(185)\,(600)}{(3960)\,(84.0)}$$

$$= 33.4 \text{ bowl bhp}$$

Example 15.13

Problem:

The bowl brake horsepower is 51.8 bhp. If the 1-in. diameter shaft is 170 ft long and is rotating at 960 rpm with a shaft fiction loss of 0.29 hp loss per 100 ft, what is the field bhp?

Solution:

Before you can calculate the field bhp, factor in the shaft loss:

$$\frac{(0.29 \text{ hp loss})\,(170 \text{ ft})}{100} = 0.5 \text{ hp loss}$$

Now determine the field bhp:

$$\text{Field bhp} = \text{Bowl bhp} + \text{Shaft Loss, hp}$$

$$= 51.8 \text{ bhp} + 0.5 \text{ hp}$$

$$= 52.3 \text{ bhp}$$

Example 15.14

Problem:

The field horsepower for a deep-well turbine pump is 62 bhp. If the thrust bearing loss is 0.5 hp and the motor efficiency is 88%, what is the motor input horsepower? (Use Equation 15.8.)

Solution:

$$Mhp = \frac{Field\ (total)\ bhp}{\dfrac{Motor\ Efficiency}{100}}$$

$$= \frac{62\ bhp\ +\ .0.5\ hp}{0.88}$$

$$= 71\ mhp$$

When we speak of the *efficiency* of any machine, we speak primarily of a comparison of what is put out by the machine (energy output) compared to its input (energy input). Horsepower efficiency, for example, is a comparison of horsepower output of the unit or system with horsepower input to that unit or system — the unit's efficiency. For vertical turbine pumps, four efficiency types are considered:

- Bowl efficiency
- Field efficiency
- Motor efficiency
- Overall efficiency

The general equation used in calculating percent efficiency is:

$$\% = \frac{Part}{Whole} \times 100 \qquad (15.13)$$

Vertical turbine pump bowl efficiency is easily determined using a pump performance curve chart — usually provided by the pump manufacturer.

Field efficiency is determined using Equation 15.14.

$$Field\ Efficiency,\ \% = \frac{\dfrac{(Field\ Head,\ ft)\ (Capacity,\ gpm)}{3960}}{Total\ bhp} \times 100 \qquad (15.14)$$

Example 15.15

Problem:

Given the following data, calculate the field efficiency of the deep-well turbine pump:

- Field head — 180 ft
- Capacity — 850 gpm
- Total bhp — 61.3 bhp

Solution:

$$\text{Field Efficiency, \%} = \frac{(\text{Field Head, ft}) (\text{Capacity, gpm})}{(3960) (\text{Total bhp})} \times 100$$

$$= \frac{(180 \text{ ft}) (850 \text{ gpm})}{(3960) (61.3 \text{ bhp})} \times 100$$

$$= 63\%$$

Overall efficiency is a comparison of the horsepower output of the system with that entering the system. Equation 15.15 is used to calculate overall efficiency:

$$\text{Overall Efficiency, \%} = \frac{\text{Field Efficiency, \% (Motor Efficiency, \%)}}{100} \qquad (15.15)$$

Example 15.16

Problem:
    The efficiency of a motor is 90%. If the field efficiency is 83%, what is the overall efficiency of the unit?

Solution:

$$\text{Overall Efficiency, \%} = \frac{(\text{Field Efficiency, \%})(\text{Motor Efficiency, \%})}{100} \times 100\%$$

$$= \frac{(83) (90)}{100}$$

$$= 74.9\%$$

15.3 WATER STORAGE

Water storage facilities for water distribution systems are required primarily to provide for fluctuating demands of water usage; to provide a sufficient amount of water to average; or to equalize daily demands on the water supply system. Other functions of water storage facilities include

- Increasing operating convenience
- Leveling pumping requirements (to keep pumps from running 24 h a day)
- Decreasing power costs
- Providing water during power source or pump failure
- Providing large quantities of water to meet fire demands
- Providing surge relief (to reduce the surge associated with stopping and starting pumps)
- Increasing detention time (to provide chlorine contact time and satisfy the desired CT (contact time) value requirements)
- Blending water sources

15.3.1 Water Storage Calculations

The storage capacity, in gallons, of a reservoir, pond, or small lake can be estimated using Equation 15.16:

$$\text{Still Water Body, gal} = (\text{Av. Length, ft}) (\text{Av. Width, ft}) (\text{Av. Depth, ft}) (7.48 \text{ gal/ft}^3) \qquad (15.16)$$

Example 15.17

Problem:
A pond has an average length of 250 ft, an average width of 110 ft, and an estimated average depth of 15 ft. What is the estimated volume of the pond in gallons?

Solution:

$$\text{Vol, gal} = (\text{Av. Length, ft}) (\text{Av. Width, ft}) (\text{Av. Depth, ft}) (7.48 \text{ gal/ft}^3)$$

$$= (250 \text{ ft}) (110 \text{ ft}) (15 \text{ ft}) (7.48 \text{ gal/ft}^3)$$

$$= 3,085,500 \text{ gal}$$

Example 15.18

Problem:
A small lake has an average length of 300 ft and an average width of 95 ft. If the maximum depth of the lake is 22 ft, what is the estimated gallons volume of the lake?

> *Note*: For small ponds and lakes, the average depth is generally about 0.4 times the greatest depth. Therefore, to estimate the average depth, measure the greatest depth and then multiply that number by 0.4.

Solution:
First, estimate the average depth of the lake:

$$\text{Estimated Av. Depth, ft} = (\text{Greatest Depth, ft}) (0.4 \text{ Depth, ft})$$

$$= (22 \text{ ft}) (0.4 \text{ ft})$$

$$= 8.8 \text{ ft}$$

Then determine the lake volume:

$$\text{Volume, gal} = (\text{Av. Length, ft}) (\text{Av. Width, ft}) (\text{Av. Depth, ft}) (7.48 \text{ gal/ft}^3)$$

$$= (300 \text{ ft}) (95 \text{ ft}) (8.8 \text{ ft}) (7.48 \text{ ft}^3)$$

$$= 1,875,984 \text{ gal}$$

15.3.2 Copper Sulfate Dosing

Algal control is perhaps the most common *in situ* treatment of lakes, ponds, and reservoirs; this is usually accomplished by the application of copper sulfate — the copper ions in the water kill the algae. Copper sulfate application methods and dosages vary depending on the specific surface water body being treated. The desired copper sulfate dosage may be expressed in milligrams per liter of copper, pounds of copper sulfate per acre-foot, or pounds of copper sulfate per acre.

For a dose expressed as milligrams per liter of copper, the following equation is used to calculate pounds of copper sulfate required:

$$\text{Copper Sulfate, lb} = \frac{\text{Copper (mg/L) (Volume, MG) (8.34 lb/gal)}}{\dfrac{\text{\% Available Copper}}{100}} \qquad (15.17)$$

Example 15.19

Problem:

For algal control in a small pond, a dosage of 0.5-mg/L copper is desired. The pond has a volume of 15 MG. How many pounds of copper sulfate are required? (Copper sulfate contains 25% available copper.)

Solution:

$$\text{Copper Sulfate, lb} = \frac{\text{Copper (mg/L) (Volume, MG) (8.34 lb/gal)}}{\dfrac{\text{\% Available Copper}}{100}}$$

$$= \frac{(0.5 \text{ mg/L}) (15 \text{ MG}) (8.34 \text{ lb/gal})}{\dfrac{25}{100}}$$

$$= 250 \text{ lb Copper Sulfate}$$

For calculating pounds of copper sulfate per acre-foot, use the following equation (assume the desired copper sulfate dosage is 0.9 lb/acre-ft):

$$\text{Copper Sulfate, lb} = \frac{(0.9 \text{ lb Copper Sulfate) (acre-ft)}}{1 \text{ acre-ft}} \qquad (15.18)$$

Example 15.20

Problem:

A pond has a volume of 35 acre-ft. If the desired copper sulfate dose is 0.9 lb/acre-ft, how many pounds of copper sulfate are required?

Solution:

$$\text{Copper Sulfate, lb} = \frac{(0.9 \text{ lb Copper Sulfate}) (\text{acre-ft})}{1 \text{ acre-ft}}$$

$$\text{Copper Sulfate, lb} = \frac{(0.9 \text{ lb Copper Sulfate})}{1 \text{ acre-ft}} \times \frac{x \text{ lb Copper Sulfate}}{35 \text{ acre-ft}}$$

Then solve for *x*:

$$(0.9) (35) = x$$

$$31.5 \text{ lb Copper Sulfate}$$

The desired copper sulfate dosage may also be expressed in terms of pounds of copper sulfate per acre. The following equation is used to determine pounds of copper sulfate (assume a desired dose of 5.2 lb of copper sulfate per acre):

$$\text{Copper Sulfate, lb} = \frac{(5.2 \text{ lb Copper Sulfate}) (\text{acre})}{1 \text{ acre}} \qquad (15.19)$$

Example 15.21

Problem:
A small lake has a surface area of 6.0 acres. If the desired copper sulfate dose is 5.2 lb/acre, how many pounds of copper sulfate are required?

Solution:

$$\text{Copper Sulfate, lb} = \frac{(5.2 \text{ lb Copper Sulfate}) (\text{acre})}{1 \text{ acre}}$$

$$= 31.2 \text{ lb Copper Sulfate}$$

15.4   COAGULATION, MIXING, AND FLOCCULATION

15.4.1 Coagulation

After screening and the other pretreatment processes, the next unit process in a conventional water treatment system is a mixer, in which the first chemicals are added in what is known as coagulation. The exception to this situation occurs in small systems using groundwater, when chlorine or other taste and odor control measures are introduced at the intake and are the extent of treatment.

The term *coagulation* refers to the series of chemical and mechanical operations by which coagulants are applied and made effective. These operations comprise two distinct phases: (1) rapid mixing to disperse coagulant chemicals by violent agitation into the water being treated; and (2) flocculation to agglomerate small particles into well-defined floc by gentle agitation for a much longer time. The coagulant must be added to the raw water and perfectly distributed into the liquid; such uniformity of chemical treatment is reached through rapid agitation or mixing.

Coagulation results from adding salts of iron or aluminum to the water. (Coagulation is the reaction between one of these salts and water.) Common coagulants (salts) include:

- Alum — aluminum sulfate
- Sodium aluminate
- Ferric sulfate
- Ferrous sulfate
- Ferric chloride
- Polymers

15.4.2 Mixing

To ensure maximum contact between the reagent and suspended particles, coagulants and coagulant aids must be rapidly dispersed (mixed) throughout the water; otherwise, the coagulant will react with water and dissipate some of its coagulating power. To ensure complete mixing and optimum plug flow reactor operation, proper detention time in the basin is required. Detention time can be calculated using the following procedures:

For complete mixing:

$$t = \frac{V}{Q} = \frac{1}{K} = \frac{C_i - C_e}{C_e} \tag{15.20}$$

For plug flow:

$$t = \frac{V}{Q} = \frac{L}{v} = \frac{1}{K} = \frac{C_i}{C_e} \tag{15.21}$$

where
t = detention time of the basin, minute
V = volume of basin, cubic meters or cubic feet
Q = flow rate, cubic meters per second or cfs
K = rate constant
C_i = influent reactant concentration, milligrams per liter
C_e = effluent reactant concentration, milligrams per liter
L = length of rectangular basin, meters or feet
v = horizontal velocity of flow, meters per second or feet per second

Example 15.22

Problem:
Alum dosage is 40 mg/L K = 90 per day based on lab tests. Compute the detention times for complete mixing and plug flow reactor for 90% reduction.

Solution:

Step 1. Find C_e:

$$C_e = (1\ 0.9)C_i = 0.1 \times C_i = 0.1 \times 40 \text{ mg/L}$$

$$= 4 \text{ mg/L}$$

Step 2. Calculate t for complete mixing (Equation 15.20):

$$t = \frac{1}{K}(C_i - C_e)\frac{1}{9/d}\left(\frac{40 \text{ mg/L} - 4 \text{ mg/L}}{4 \text{ mg/L}}\right)$$

$$t = \frac{1d}{90}\left(\frac{1440 \text{ min}}{1d}\right)$$

$$= 144 \text{ minutes}$$

Step 3. Calculate t for plug flow using the following formula:

$$t = \frac{1}{K}\left(\text{In } \frac{C_i}{C_e}\right) = \frac{1440}{90}\left(\text{In } \frac{40}{4}\right)$$

$$= 36.8 \text{ minutes}$$

15.4.3 Flocculation

Flocculation follows coagulation in the conventional water treatment process; it is the physical process of slowly mixing the coagulated water to increase the probability of particle collision. Through experience, we see that effective mixing reduces the required amount of chemicals and greatly improves the sedimentation process, which results in longer filter runs and higher quality finished water. Flocculation's goal is to form a uniform, feather-like material similar to snowflakes — a dense, tenacious floc that entraps the fine, suspended, and colloidal particles and carries them down rapidly in the settling basin. To increase the speed of floc formation and the strength and weight of the floc, polymers are often added.

15.4.4 Coagulation and Flocculation General Calculations

In the proper operation of the coagulation and flocculation unit processes, calculations are performed to determine chamber or basin volume, chemical feed calibration, chemical feeder settings, and detention time.

15.4.4.1 Chamber and Basin Volume Calculations

To determine the volume of a square or rectangular chamber or basin, use Equation 15.22 or 15.23:

$$\text{Volume, ft}^3 = \text{(length, ft) (width, ft) (depth, ft)} \qquad (15.22)$$

$$\text{Volume, gal} = \text{(length, ft) (width, ft) (depth, ft) } (7.48 \text{ gal/ft}^3) \qquad (15.23)$$

Example 15.23

Problem:

A flash mix chamber is 4 ft² with water to a depth of 3 ft. What is the volume of water (in gallons) in the chamber?

Solution:

$$\text{Volume, gal} = \text{(length, ft) (width, ft) (depth, ft) } (7.48 \text{ gal/ft}^3)$$

$$= (4 \text{ ft}) (4 \text{ ft}) (3 \text{ ft}) (7.48 \text{ gal/ft}^3)$$

$$= 359 \text{ gal}$$

Example 15.24

Problem:

A flocculation basin is 40 ft long and 12 ft wide, with water to a depth of 9 ft. What is the volume of water (in gallons) in the basin?

Solution:

$$\text{Volume, gal} = \text{(length, ft) (width, ft) (depth, ft) } (7.48 \text{ gal/ft}^3)$$

$$= (40 \text{ ft}) (12 \text{ ft}) (9 \text{ ft}) (7.48 \text{ gal/ft}^3)$$

$$= 32,314 \text{ gal}$$

Example 15.25

Problem:

A flocculation basin is 50 ft long, 22 ft wide, and contains water to a depth of 11 ft 6 in. How many gallons of water are in the tank?

Solution:

First, convert the 6 in. of the depth measurement to feet:

$$\frac{6 \text{ in.}}{12 \text{ in./ft}} = 0.5 \text{ ft}$$

Then calculate basin volume:

$$\text{Volume, gal} = (\text{length, ft}) \, (\text{width, ft}) \, (\text{depth, ft}) \, (7.48 \text{ gal/ft}^3)$$

$$= (50 \text{ ft}) \, (22 \text{ ft}) \, (11.5 \text{ ft}) \, (7.48 \text{ gal/ft}^3)$$

$$= 94{,}622 \text{ gal}$$

15.4.4.2 Detention Time

Because coagulation reactions are rapid, detention time for flash mixers is measured in seconds, whereas the detention time for flocculation basins is generally between 5 and 30 min. The equation used to calculate detention time is

$$\text{Detention Time, min} = \frac{\text{Volume of Tank, gal}}{\text{Flow Rate, gpm}} \tag{15.24}$$

Example 15.26

Problem:
The flow to a flocculation basin 50 ft long, 12 ft wide, and 10 ft deep is 2100 gpm. What is the detention time in the tank, in minutes?

Solution:

$$\text{Tank Volume, gal} = (50 \text{ ft}) \, (12 \text{ ft}) \, (10 \text{ ft}) \, (7.48 \text{ gal/ft}^3)$$

$$= 44{,}880 \text{ gal}$$

$$\text{Detention Time, min} = \frac{\text{Volume of Tank, gal}}{\text{Flow Rate, gpm}}$$

$$= \frac{44{,}880 \text{ gal}}{2100 \text{ gpm}}$$

$$= 21.4 \text{ min}$$

Example 15.27

Problem:
A flash mix chamber is 6 ft long and 4 ft wide, with water to a depth of 3 ft. Assuming the flow is steady and continuous, if the flow to the flash mix chamber is 6 MGD, what is the chamber detention time in seconds?

Solution:
First, convert the flow rate from gallons per day to gallons per second so that time units will match:

$$\frac{6,000,000}{(1440 \text{ min/day}) (60 \text{ sec/min})} = 69 \text{ gps}$$

Then calculate detention time:

$$\text{Detention Time, sec} = \frac{\text{Volume of Tank, gal}}{\text{Flow Rate, gps}}$$

$$= \frac{(6 \text{ ft}) (4 \text{ ft}) (3 \text{ ft}) (7.48 \text{ gal/ft}^3)}{69 \text{ gps}}$$

$$= 7.8 \text{ sec}$$

15.4.4.3 Determining Dry Chemical Feeder Setting (Pounds per Day)

Adding (dosing) chemicals to the water flow calls for a measured amount of chemical. The amount of chemical required depends on such factors as the type of chemical used, the reason for dosing, and the flow rate being treated. To convert from milligrams per liter to pounds per day, use the following equation:

$$\text{Chem. added, lb/day} = (\text{Chemical, mg/L}) (\text{Flow, MGD}) (8.34 \text{ lb/gal}) \qquad (15.25)$$

Example 15.28

Problem:

Jar tests indicate that the best alum dose for a water sample is 8 mg/L. If the flow to be treated is 2,100,000 gpd, what should the pounds per day setting be on the dry alum feeder?

Solution:

$$\text{Lb/day} = (\text{Chemical, mg/L}) (\text{Flow, MGD}) (8.34 \text{ lb/gal})$$

$$= (8 \text{ mg/L}) (2.10 \text{ MGD}) (8.34 \text{ lb/gal})$$

$$= 140 \text{ lb/day}$$

Example 15.29

Problem:

Determine the desired pounds per day setting on a dry chemical feeder if jar tests indicate an optimum polymer dose of 12 mg/L, and the flow to be treated is 4.15 MGD.

Solution:

$$\text{Polymer, lb/day} = (12 \text{ mg/L}) (4.15 \text{ MGD}) (8.34 \text{ lb/gal})$$

$$= 415 \text{ lb/day}$$

15.4.4.4 Determining Chemical Solution Feeder Setting (Gallons per Day)

When solution concentration is expressed as pounds of chemical per gallon of solution, the required feed rate can be determined using the following equations:

Solution:

$$\text{Chem., lb/day} = (\text{Chemical, mg/L}) (\text{Flow, MGD}) (8.34 \text{ lb/gal}) \tag{15.26}$$

Then, convert the pounds per day dry chemical to gallons per day solution:

$$\text{Solution, gpd} = \frac{\text{Chemical, lb/day}}{\text{lb Chemical/gal Solution}} \tag{15.27}$$

Example 15.30

Problem:

Jar tests indicate that the best alum dose for a water sample is 7 mg/L. The amount to be treated is 1.52 MGD. Determine the gallons-per-day setting for the alum solution feeder if the liquid alum contains 5.36 lb of alum per gallon of solution.

Solution:

First calculate the pounds per day of dry alum required, using the milligrams-per-liter to pounds-per-day equation:

$$\text{Dry Alum, lb/day} = (\text{mg/L}) (\text{Flow, MGD}) (8.34 \text{ lb/gal})$$

$$= (7 \text{ mg/L}) (1.52 \text{ MGD}) (8.34 \text{ lb/gal})$$

$$= 89 \text{ lb/day}$$

Then, calculate the gallons per day solution required:

$$\text{Alum Solution, gpd} = \frac{89 \text{ lb/day}}{5.36 \text{ lb Alum/gal Solution}}$$

$$= 16.6 \text{ gpd}$$

15.4.4.5 Determining Chemical Solution Feeder Setting (Milliliters per Minute)

Some solution chemical feeders dispense chemical as milliliters per minute (mL/min). To calculate the milliliters per minute solution required, use the following procedure:

$$\text{Solution, mL/min} = \frac{(\text{gpd}) (3785 \text{ mL/gal})}{1440 \text{ min/day}} \tag{15.28}$$

Example 15.31

Problem:

The desired solution feed rate was calculated at 9 gpd. What is this feed rate expressed as milliliters per minute?

Solution:

$$mL/min = \frac{(gpd)\,(3785\ mL/gal)}{1440\ min/day}$$

$$= \frac{(9\ gpd)\,(3785\ mL/gal)}{1440\ min/day}$$

$$= 24\ mL/min\ Feed\ Rate$$

Example 15.32

Problem:

The desired solution feed rate has been calculated at 25 gpd. What is this feed rate expressed as milliliters per minute?

Solution:

$$mL/min = \frac{(gpd)\,(3785\ mL/gal)}{1440\ min/day}$$

$$= \frac{(25\ gpd)\,(3785\ mL/gal)}{1440\ min/day}$$

$$= 65.7\ mL/min\ Feed\ Rate$$

Sometimes we need to know milliliters per minute solution feed rate when we do not know the gallons per day solution feed rate. In such cases, calculate the gallons per day solution feed rate first, using the following equation:

$$gpd = \frac{(Chemical,\ mg/L)\,(Flow,\ MGD)\,(8.34\ lb/gal)}{Chemical,\ lb/Solution,\ gal} \tag{15.29}$$

15.4.5 Determining Percent of Solutions

The strength of a solution is a measure of the amount of chemical solute dissolved in the solution. We use the following equation to determine percent strength of solution:

$$\%\ Strength = \frac{Chemical,\ lb}{Water,\ lb + Chemical,\ lb} \times 100 \tag{15.30}$$

Example 15.33

Problem:
  If a total of 10 oz of dry polymer are added to 15 gal of water, what is the percent strength (by weight) of the polymer solution?

Solution:
  Before calculating percent strength, the ounces of chemical must be converted to pounds of chemical:

$$\frac{10 \text{ oz}}{16 \text{ oz/lb}} = 0.625 \text{ lb Chemical}$$

Now calculate percent strength:

$$\% \text{ Strength} = \frac{\text{Chemical, lb}}{\text{Water, lb + Chemical, lb}} \times 100$$

$$= \frac{= 0.625 \text{ lb Chemical}}{(15 \text{ gal}) (8.34 \text{ lb/gal}) + 0.625 \text{ lb}} \times 100$$

$$= \frac{0.625 \text{ lb Chemical}}{125.7 \text{ lb Solution}} \times 100$$

$$= 0.5\%$$

Example 15.34

Problem:
  If 90 g (1 g = 0.0022 lb) of dry polymer are dissolved in 6 gal of water, what percent strength is the solution?

Solution:
  First, convert grams of chemical to pounds of chemical. Because 1 g = 0.0022 lb, 90 g = 90 times 0.0022 lb:

$$(90 \text{ g Polymer}) (0.0022 \text{ lb/g}) = 0.198 \text{ lb Polymer}$$

Now calculate percent strength of the solution:

$$\% \text{ Strength} = \frac{\text{lb Polymer}}{\text{lb Water + lb Polymer}} \times 100$$

$$= \frac{0.198 \text{ lb Polymer}}{(6 \text{ gal}) (8.34 \text{ lb/gal}) + 0.198 \text{ lb}} \times 100$$

$$= 4\%$$

15.4.5.1 Determining Percent Strength of Liquid Solutions

When using liquid chemicals to make up solutions (liquid polymer, for example), a different calculation is required:

$$\text{Liq. Poly., lb} = \frac{\text{Liq. Poly (\% Strength)}}{100} = \text{Poly. Sol., lb} \frac{\text{Poly. Sol. (\% Strength)}}{100} \quad (15.31)$$

Example 15.35

Problem:

A 12% liquid polymer is used in making up a polymer solution. How many pounds of liquid polymer should be mixed with water to produce 120 lb of a 0.5% polymer solution?

Solution:

$$\frac{(\text{Liq. Poly., lb}) (\text{Liq. Poly \% Strength})}{100} = \text{Poly. Sol., lb} \frac{\text{Poly. Sol. (\% Strength)}}{100}$$

$$\frac{(x \text{ lb}) (12)}{100} = \frac{(120 \text{ lb}) (0.5)}{100}$$

$$x = \frac{(120) (0.005)}{0.12}$$

$$x = 5 \text{ lb}$$

15.4.5.2 Determining Percent Strength of Mixed Solutions

The percent strength of solution mixture is determined using the following equation:

$$\% \text{ Strength of Mix.} = \frac{\dfrac{(\text{Sol. 1, lb})(\% \text{ Strength, Sol.1})}{100} + \dfrac{(\text{Sol. 2, lb})(\% \text{ Strength, Sol. 2})}{100}}{\text{lb Solution 1} + \text{lb Solution 2}} \times 100 \quad (15.32)$$

Example 15.36

Problem:

If 12 lb of a 10% strength solution are mixed with 40 lb of 1% strength solution, what is the percent strength of the solution mixture?

Solution:

$$\% \text{ Strength of Mix.} = \frac{\dfrac{(\text{Sol. 1, lb})(\% \text{ Strength, Sol.1})}{100} + \dfrac{(\text{Sol. 2, lb})(\% \text{ Strength, Sol. 2})}{100}}{\text{lb Solution 1} \quad + \quad \text{lb Solution 2}} \times 100$$

$$= \frac{(12 \text{ lb}) (0.1) + (40 \text{ lb}) (0.01)}{12 \text{ lb} + 40 \text{ lb}} \times 100$$

$$= \frac{1.2 \text{ lb} + 0.40}{52 \text{ lb}} \times 100$$

$$= 3.1\%$$

15.4.6 Dry Chemical Feeder Calibration

Occasionally we need to perform a calibration calculation to compare the actual chemical feed rate with the feed rate indicated by the instrumentation. To calculate the actual feed rate for a dry chemical feeder, place a container under the feeder; weigh the container when it is empty and then weigh it again after a specified length of time (30 min, for example).

The actual chemical feed rate can be calculated using the following equation:

$$\text{Chemical Feed Rate, lb/min} = \frac{\text{Chemical Applied, lb}}{\text{Length of Application, min}} \qquad (15.33)$$

If desired, the chemical feed rate can be converted to pounds per day:

$$\text{Feed Rate, lb/day} = \frac{(\text{Feed Rate, lb/min}) (1440 \text{ min})}{\text{day}} \qquad (15.34)$$

Example 15.37

Problem:

Calculate the actual chemical feed rate in pounds per day, if a container placed under a chemical feeder collects a total of 2 lb during a 30-min period.

Solution:

First, calculate the pounds per minute feed rate:

$$\text{Chemical Feed Rate, lb/min} = \frac{\text{Chemical Applied, lb}}{\text{Length of Application, min}}$$

$$= \frac{2 \text{ lb}}{30 \text{ min}}$$

$$= 0.06 \text{ lb/min Feed Rate}$$

Then, calculate the pounds per day feed rate:

$$\text{Chemical Feed Rate, lb/day} = (0.06 \text{ lb/min}) (1440 \text{ min/day})$$

$$= 86.4 \text{ lb/day Feed Rate}$$

Example 15.38

Problem:

Calculate the actual chemical feed rate in pounds per day, if a container placed under the chemical feeder collects a total of 1.6 lb during a 20-min period.

Solution:

First, calculate the pounds per minute feed rate:

$$\text{Chemical Feed Rate, lb/min} = \frac{\text{Chemical Applied, lb}}{\text{Length of Application, min}}$$

$$= \frac{1.6 \text{ lb}}{20 \text{ min}}$$

$$= 0.08 \text{ lb/min Feed Rate}$$

Then, calculate the pounds per day feed rate:

$$\text{Chemical Feed Rate, lb/day} = (0.08 \text{ lb/min}) (1440 \text{ min/day})$$

$$= 115 \text{ lb/day Feed Rate}$$

15.4.6.1 Solution Chemical Feeder Calibration

As with other calibration calculations, the actual solution chemical feed rate is determined and then compared with the feed rate indicated by the instrumentation. To calculate the actual solution chemical feed rate, first express the solution feed rate in million gallons per day (MGD). Once the million gallons per day solution flow rate has been calculated, use the milligrams per liter equation to determine chemical dosage in pounds per day.

If solution feed is expressed as milliliters per minute, first convert milliliters per minute flow rate to gallons per day flow rate.

$$\text{gpd} = \frac{(\text{mL/min})(1440 \text{ min/day})}{3785 \text{ mL/gal}} \tag{15.35}$$

Then, calculate chemical dosage, pounds per day.

$$\text{Chemical, lb/day} = (\text{mg/L Chemical}) (\text{MGD Flow}) (8.34 \text{ lb/day}) \tag{15.36}$$

Example 15.39

Problem:

A calibration test is conducted for a solution chemical feeder. During a 5-min test, the pump delivered 940 mg/L of the 1.20% polymer solution. (Assume the polymer solution weighs 8.34 lb/gal.) What is the polymer dosage rate in pounds per day?

Solution:

The flow rate must be expressed as million gallons per day; therefore, the milliliters per minute solution flow rate must first be converted to gallons per day and then million gallons per day. The milliliters per minute flow rate are calculated as:

$$\frac{940 \text{ mL}}{5 \text{ min}} = 188 \text{ mL/min}$$

Next, convert the milliliters per minute flow rate to gallons per day flow rate:

$$\frac{(188 \text{ mL/min}) (1440 \text{ min/day})}{3785 \text{ mL/gal}} = 72 \text{ gpd flow rate}$$

Then, calculate the pounds per day polymer feed rate:

$$(12{,}000 \text{ mg/L}) (0.000072 \text{ MGD}) (8.34 \text{ lb/day}) = 7.2 \text{ lb/day Polymer}$$

Example 15.40

Problem:

During a calibration test conducted for a solution chemical feeder, over a 24-h period, the solution feeder delivers a total of 100 gal of solution. The polymer solution is a 1.2% solution. What is the pounds-per-day feed rate? (Assume the polymer solution weighs 8.34 lb/gal.)

Solution:

The solution feed rate is 100 gal/day or 100 gpd. Expressed as million gallons per day, this is 0.000100 MGD. Use the milligrams-per-liter to pounds-per-day equation to calculate actual feed rate, pounds per day:

$$\text{lb/day Chemical} = (\text{Chemical, mg/L}) (\text{Flow, MGD}) (8.34 \text{ lb/day})$$

$$= (12{,}000 \text{ mg/L}) (0.000100 \text{ MGD}) (8.34 \text{ lb/day})$$

$$= 10 \text{ lb/day Polymer}$$

The actual pumping rates can be determined by calculating the volume pumped during a specified time frame. For example, if 60 gal are pumped during a 10-min test, the average pumping rate during the test is 6 gpm.

Actual volume pumped is indicated by drop in tank level. Using the following equation, we can determine the flow rate in gallons per minute:

$$\text{Flow Rate, gpm} = \frac{(0.785)\,(D^2)\,(\text{Drop in Level, ft})\,(7.48\ \text{gal/ft}^3)}{\text{Duration of Test, min}} \qquad (15.37)$$

Example 15.41

Problem:

For a pumping rate calibration test conducted for a 15-min period, the liquid level in the 4-ft diameter solution tank is measured before and after the test. If the level drops 0.5 ft during the 15-min test, what is the pumping rate in gallons per minute?

Solution:

$$\text{Flow Rate, gpm} = \frac{(0.785)\,(D^2)\,(\text{Drop in Level, ft})\,(7.48\ \text{gal/ft}^3)}{\text{Duration of Test, min}}$$

$$= \frac{(0.785)\,(4\ \text{ft})\,(4\ \text{ft})\,(0.5\ \text{ft})\,(7.48\ \text{gal/ft}^3)}{15\ \text{min}}$$

$$= 3.1\ \text{gpm Pumping Rate}$$

15.4.7 Determining Chemical Usage

One of the primary functions performed by water operators is the recording of data. The pounds-per-day or gallons-per-day chemical use is part of these data. From them, the average daily use of chemicals and solutions can be determined. This information is important in forecasting expected chemical use, comparing it with chemicals in inventory, and determining when additional chemicals will be required.

To determine average chemical use, use Equation 15.38 (pounds per day) or Equation 15.39 (gallons per day):

$$\text{Average Use, lb/day} = \frac{\text{Total Chemical Used, lb}}{\text{Number of Days}} \qquad (15.38)$$

or

$$\text{Average Use, gpd} = \frac{\text{Total Chemical Used, gal}}{\text{Number of Days}} \qquad (15.39)$$

Then we can calculate days' supply in inventory:

$$\text{Days Supply in Inventory} = \frac{\text{Total Chemical in Inventory, lb}}{\text{Average Use, lb/day}} \qquad (15.40)$$

or

$$\text{Days Supply in Inventory} = \frac{\text{Total Chemical in Inventory, gal}}{\text{Average Use, gpd}} \qquad (15.41)$$

Example 15.42

Problem:
   The chemical used for each day during a week is given in the following table. Based on these data, what was the average pounds-per-day chemical use during the week?

Day of week	Amount of chemical (lb/day)
Monday	88
Tuesday	93
Wednesday	91
Thursday	88
Friday	96
Saturday	92
Sunday	86

Solution:

$$\text{Average Use, lb/day} = \frac{\text{Total Chemical Used, lb}}{\text{Number of Days}}$$

$$= \frac{634 \text{ lb}}{7 \text{ days}}$$

$$= 90.6 \text{ lb/day Average Use}$$

Example 15.43

Problem:
   The average chemical use at a plant is 77 lb/day. If the chemical inventory is 2800 lb, how many days' supply is this?

Solution:

$$\text{Days Supply in Inventory} = \frac{\text{Total Chemical in Inventory, lb}}{\text{Average Use, lb/day}}$$

$$= \frac{2800 \text{ lb in Inventory}}{77 \text{ lb/day Average Use}}$$

$$= 36.4 \text{ days Supply in Inventory}$$

15.4.7.1 Paddle Flocculator Calculations

The gentle mixing required for flocculation is accomplished by a variety of devices. Probably the most common device in use is the basin equipped with mechanically driven paddles. Paddle flocculators have individual compartments for each set of paddles. The useful power input imparted by a paddle to the water depends on the drag force and the relative velocity of the water with respect to the paddle (Droste, 1997).

For paddle flocculator design and operation, environmental engineers are mainly interested in determining the velocity of a paddle at a set distance, the drag force of the paddle on the water, and the power input imparted to the water by the paddle.

Because of slip (factor k), the velocity of the water will be less than the velocity of the paddle. If baffles are placed along the walls in a direction perpendicular to the water movement, the value of k decreases because the baffles obstruct the movement of the water (Droste, 1997). The frictional dissipation of energy depends on the relative velocity, v. The relative velocity can be determined using Equation 15.42:

$$v = v_p - v_t = v_p - kv_p = v_p (1 - k) \qquad (15.42)$$

where
v_t = water velocity
v_p = paddle velocity

To determine the velocity of the paddle at a distance r from the shaft, use Equation 15.43:

$$v_p = \frac{2\pi N}{60} r \qquad (15.43)$$

where N = rate of revolution of the shaft (rpm).

To determine the drag force of the paddle on the water, use Equation 15.44:

$$F_D = 1/2 p C_D A v^2 \qquad (15.44)$$

where
A = area of the paddle
F_D = drag force
C_D = drag coefficient

To determine the power input imparted to the water by an elemental area of the paddle, the equation typically used is 15.45:

$$dP = dF_D v = 1/2 p C_D v^3 \, dA \qquad (15.45)$$

15.5 SEDIMENTATION CALCULATIONS

Sedimentation (solid from liquid separation by gravity) is one of the most basic processes of water and wastewater treatment. In water treatment, plain sedimentation, such as the use of a presedimentation basin for grit removal and sedimentation basin following coagulation–flocculation, is the most commonly used.

15.5.1 Tank Volume Calculations

The two common tank shapes of sedimentation tanks are rectangular and cylindrical. The equations for calculating the volume for each type of tank are shown next.

15.5.1.1 Calculating Tank Volume

For rectangular sedimentation basins, we use Equation 15.46:

$$\text{Volume, gal} = (\text{length, ft}) (\text{width, ft}) (\text{depth, ft}) (7.48 \text{ gal/ft}^3) \qquad (15.46)$$

For circular clarifiers, we use Equation 15.47:

$$\text{Volume, gal} = (0.785) (\text{D2}) (\text{depth, ft}) (7.48 \text{ gal/ft}^3) \qquad (15.47)$$

Example 15.44

Problem:
A sedimentation basin is 25 ft wide, 80 ft long, and contains water to a depth of 14 ft. What is the volume of water in the basin in gallons?

Solution:

$$\text{Volume, gal} = (\text{length, ft}) (\text{width, ft}) (\text{depth, ft}) (7.48 \text{ gal/ft}^3)$$

$$= (80 \text{ ft}) (25 \text{ ft}) (14 \text{ ft}) (7.48 \text{ gal/ft}^3)$$

$$= 209{,}440 \text{ gal}$$

Example 15.45

Problem:
A sedimentation basin is 24 ft wide and 75 ft long. When the basin contains 140,000 gal, what will the water depth be?

Solution:

$$\text{Volume, gal} = (\text{length, ft}) (\text{width, ft}) (\text{depth, ft}) (7.48 \text{ gal/ft}^3)$$

$$140{,}000 \text{ gal} = (75 \text{ ft}) (24 \text{ ft}) (x \text{ ft}) (7.48 \text{ gal/ft}^3)$$

$$x \text{ ft} = \frac{140{,}000}{(75)(24)(7.48)}$$

$$x \text{ ft} = 10.4 \text{ ft}$$

15.5.2 Detention Time

Detention time for clarifiers varies from 1 to 3 h. The equations used to calculate detention time are shown here.

- Basic detention time equation:

$$\text{Detention Time, h} = \frac{\text{Volume of Tank, gal}}{\text{Flow Rate, gph}} \qquad (15.48)$$

- Rectangular sedimentation basin equation:

$$\text{Detention Time, h} = \frac{(\text{Length, ft}) (\text{width, ft}) (\text{Depth, ft}) (7.48 \text{ gal/ft}^3)}{\text{Flow Rate, gph}} \qquad (15.49)$$

- Circular basin equation

$$\text{Detention Time, h} = \frac{(0.785) (D^2) (\text{Depth, ft}) (7.48 \text{ gal/ft}^3)}{\text{Flow Rate, gph}} \qquad (15.50)$$

Example 15.46

Problem:

A sedimentation tank has a volume of 137,000 gal. If the flow to the tank is 121,000 gph, what is the detention time in the tank, in hours?

Solution:

$$\text{Detention Time, h} = \frac{\text{Volume of Tank, gal}}{\text{Flow Rate, gph}}$$

$$= \frac{137,000 \text{ gal}}{121,000 \text{ gph}}$$

$$= 1.1 \text{ h}$$

Example 15.47

Problem:

A sedimentation basin is 60 ft long, 22 ft wide, and has water to a depth of 10 ft. If the flow to the basin is 1,500,000 gpd, what is the sedimentation basin detention time in hours?

Solution:

First, convert the flow rate from gallons per day to gallons per hour so that time units will match (1,500,000 gpd ÷ 24 h/day = 62,500 gph). Then calculate detention time:

$$\text{Detention Time, h} = \frac{\text{Volume of Tank, gal}}{\text{Flow Rate, gph}}$$

$$= \frac{(60 \text{ ft}) (22 \text{ ft}) (10 \text{ ft}) (7.48 \text{ gal/ft}^3)}{62,500 \text{ gph}}$$

$$= 1.6 \text{ h}$$

15.5.3 Surface Overflow Rate

Surface loading rate — similar to hydraulic loading rate (flow per unit area) — is used to determine loading on sedimentation basins and circular clarifiers. Hydraulic loading rate, however, measures the total water entering the process, whereas surface overflow rate measures only the water overflowing the process (plant flow only).

> *Note*: Surface overflow rate calculations do not include recirculated flows. Other terms used synonymously with surface overflow rate are surface loading rate and surface settling rate.

Surface overflow rate is determined using the following equation:

$$\text{Surface Overflow Rate} = \frac{\text{Flow, gpm}}{\text{Area, ft}^2} \qquad (15.51)$$

Example 15.48

Problem:

A circular clarifier has a diameter of 80 ft. If the flow to the clarifier is 1800 gpm, what is the surface overflow rate in gallons per minute per square foot?

Solution:

$$\text{Surface Overflow Rate} = \frac{\text{Flow, gpm}}{\text{Area, ft}^2}$$

$$= \frac{1800 \text{ gpm}}{(0.785)\,(80 \text{ ft})\,(80 \text{ ft})}$$

$$= 0.36 \text{ gpm/ft}^2$$

Example 15.49

Problem:

A sedimentation basin 70 ft by 25 ft receives a flow of 1000 gpm. What is the surface overflow rate in gallons per minute per square foot?

Solution:

$$\text{Surface Overflow Rate} = \frac{\text{Flow, gpm}}{\text{Area, ft}^2}$$

$$= \frac{1{,}000 \text{ gpm}}{(70 \text{ ft})\,(25 \text{ ft})}$$

$$= 0.6 \text{ gpm/ft}^2$$

15.5.4 Mean Flow Velocity

The measure of average velocity of the water as it travels through a rectangular sedimentation basin is known as mean flow velocity and is calculated using Equation 15.52:

$$Q \text{ (Flow), ft}^3/\text{min} = A \text{ (Cross-Sectional Area), ft}^2 \times V \text{ (Vol.) ft/min} \qquad (15.52)$$

$$(Q = A \times V)$$

Example 15.50

Problem:
   A sedimentation basin 60 ft long and 18 ft wide has water to a depth of 12 ft. When the flow through the basin is 900,000 gpd, what is the mean flow velocity in the basin in feet per minute?

Solution:
   Because velocity is desired in feet per minute, the flow rate in the $Q = AV$ equation must be expressed in cubic feet per minute (cfm):

$$\frac{900,000 \text{ gpd}}{(1440 \text{ min/day}) (7.48 \text{ gal/ft}^3)} = 84 \text{ cfm}$$

Then the $Q = AV$ equation can be used to calculate velocity:

$$(Q = A \, V)$$

$$84 \text{ cfm} = (18 \text{ ft}) (12 \text{ ft}) (x \text{ fpm})$$

$$x = \frac{84}{(18) (12)}$$

$$= 0.4 \text{ fpm}$$

Example 15.51

Problem:
   A rectangular sedimentation basin 50 ft long and 20 ft wide has a water depth of 9 ft. If the flow to the basin is 1,880,000 gpd, what is the mean flow velocity in feet per minute?

Solution:
   Because velocity is desired in feet per minute, the flow rate in the $Q = AV$ equation must be expressed in cubic feet per minute (cfm):

$$\frac{1,880,000 \text{ gpd}}{(1440 \text{ min/day}) (7.48 \text{ gal/ft}^3)} = 175 \text{ cfm}$$

The $Q = AV$ equation can be used to calculate velocity:

$$(Q = A V)$$

$$175 \text{ cfm} = (20 \text{ ft}) (9 \text{ ft}) (x \text{ fpm})$$

$$x = \frac{175 \text{ cfm}}{(20)(9)}$$

$$x = 0.97 \text{ fpm}$$

15.5.5 Weir Loading Rate (Weir Overflow Rate)

Weir loading rate (weir overflow rate) is the amount of water leaving the settling tank per linear foot of weir. The result of this calculation can be compared with design. Normally, weir overflow rates of 10,000 to 20,000 gal/day/ft are used in the design of a settling tank. Typically, weir-loading rate is a measure of the gallons per minute (gpm) flow over each foot of weir. Weir loading rate is determined using the following equation:

$$\text{Weir Loading Rate, gpm/ft} = \frac{\text{Flow, gpm}}{\text{Weir Length, ft}} \qquad (15.53)$$

Example 15.52

Problem:
 A rectangular sedimentation basin has a total of 115 ft of weir. What is the weir loading rate in gallons per minute per foot when the flow is 1,110,000 gpd?

Solution:

$$\frac{1,110,000 \text{ gpd}}{1440 \text{ min/day}} = 771 \text{ gpm}$$

$$\text{Weir Loading Rate, gpm/ft} = \frac{\text{Flow, gpm}}{\text{Weir Length, ft}}$$

$$= \frac{771 \text{ gpm}}{115 \text{ ft}}$$

$$= 6.7 \text{ gpm/ft}$$

Example 15.53

Problem:
 A circular clarifier receives a flow of 3.55 MGD. If the diameter of the weir is 90 ft, what is the weir-loading rate in gallons per minute per foot?

Solution:

$$\frac{3{,}550{,}000 \text{ gpd}}{1440 \text{ min/day}} = 2465 \text{ gpm}$$

$$\text{ft of weir} = (3.14)\,(90 \text{ ft})$$

$$= 283 \text{ ft}$$

$$\text{Weir Loading Rate, gpm/ft} = \frac{\text{Flow, gpm}}{\text{Weir Length, ft}}$$

$$= \frac{2465 \text{ gpm}}{283 \text{ ft}}$$

$$= 8.7 \text{ gpm/ft}$$

15.5.6 Percent Settled Biosolids

The percent settled biosolids test (a.k.a. "volume over volume" test, or V/V test) is conducted by collecting a 100-mL slurry sample from the solids contact unit and allowing it to settle for 10 min. After 10 min, the volume of settled biosolids at the bottom of the 100-mL graduated cylinder is measured and recorded. The equation used to calculate percent settled biosolids is

$$\% \text{ Settled Biosolids} = \frac{\text{Settled Biosolids Volume, mL}}{\text{Total Sample Volume, mL}} \times 100 \qquad (15.54)$$

Example 15.54

Problem:

A 100-mL sample of slurry from a solids contact unit is placed in a graduated cylinder and allowed to settle for 10 min. The settled biosolids at the bottom of the graduated cylinder after 10 min is 22 mL. What is the percent of settled biosolids of the sample?

Solution:

$$\% \text{ Settled Biosolids} = \frac{\text{Settled Biosolids Volume, mL}}{\text{Total Sample Volume, mL}} \times 100$$

$$= \frac{22 \text{ mL}}{100 \text{ mL}} \times 100$$

$$= 19\% \text{ Settled Biosolids}$$

Example 15.55

Problem:

A 100-mL sample of slurry from a solids contact unit is placed in a graduated cylinder. After 10 min, a total of 21 mL of biosolids settles to the bottom of the cylinder. What is the percent settled biosolids of the sample?

Solution:

$$\% \text{ Settled Biosolids} = \frac{\text{Settled Biosolids Volume, mL}}{\text{Total Sample Volume, mL}} \times 100$$

$$= \frac{21 \text{ mL}}{100 \text{ mL}} \times 100$$

$$= 21\% \text{ Settled Biosolids}$$

15.5.7 Determining Lime Dosage (Milligrams per Liter)

During the alum dosage process, lime is sometimes added to provide adequate alkalinity (HCO_3) in the solids contact clarification process for the coagulation and precipitation of the solids. Determining the required lime dose in milligrams per liter uses three steps.

In *Step 1*, the total alkalinity required is calculated. Total alkalinity required to react with the alum to be added and provide proper precipitation is determined using the following equation:

$$\text{Tot. Alk. Req., mg/L} = \underset{\underset{(1 \text{ mg/L Alum Reacts w/0.45 mg/L Alk.})}{\uparrow}}{\text{Alk. Reacting w/Alum, mg/L}} + \text{Alk. in the Water, mg/L} \qquad (15.55)$$

Example 15.56

Problem:

Raw water requires an alum dose of 45 mg/L, as determined by jar testing. If 30-mg/L residual alkalinity levels must be present in the water to ensure complete precipitation of alum added, what is the total alkalinity required in milligrams per liter?

Solution:

First, calculate the alkalinity that will react with 45 mg/L of alum:

$$\frac{0.45 \text{ mg/L Alk.}}{1 \text{ mg/L Alum}} = \frac{x \text{ mg/L Alk}}{45 \text{ mg/L Alum}}$$

$$(0.45)(45) = x$$

$$= 20.25 \text{ mg/L Alk.}$$

Next, calculate the total alkalinity required:

$$\text{Total Alk. Required, mg/L} = \text{Alk to React w/Alum, mg/L} + \text{Residual Alk, mg/L}$$

$$= 20.25 \text{ mg/L} + 30 \text{ mg/L}$$

$$= 50.25 \text{ mg/L}$$

Example 15.57

Problem:

Jar tests indicate that 36 mg/L of alum is optimum for particular raw water. If a residual 30-mg/L alkalinity must be present to promote complete precipitation of the alum added, what is the total alkalinity required in milligrams per liter?

Solution:

First, calculate the alkalinity that will react with 36 mg/L of alum:

$$\frac{0.45 \text{ mg/L Alk.}}{1 \text{ mg/L Alum}} = \frac{x \text{ mg/L Alk}}{36 \text{ mg/L Alum}}$$

$$(0.45)(36) = x$$

$$= 16.2$$

Then, calculate the total alkalinity required:

$$\text{Total Alk. Required, mg/L} = A = 16.2 \text{ mg/L} + 30 \text{ mg/L}$$

$$= 46.2 \text{ mg/L}$$

In *Step 2*, we make a comparison between required alkalinity and alkalinity already in the raw water to determine how many milligrams per liter of alkalinity should be added to the water. The equation used to make this calculation is:

$$\text{Alk. to be Added to the Water, mg/L} = \text{Tot. Alk. Req., mg/L}^- \qquad (15.56)$$
$$\text{Alk. Present in the Water, mg/L}$$

Example 15.58

Problem:

A total of 44-mg/L alkalinity is required to react with alum and ensure proper precipitation. If the raw water has an alkalinity of 30 mg/L as bicarbonate, how many milligrams per liter alkalinity should be added to the water?

Solution:

$$\text{Alk. to be Added, mg/L} = \text{Tot. Alk. Req., mg/L-Alk. Present in the Water, mg/L}$$

$$= 44 \text{ mg/L} - 30 \text{ mg/L}$$

$$= 14 \text{ mg/L Alkalinity to be Added}$$

In *Step 3*, after determining the amount of alkalinity to be added to the water, we determine how much lime (the source of alkalinity) needs to be added. We accomplish this by using the ratio shown in Example 15.59.

Example 15.59

Problem:

If 16-mg/L alkalinity must be added to a raw water, how many milligrams-per-liter lime will be required to provide this amount of alkalinity? (1 mg/L alum reacts with 0.45 mg/L alkalinity and 1 mg/L alum reacts with 0.35 mg/L lime.)

Solution:

First, we determine the milligrams per liter of lime required by using a proportion that relates bicarbonate alkalinity to lime:

$$\frac{0.45 \text{ mg/L Alk.}}{0.35 \text{ mg/L Lime}} = \frac{16 \text{ mg/L Alk.}}{x \text{ mg/L Lime}}$$

Next, we cross-multiply:

$$0.45\, x = (16)(0.35)$$

$$x = \frac{(16)(0.35)}{0.45}$$

$$x = 12.4 \text{ mg/L Lime}$$

In Example 15.60, we use all three steps to determine a required lime dosage (milligrams per liter).

Example 15.60

Problem:

Given the following data, calculate the lime dose required, in milligrams per liter:

- Alum dose required (determined by jar tests) — 52 mg/L
- Residual alkalinity required for precipitation — 30 mg/L
- 1 mg/L alum reacts with 0.35 mg/L lime

- 1 mg/L alum reacts with 0.45 mg/L alkalinity
- Raw water alkalinity — 36 mg/L

Solution:

To calculate the total alkalinity required, we must first calculate the alkalinity that will react with 52 mg/L of alum:

$$\frac{0.45 \text{ mg/L Alk.}}{1 \text{ mg/L Alum}} = \frac{x \text{ mg/L Alk.}}{52 \text{ mg/L Lime}}$$

$$(0.45)\,(52) = x$$

$$23.4 \text{ mg/L Alk.} = x$$

The total alkalinity requirement can now be determined:

$$\text{Total Alk. Required, mg/L} = \text{Alk. to React w/Alum, mg/L} + \text{Residual Alk, mg/L}$$

$$= 23.4 \text{ mg/L} + 30 \text{ mg/L}$$

$$= 53.4 \text{ mg/L Total Alkalinity Required}$$

Next, calculate how much alkalinity must be *added* to the water:

$$\text{Alk. to be Added, mg/L} = \text{Tot. Alk. Req., mg/L}^- \text{ Alk. Present, mg/L}$$

$$= 53.4 \text{ mg/L} - 36 \text{ mg/L}$$

$$= 17.4 \text{ mg/L Alk to be added to the Water}$$

Finally, calculate the lime required to provide this additional alkalinity:

$$\frac{0.45 \text{ mg/L Alk.}}{0.35 \text{ mg/L Alum}} = \frac{17.4 \text{ mg/L Alk.}}{x \text{ mg/L Lime}}$$

$$0.45 \, x = (17.4)\,(0.35)$$

$$x = \frac{(17.4)\,(0.35)}{0.45}$$

$$x = 13.5 \text{ mg/L Lime}$$

15.5.8 Determining Lime Dosage (Pounds per Day)

After the lime dose has been determined in terms of milligrams per liter, calculating the lime dose in pounds per day is a fairly simple matter. This is one of the most common calculations in water and wastewater treatment. To convert from milligrams per liter to pounds per day of lime dose, use the following equation:

$$\text{Lime, lb/day} = \text{Lime (mg/L) (Flow, MGD) (8.34 lb/gal)} \tag{15.57}$$

Example 15.61

Problem:
    The lime dose for a raw water is calculated at 15.2 mg/L. If the flow to be treated is 2.4 MGD, how many pounds per day of lime are required?

Solution:

$$\text{Lime, lb/day} = \text{Lime (mg/L) (Flow, MGD) (8.34 lb/gal)}$$

$$= (15.2 \text{ mg/L}) (2.4 \text{ MGD}) (8.34 \text{ lb/gal})$$

$$= 304 \text{ lb/day Lime}$$

Example 15.62

Problem:
    The flow to a solids contact clarifier is 2,650,000 gpd. If the lime dose required is 12.6 mg/L, how many pounds per day of lime are required?

Solution:

$$\text{Lime, lb/day} = \text{Lime (mg/L) (Flow, MGD) (8.34 lb/gal)}$$

$$= (12.6 \text{ mg/L}) (2.65 \text{ MGD}) (8.34 \text{ lb/gal})$$

$$= 278 \text{ lb/day Lime}$$

15.5.9 Determining Lime Dosage (Grams per Minute)

In converting from milligrams per liter of lime to grams per minute (g/min) of lime, use Equation 15.58.

    Key point: 1 lb = 453.6 g.

$$\text{Lime, g/min} = \frac{(\text{Lime, lb/day}) (453.6 \text{ g/lb})}{1440 \text{ min/day}} \tag{15.58}$$

Example 15.63

Problem:

A total of 275 lb/day of lime is required to raise the alkalinity of the water passing through a solids-contact clarification process. How many grams per minute of lime does this represent?

Solution:

$$\text{Lime, g/min} = \frac{(\text{lb/day}) \, (453.6 \text{ g/lb})}{1440 \text{ min/day}}$$

$$= \frac{(275 \text{ lb/day}) \, (453.6 \text{ g/lb})}{1440 \text{ min/day}}$$

$$= 86.6 \text{ g/min Lime}$$

Example 15.64

Problem:

A lime dose of 150 lb/day is required for a solids-contact clarification process. How many grams per minute of lime does this represent?

Solution:

$$\text{Lime, g/min} = \frac{(\text{lb/day}) \, (453.6 \text{ g/lb})}{1440 \text{ min/day}}$$

$$= \frac{(150 \text{ lb/day}) \, (453.6 \text{ g/lb})}{1440 \text{ min/day}}$$

$$= 47.3 \text{ g/min Lime}$$

15.5.10 Particle Settling (Sedimentation)

> *Note*: Much of the information presented in the following subsection is based on USEPA's Turbidity Provisions, EPA Guidance Manual, April 1999.

Particle settling (sedimentation) may be described for a singular particle by the Newton equation (Equation 15.64) for terminal settling velocity of a spherical particle. For the engineer, knowledge of this velocity is basic to the design and performance of a sedimentation basin.

The rate at which discrete particles will settle in a fluid of constant temperature is given by the equation:

$$u = \frac{4g(p_P - p)d}{3C_D p} \,^{0.5} \tag{15.59}$$

where

u = settling velocity of particles, meters per second or feet per second
g = gravitational acceleration, meters per square second or feet per square second
p_p = density of particles, kilogram per cubic meter or pounds per cubic foot
p = density of water, kilograms per cubic meter or pounds per cubic feet
d = diameter of particles, meters or feet
C_D = coefficient of drag

The terminal settling velocity is derived by equating the drag, buoyant, and gravitational forces acting on the particle. At low settling velocities, the equation is not dependent on the shape of the particle, and most sedimentation processes are designed to remove small particles ranging from 1.0 to 0.5 μm, which settle slowly. Larger particles settle at higher velocity and are removed whether or not they follow Newton's law or Stokes' law — the governing equations when the drag coefficient is sufficiently small (0.5 or less) as is the case for colloidal products (McGhee, 1991).

Typically, a large range of particle sizes exists in raw water supplies. There are four types of sedimentation:

- Type 1 — discrete particle settling (particles of various sizes, in a dilute suspension, which settle without flocculating
- Type 2 — flocculant settling (heavier particles coalesced with smaller and lighter particles)
- Type 3 — hindered settling (high densities of particles in suspension resulting in an interaction of particles)
- Type 4—compression settling (Gregory and Zabel, 1990).

The values of drag coefficient depend on the density of water (p); relative velocity (u); particle diameter (d); and viscosity of water (μ), which gives the Reynolds number **R** as:

$$R = \frac{pud}{\mu} \qquad (15.60)$$

As the Reynolds number increases, the value of C_D increases. For **R** of less than 2, C_D is related to **R** by the following linear expression:

$$C_D = \frac{24}{R} \qquad (15.61)$$

At low levels of **R**, the Stokes equation for laminar flow conditions is used (Equation 15.60 and Equation 15.61 substituted into Equation 15.59).

$$u = \frac{g(p_P - p)d^2}{18\mu} \qquad (15.62)$$

In the region of higher Reynolds numbers ($2 < R < 500$ to 1000), C_D becomes (Fair et al., 1968):

$$C_D = \frac{24}{R} + \frac{3}{\sqrt{R}} + 0.34 \qquad (15.63)$$

Key point: In the region of turbulent flow (500–$1000 < R < 200{,}000$), the C_D remains approximately constant at 0.44.

The velocity of settling particles results in Newton's equation (AWWA and ASCE, 1990):

$$u = 1.74 \left[\frac{(p_p - p)gd}{p} \right]^{0.5}$$

(15.64)

Key point: When the Reynolds number is greater than 200,000, the drag force decreases substantially, and C_D becomes 0.10. No settling occurs at this condition.

Example 15.65

Problem:

Estimate the terminal settling velocity in water at a temperature of 21°C, for spherical particles with specific gravity of 2.40 and average diameter of (a) 0.006 mm and (b) 1.0 mm.

Solution:

Step 1. Use Equation 15.62 for (a).
Given: at temperature $(T) = 21°C$
$p = 998$ kg/m³
$\mu = 0.00098$ N s/m²
$d = 0.06$ mm $= 6 \times 10^{-5}$ m
$g = 9.81$ m/sec²

$$u = \frac{g(p_P - p)d^2}{18\mu}$$

$$u = \frac{9.81 \text{ m/sec}^2 \, (2400 - 998) \text{ kg/m}^3 \, (6 \times 10^{-5} \text{ m})^2}{18 \times 0.00098 \text{ N sec/m}^2}$$

$$= 0.00281 \text{ m/sec}$$

Step 2. Use Equation 15.60 to check the Reynolds number:

$$R = \frac{pud}{\mu} = \frac{998 \times 0.00281 \times 6 \times 10^{-5}}{0.00098}$$

$$= 0.172$$

(a) Stokes' law applies because **R** < 2.
Step 3. Using Stokes' law for (b):

$$u = \frac{9.81 \, (2400 - 998) \, (0.001)^2}{18 \times 0.00098}$$

$$= \frac{0.0137536}{0.01764}$$

$$= 0.779 \text{ m/sec}$$

Step 4. Check the Reynolds number:
Assume the irregularities of the particles $\Phi = 0.80$

$$= \frac{\Phi p u d}{\mu} = \frac{0.80 \times 998 \times 0.779 \times 0.001}{0.00098}$$

$$= 635$$

Because $\mathbf{R} > 2$, Stokes' law does not apply. Use Equation 15.59 to calculate u.
Step 5. Using Equation 15.63 and Equation 15.59, calculate u.

$$C_D = \frac{24}{R} + \frac{3}{\sqrt{R}} + 0.34 = \frac{24}{635} + \frac{3}{\sqrt{635}} = 0.34$$

$$= 0.50$$

$$u^2 = \frac{4g(p_p - p)d}{3C_D p}$$

$$u^2 = \frac{4 \times 9.81 \times (2400 - 998) \times 0.001}{3 \times 0.50 \times 998}$$

$$u = 0.192 \text{ m/sec}$$

Step 6. Recheck $\mathbf{R}$:

$$R = \frac{\Phi p u d}{\mu} = \frac{0.80 \times 998 \times 0.192 \times 0.001}{0.00098}$$

$$= 156$$

Step 7. Repeat Step 5 with new $\mathbf{R}$:

$$C_D = \frac{24}{156} + \frac{3}{\sqrt{156}} + 0.34$$

$$= 0.73$$

$$u^2 = \frac{4 \times 9.81 \times 1402 \times 0.001}{3 \times 0.73 \times 998}$$

$$= 159 \text{ m/sec}$$

Step 8. Recheck **R:**

$$R = \frac{0.80 \times 998 \times 0.159 \times 0.001}{0.00098}$$

$$= 130$$

Step 9. Repeat Step 7:

$$C_D = \frac{24}{130} + \frac{3}{\sqrt{130}} + 0.34$$

$$= 0.79$$

$$u^2 = \frac{4 \times 9.81 \times 1402 \times 0.001}{3 \times 0.79 \times 998}$$

$$u = 0.152 \text{ m/sec}$$

(b) The estimated velocity is approximately 0.15 m/sec.

15.5.11 Overflow Rate (Sedimentation)

Overflow rate, along with detention time, horizontal velocity, and weir loading rate, is the parameter typically used for sizing sedimentation basin. The theoretical detention time (plug flow theory) is computed from the volume of the basin, divided by average daily flow.

$$t = \frac{24 \, V}{Q} \tag{15.65}$$

where
t = detention time, hours
24 = 24 hours per day
V = volume of basin, cubic meters or million gallons per day
Q = average daily flow, cubic meters per day or million gallons per day

The overflow rate is a standard design parameter that can be determined from discrete particle settling analysis. The overflow rate or surface loading rate is calculated by dividing the average daily flow by the total area of the sedimentation basin.

$$u = \frac{Q}{A} = \frac{Q}{lw} \tag{15.66}$$

where
u = overflow rate, cubic meters (square meters per day) or gallons per day per foot
Q = average daily flow, cubic meters per day or gallons per day
A = total surface area of basin, square meters or square feet
l and w = length and width of basin, meters or feet

Key point: All particles with a settling velocity greater than the overflow rate will settle and be removed.

Hudson points out that rapid particle density changes from temperature, solid concentration, or salinity can induce density current, which can cause severe short-circuiting in horizontal tanks (Hudson, 1989).

Example 15.66

Problem:
A water treatment plant has two clarifiers treating 2.0 MGD of water. Each clarifier is 14 ft wide, 80 ft long, and 17 ft deep. Determine: (a) detention time; (b) overflow; (c) horizontal velocity; and (d) weir loading rate, assuming the weir length is 2.5 times the basin width.

Solution:

Step 1. Compute detention time (t) for each clarifier:

$$Q = \frac{2 \text{ mgd}}{2} = \frac{1,000,000 \text{ gal}}{d} \times \frac{1 \text{ ft}}{7.48 \text{ gal}} \times \frac{1 \text{ d}}{24 \text{ h}}$$

$$= 5570 \text{ ft}^3/h$$

$$= 92.8 \text{ ft}^3/min$$

(a)
$$t = \frac{V}{Q} = \frac{14 \text{ ft} \times 80 \text{ ft} \times 17 \text{ ft}}{5570 \text{ ft}^3/h} 3.42 \text{ h}$$

Step 2. Compute overflow rate u:

(b)
$$u = \frac{Q}{Iw} = \frac{1,000,000 \text{ gpd}}{14 \text{ ft} \times 80 \text{ ft}} = 893 \text{ gpd/ft}$$

Step 3. Compute horizontal velocity V:

(c)
$$V = \frac{Q}{wd} = \frac{92.8 \text{ ft}^3/min}{14 \text{ ft} \times 17 \text{ ft}} = 0.39 \text{ ft/min}$$

Step 4. Compute weir loading rate u_w:

(d)
$$u_w = \frac{Q}{2.5 \text{ w}} = \frac{1,000,000 \text{ gpd}}{2.5 \times 14 \text{ ft}} = 28,571 \text{ gpd/ft}$$

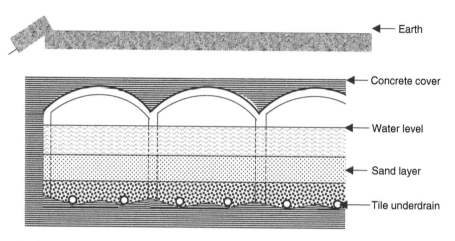

Figure 15.3 Slow sand filter.

15.6 WATER FILTRATION CALCULATIONS

Water filtration is the physical process of separating suspended and colloidal particles from waste by passing the water through a granular material. The process of filtration involves straining, settling, and adsorption. As floc passes into the filter, the spaces between the filter grains become clogged, reducing openings and increasing removal. Some material is removed merely because it settles on a media grain. One of the most important processes is adsorption of the floc onto the surface of individual filter grains. In addition to removing silt and sediment, flock, algae, insect larvae and any other large elements, filtration also contributes to the removal of bacteria and protozoans such as Giardia lamblia and Cryptosporidium. Some filtration processes are also used for iron and manganese removal.

The surface water treatment rule (SWTR) specifies four filtration technologies, although SWTR also allows the use of alternate filtration technologies (for example, cartridge filters). These include slow sand filtration (see Figure 15.3); rapid sand filtration; pressure filtration; diatomaceous earth filtration; and direct filtration. Of these, all but rapid sand filtration are commonly employed in small water systems that use filtration. Each type of filtration system has advantages and disadvantages. Regardless of the type of filter, however, filtration involves the processes of *straining* (where particles are captured in the small spaces between filter media grains); *sedimentation* (where the particles land on top of the grains and stay there); and *adsorption* (where a chemical attraction occurs between the particles and the surface of the media grains.

15.6.1 Flow Rate through a Filter (Gallons per Minute)

Flow rate in gallons per minute through a filter can be determined by simply converting the gallons per day flow rate, as indicated on the flow meter. Calculate the flow rate (gallons per minute) by taking the meter flow rate (gallons per day) and dividing by 1440 min/day, using Equation 15.67.

$$\text{Flow Rate, gpm} = \frac{\text{Flow Rate, gpd}}{1440 \text{ min/day}} \qquad (15.67)$$

Example 15.67

Problem:

The flow rate through a filter is 4.25 MGD. What is this flow rate expressed as gallons per minute?

Solution:

$$\text{Flow Rate, gpm} = \frac{4.25 \text{ gpd}}{1440 \text{ min/day}}$$

$$= \frac{4,250,000 \text{ gpd}}{1440 \text{ min/day}}$$

$$= 2951 \text{ gpm}$$

Example 15.68

Problem:

During a 70-h filter run, a total of 22.4 million gal of water are filtered. What is the average flow rate through the filter in gallons per minute during this filter run?

Solution:

$$\text{Flow Rate, gpm} = \frac{\text{Total Gallons Produced}}{\text{Filter Run, min}}$$

$$= \frac{22,400,000 \text{ gal}}{(70 \text{ h}) (60 \text{ min/h})}$$

$$= 5333 \text{ gpm}$$

Example 15.69

Problem:

At an average flow rate of 4000 gpm, how long a filter run (in hours) would be required to produce 25 MG of filtered water?

Solution:

Write the equation as usual, filling in known data:

$$\text{Flow Rate, gpm} = \frac{\text{Total Gallons Produced}}{\text{Filter Run, min}}$$

$$4000 \text{ gpm} = \frac{25,000,000 \text{ gal}}{(x \text{ h}) (60 \text{ min/h})}$$

Then solve for *x:*

$$= \frac{25,000,000 \text{ gal}}{(4000) (60)}$$

$$= 104 \text{ h}$$

Example 15.70

Problem:

A filter box is 20 ft × 30 ft (including the sand area). If the influent valve is shut, the water drops 3.0 inches per minute. What is the rate of filtration in million gallons per day?

Solution:

Given:

$$\text{Filter Box} = 20 \text{ ft} \times 30 \text{ ft}$$

$$\text{Water drop} = 3.0 \text{ inches per minute}$$

Find the volume of water passing through the filter:

$$\text{Volume} = \text{Area} \times \text{Height}$$

$$\text{Area} = \text{Width} \times \text{Length}$$

Note: The best way to perform calculations for this type of problem is step by step, breaking down the problem into what is given and what must be found.

Step 1:

$$\text{Area} = 20 \text{ ft} \times 30 \text{ ft} = 600 \text{ ft}^2$$

Convert 3.0 in. into feet.
Divide 3.0 by 12 to find feet.

$$3.0/12 = 0.25 \text{ feet}$$

$$\text{Volume} = 600 \text{ ft}^2 \times 0.25 \text{ ft}$$

$$= 150 \text{ ft}^3 \text{ of water passing through the filter in one minute}$$

Step 2. Convert cubic feet to gallons:

$$150 \text{ ft}^3 \times 7.48 \text{ gal/ft}^3 = 1,122 \text{ gal/min}$$

Step 3. The problem asks for the rate of filtration in million gallons per day. To find this, multiply the number of gallons per minute by the number of minutes per day.

$$1122 \text{ gal/min} \times 1440 \text{ min/day} = 1.62 \text{ MGD}$$

Example 15.71

Problem:

The influent valve to a filter is closed for 5 min. During this time, the water level in the filter drops 0.8 ft (10 in.). If the filter is 45 ft long and 15 ft wide, what is the gallons per minute flow rate through the filter? Water drop equals 0.16 ft/min.

Solution:

First, calculate cubic feet per minute flow rate using the $Q = AV$ equation:

$$Q, \text{cfm} = (\text{Length, ft}) \, (\text{Width, ft}) \, (\text{Drop Velocity, ft/min})$$

$$= (45 \text{ ft}) \, (15 \text{ ft}) \, (0.16 \text{ ft/min})$$

$$= 108 \text{ cfm}$$

Then, convert cubic feet per minute flow rate to gallons per minute flow rate:

$$(108 \text{ cfm}) \, (7.48 \text{ gal/ft}^3) = 808 \text{ gpm}$$

15.6.2 Filtration Rate

One measure of filter production is filtration rate (generally a range of from 2 to 10 gpm/ft²). Along with filter run time, filtration rate — the gallons per minute of water filtered through each square foot of filter area — provides valuable information for filter operation. Determine filtration rate using Equation 15.68:

$$\text{Filtration Rate, gpm/ft}^2 = \frac{\text{Flow Rate, gpm}}{\text{Filter Surface Area, ft}^2} \qquad (15.68)$$

Example 15.72

Problem:

A filter 18 ft by 22 ft receives a flow of 1750 gpm. What is the filtration rate in gallons per minute per square foot?

Solution:

$$\text{Filtration Rate, gpm/ft}^2 = \frac{\text{Flow Rate, gpm}}{\text{Filter Surface Area, ft}^2}$$

$$= \frac{1750 \text{ gpm}}{(18 \text{ ft}) \, (22 \text{ ft})}$$

$$= 4.4 \text{ gpm/ft}^2$$

Example 15.73

Problem:

A filter 28 ft long and 18 ft wide treats a flow of 3.5 MGD. What is the filtration rate in gallons per minute per square foot?

Solution:

$$\text{Flow rate} = \frac{3,500,000 \text{ gpd}}{1440 \text{ min/day}} = 2431 \text{ gpm}$$

$$\text{Filtration Rate, gpm/ft}^2 = \frac{\text{Flow Rate, gpm}}{\text{Filter Surface Area, ft}^2}$$

$$= \frac{2431 \text{ gpm}}{(28 \text{ ft})(18 \text{ ft})}$$

$$= 4.8 \text{ gpm/ft}^2$$

Example 15.74

Problem:

A filter 45 ft long and 20 ft wide produces a total of 18 MG during a 76-h filter run. What is the average filtration rate in gallons per minute per square foot for this filter run?

Solution:

First, calculate the gallons per minute flow rate through the filter:

$$\text{Flow rate} = \frac{\text{Total Gallons Produced}}{\text{Filter Run, min}}$$

$$\text{Flow rate} = \frac{18,000,000 \text{ gal}}{(76 \text{ h})(60 \text{ min/h})}$$

$$= 3947 \text{ gpm}$$

Then, calculate filtration rate:

$$\text{Filtration Rate, gpm/ft}^2 = \frac{\text{Flow Rate, gpm}}{\text{Filter Surface Area, ft}^2}$$

$$\text{Flow rate} = \frac{3947 \text{ gpm}}{(45 \text{ ft})(20 \text{ ft})}$$

$$= 4.4 \text{ gpm/ft}^2$$

Example 15.75

Problem:
A filter is 40 ft long and 20 ft wide. During a test of flow rate, the influent valve to the filter is closed for 6 min. The water level drop during this period is 16 in. What is the filtration rate for the filter in gallons per minute per square foot?

Solution:
First, calculate gallons per minute flow rate, using the $Q = AV$ equation:

$$Q, \text{gpm} = (\text{Length, ft}) \, (\text{Width, ft}) \, (\text{Drop Velocity, ft/min}) \, (7.48 \text{ gal/ft}^3)$$

$$= \frac{(40 \text{ ft}) \, (20 \text{ ft}) \, (1.33 \text{ ft}) \, (7.48 \text{ gal/ft}^3)}{6 \text{ min}}$$

$$= 1316 \text{ gpm}$$

Then, calculate filtration rate:

$$\text{Filtration Rate} = \frac{\text{Flow Rate, gpm}}{\text{Filter Area, ft}^2}$$

$$= \frac{1316 \text{ gpm}}{(40 \text{ ft}) \, (20 \text{ ft})}$$

$$= 1.6 \text{ gpm/ft}^2$$

15.6.3 Unit Filter Run Volume (UFRV)

The unit filter run volume (UFRV) calculation indicates the total gallons passing through each square foot of filter surface area during an entire filter run. This calculation is used to compare and evaluate filter runs. UFRVs are usually at least 5000 gal/ft² and generally in the range of 10,000 gpd/ft². The UFRV value will begin to decline as the performance of the filter begins to deteriorate. The equation used in these calculations is

$$\text{UFRV} = \frac{\text{Total Gallons Filtered}}{\text{Filter Surface Area, ft}^2} \qquad (15.69)$$

Example 15.76

Problem:
The total water filtered during a filter run (between backwashes) is 2,220,000 gal. If the filter is 18 ft by 18 ft, what is the unit filter run volume (UFRV) in gallons per square foot?

Solution:

$$UFRV = \frac{\text{Total Gallons Filtered}}{\text{Filter Surface Area, ft}^2}$$

$$= \frac{2,220,000 \text{ gal}}{(18 \text{ ft}) (18 \text{ ft})}$$

$$= 6852 \text{ gal/ft}^2$$

Example 15.77

Problem:

The total water filtered during a filter run is 4,850,000 gal. If the filter is 28 ft by 18 ft, what is the unit filter run volume in gallons per square foot?

Solution:

$$UFRV = \frac{\text{Total Gallons Filtered}}{\text{Filter Surface Area, ft}^2}$$

$$= \frac{4,850,000 \text{ gal}}{(28 \text{ ft}) (18 \text{ ft})}$$

$$= 9623 \text{ gal/ft}^2$$

Equation 15.69 can be modified as shown in Equation 15.70 to calculate the unit filter run volume, given filtration rate and filter run data.

$$UFRV = (\text{Filtration Rate, gpm, ft}^2) (\text{Filter Run Time, min}) \qquad (15.70)$$

Example 15.78

Problem:

The average filtration rate for a filter is determined at 2.0 gpm/ft². If the filter run time was 4250 min, what was the unit filter run volume in gallons per square foot?

Solution:

$$UFRV = (\text{Filtration Rate, gpm, ft}^2) (\text{Filter Run Time, min})$$

$$= 8500 \text{ gal/ft}^2$$

The problem indicates an average filtration rate of 2.0 gal entering each square foot of filter each minute. The total gallons entering during the total filter run is 4250 times that amount.

Example 15.79

Problem:
   The average filtration rate during a particular filter run was determined at 3.2 gpm/ft². If the filter run time was 61.0 h, what was the UFRV in gallons per square foot for the filter run?

Solution:

$$UFRV = (\text{Filtration Rate, gpm, ft}^2)\,(\text{Filter Run, h})\,(60 \text{ min/h})$$

$$= (3.2 \text{ gpm/ft}^2)\,(61.0 \text{ h})\,(60 \text{ min/h})$$

$$= 11{,}712 \text{ gal/ft}^2$$

15.6.4 Backwash Rate

In filter backwashing, one of the most important operational parameters to be determined is the amount of water in gallons required for each backwash. This amount depends on the design of the filter and the quality of the water being filtered. The actual washing typically lasts 5 to 10 min and uses amounts from 1 to 5% of the flow produced.

Example 15.80

Problem:
   A filter has the following dimensions:

- Length = 30 ft
- Width = 20 ft
- Depth of filter media = 24 in.

Assuming that a backwash rate of 15 gal/ft²/min is recommended and 10 min of backwash is required, calculate the amount of water in gallons required for each backwash.

Solution:

   Step 1. Area of filter = 30 ft × 20 ft = 600 ft²
   Step 2. Gallons of water used per square foot of filter = 15 gal/ft²/min × 10 min = 150 gal/ft²
   Step 3. Gallons required = 150 gal/ft² × 600 ft² = 90,000 gal required for backwash.

Typically, backwash rates range from 10 to 25 gpm/ft². The backwash rate is determined by using Equation 15.71:

$$\text{Backwash} = \frac{\text{Flow Rate, gpm}}{\text{Filter Area, ft}^2} \qquad (15.71)$$

Example 15.81

Problem:
   A filter 30 ft by 10 ft has a backwash rate of 3120 gpm. What is the backwash rate in gallons per minute per square foot?

Solution:

$$\text{Backwash rate} = \frac{\text{Flow Rate, gpm}}{\text{Filter Area, ft}^2}$$

$$= \frac{3120 \text{ gpm}}{(30 \text{ ft}) (10 \text{ ft})}$$

$$= 10.4 \text{ gpm/ft}^2$$

Example 15.82

Problem:

A filter 20 ft long and 20 ft wide has a backwash flow rate of 4.85 MGD. What is the filter backwash rate in gallons per minute per square foot?

Solution:

$$\text{Backwash rate} = \frac{\text{Flow Rate, gpm}}{\text{Filter Area, ft}^2}$$

$$= \frac{4,850,000 \text{ gpd}}{1440 \text{ min/day}}$$

$$= 3368 \text{ gpm}$$

$$= \frac{3368 \text{ gpm}}{(20 \text{ ft})(20 \text{ ft})}$$

$$= 8.42 \text{ gpm/ft}^2$$

15.6.5 Backwash Rise Rate

Backwash rate is occasionally measured as the upward velocity of the water during backwashing expressed as inches per minute of rise. To convert from a backwash rate of gallons per minute per square foot to a rise rate of inches per minute, use Equation 15.72 or Equation 15.73:

$$\text{Backwash Rate, in./min} = \frac{(\text{Backwash Rate, gpm/ft}^2) \, (12 \text{ in./ft})}{7.48 \text{ gal/ft}^3} \qquad (15.72)$$

$$\text{Backwash Rate, in./min} = (\text{Backwash Rate, gpm/ft}^2) \, (1.6) \qquad (15.73)$$

Example 15.83

Problem
    A filter has a backwash rate of 16 gpm/ft². What is this backwash rate expressed as inches per minute of rise rate?

Solution:

$$\text{Backwash Rate, in/min} = \frac{(\text{Backwash Rate, gpm/ft}^2)\,(12\text{ in./ft})}{7.48\text{ gal/ft}^3}$$

$$= \frac{(16\text{ gpm/ft}^2)\,(12\text{ in./ft})}{7.48\text{ gal/ft}^3}$$

$$= 25.7\text{ in./min}$$

Example 15.84

Problem:
    A filter 22 ft long and 12 ft wide has a backwash rate of 3260 gpm. What is this backwash rate expressed as inches per minute of rise?

Solution:
    First, calculate the backwash rate as gallons per minute per square foot:

$$\text{Backwash Rate} = \frac{\text{Flow Rate, gpm}}{\text{Filter Area, ft}^2}$$

$$= \frac{3260\text{ gpm}}{(22\text{ ft})\,(12\text{ ft})}$$

$$= 12.3\text{ gpm/ft}^2$$

Then, convert gallons per minute per square foot to inches per minute of rise rate:

$$= \frac{(12.3\text{ gpm/ft}^2)\,(12\text{ in./ft})}{7.48\text{ gal/ft}^3}$$

$$= 19.7\text{ in./min}$$

15.6.6 Volume of Backwash Water Required (Gallons)

To determine the volume of water required for backwashing, we must know the desired backwash flow rate (gallons per minute) and the duration of backwash (minutes):

$$\text{Backwash Water Vol., gal} = (\text{Backwash, gpm}) (\text{Duration of Backwash, min}) \qquad (15.74)$$

Example 15.85

Problem:

For a backwash flow rate of 9000 gpm and a total backwash time of 8 min, how many gallons of water are required for backwashing?

Solution:

$$\text{Backwash Water Vol., gal} = (\text{Backwash, gpm}) (\text{Duration of Backwash, min})$$

$$= (9000 \text{ gpm}) (8 \text{ min})$$

$$= 72,000 \text{ gal}$$

Example 15.86

Problem:

How many gallons of water are required to provide a backwash flow rate of 4850 gpm for a total of 5 min?

Solution:

$$\text{Backwash Water Vol., gal} = (\text{Backwash, gpm}) (\text{Duration of Backwash, min})$$

$$= (4850 \text{ gpm}) (7 \text{ min})$$

$$= 33,950 \text{ gal}$$

15.6.7 Required Depth of Backwash Water Tank (Feet)

The required depth of water in the backwash water tank is determined from the volume of water required for backwashing. To make this calculation, simply use Equation 15.75:

$$\text{Volume, gal} = (0.785) (D^2) (\text{Depth, ft}) (7.48 \text{ gal/ft}^3) \qquad (15.75)$$

Example 15.87

Problem:

The volume of water required for backwashing has been calculated at 85,000 gal. What is the required depth of water in the backwash water tank to provide this amount of water if the diameter of the tank is 60 ft?

Solution:

Use the volume equation for a cylindrical tank, filling in known data; then solve for *x:*

$$\text{Volume, gal} = (0.785)\,(D^2)\,(\text{Depth, ft})\,(7.48\ \text{gal/ft}^3)$$

$$85{,}000\ \text{gal} = (0.785)\,(60\ \text{ft})\,(60\ \text{ft})\,(x\ \text{ft})\,(7.48\ \text{gal/ft}^3)$$

$$= \frac{85{,}000}{(0.785)\,(60)\,(60)\,(7.48)}$$

$$x = 4\ \text{ft}$$

Example 15.88

Problem:

A total of 66,000 gal of water is required for backwashing a filter at a rate of 8000 gpm for a 9-min period. What depth of water is required in the backwash tank with a diameter of 50 ft?

Solution:

Use the volume equation for cylindrical tanks:

$$\text{Volume, gal} = (0.785)\,(D^2)\,(\text{Depth, ft})\,(7.48\ \text{gal/ft}^3)$$

$$66{,}000\ \text{gal} = (0.785)\,(50\ \text{ft})\,(50\ \text{ft})\,(x\ \text{ft})\,(7.48\ \text{gal/ft}^3)$$

$$x = \frac{66{,}000}{(0.785)\,(50)\,(50)\,(7.48)}$$

$$x = 4.5\ \text{ft}$$

15.6.8 Backwash Pumping Rate (Gallons per Minute)

The desired backwash-pumping rate (gallons per minute) for a filter depends on the desired backwash rate in gallons per minute per square foot, as well as the square foot area of the filter. The backwash pumping rate gallons per minute can be determined by using Equation 15.76:

$$\text{Backwash Pumping Rate, gpm} = (\text{Desired Backwash Rate, gpm/ft}^2)\,(\text{Filter Area, ft}^2) \qquad (15.76)$$

Example 15.89

Problem:

A filter is 25 ft long and 20 ft wide. If the desired backwash rate is 22 gpm/ft², what backwash pumping rate (gallons per minute) will be required?

Solution:

The desired backwash flow through each square foot of filter area is 20 gpm. The total gallons-per-minute flow through the filter is therefore 20 gpm times the entire square foot area of the filter:

$$\text{Backwash Pumping Rate, gpm} = (\text{Desired Backwash Rate, gpm/ft}^2) (\text{Filter Area, ft}^2)$$

$$= 20 \text{ gpm/ft}^2 \ (25 \text{ ft}) \ (20 \text{ ft})$$

$$= 10,000 \text{ gpm}$$

Example 15.90

Problem:

The desired backwash-pumping rate for a filter is 12-gpm/ft². If the filter is 20 ft long and 20 ft wide, what backwash pumping rate (gallons per minute) is required?

Solution:

$$\text{Backwash Pumping Rate, gpm} = (\text{Desired Backwash Rate, gpm/ft}^2) (\text{Filter Area, ft}^2)$$

$$= (12 \text{ gpm/ft}^2) (20 \text{ ft}) (20 \text{ ft})$$

$$= 4800 \text{ gpm}$$

15.6.9 Percent Product Water Used for Backwashing

Along with measuring filtration rate and filter run time, another aspect of filter operation that is monitored for filter performance is the percent of product water used for backwashing. The equation for percent of product water used for backwashing calculations is

$$\text{Backwash Water, \%} = \frac{\text{Backwash Water, gal}}{\text{Water Filtered, gal}} \times 100 \tag{15.77}$$

Example 15.91

Problem:

A total of 18,100,000 gal of water were filtered during a filter run. If 74,000 gal of this product water were used for backwashing, what percent of the product water was used for backwashing?

Solution:

$$\text{Backwash Water, \%} = \frac{\text{Backwash Water, gal}}{\text{Water Filtered, gal}} \times 100$$

$$= \frac{74,000 \text{ gal}}{18,100,000 \text{ gal}} \times 100$$

$$= 0.4\%$$

Example 15.92

Problem:

A total of 11,400,000 gal of water are filtered during a filter run. If 48,500 gal of product water are used for backwashing, what percent of the product water is used for backwashing?

Solution:

$$\text{Backwash Water, \%} = \frac{\text{Backwash Water, gal}}{\text{Water Filtered, gal}} \times 100$$

$$= \frac{48,500 \text{ gal}}{11,400,000 \text{ gal}} \times 100$$

$$= 0.43 \text{ \% Backwash Water}$$

15.6.10 Percent Mud Ball Volume

Mud balls occur when heavy deposits of solids near the top surface of the medium break into pieces during backwash. This results in spherical accretions (usually less than 12 in. in diameter) of floc and sand. The filter media must be checked periodically for the presence of mud balls, which diminish the effective filter area. To calculate the percent of mud ball volume, use Equation 15.78:

$$\text{\% Mud Ball Volume} = \frac{\text{Mud Ball Vol., mL}}{\text{Total Sample, Vol., mL}} \times 100 \qquad (15.78)$$

Example 15.93

Problem:

A 3350-mL sample of filter media was taken for mud ball evaluation. The volume of water in the graduated cylinder rose from 500 to 525 mL when mud balls were placed in the cylinder. What is the percent mud ball volume of the sample?

Solution:

First, determine the volume of mud balls in the sample:

$$525 \text{ mL} - 500 \text{ mL} = 25 \text{ mL}$$

Then, calculate the percent of mud ball volume:

$$\% \text{ Mud Ball Volume } = \frac{\text{Mud Ball Vol., mL}}{\text{Total Sample, Vol., mL}} \times 100$$

$$= \frac{25 \text{ mL}}{3350 \text{ mL}} \times 100$$

$$= 0.75\%$$

Example 15.94

Problem:

A filter is tested for the presence of mud balls. The mud ball sample has a total sample volume of 680 mL. Five samples are taken from the filter. When the mud balls are placed in 500 mL of water, the water level rises to 565 mL. What is the percent of mud ball volume of the sample?

Solution:

$$\% \text{ Mud Ball Volume } = \frac{\text{Mud Ball Vol., mL}}{\text{Total Sample, Vol., mL}} \times 100$$

The mud ball volume is the volume that the water rose:

$$565 \text{ mL} - 500 \text{ mL} = 65 \text{ mL}$$

Because five samples of media were taken, the total sample volume is five times the sample volume.

$$(5)(680 \text{ mL}) = 3400 \text{ mL}$$

$$\% \text{ Mud Ball Volume } = \frac{65 \text{ mL}}{3400 \text{ mL}} \times 100$$

$$= 1.9\%$$

15.6.11 Filter Bed Expansion

Expanding the filter media during the wash maximizes the removal of particles held in the filter or by the media; the efficiency of the filter wash operation depends on the expansion of the sand bed. Bed expansion is determined by measuring the distance from the top of the unexpanded media to a reference point (for example, the top of the filter wall) and from the top of the expanded media to the same reference. A proper backwash rate should expand the filter 20 to 25% (AWWA and ASCE, 1990). Percent bed expansion is given by dividing the bed expansion by the total depth of expandable media (media depth less support gravels) and multiplying by 100:

Expanded Measurement = depth to top of media during backwash (inches)

Unexpanded Measurement = depth to top of media before backwash (inches)

$$\text{Bed Exp.} = \text{unexpanded measurement (inches) expanded measurement (inches)}$$

$$\text{Bed Expansion, \%} = \frac{\text{Bed expansion measurement (inches)}}{\text{Total depth of expandable media (inches)}} \times 100 \qquad (15.79)$$

Example 15.95 (USEPA, 1999)

Problem:

The backwashing practices are being evaluated for a filter with 30 in. of anthracite and sand. While at rest, the distance from the top of the media to the concrete floor surrounding the top of filter is measured at 41 in. After the backwash has been started and the maximum backwash rate is achieved, a probe containing a white disk is slowly lowered into the filter bed until anthracite is observed on the disk. The distance from the expanded media to the concrete floor is measured at 34.5 in. What is the percent of bed expansion?

Solution:

Given:
Unexpanded measurement = 41 in.
Expanded measurement = 34.5 in.

$$\text{Bed expansion} = 6.5 \text{ in.}$$
$$\text{Bed expansion (percent)} = (6.5 \text{ in.}/30 \text{ in.}) \times 100 = 22\%$$

15.6.12 Filter Loading Rate

Filter loading rate is the flow rate of water applied to the unit area of the filter — the same value as the flow velocity approaching the filter surface. The rate can be determined by using Equation 15.80:

$$u = Q/A \qquad (15.80)$$

where
u = loading rate, cubic meters/(square meters per day) or gallons per minute per square foot
Q = flow rate, cubic meters per day or cubic feet per day of gallons per minute
A = surface area of filter, square meters or square feet

On the basis of loading rate, filters are classified as slow sand filters, rapid sand filters, and high-rate sand filters. Typically, the loading rate for rapid sand filters is 120 $m^3/(m^2 \text{ d})$(83 $L/(m^2$ min) or 2 $gal/min/ft^2$. The loading rate may be up to five times this rate for high-rate filters.

Example 15.96

Problem:

A sanitation district is to install rapid sand filters downstream of the clarifiers. The design-loading rate is selected at 150 m^3/m^2. The design capacity of the waterworks is 0.30 m^3/sec (6.8 MGD). The maximum surface per filter is limited to 45 m^2. Design the number and size of filters; calculate the normal filtration rate.

Solution:

Step 1. Determine the total surface area required:

$$A = \frac{Q}{u} \quad \frac{0.30 \text{ m}^3/\text{sec } (85,400 \text{ sec/day})}{150 \text{ m}^3/\text{m}^2 \text{ day}}$$

$$= \frac{25,920}{150}$$

$$= 173 \text{ m}^2$$

Step 2. Determine the number of filters:

$$= \frac{173 \text{ m}^2}{45 \text{ m}} = 3.8$$

Select 4 m. The surface area (a) for each filter is:

$$a = 173 \text{ m}^2/4 = 43.25 \text{ m}^2$$

We can use 6 m × 7 m; 6.4 m × 7 m; or 6.42 m × 7 m.
Step 3. If a 6-m × 7-m filter is installed, the normal filtration rate is:

$$u = \frac{Q}{A} \quad \frac{0.30 \text{ m}^3/\text{sec } (85,400 \text{ sec/day})}{4 \times 6 \times 7 \text{ m}}$$

$$= 154.3 \text{ m}^3/(\text{m}^2\text{day})$$

15.6.13 Filter Medium Size

Filter medium grain size has an important effect on the filtration efficiency and on backwashing requirements for the medium. The actual medium selected is typically determined by performing a grain size distribution analysis — sieve size and percentage passing by weight relationships are plotted on logarithmic-probability paper. The most common parameters used in the U.S. to characterize the filter medium are effective size (ES) and uniformity coefficient (UC) of medium size distribution. The ES is the grain size for which 10% of the grains are smaller by weight; it is often abbreviated by d^{10}. The UC is the ratio of the 60 percentile (d^{60}) to the 10 percentile. The 90 percentile, d_{90}, is the size for which 90% of the grains are smaller by weight. The d_{90} size is used for computing the required filter backwash rate for a filter medium.

Values of d_{10}, d_{60}, and d_{90} can be read from an actual sieve analysis curve. If such a curve is not available and if a linear log-probability plot is assumed, the values can be interrelated by Equation 15.81 (Cleasby, 1990).

$$d_{90} = d_{10}(10^{1.67 \log UC}) \tag{15.81}$$

Example 15.97

Problem:
   A sieve analysis curve of a typical filter sand gives $d_{10} = 0.52$ mm and $d_{60} = 0.70$ mm. What are its uniformity coefficient and d_{90}?

Solution:

Step 1.

$$UC = d_{90}/d_{10} = 0.70 \text{ mm}/0.52 \text{ mm}$$

$$= 1.35$$

Step 2. Find d_{90} using Equation 15.81.

$$d_{90} = d_{10}(10^{1.67 \log UC})$$

$$= 0.52 \text{ mm}(10^{1.67 \log 1.35})$$

$$= 0.52 \text{ mm}(10^{0.218})$$

$$= 0.86 \text{ mm}$$

15.6.14 Mixed Media

Recently, an innovation in filtering systems has offered a significant improvement and economic advantage over rapid rate filtration: the mixed media filter bed. Mixed media filter beds offer specific advantages in specific circumstances and will give excellent operating results at a filtering rate of 5 gal/ft²/min. The mixed media filtering unit is more tolerant of handling higher turbidities in the settled water. For improved process performance, activated carbon or anthracite is added to the top of the sand bed. The approximate specific gravity (s) of ilmenite (Chavara, <60% TiO_2); silica sand; anthracite; and water are 4.2; 2.6; 1.5; and 1.0, respectively. The economic advantage of the mixed bed media filter is based upon filter area; it will safely produce 2½ times as much filtered water as a rapid sand filter.
   When settling velocities are equal, the particle sizes for media of different specific gravities can be computed by using Equation 15.82:

$$= \frac{d_1}{d_2} = \left(\frac{s_2 - s}{s_1 - s} \right)^{2/3} \tag{15.82}$$

where
d_1, d_2 = diameters of particles 1 and 2, and water, respectively
s_1, s_2 = specific gravity of particles 1 and 2, and water, respectively

Example 15.98

Problem:

Estimate the particle size of ilmenite sand (specific gravity = 4.2) that has the same settling velocity of silica sand, 0.60 mm in known diameter (specific gravity = 2.6).

Solution:

Find the diameter on ilmenite sand by Equation 15.82.

$$d = (0.6 \text{ mm}) = \frac{2.6 - 1}{4.2 - 1}^{2/3}$$

$$= 0.38 \text{ mm settling size}$$

15.6.15 Head Loss for Fixed Bed Flow

When water is pumped upward through a bed of fine particles at a very low flow rate, the water percolates through the pores (void spaces) without disturbing the bed. This is a fixed bed process. The head loss (pressure drop) through a clean granular-media filter is generally less than 0.9 m (3 ft). With the accumulation of impurities, head loss gradually increases until the filter is backwashed. The Kozeny equation, shown next, is typically used to calculate head loss through a clean fixed-bed flow filter.

$$\frac{h}{L} = \frac{k\mu(1-\varepsilon)^2}{gpe^3}\left(\frac{A}{V}\right)^2 u \tag{15.83}$$

where
h = head loss in filter depth L, meters or feet
k = dimensionless Kozeny constant, 5 for sieve openings, 6 for size of separation
g = acceleration of gravity, 9.81 m/sec or 32.2 ft/sec
μ = absolute viscosity of water, N sec/m^2 or lb-sec/ft^2
p = density of water, kilograms per cubic meters or pounds per cubic foot
ε = porosity, dimensionless
A/V = grain surface area per unit volume of grain
  = specific surface S (or shape factor = 6.0 to 7.7)
  = 6/day for spheres
  = 6/ ψd_{eq} for irregular grains
ψ = grain sphericity or shape factor
d_{eq} = grain diameter of spheres of equal volume
u = filtration (superficial) velocity, meters per second or feet per second

Example 15.99

Problem:

A dual medium filter is composed of 0.3 m anthracite (mean size of 2.0 mm) placed over a 0.6-m layer of sand (mean size 0.7 mm) with a filtration rate of 9.78 m/h. Assume the grain sphericity is $\psi = 0.75$ and a porosity for both is 0.42. Although normally taken from the appropriate table at 15°C, we provide the head loss data of the filter at 1.131×10^{-6} m^2 sec.

Solution:

Step 1. Determine head loss through anthracite layer using the Kozeny equation (Equation 15.83).

$$\frac{h}{L} = \frac{k\mu(1-\varepsilon)^2}{gp\varepsilon^3}\left(\frac{A}{V}\right)^2 u$$

where
k   = 6
g   = 9.81 m/sec²
μ_p = v = 1.131 × 10⁻⁶ m² sec (from the appropriate table)
ε = 0.40
A/V = 6/0.75 day = 8/day = 8/0.002
u   = 9.78 m/h = 0.00272 m/sec
L   = 0.3 m

then

$$h = 6 \times \frac{1.131\times10^{-6}}{9.81} \times \frac{1-0.42^2}{0.42^3} \times \left(\frac{8}{0.002}\right)^2 (0.00272)\,(0.2)$$

$$= 0.0410 \text{ m}$$

Step 2. Compute the head loss passing through the sand. Use data in Step 1, except insert:
    k = 5
    d = 0.0007 m
    L = 0.6 m

$$h = 5 \times \frac{1.131\times10^{-6}}{9.81} \times \frac{0.58^2}{0.42^3} \times \left(\frac{8}{d}\right)^2 (0.00272)\,(0.4)$$

$$= 0.5579 \text{ m}$$

Step 3. Compute total head loss:

$$h = 0.0410 \text{ m} + 0.5579 \text{ m}$$

$$= 0.599 \text{ m}$$

15.6.16 Head Loss through a Fluidized Bed

If the upward water flow rate through a filter bed is very large, the bed mobilizes pneumatically and may be swept out of the process vessel. At an intermediate flow rate, the bed expands and is in what we call an *expanded* state. In the fixed bed, the particles are in direct contact with each other, supporting each other's weight. In the expanded bed, the particles have a mean free distance between particles and the drag force of the water supports the particles. The expanded bed has some of the properties of the water (of a fluid) and is called a fluidized bed (Chase, 2002). Simply, *fluidization* is defined as upward flow through a granular filter bed at sufficient velocity to suspend the grains in the water. Minimum fluidizing velocity (U_{mf}) is the superficial fluid velocity needed to start fluidization and is used in determining the required minimum backwashing flow rate. Wen and Yu proposed the U_{mf} equation including the near constants (over a wide range of particles) 33.7 and 0.0408, but excluding porosity of fluidization and shape factor (Wen and Yu, 1966):

$$U_{mf} \; \frac{\mu}{pd_{eq}}(1135.69 + 0.0408G_n)^{0.5} - \frac{33.7\mu}{pd_{eq}} \tag{15.84}$$

where
μ = absolute viscosity of water, N seconds per square meter or pounds seconds per square foot
p = density of water, kilograms per cubic meters or pounds per cubic foot
$d_{eq} = d_{90}$ sieve size is used instead of d_{eq}
G_n = Galileo number

$$= d_{eq}^3 p(p_s - p)g/\mu^2 \tag{15.85}$$

Other variables used are expressed in Equation 15.83.

Note: Based on the studies of Cleasby and Fan (1981), we use a safety factor of 1.3 to ensure adequate movement of the grains.

Example 15.100

Problem:
Estimate the minimum fluidized velocity and backwash rate for the sand filter. The d_{90} m size of sand is 0.90 mm. The density of sand is 2.68 g/cm³.

Solution:

Step 1. Compute the Galileo number:
From given data and the applicable table, at 15°C:

p = 0.999 g/cm³
μ = 0.0113 N sec/m² = 0.00113 kg/msec = 0.0113 g/cm-sec
μp = 0.0113 cm²/sec
g = 981 cm/sec²
d = 0.090 cm
p_s = 2.68 g/cm³

Using Equation 15.85:

$$G_n = d_{eq}^3 p(p_s - p)g/m^2$$

$$= (0.090)^3 (0.999)(2.68 - 0.999)(981)/(0.0113)^2$$

$$= 9405$$

Step 2. Compute U_{mf} using Equation 15.84:

$$U_{mf} \quad \frac{0.0113}{0.999 \times 0.090}(1135.69 + 0.0408 \times 9405)^{0.5} - \frac{33.7 \times 0.0113}{0.999 \times 0.090}$$

$$= 0.660 \text{ cm/sec}$$

Step 3. Compute backwash rate. Apply a safety factor of 1.3 to U_{mf} as backwash rate:

$$\text{Backwash rate} = 1.3 \times 0.660 \text{ cm/sec} = 0.858 \text{ cm/sec}$$

$$0.858 \frac{cm^3}{cm^2 sec} \times \frac{L}{1000 \text{ cm}^3} \times \frac{1}{3.785} \times \frac{gal}{L} \times 929 \times \frac{cm^2}{ft^2} \times \frac{60 \text{ sec}}{min} =$$

$$= 12.6 \text{ gpm/ft}^2$$

15.6.17 Horizontal Washwater Troughs

Wastewater troughs are used to collect backwash water as well as to distribute influent water during the initial stages of filtration. These troughs are normally placed above the filter media in the U.S. Proper placement ensures that the filter media are not carried into the troughs during the backwash and removed from the filter. These backwash troughs are constructed from concrete, plastic, fiberglass, or other corrosion-resistant materials. The total rate of discharge in a rectangular trough with free flow can be calculated by using Equation 15.86:

$$Q = Cwh^{1.5} \qquad\qquad (15.86)$$

where
Q = flow rate, cfs
C = constant (2.49)
w = trough width, feet
h = maximum water depth in trough, feet

Example 15.101

Problem:
    Troughs are 18 ft long, 18 in. wide, and 8 ft to the center with a horizontal flat bottom. The backwash rate is 24 in./min. Estimate (1) the water depth of the troughs with free flow into the gullet; and (2) the distance between the top of the troughs and the 30-in. sand bed. Assume a 40% expansion, 6 in. of freeboard in the troughs, and 6 in. of thickness.

Solution:

Step 1. Estimate the maximum water depth (h) in trough:

$$v = 24 \text{ in./min} = 2 \text{ ft/60 sec} = 1/30 \text{ }1$$

$$A = 18 \text{ ft} \times 8 \text{ ft} = 144 \text{ ft}^2$$

$$Q = VA = 144/30 \text{ cfs}$$

$$= 4.8 \text{ cfs}$$

Using Equation 15.86:

$$Q = 2.49 \text{ wh}^{1.5}, \text{ w} = 1.5 \text{ ft}$$

$$h = (Q/2.49 \text{ w})^{2/3}$$

$$= [4.8/(2.49 \times 1.5)]^{2/3}$$

$$= 1.18 \text{ ft (or approximately 14 in.} = 1.17 \text{ ft)}$$

Step 2. Determine the distance (y) between the sand bed surface and the top troughs:

$$\text{Freeboard} = 6 \text{ in.} = 0.5 \text{ ft}$$

$$\text{Thickness} = 8 \text{ in.} = 0.67 \text{ ft (the bottom of trough)}$$

$$y = 2.5 \text{ ft} \times 0.4 + 1.17 \text{ ft} + 0.5 \text{ ft} + 0.5 \text{ ft}$$

$$= 3.2 \text{ ft}$$

15.6.18 Filter Efficiency

Water treatment filter efficiency is defined as the effective filter rate divided by the operation filtration rate as shown in Equation 15.87(AWWA and ASCE, 1998).

$$E = \frac{R_e}{R_o} = \frac{\text{UFRV} - \text{UBWU}}{\text{UFRV}} \tag{15.87}$$

where
E = filter efficiency, percent
R_e = effective filtration rate, gallons per minute per square foot
R_o = operating filtration rate, gallons per minute per square foot
UFRV = unit filter run volume, gallons per square foot
UBWV = unit backwash volume, gallons per square foot

Example 15.102

Problem:
    A rapid sand filter operates at 3.9 gal/min/ft² for 48 h. Upon completion of the filter run, 300 gal/ft² of backwash water is used. Find the filter efficiency.

Solution:

Step 1. Calculate operating filtration rate, R_o

$$R_o = 3.9 \text{ gpm/ft}^2 \times 60 \text{ min/h} \times 48 \text{ h}$$

$$= 11{,}232 \text{ gal/ft}^2$$

Step 2. Calculate effective filtration rate, R_e

$$R_e = (11{,}232 - 300) \text{ gal/ft}^2$$

$$= 10{,}932 \text{ gal/ft}^2$$

Step 3. Calculate filter efficiency, E, using Equation 15.87.

$$E = 10{,}932/11{,}232$$

$$= 97.3\%$$

15.7 WATER CHLORINATION CALCULATIONS

Chlorine is the most commonly used substance for disinfection of water in the U.S. The addition of chlorine or chlorine compounds to water is called chlorination. Chlorination is considered the single most important process for preventing the spread of waterborne disease.

15.7.1 Chlorine Disinfection

Chlorine deactivates microorganisms through several mechanisms and can destroy most biological contaminants:

- It causes damage to the cell wall.
- It alters the permeability of the cell (the ability to pass water in and out through the cell wall).
- It alters the cell protoplasm.
- It inhibits the enzyme activity of the cell so that it is unable to use its food to produce energy.
- It inhibits cell reproduction.

Chlorine is available in a number of different forms: (1) as pure elemental gaseous chlorine (a greenish-yellow gas heavier than air, nonflammable, and nonexplosive, with a pungent and irritating odor); when released to the atmosphere, this form is toxic and corrosive; (2) as solid calcium hypochlorite (in tablets or granules); or (3) as a liquid sodium hypochlorite solution (in various strengths). The strengths of one form of chlorine over the others for a given water system depend on the amount of water to be treated, the configuration of the water system, the local availability of the chemicals, and the skill of the operator.

One of the major advantages of using chlorine is the effective residual that it produces. A residual indicates that disinfection is completed and that the system has an acceptable bacteriological quality. Maintaining a residual in the distribution system helps to prevent regrowth of those microorganisms that were injured but not killed during the initial disinfection stage.

15.7.2 Determining Chlorine Dosage (Feed Rate)

The expressions *milligrams per liter* (mg/L) and *pounds per day* (lb/day) are most often used to describe the amount of chlorine added or required. Equation 15.88 can be used to calculate milligrams per liter or pounds per day of chlorine dosage.

$$\text{Chlorine Feed Rate, lb/day} = (\text{Cl., mg/L}) (\text{flow, MGD}) (8.34, \text{lb/gal}) \qquad (15.88)$$

Example 15.103

Problem:
    Determine the chlorinator setting (pounds per day) needed to treat a flow of 4 MGD with a chlorine dose of 5 mg/L.

Solution:

$$\text{Chlorine Feed Rate, lb/day} = (\text{Cl., mg/L}) (\text{flow, MGD}) (8.34, \text{lb/gal})$$

$$= (5 \text{ mg/L}) (4 \text{ MGD}) (8.34 \text{ lb/gal})$$

$$= 167 \text{ lb/day}$$

Example 15.104

Problem:
    A pipeline 12 in. in diameter and 1400 ft long must be treated with a chlorine dose of 48 mg/L. How many pounds of chlorine will this require?

Solution:
    Determine the gallon volume of the pipeline.

$$\text{Volume, gal} = (0.785) (D^2) (\text{Length, ft}) (7.48 \text{ gal/ft}^3)$$

$$= (0.785) (1 \text{ ft}) (1 \text{ ft}) (1400 \text{ ft}) (7.48 \text{ gal/ft}^3)$$

$$= 8221 \text{ gal}$$

Now calculate the pounds chlorine required.

$$\text{lb Chlorine} = (\text{Chlorine, mg/L}) (\text{MG Volume}) (8.34 \text{ lb/gal})$$

$$= (48 \text{ mg/L}) (0.008221 \text{ MG}) (8.34 \text{ lb/gal})$$

$$= 3.3 \text{ lb}$$

Example 15.105

Problem:

A chlorinator setting is 30 lb/24h. If the flow being chlorinated is 1.25 MGD, what is the chlorine dosage expressed as milligrams per liter?

Solution:

$$\text{Chlorine lb/day} = (\text{Chlorine, mg/L}) (\text{MGD flow}) (8.34, \text{lb/gal})$$

$$30 \text{ lb/day} = (\text{x mg/L}) (\text{flow, MGD}) (8.34 \text{ lb/gal})$$

$$x = \frac{30}{(1.25)(8.34)}$$

$$x = 2.9 \text{ mg/L}$$

Example 15.106

Problem:

A flow of 1600 gpm must be chlorinated. At a chlorinator setting of 48 lb/24 h, what is the chlorine dosage in milligrams per liter?

Solution:

Convert the gallons per minute flow rate to million gallons per day flow rate:

$$(1600 \text{ gpm}) (1440 \text{ min/day}) = 2{,}304{,}000 \text{ gpd}$$

$$= 2.304 \text{ MGD}$$

Calculate the chlorine dosage in milligrams per liter:

$$\text{Chlorine, lb/day} = (\text{Chlorine, mg/L}) (\text{Flow, MGD})$$

$$(\text{x mg/L}) (2.304 \text{ MGD}) (8.34 \text{ lb/gal}) = 48 \text{ lb/day}$$

$$x = \frac{48}{(2.304)(8.34)}$$

$$x = 2.5 \text{ mg/L}$$

15.7.3 Calculating Chlorine Dose, Demand, and Residual

Common terms used in chlorination include:

- *Chlorine dose* — the amount of chlorine added to the system, determined by adding the desired residual for the finished water to the chlorine demand of the untreated water. Dosage can be milligrams per liter or pounds per day. The most common is milligrams per liter.

Chlorine dose, milligrams per liter = chlorine demand, milligrams per liter + chlorine, milligrams per liter residual, milligrams per liter

- *Chlorine demand* — the amount of chlorine used by iron, manganese, turbidity, algae, and microorganisms in the water. Because the reaction between chlorine and microorganisms is not instantaneous, demand is relative to time. For instance, the demand 5 min after applying chlorine will be less than the demand after 20 min. Demand, like dosage, is expressed in milligrams per liter. The chlorine demand is as follows:

$$\text{Chlorine Demand} = \text{Chlorine Dose} - \text{Chlorine Residual}$$

- *Chlorine residual* — the amount of chlorine (determined by testing) remaining after the demand is satisfied. Residual, like demand, is based on time. The longer the time after dosage is, the lower the residual will be, until all of the demand has been satisfied. Residual, like dosage and demand, is expressed in milligrams per liter. The presence of a *free residual* of at least 0.2 to 0.4 ppm usually provides a high degree of assurance that the disinfection of the water is complete. *Combined residual* is the result of combining free chlorine with nitrogen compounds. Combined residuals are also called chloramines. *Total chlorine residual* is the mathematical combination of free and combined residuals. Total residual can be determined directly with standard chlorine residual test kits.

$$\text{Chlorine Dose, mg/L} = \text{Cl. Demand, mg/L} + \text{Cl. Residual, mg/L} \qquad (15.89)$$

The following examples show the calculation of chlorine dose, demand, and residual.

Example 15.107

Problem:
    A water sample is tested and has a chlorine demand of 1.7 mg/L. If the desired chlorine residual is 0.9 mg/L, what is the desired chlorine dose in milligrams per liter?

Solution:

$$\text{Chlorine Dose, mg/L} = \text{Cl. Demand, mg/L} + \text{Cl. Residual, mg/L}$$

$$= 1.7 \text{ mg/L} + 0.9 \text{ mg/L}$$

$$= 2.6 \text{ mg/L Chlorine Dose}$$

Example 15.108

Problem:
    The chlorine dosage for water is 2.7 mg/L. If the chlorine residual after 30 min of contact time is 0.7 mg/L, what is the chlorine demand expressed in milligrams per liter?

Solution:

$$\text{Chlorine Dose, mg/L} = \text{Cl. Demand, mg/L} + \text{Cl. Residual, mg/L}$$

$$2.7 \text{ mg/L} = x \text{ mg/L} + 0.6 \text{ mg/L}$$

$$2.7 \text{ mg/L} - 0.7 \text{ mg/L} = x \text{ mg/L}$$

$$x \text{ Chlorine Demand, mg/L} = 2.0 \text{ mg/L}$$

Example 15.109

Problem:

What should the chlorinator setting be (pounds per day) to treat a flow of 2.35 MGD if the chlorine demand is 3.2 mg/L and a chlorine residual of 0.9 mg/L is desired?

Solution:

Determine the chlorine dosage in milligrams per liter:

$$\text{Chlorine Dose, mg/L} = \text{Cl. Demand, mg/L} + \text{Cl. Residual, mg/L}$$

$$= 3.2 \text{ mg/L} + 0.9 \text{ mg/L}$$

$$= 4.1 \text{ mg/L}$$

Calculate the chlorine dosage (feed rate) in pounds per day:

$$\text{Chlorine, lb/day} = (\text{Chlorine, mg/L}) (\text{Flow, MGD}) (8.34 \text{ lb/gal})$$

$$= (4.1 \text{ mg/L}) (2.35 \text{ MGD}) (8.34 \text{ lb/gal})$$

$$= 80.4 \text{ lb/day Chlorine}$$

15.7.4 Breakpoint Chlorination Calculations

To produce a free chlorine residual, enough chlorine must be added to the water to produce what is referred to as *breakpoint chlorination* (the point at which near complete oxidation of nitrogen compounds is reached; any residual beyond breakpoint is mostly free chlorine [see Figure 15.4)]). When chlorine is added to natural waters, the chlorine begins combining with and oxidizing the chemicals in the water before it begins disinfecting. Although residual chlorine will be detectable in the water, the chlorine will be in the combined form with a weak disinfecting power. As we see in Figure 15.4, adding more chlorine to the water at this point actually decreases the chlorine residual as the additional chlorine destroys the combined chlorine compounds. At this stage, water may have a strong swimming pool or medicinal taste and odor. To avoid this taste and odor, add still more chlorine to produce a free residual chlorine. Free chlorine has the highest disinfecting

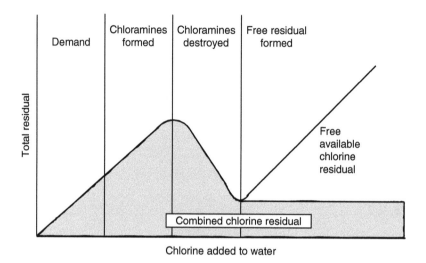

Figure 15.4 Breakpoint chlorination curve.

power. The point at which most of the combined chlorine compounds have been destroyed and the free chlorine starts to form is the *breakpoint*.

Key point: The actual chlorine breakpoint of water can only be determined by experimentation.

To calculate the actual increase in chlorine residual that would result from an increase in chlorine dose, we use the milligrams-per-liter to pounds-per-day equation:

$$\text{Increase in Cl dose, lb/day} = (\text{Expected Increase, mg/L}) (\text{Flow, MGD}) (8.34 \text{ lb/gal}) \qquad (15.90)$$

Key point: The actual increase in residual is simply a comparison of new and old residual data.

Example 15.110

Problem:

A chlorinator setting is increased by 2 lb/day. The chlorine residual before the increased dosage was 0.2 mg/L. After the increased chlorine dose, the chlorine residual was 0.5 mg/L. The average flow rate being chlorinated is 1.25 MGD. Is the water chlorinated beyond the breakpoint?

Solution:

Calculate the expected increase in chlorine residual. Use the milligrams-per-liter to pounds-per-day equation:

$$\text{Lb/day Increase} = (\text{mg/L Increase}) (\text{Flow, MGD}) (8.34 \text{ lb/gal})$$

$$2 \text{ lbs/day} = (x \text{ mg/L}) (1.25 \text{ MGD}) (8.34 \text{ lbs/gal})$$

$$x = \frac{2}{(1.25)(8.34)}$$

$$x = 0.19 \text{ mg/L}$$

Actual increase in residual:

$$0.5 \text{ mg/L} - 0.19 \text{ mg/L} = 0.31 \text{ mg/L}$$

Example 15.111

Problem:

A chlorinator setting of 18 lb of chlorine per 24 h results in a chlorine residual of 0.3 mg/L. The chlorinator setting is increased to 22 lb/24 h. The chlorine residual increases to 0.4 mg/L at this new dosage rate. The average flow being treated is 1.4 MGD. On the basis of these data, is the water being chlorinated past the breakpoint?

Solution:

Calculate the expected increase in chlorine residual:

$$\text{Lb/day Increase} = (\text{mg/L Increase})\,(\text{Flow, MGD})\,(8.34 \text{ lb/gal})$$

$$4 \text{ lb/day} = (\text{x mg/L})\,(1.4 \text{ MGD})\,(8.34 \text{ lb/gal})$$

$$\text{x} = \frac{4}{(1.4 \text{ MGD})\,(8.34)}$$

$$\text{x} = 0.34 \text{ mg/L}$$

Actual increase in residual:

$$(0.4 \text{ mg/L} - 0.3 \text{ mg/L} = 0.1 \text{ mg/L}$$

15.7.5 Calculating Dry Hypochlorite Feed Rate

The most commonly used dry hypochlorite, calcium hypochlorite, contains about 65 to 70% available chlorine, depending on the brand. Because hypochlorites are not 100% pure chorine, more pounds per day than with liquid chlorine must be fed into the system to obtain the needed levels of chlorine for disinfection. To calculate the pounds per day of hypochlorite needed, use Equation 15.91:

$$\text{Hypochlorite, lb/day} = \frac{\text{Lb/day Chlorine}}{\dfrac{\% \text{ Available Chlorine}}{100}} \qquad (15.91)$$

Example 15.112

Problem:

A chlorine dosage of 110 lb/day is required to disinfect a flow of 1,550,000 gpd. If the calcium hypochlorite used contains 65% available chlorine, how many pounds per day of hypochlorite are required for disinfection?

Solution:

Because only 65% of the hypochlorite is chlorine, more than 110 lb of hypochlorite will be required:

$$\text{Hypochlorite, lb/day} = \frac{\text{Lb/day Chlorine}}{\dfrac{\text{\% Available Chlorine}}{100}}$$

$$= \frac{110 \text{ Lb/day}}{\dfrac{65}{100}}$$

$$= \frac{110}{0.65}$$

$$= 169 \text{ lb/day Hypochlorite}$$

Example 15.113

Problem:

A water flow of 900,000 gpd requires a chlorine dose of 3.1 mg/L. If calcium hypochlorite (65% available chlorine) is used, how many pounds per day of hypochlorite are required?

Solution:

Calculate the pounds per day chlorine required:

$$\text{Chlorine, lb/day} = (\text{Chlorine, mg/L})\,(\text{Flow, MGD})\,(8.34 \text{ lb/gal})$$

$$= (3.1 \text{ mg/L})\,(0.90 \text{ MGD})\,(8.34 \text{ lb/gal})$$

$$= 23 \text{ lb/day}$$

The pounds per day of hypochlorite:

$$\text{Hypochlorite, lb/day} = \frac{\text{Lb/day Chlorine}}{\dfrac{\text{\% Available Chlorine}}{100}}$$

$$= \frac{23 \text{ lb/day Chlorine}}{0.65 \text{ Available Chlorine}}$$

$$= 35 \text{ lb/day Available Hypochlorite}$$

Example 15.114

Problem:

A tank contains 550,000 gal of water and will receive a chlorine dose of 2.0 mg/L. How many pounds of calcium hypochlorite (65% available chlorine) are required?

Solution:

$$\text{Hypochlorite, lb/day} = \frac{\text{Lb/day Chlorine}}{\dfrac{\%\ \text{Available Chlorine}}{100}}$$

$$= \frac{(2.0\ \text{mg/L})\,(0.550\ \text{MG})\,(8.34\ \text{lb/gal})}{\dfrac{65}{100}}$$

$$= \frac{9.2\ \text{lb}}{0.65}$$

$$= 14.2\ \text{lb Hypochlorite}$$

Example 15.115

Problem:

A total of 40 lb of calcium hypochlorite (65% available chlorine) are used in a day. If the flow rate treated is 1,100,000 gpd, what is the chlorine dosage in milligrams per liter?

Solution:

Calculate the pounds per day of chlorine dosage:

$$\text{Hypochlorite, lb/day} = \frac{\text{Lb/day Chlorine}}{\dfrac{\%\ \text{Available Chlorine}}{100}}$$

$$40\ \text{lb/day Hypochlorite} = \frac{\text{x lb/day Chlorine}}{0.65}$$

$$(0.65)\,(40) = \text{x}$$

$$26\ \text{lb/day Chlorine} = \text{x}$$

Then, calculate milligrams per liter of chlorine, using the milligrams-per-liter to pounds-per-day equation and filling in the new information:

$$26\ \text{lb/day Chlorine} = (\text{x mg/L Chlorine})\,(1.10\ \text{MGD})\,(8.34\ \text{lb/gal})$$

$$x = \frac{26 \text{ lb/day}}{(1.10 \text{ MGD}) (8.34 \text{ lb/gal})}$$

$$= 2.8 \text{ mg/L Chlorine}$$

Example 15.116

Problem:

A flow of 2,550,000 gpd is disinfected with calcium hypochlorite (65% available chlorine). If 50 lb of hypochlorite are used in a 24-h period, what is the milligrams-per-liter chlorine dosage?

Solution:

Pounds per day chlorine dosage:

$$50 \text{ lb/day Hypochlorite} = \frac{x \text{ lb/day Chlorine}}{0.65}$$

$$x = 32.5 \text{ Chlorine}$$

Milligrams per liter of chlorine:

$$(x \text{ mg/L Chlorine}) (2.55 \text{ MGD}) (8.34 \text{ lb/gal}) = 32.5 \text{ lb/day}$$

$$x = 1.5 \text{ mg/L Chlorine}$$

15.7.6 Calculating Hypochlorite Solution Feed Rate

Liquid hypochlorite (sodium hypochlorite) is supplied as a clear, greenish-yellow liquid in strengths from 5.25 to 16% available chlorine. Often referred to as "bleach," it is, in fact, used for bleaching; common household bleach is a solution of sodium hypochlorite containing 5.25% available chlorine.

When calculating gallons per day of liquid hypochlorite, the pounds per day of hypochlorite required must be converted to gallons per day of required hypochlorite. This conversion is accomplished using Equation 15.92:

$$\text{Hypochlorite, gpd} = \frac{\text{Hypochlorite, lb/day}}{8.34 \text{ lb/gal}} \qquad (15.92)$$

Example 15.117

Problem:

A total of 50 lb/day of sodium hypochlorite are required for disinfection of a 1.5-MGD flow. How many gallons per day of hypochlorite is this?

Solution:

Because pounds per day of hypochlorite has already has been calculated, we simply convert pounds per day to gallons per day:

$$\text{Hypochlorite, gpd} = \frac{\text{Hypochlorite, lb/day}}{8.34 \text{ lb/gal}}$$

$$= \frac{50 \text{ lb/day}}{8.34 \text{ lb/gal}}$$

$$= 6.0 \text{ gpd Hypochlorite}$$

Example 15.118

Problem:

A hypochlorinator is used to disinfect the water pumped from a well. The hypochlorite solution contains 3% available chlorine. A chlorine dose of 1.3 mg/L is required for adequate disinfection throughout the system. If the flow being treated is 0.5 MGD, how many gallons per day of the hypochlorite solution are required?

Solution:

Calculate the pounds per day of chlorine required:

$$(1.3 \text{ mg/L}) (0.5 \text{ MGD}) (8.34 \text{ lb/gal}) = 5.4 \text{ lb/day Chlorine}$$

Calculate the pounds per day hypochlorite solution required:

$$\text{Hypochlorite, lb/day} = \frac{5.4 \text{ lb/day Chlorine}}{0.03}$$

$$= 180 \text{ lb/day Hypochlorite}$$

Calculate the gallons per day of hypochlorite solution required:

$$= \frac{180 \text{ lb/day}}{8.34 \text{ lb/gal}}$$

$$= 21.6 \text{ gpd Hypochlorite}$$

15.7.7 Calculating Percent Strength of Solutions

If a teaspoon of salt is dropped into a glass of water, it gradually disappears; the salt dissolves in the water. A microscopic examination of the water would not show the salt. Only examination at the molecular level, which is not easily done, would show salt and water molecules intimately mixed. If we taste the liquid, of course, we know the salt is there, and we could recover the salt by evaporating the water. In a solution, the molecules of the salt, the *solute*, are homogeneously dispersed among the molecules of water, the *solvent*. This mixture of salt and water is homogenous on a molecular level. Such a homogenous mixture is called a *solution*. The composition of a solution can be varied within certain limits. There are three common states of matter: gas, liquid, and solids.

In this discussion, at the moment, we are only concerned with solids (calcium hypochlorite) and liquid (sodium hypochlorite).

15.7.8 Calculating Percent Strength Using Dry Hypochlorite

To calculate the percent strength of a chlorine solution, use Equation 15.93:

$$\% \text{ Chlorine Strength} = \frac{(\text{Hypochlorite, lb})\dfrac{(\% \text{ Available Chlorine})}{100}}{\text{Water, lb} + (\text{Hypochlorite, lb})\dfrac{(\% \text{ Available Chlorine})}{100}} \times 100 \quad (15.93)$$

Example 15.119

Problem:

If a total of 72 oz of calcium hypochlorite (65% available chlorine) are added to 15 gal of water, what is the percent chlorine strength (by weight) of the solution?

Solution:

Convert the ounces of hypochlorite to pounds of hypochlorite:

$$\frac{72 \text{ oz}}{16 \text{ oz/lb}} = 4.5 \text{ lb chemical}$$

$$\% \text{ Chlorine Strength} = \frac{(\text{Hypochlorite, lb})\dfrac{(\% \text{ Available Chlorine})}{100}}{\text{Water, lbs} + (\text{Hypochlorite, lb})\dfrac{(\% \text{ Available Chlorine})}{100}} \times 100$$

$$= \frac{(4.5 \text{ lb})(0.65)}{(15 \text{ gal})(8.34 \text{ lb/gal}) + (4 \text{ lb})(0.65)} \times 100$$

$$= \frac{2.9 \text{ lb}}{125.1 \text{ lb} + 2.9 \text{ lb}} \times 100$$

$$= \frac{(2.9)(100)}{126}$$

$$= 2.3 \text{ Chlorine Strength}$$

15.7.9 Calculating Percent Strength Using Liquid Hypochlorite

To calculate percent strength using liquid solutions (including liquid hypochlorite), use Equation 15.94:

$$(\text{Liq. Hypo., gal})(8.34 \text{ lb/gal})\frac{(\% \text{ Strength of Hypo.})}{100}$$

$$= (\text{Hypo. Sol., gal})(8.34 \text{ lb/gal})\frac{(\% \text{ Strength of Hypo.})}{100} \quad (15.94)$$

Example 15.120

Problem:

In making up a hypochlorite solution, 12% of liquid hypochlorite solution is used. If 3.3 gal of liquid hypochlorite are mixed with water to produce 25 gal of hypochlorite solution, what is the percent strength of the solution?

Solution:

$$(\text{Liq. Hypo., gal}) \, (8.34 \, \text{lb/gal}) \, \frac{(\% \, \text{Strength of Hypo.})}{100}$$

$$= (\text{Hypo. Sol., gal}) \, (8.34 \, \text{lb/gal}) \, \frac{(\% \, \text{Strength of Hypo.})}{100}$$

$$(3.3 \, \text{gal}) \, (8.34 \, \text{lb/gal}) \, \frac{12}{100} = (25 \, \text{gal}) \, (8.34 \, \text{lb/gal}) \, \frac{(x)}{100}$$

$$= \frac{(3.3) \, (12)}{(25)}$$

$$x = 1.6\%$$

15.8 CHEMICAL USE CALCULATIONS

In typical plant operation, chemical use, in pounds per day or gallons per day, is recorded each day. These data provide a record of daily use from which the average daily use of the chemical or solution can be calculated. To calculate average use in pounds per day, use Equation 15.95. To calculate average use in gallons per day, use Equation 15.96.

$$\text{Average Use, lb/day} = \frac{\text{Total Chemical Used, lb}}{\text{Number of Days}} \tag{15.95}$$

$$\text{Average Use, gpd} = \frac{\text{Total Chemical Used, gal}}{\text{Number of Days}} \tag{15.96}$$

To calculate the day's supply in inventory, we use Equation 15.97 or Equation 15.98:

$$\text{Days Supply in Inventory} = \frac{\text{Total Chemical in Inventory, lb}}{\text{Average Use, lb/day}} \tag{15.97}$$

$$\text{Days Supply in Inventory} = \frac{\text{Total Chemical in Inventory, gal}}{\text{Average Use, gpd}} \tag{15.98}$$

Example 15.121

Problem:

The pounds of calcium hypochlorite used for each day during a week are given in the following table. Based on these data, what were the average pounds per day of hypochlorite chemical use during the week?

Day of week	Pounds per day
Monday	50
Tuesday	55
Wednesday	51
Thursday	46
Friday	56
Saturday	51
Sunday	48

$$\text{Average Use, lb/day} = \frac{\text{Total Chemical Used, lb}}{\text{Number of Days}}$$

$$= \frac{357}{7}$$

$$= 51 \text{ lb/day Average Use}$$

Example 15.122

Problem:

The average calcium hypochlorite use at a plant is 40 lb/day. If the chemical inventory in stock is 1100 lb, how many days' supply is this?

Solution:

$$\text{Days Supply in Inventory} = \frac{\text{Total Chemical in Inventory, lb}}{\text{Average Use, lb/day}}$$

$$\text{Days Supply in Inventory} = \frac{1100 \text{ lb in Inventory}}{40 \text{ lb/day Average Use}}$$

$$= 27.5 \text{ days Supply in Inventory}$$

15.8.1 Chlorination Chemistry

As mentioned, chlorine is used in the form of free elemental chlorine or as hypochlorites. Temperature, pH, and organic content in the water influence its chemical form in water. When chlorine gas is dissolved in water, it rapidly hydrolyzes to hydrochloric acid (HCl) and hypochlorous acid (HOCl):

$$Cl_2 + H_2O \leftrightarrow H^+ + Cl^- + HOCl \tag{15.99}$$

According to White (1972), the equilibrium constant is:

$$K_H = \frac{[H^+][Cl^-][HOCl]}{[Cl_{2(aq)}]} \tag{15.100}$$

$$= 4.48 \times 10^4 \times @ \ 25°C$$

Henry's law is used to explain the dissolution of gaseous chlorine, $Cl_{2(aq)}$; it describes the effect of the pressure on the solubility of the gases: A linear relationship exists between the partial pressure of gas above a liquid and the mole fraction of the gas dissolved in the liquid (Fetter, 1998).

The Henry's law constant K_H (shown in Equation 15.100) is the measure unit of the compound transfer between the gaseous and aqueous phases. K_H is presented as a ratio of the compounds concentration in the gaseous phase to that in the aqueous phase at equilibrium:

$$K_H = \frac{P}{C_{water}} \tag{15.101}$$

where
K_H = Henry's law constant
P = compounds partial pressure in the gaseous phase
C_{water} = compounds concentration in the aqueous solution

> *Note*: The unit of the Henry's law constant is dependent on the choice of measure; however, it can also be dimensionless. For our purposes, Henry's law can be expressed as (Downs and Adams, 1973):

$$Cl_{2(aq)} = \frac{Cl_{2(aq)}}{H(mol/L \ atm)} = \frac{[Cl_{2(aq)}]}{P_{Cl2}} \tag{15.102}$$

where
$[Cl_{2(aq)}]$ = molar concentration of Cl_2
Pcl_2 = partial pressure of chlorine in atmosphere

Water (1978) points out that the disinfection capabilities of hypochlorous acid (HOCl) are generally higher than that of hypochlorite ions (OCl).

$$\text{Henry's (H) Law constant, mol/L atm} = 4.805 \times 10^{-6} \exp\left(\frac{2818.48}{T}\right) \tag{15.103}$$

Hypochlorous acid is a weak acid and subject to further dissociation to hypochlorite ions (OCl⁻) and hydrogen ions:

$$HOCl \leftrightarrow OCl^- + H^+ \tag{15.104}$$

Its acid dissociation constant K_a is

$$K_A = \frac{[OCl^-][H^+]}{[HOCl]} \tag{15.105}$$

$$3.7 \times 10^{-8} @ 25°C$$

$$= 2.61 \times 10^{-8} @ 20°C$$

Morris (1966) points out that the values of K_a for hypochlorous acid is a function of temperature in Kelvin (K) as follows:

$$\ln K_a = 23.184 - 0.058T - 6908/T \tag{15.106}$$

REFERENCES

AWWA (American Water Works Association) and ASCE (American Society of Civil Engineers) (1990). *Water Treatment Plant Design*, 2nd ed. American Water Works Association and American Society of Civil Engineers. New York: McGraw-Hill.

AWWA (American Water Works Association) and ASCE (American Society of Civil Engineers) (1998). *Water Treatment Plant Design*, 3rd ed. American Water Works Association and American Society of Civil Engineers. New York: McGraw-Hill.

Chase, G.L. (2002). *Solids Notes: Fluidization*. Akron, Ohio: The University of Akron.

Cleasby, J.L. (1990). Filtration. In AWWA, *Water Quality and Treatment*. New York: McGraw-Hill.

Cleasby, J.L. and Fan, K.S. (1981). Predicting fluidization and expansion of filter media. *J. Environ. Eng. Div.* ASCE 107(EE3), 355–471.

Downs, A.J. and Adams, C.J. (1973). *The Chemistry of Chlorine, Bromine, Iodine and Astatine*. Oxford: Pergamon Press.

Droste, R.L. (1997). *Theory and Practice of Water and Wastewater Treatment*. New York: John Wiley & Sons.

Fair, G.M., Geyer, J.C., and Okun, D.A. (1968). *Water and Wastewater Engineering*, Vol. 2: *Water Purification and Wastewater Treatment and Disposal*. New York: John Wiley & Sons.

Fetter, C.W. (1998). *Handbook of Chlorination*. New York: Litton Educational.

Gregory, R. and Zabel, T.R. (1990). Sedimentation and flotation, in *Water Quality and Treatment, A Handbook of Community Water Supplies*, 4th ed. AWWA. Ed. F.W. Pontius. New York: McGraw-Hill.

Gupta, R.S. (1997). *Environmental Engineering and Science: An Introduction*. Rockville, Maryland: Government Institutes.

Hudson, H.E., Jr. (1989). Density considerations in sedimentation. *J. Am. Water Works Assoc.* 64(6): 382–386.

McGhee, T.J. (1991). *Water Resources and Environmental Engineering*, 6th ed. New York: McGraw-Hill.

Morris, J.C. (1966). The acid ionization constant of HOCl from 5 to 35°C. *J. Phys. Chem.* 70(12),: 3789.

USEPA (1999). Individual filter self assessment. EPA Guidance Manual, Turbidity provisions, 5–12.

Water, G.C. (1978). *Disinfection of Wastewater and Water for Reuse*. New York: Van Nostrand Reinhold.

Wen, C.Y. and Yu, Y.H. (1966). Minimum fluidization velocity. *AIChE J.* 12(3), 610–612.

White, G.C. (1972). *Handbook of Chlorination*. New York: Litton Education.

PART V

Math Concepts: Wastewater Engineering

CHAPTER 16

Wastewater Calculations

In the glory days of Empire Textiles, the solvents and dyes used on the fabrics were dumped directly into the river, staining the banks below the falls red and green and yellow, according to the day of the week and the size of the batch. The sloping banks contained rings, like those in a tree trunk, except these were in rainbow colors, they recorded not the years but the rise and fall of the river. Even now, fifty years later, only the hardiest weeds and scrub trees grew south of the pavement on Front Street, and when the brush was periodically cleared, surprising patches of fading chartreuse and magenta were revealed.

Richard Russo, *Empire Falls* **(2001)**

16.1 INTRODUCTION

Standard wastewater treatment consists of a series of steps or unit processes tied together (see Figure 16.1) with the ultimate purpose of taking the raw sewage influent and turning it into an effluent that is often several times cleaner than the water in the outfalled water body. As we did for the water calculations presented in Chapter 15, we present math calculations related to wastewater at the operations level as well as the engineering level. Again, our purpose in using this format is consistent with our intention to provide a single, self-contained, ready reference source.

16.2 PRELIMINARY TREATMENT CALCULATIONS

The initial stage of treatment in the wastewater treatment process (following collection and influent pumping) is *preliminary treatment.* Process selection is normally based upon the expected characteristics of the influent flow. Raw influent entering the treatment plant may contain many kinds of materials (trash); preliminary treatment protects downstream plant equipment by removing these materials, which could cause clogs, jams, or excessive wear in plant machinery. In addition, the removal of various materials at the beginning of the treatment train saves valuable space within the treatment plant.

Two of the processes used in preliminary treatment include screening and grit removal. However, preliminary treatment may also include other processes, each designed to remove a specific type of material that could present a potential problem for downstream unit treatment processes. These processes exclude shredding, flow measurement, preaeration, chemical addition, and flow equalization. Except in extreme cases, plant design will not include all of these items. In this chapter, we focus on and describe typical calculations used in two of these processes: screening and grit removal.

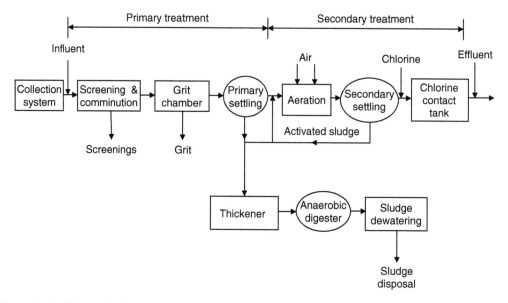

Figure 16.1 Schematic of an example wastewater treatment process providing primary and secondary treatment using activated sludge process. (From Spellman, F.R., 1999, *Spellman's Standard Handbook for Wastewater Operators,* Vol. 1, Lancaster, PA: Technomic Publishing Company.)

16.2.1 Screening

Screening removes large solids (rags, cans, rocks, branches, leaves, and roots, for example) from the flow before the flow moves on to downstream processes.

16.2.2 Screenings Removal Calculations

Wastewater operators responsible for screenings disposal are typically required to keep a record of the amount of screenings removed from the flow (the plant engineer, obviously, is responsible for ensuring the accuracy of these records). To keep and maintain accurate screening records, the volume of screenings withdrawn must be determined. Two methods are commonly used to calculate the volume of screenings withdrawn:

$$\text{Screenings Removed, ft}^3/\text{day} = \frac{\text{Screenings, ft}^3}{\text{days}} \tag{16.1}$$

$$\text{Screenings Removed, ft}^3/\text{MG} = \frac{\text{Screenings, ft}^3}{\text{Flow, MG}} \tag{16.2}$$

Example 16.1

Problem:

A total of 65 gal of screenings is removed from the wastewater flow during a 24-h period. What is the screenings removal reported as cubic feet per day?

Solution:
First, convert gallon screenings to cubic feet:

$$\frac{65 \text{ gal}}{7.48 \text{ gal/ft}^3} = 8.7 \text{ ft}^3 \text{ screenings}$$

Next, calculate screenings removed as cubic feet per day:

$$\text{Screenings Removed, ft}^3/\text{day} = \frac{8.7 \text{ ft}^3}{1 \text{ day}} = 8.7 \text{ ft}^3/\text{day}$$

Example 16.2

Problem:
During 1 week, a total of 310 gal of screenings was removed from the wastewater screens. What is the average removal in cubic feet per day?

Solution:
First, gallon screenings must be converted to cubic feet screenings:

$$\frac{310 \text{ gal}}{7.48 \text{ gal/ft}^3} = 41.4 \text{ ft}^3 \text{ screenings}$$

Next, the screenings removal calculation is completed:

$$\text{Screenings Removed, ft}^3/\text{day} = \frac{41.4 \text{ ft}^3}{7 \text{ day}} = 5.9 \text{ ft}^3/\text{day}$$

16.2.3 Screenings Pit Capacity Calculations

Recall that detention time may be considered the time required to flow through a basin or tank, or the time required to fill a basin or tank at a given flow rate. In screenings pit capacity problems, the time required to fill a screenings pit is calculated. The equation used for these types of problems is:

$$\text{Screenings Pit Fill Time, days} = \frac{\text{Volume of Pit, ft}^3}{\text{Screenings Removed, ft}^3/\text{day}} \qquad (16.3)$$

Example 16.3

Problem:
A screenings pit has a capacity of 500 ft³. (The pit is actually larger than 500 ft³ to accommodate soil for covering.) If an average of 3.4 ft³ of screenings is removed daily from the wastewater flow, in how many days will the pit be full?

Solution:

$$\text{Screenings Pit Fill Time, days} = \frac{\text{Volume of Pit, ft}^3}{\text{Screenings Removed, ft}^3/\text{day}}$$

$$\frac{500 \text{ ft}^3}{3.4 \text{ ft}^3/\text{day}} = 147.1 \text{ days}$$

Example 16.4

Problem:

A plant has been averaging a screenings removal of 2 ft³/MG. If the average daily flow is 1.8 MGD, how many days will it take to fill the pit with an available capacity of 125 ft³?

Solution:

The filling rate must first be expressed as cubic feet per day:

$$\frac{(2 \text{ ft}^3)(1.8 \text{ MGD})}{\text{MG}} = 3.6 \text{ ft}^3/\text{day}$$

$$\text{Screenings Pit Fill Time, days} = \frac{125 \text{ ft}^3}{3.6 \text{ ft}^3/\text{day}}$$

$$= 34.7 \text{ days}$$

Example 16.5

Problem:

A screenings pit has a capacity of 12 yd³ available for screenings. If the plant removes an average of 2.4 ft³ of screenings per day, in how many days will the pit be filled?

Solution:

Because the filling rate is expressed as cubic feet per day, the volume must be expressed as cubic feet:

$$(12 \text{ yd}^3)(27 \text{ ft}^3/\text{yd}^3) = 324 \text{ ft}^3$$

Now calculate fill time:

$$\text{Screenings Pit Fill Time, days} = \frac{\text{Volume of Pit, ft}^3}{\text{Screenings Removed, ft}^3/\text{day}}$$

$$\frac{324 \text{ ft}^3}{2.4 \text{ ft}^3/\text{day}}$$

$$135 \text{ days}$$

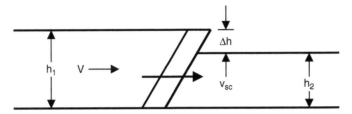

Figure 16.2 Water profile through a screen.

16.2.4 Headloss through Bar Screen

Headloss through a bar screen is determined by using Bernoulli's equation (see Figure 16.2):

$$h_1 + \frac{v^2}{2g} = h_2 + \frac{v_{sc}^{2}}{2g} + \text{Losses} \qquad (16.4)$$

where
h_1 = upstream depth of flow
h_2 = downstream depth of flow
g = acceleration of gravity
v = upstream velocity
v_{sc} = velocity of flow through the screen

The losses can be incorporated into a coefficient.

$$\Delta h = h_1 - h_2 = \frac{1}{2gC_d^{2}} \; (v_{sc}^{2} - v^2) \qquad (16.5)$$

where C_d = discharge coefficient (typical value = 0.84), a value usually supplied by manufacturer or determined through experimentation.

16.2.5 Grit Removal

The purpose of grit removal is to remove inorganic solids (sand, gravel, clay, egg shells, coffee grounds, metal filings, seeds, and other similar materials) that could cause excessive mechanical wear. Several processes or devices are used for grit removal, all based on the fact that grit is heavier than the organic solids, which should be kept in suspension for treatment in unit processes that follow grit removal. Grit removal may be accomplished in grit chambers or by the centrifugal separation of biosolids. Processes use gravity/velocity, aeration, or centrifugal force to separate the solids from the wastewater.

16.2.6 Grit Removal Calculations

Wastewater systems typically average 1 to 15 ft³ of grit per million gallons of flow (sanitary systems: 1 to 4 ft³/MG; combined wastewater systems average from 4 to 15 ft³/million gals of flow), with higher ranges during storm events. Generally, grit is disposed of in sanitary landfills. Because of this process, for planning purposes, operators must keep accurate records of grit removal. Most often, the data are reported as cubic feet of grit removed per million gallons of flow:

$$\text{Grit Removed, ft}^3/\text{MG} = \frac{\text{Grit Volume, ft}^3}{\text{Flow, MG}} \qquad (16.6)$$

Over a given period, the average grit removal rate at a plant (at least a seasonal average) can be determined and used for planning purposes. Typically, grit removal is calculated as cubic yards because excavation is normally expressed in terms of cubic yards:

$$\text{Cubic Yards Grit} = \frac{\text{Total Grit, ft}^3}{27 \text{ ft}^3/\text{yd}^3} \qquad (16.7)$$

Example 16.6

Problem:
A treatment plant removes 10 ft³ of grit in 1 day. How many cubic feet of grit are removed per million gallons if the plant flow is 9 MGD?

Solution:

$$\text{Grit Removed, ft}^3/\text{MG} = \frac{\text{Grit Volume, ft}^3}{\text{Flow, MG}}$$

$$\frac{10 \text{ ft}^3}{9 \text{ MG}} = 1.1 \text{ft}^3/\text{MG}$$

Example 16.7

Problem:
The total daily grit removed for a plant is 250 gal. If the plant flow is 12.2 MGD, how many cubic feet of grit are removed per million gallons of flow?

Solution:
First, convert gallon grit removed to cubic feet:

$$\frac{250 \text{ gal}}{7.48 \text{ gal/ft}^3} = 33 \text{ ft}^3$$

Next, complete the calculation of cubic feet per million gallons:

$$\text{Grit Removed, ft}^3/\text{MG} = \frac{\text{Grit Volume, ft}^3}{\text{Flow, MG}}$$

$$\frac{33 \text{ ft}^3}{12.2 \text{ MGD}} = 2.7 \text{ ft}^3/\text{MGD}$$

Example 16.8

Problem:

The monthly average grit removal is 2.5 ft³/MG. If the monthly average flow is 2,500,000 gpd, how many cubic yards must be available for grit disposal if the disposal pit has a 90-day capacity?

Solution:

First, calculate the grit generated each day:

$$\frac{(2.5 \text{ ft}^3)}{\text{MGD}}(2.5 \text{ MGD}) = 6.25 \text{ ft}^3 \text{ each day}$$

The cubic feet of grit generated for 90 days would be

$$\frac{(6.25 \text{ ft}^3)}{\text{day}}(90 \text{ days}) = 562.5 \text{ ft}^3$$

Convert cubic feet to cubic yards of grit:

$$\frac{562.5 \text{ ft}^3}{27 \text{ ft}^3/\text{yd}^3} = 21 \text{ yd}^3$$

16.2.7 Grit Channel Velocity Calculation

The optimum velocity in sewers is approximately 2 ft/sec at peak flow because this velocity normally prevents solids from settling from the lines. However, when the flow reaches the grit channel, the velocity should decrease to about 1 ft/sec to permit heavy inorganic solids to settle. In the example calculations that follow, we describe how the velocity of the flow in a channel can be determined by the float and stopwatch method and by channel dimensions.

Example 16.9

Velocity by Float and Stopwatch

$$\text{Velocity, fps } = \frac{\text{Distance Traveled, ft}}{\text{Time Required, sec}} \tag{16.8}$$

Problem:

A float takes 30 sec to travel 37 ft in a grit channel. What is the velocity of the flow in the channel?

Solution:

$$\text{Velocity, fps } = \frac{37 \text{ ft}}{30 \text{ sec}} = 1.2 \text{ fps}$$

Example 16.10

Velocity by Flow and Channel Dimensions

This calculation can be used for a single channel or tank or for multiple channels or tanks with the same dimensions and equal flow. If the flow through each unit of the unit dimensions is unequal, the velocity for each channel or tank must be computed individually.

$$\text{Velocity, fps} = \frac{\text{Flow, MGD} \times 1.55 \text{ cfs/MGD}}{\text{\# Channels in Service} \times \text{Channel Width, ft} \times \text{Water Depth, ft}} \quad (16.9)$$

Problem:

The plant is currently using two grit channels. Each channel is 3 ft wide and has a water depth of 1.3 ft. What is the velocity when the influent flow rate is 4.0 MGD?

Solution:

$$\text{Velocity, fps} = \frac{4.0 \text{ MGD} \times 1.55 \text{ cfs/MGD}}{2 \text{ Channels} \times 3 \text{ ft} \times 1.3 \text{ ft}}$$

$$\text{Velocity, fps} = \frac{6.2 \text{ cfs}}{7.8 \text{ ft}^2} = 0.79 \text{ fps}$$

Key point: Because 0.79 is within the 0.7 to 1.4 level, the operator of this unit would not make any adjustments.

Key point: The channel dimensions must always be in feet. Convert inches to feet by dividing by 12 in. per foot.

16.2.7.1 Required Settling Time

This calculation can be used to determine the time required for a particle to travel from the surface of the liquid to the bottom at a given settling velocity. To compute the settling time, settling velocity in feet per second must be provided or determined by experiment in a laboratory.

$$\text{Settling Time, sec} = \frac{\text{Liquid Depth, ft}}{\text{Settling, Velocity, fps}} \quad (16.10)$$

Example 16.11

Problem:

The plant's grit channel is designed to remove sand, which has a settling velocity of 0.080 ft/sec. The channel is currently operating at a depth of 2.3 ft. How many seconds will it take for a sand particle to reach the channel bottom?

Solution:

$$\text{Settling Time, sec} = \frac{2.3 \text{ ft}}{0.080 \text{ fps}} = 28.8 \text{ sec}$$

16.2.7.2 Required Channel Length

This calculation can be used to determine the length of channel required to remove an object with a specified settling velocity.

$$\text{Required Channel Length} = \frac{\text{Channel Depth, ft} \times \text{Flow Velocity, fps}}{0.080 \text{ fps}} \tag{16.11}$$

Example 16.12

Problem:
   The plant's grit channel is designed to remove sand, which has a settling velocity of 0.080 ft/sec. The channel is currently operating at a depth of 3 ft. The calculated velocity of flow through the channel is 0.85 ft/sec. The channel is 36 ft long; is it long enough to remove the desired sand particle size?

Solution:

$$\text{Required Channel Length} = \frac{3 \text{ ft} \times 0.85 \text{ fps}}{0.080 \text{ fps}} = 31.9 \text{ ft}$$

Yes, the channel is long enough to ensure that all the sand will be removed.

16.2.7.3 Velocity of Scour

The Camp–Shields equation (Camp, 1942) is used to estimate the velocity of scour necessary to resuspend settled organics:

$$v_s = \sqrt{\frac{8kgd}{f} \frac{p_p - p}{p}} \tag{16.12}$$

where
v_s = velocity of scour
d = nominal diameter of the particle
k = empirically determined constant
f = Darcy–Weisbach friction factor

   If the channel is rectangular and discharges over a rectangular weir, the discharge relation based on Bernoulli's equation is:

$$Q = C_d A \sqrt{2gH} = C_w H^{3/2} \tag{16.13}$$

where
w = width of the channel
A = cross-sectional area of the channel
C_d = discharge coefficient
C = equal to $C_d \sqrt{2g}$
H = depth of flow in the channel

The horizontal velocity, v_h, is related to the discharge rate and channel velocity by:

$$= \frac{Q}{A} = \frac{Q}{wH} = CH^{1/2} = C\left(\frac{Q}{C_w}\right)^{1/3} \qquad (16.14)$$

16.3   PRIMARY TREATMENT CALCULATIONS

Primary treatment (primary sedimentation or clarification) should remove organic settleable and organic floatable solids. Poor solids removal during this step of treatment may cause organic overloading of the biological treatment processes following primary treatment. Normally, each primary clarification unit can be expected to remove 90 to 95% of settleable solids; 40 to 60% of the total suspended solids; and 25 to 35% of BOD.

16.3.1   Process Control Calculations

As with many other wastewater treatment plant unit processes, several process control calculations may be helpful in evaluating the performance of the primary treatment process. Process control calculations are used in the sedimentation process to determine:

- Percent removal
- Hydraulic detention time
- Surface loading rate (surface settling rate)
- Weir overflow rate (weir loading rate)
- Biosolids pumping
- Percent total solids (% ts)
- BOD and SS removed, pounds per day

In the following subsections, we take a closer look at a few of these process control calculations and example problems.

> *Key point*: The calculations presented in the following sections allow determination of values for each function performed. Again, keep in mind that an optimally operated primary clarifier should have values in an expected range. Recall that the expected range percentage removal for a primary clarifier is

- Settleable solids: 90 to 95%
- Suspended solids: 40 to 60%
- BOD: 25 to 35%

The expected range of hydraulic detention time for a primary clarifier is 1 to 3 h. The expected range of surface loading/settling rate for a primary clarifier is 600 to 1200 gpd/ft² (ballpark estimate). The expected range of weir overflow rate for a primary clarifier is 10,000 to 20,000 gpd/ft.

16.3.2   Surface Loading Rate (Surface Settling Rate/Surface Overflow Rate)

Surface loading rate is the number of gallons of wastewater passing over 1 ft² of tank per day. This can be used to compare actual conditions with design. Plant designs generally use a surface-loading rate of 300 to 1200 gpd/ft².

$$\text{Surface Loading Rate, gpd/ft}^2 = \frac{\text{gal/day}}{\text{Surface Tank Area, ft}^2} \qquad (16.15)$$

Example 16.13

Problem:
The circular settling tank has a diameter of 120 ft. If the flow to the unit is 4.5 MGD, what is the surface loading rate in gallons per day per square foot?

Solution:

$$\text{Surface Loading Rate} = \frac{4.5\ \text{MGD} \times 1,000,000\ \text{gal/MGD}}{0.785 \times 120\ \text{ft} \times 120\ \text{ft}} = 398\ \text{gpd/ft}^2$$

Example 16.14

Problem:
A circular clarifier has a diameter of 50 ft. If the primary effluent flow is 2,150,000 gpd, what is the surface overflow rate in gallons per day per square foot?

Solution:
Key point: Remember that area = (0.785) (50 ft) (50 ft)

$$\text{Surface Overflow Rate} = \frac{\text{Flow, gpd}}{\text{Area, ft}^2}$$

$$\frac{2,150,000}{(0.785)\ (50\ \text{ft})\ (50\ \text{ft})} = 1,096\ \text{gpd/ft}^2$$

Example 16.15

Problem:
A sedimentation basin 90 ft by 20 ft receives a flow of 1.5 MGD. What is the surface overflow rate in gallons per day per square foot?

Solution:

$$\text{Surface Overflow Rate} = \frac{\text{Flow, gpd}}{\text{Area, ft}^2}$$

$$= \frac{1,500,000\ \text{gpd}}{(90\ \text{ft})\ (20\ \text{ft})}$$

$$= 833\ \text{gpd/ft}^2$$

16.3.3 Weir Overflow Rate (Weir Loading Rate)

A weir is a device used to measure wastewater flow. *Weir overflow rate (weir loading rate)* is the amount of water leaving the settling tank per linear foot of water. The result of this calculation can be compared with design. Normally, weir overflow rates of 10,000 to 20,000 gal/day/ft are used in the design of a settling tank.

$$\text{Weir Overflow Rate, gpd/ft} = \frac{\text{Flow, gal/day}}{\text{Weir Length, ft}} \qquad (16.16)$$

Key point: In calculating weir circumference, use total feet of weir = (3.14) (weir diameter, feet).

Example 16.16

Problem:

The circular settling tank is 80 ft in diameter and has a weir along its circumference. The effluent flow rate is 2.75 MGD. What is the weir overflow rate in gallons per day per foot?

Solution:

$$\text{Weir Overflow Rate, gpd/ft} = \frac{2.75 \text{ MGD} \times 1,000,000 \text{ gal}}{3.14 \times 80 \text{ ft}} = 10,947 \text{ gal/day/ft}$$

Key point: Notice that 10,947 gal/day/ft is above the recommended minimum of 10,000.

Example 16.17

Problem:

A rectangular clarifier has a total of 70 ft of weir. What is the weir overflow rate in gallons per day per square foot when the flow is 1,055,000 gpd?

Solution:

$$\text{Weir Overflow Rate, gpd/ft} = \frac{\text{Flow, gal/day}}{\text{Weir Length, ft}}$$

$$= \frac{1,055,000 \text{ gpd}}{70 \text{ ft}} = 15,071 \text{ gpd}$$

16.3.4 Primary Sedimentation Basins

Example 16.18

Problem:

Two rectangular settling tanks are each 8 m wide, 26 m long, and 2.5 m deep. Each is used alternatively to treat 1800 m³ in a 12-h period. Compute the surface overflow (settling) rate, detention time, horizontal velocity, and outlet weir-loading rate using an H-shaped weir with three times width.

Solution:

Step 1. Determine the design flow Q:

$$Q = \frac{1800 \text{ m}^3}{12 \text{ h}} \times \frac{24 \text{ h}}{1 \text{ day}}$$

$$= 3600 \text{ m}^3/\text{day}$$

Step 2. Compute surface overflow rate v_o:

$$V_o = Q/A = 3600 \text{ m}^3/\text{day} \div (8 \text{ m} \times 26 \text{ m})$$

$$= 17.3 \text{ m}^3 \ (\text{m}^2 \cdot \text{day})$$

Step 3. Compute detention time t:

$$\text{Tank Volume V} = 8 \text{ m} \times 26 \text{ m} \times 2.5 \text{ m} \times 2$$

$$= 1040 \text{ m}_3$$

$$t = V/Q = 1040 \text{ m}^3/(3600 \text{ m}^3/\text{day})$$

$$= 0.289 \text{ day}$$

$$= 6.9 \text{ h}$$

Step 4. Compute horizontal velocity v_h:

$$V_h = \frac{3600 \text{ m}^3/\text{day}}{8 \text{ m} \times 2.5 \text{ m}}$$

$$= 180 \text{ m/day}$$

$$= 0.125 \text{ m/min}$$

$$= 0.410 \text{ ft/min}$$

Step 5. Compute outlet weir loading, wl:

$$wl = \frac{3600 \text{ m}^3/\text{day}}{8 \text{ m} \times 3 \text{ m}} = 150 \text{ m}^3/(\text{day} \cdot \text{m})$$

$$= 12{,}100 \text{ gal/(day)}$$

16.4 BIOSOLIDS PUMPING

Determination of biosolids pumping (the quantity of solids and volatile solids removed from the sedimentation tank) provides accurate information needed for process control of the sedimentation process.

$$\text{Solids Pumped} = \text{Pump Rt., gpm} \times \text{Pump Time, min/day} \times 8.34 \text{ lb/gal} \times \% \text{ Solid} \quad (16.17)$$

$$\text{Volatile Solids/lb/day} = \text{Pump Rt.} \times \text{Pump Time } 8.34 \times \% \text{ Solids} \times \% \text{ Vol. Matter} \quad (16.18)$$

Example 16.19

Problem:

The biosolids pump operates 30 min/h and delivers 25 gal/min of biosolids. Laboratory tests indicate that the biosolids are 5.3% solids and 68% volatile matter. Assuming a 24-h period, how many pounds of volatile matter are transferred from the settling tank to the digester?

Solution:

$$\text{Pump Time} = 30 \text{ min/hr}$$

$$\text{Pump Rate} = 25 \text{ gpm}$$

$$\% \text{ Solids} = 5.3\%$$

$$\% \text{ V.M.} = 68\%$$

$$\text{Volatile Solids, lb/day} = 25 \text{ gpm} \times (30 \text{ min/h} \times 24 \text{ h/day}) \times 8.34 \text{ lb/gal} \times 0.053 \times 0.68$$

$$= 5410 \text{ lb/day}$$

16.4.1 Percent Total Solids (% TS)

Problem:

A settling tank biosolids sample is tested for solids. The sample and dish weigh 73.79 g. The dish alone weighs 21.4 g. After drying, the dish with dry solids weighs 22.4 g. What is the percent total solids (% TS) of the sample?

Solution:

$$
\begin{array}{rr}
\text{Sample + Dish} & 73.79 \text{ g} \\
\text{Dish alone} & -21.40 \text{ g} \\
\hline
 & 52.39 \text{ g}
\end{array}
$$

$$
\begin{array}{rr}
\text{Dish + Dry Solids} & 22.4 \text{ g} \\
\text{Dish alone} & -21.4 \text{ g} \\
\hline
 & 1.0 \text{ g}
\end{array}
$$

$$\frac{1.0 \text{ g}}{52.39 \text{ g}} \times 100\% = 1.9\%$$

16.4.2 BOD and SS Removed, Pounds per Day

To calculate the pounds of BOD or suspended solids removed each day, we need to know the milligrams per liter of BOD or SS removed and the plant flow. Then, we can use the milligrams-per-liter to pounds-per-day equation:

$$\text{SS Removed} = \text{mg/L} \times \text{MGD} \times 8.34 \text{ lb/gal} \tag{16.19}$$

Example 16.20

Problem:
   If 120 mg/L suspended solids are removed by a primary clarifier, how many pounds per day of suspended solids are removed when the flow is 6,250,000 gpd?

Solution:

$$\text{SS Removed} = 120 \text{ mg/L} \times 6.25 \text{ MGD} \times 8.34 \text{ lb/gal} = 6255 \text{ lb/day}$$

Example 16.21

Problem:
   The flow to a secondary clarifier is 1.6 MGD. If the influent BOD concentration is 200 mg/L and the effluent BOD concentration is 70 mg/L, how many pounds of BOD are removed daily?

Solution:

$$\text{lb/day BOD removed} = 200 \text{ mg/L} - 70 \text{ mg/L} = 130 \text{ mg/L}$$

   After calculating milligrams per liter of BOD removed, calculate pounds per day of BOD removed:

$$\text{BOD removed, lb/day} = (130 \text{ mg/L})(1.6 \text{ MGD})(8.34 \text{ lb/gal}) = 1735 \text{ lb/day}$$

16.5 TRICKLING FILTER CALCULATIONS

The trickling filter process (see Figure 16.3) is one of the oldest forms of dependable biological treatment for wastewater. By its very nature, the trickling filter has its advantages over other unit processes. It is a very economical and dependable process for treatment of wastewater prior to discharge and is capable of withstanding periodic shock loading; furthermore, process energy demands are low because aeration is a natural process.

   As shown in Figure 16.4, trickling filter operation involves spraying wastewater over solid media such as rock, plastic, or redwood slats (or laths). As the wastewater trickles over the surface of the media, a growth of microorganisms (bacteria, protozoa, fungi, algae, helminths or worms, and larvae) develops. This growth is visible as a shiny slime similar to the slime found on rocks in a stream. As wastewater passes over this slime, the slime adsorbs the organic (food) matter. This organic matter is used for food by the microorganisms. At the same time, air moving through the open spaces in the filter transfers oxygen to the wastewater. This oxygen is then transferred to the slime to keep the outer layer aerobic. As the microorganisms use the food and oxygen, they produce

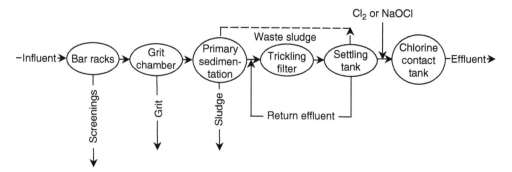

Figure 16.3 Simplified flow diagram of trickling filter used for wastewater treatment. (From Spellman, F.R., 1999, *Spellman's Standard Handbook for Wastewater Operators,* Vol. 1, Lancaster, PA: Technomic Publishing Company.)

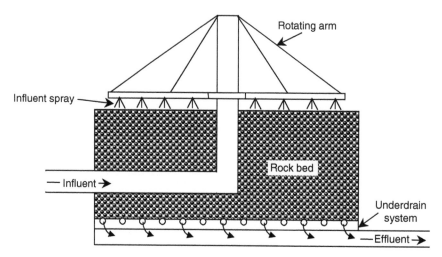

Figure 16.4 Schematic of cross-section of a trickling filter. (From Spellman, F.R., 1999, *Spellman's Standard Handbook for Wastewater Operators,* Vol. 1, Lancaster, PA: Technomic Publishing Company.)

more organisms, carbon dioxide, sulfates, nitrates, and other stable by-products; these materials are then discarded from the slime back into the wastewater flow and carried out of the filter.

16.5.1 Trickling Filter Process Calculations

Several calculations are useful in the operation of trickling filters; these include hydraulic loading, organic loading, and biochemical oxygen demand (BOD) and suspended solids (SS) removal. Each type of trickling filter is designed to operate with specific loading levels, which depend on the filter classification. To operate the filter properly, filter loading must be within the specified levels. The main three loading parameters for the trickling filter are hydraulic loading, organic loading, and recirculation ratio.

16.5.2 Hydraulic Loading

Calculating the hydraulic loading rate is important in accounting for the primary effluent as well as the recirculated trickling filter effluent. These are combined before they are applied to the filter surface. The hydraulic loading rate is calculated based on filter surface area. The normal hydraulic loading-rate ranges for standard- and high-rate trickling filters are:

$$\text{Standard Rate} = 25 - 100 \text{ gpd/ft}^2 \text{ or } 1 - 40 \text{ MGD/acre}$$

$$\text{High Rate} = 100 - 1000 \text{ gpd/ft}^2 \text{ or } 4 - 40 \text{ MGD/acre}$$

Key point: If the hydraulic loading rate for a particular trickling filter is too low, septic conditions begin to develop.

Example 16.22

Problem:
A trickling filter 80 ft in diameter is operated with a primary effluent of 0.588 MGD and a recirculated effluent flow rate of 0.660 MGD. Calculate the hydraulic loading rate on the filter in units of gallons per day per square foot.

Solution:
The primary effluent and recirculated trickling filter effluent are applied together across the surface of the filter; therefore, 0.588 MGD + 0.660 MGD = 1.248 MGD = 1,248,000 gpd.

$$\text{Circular surface area} = 0.785 \times (\text{diameter})^2$$

$$= 0.785 \times (80 \text{ ft})^2$$

$$= 5,024 \text{ ft}^2$$

$$\frac{1,248,000 \text{ gpd}}{5,024 \text{ ft}^2} = 248.4 \text{ gpd/ft}^2$$

Example 16.23

Problem:
A trickling filter 80 ft in diameter treats a primary effluent flow of 550,000 gpd. If the recirculated flow to the clarifier is 0.2 MGD, what is the hydraulic loading on the trickling filter?

Solution:

$$\text{Hydraulic loading rate} = \frac{\text{Total Flow, gpd}}{\text{Area, ft}^2}$$

$$\frac{750,000 \text{ gpd total flow}}{(0.785)(80 \text{ ft})(80 \text{ ft})}$$

$$= 149 \text{ gpd/ft}^2$$

Example 16.24

Problem:

A high-rate trickling filter receives a daily flow of 1.8 MGD. What is the dynamic loading rate in MGD per acre if the filter is 90 ft in diameter and 5 ft deep?

Solution:

$$(0.785)\ (90\ \text{ft})\ (90\ \text{ft})\ =\ 6359\ \text{ft}^2$$

$$\frac{6359\ \text{ft}^2}{43{,}560\ \text{ft}^2/\text{acre}}\ =\ 0.146\ \text{acre}$$

$$\text{Hydraulic Loading Rate}\ =\ \frac{1.8\ \text{MGD}}{0.146\ \text{acre}} = 12.3\ \text{MGD/acre}$$

Key point: When hydraulic loading rate is expressed as MGD per acre, this is still an expression of gallon flow over surface area of trickling filter.

16.5.3 Organic Loading Rate

Trickling filters are sometimes classified by the organic loading rate applied. This rate is expressed as a certain amount of BOD applied to a certain volume of media. In other words, the organic loading is defined as the pounds of BOD or chemical oxygen demand (COD) applied per day per 1000 ft³ of media — a measure of the amount of food applied to the filter slime. To calculate the organic loading on the trickling filter, two things must be known: (1) the pounds of BOD or COD applied to the filter media per day; and (2) the volume of the filter media in 1000 ft³-units. The BOD and COD contribution of the recirculated flow is not included in the organic loading.

Example 16.25

Problem:

A trickling filter 60 ft in diameter receives a primary effluent flow rate of 0.440 MGD. Calculate the organic loading rate in units of pounds of BOD applied per day per 1000 ft³ of media volume. The primary effluent BOD concentration is 80 mg/L. The media depth is 9 ft.

Solution:

$$0.440\ \text{MGD}\ \times\ 80\ \text{mg/L}\ \times\ 8.34\ \text{lb/gal}\ =\ 293.6\ \text{lb of BOD applied/day}$$

$$\text{Surface Area}\ =\ 0.785\ \times\ (60)^2\ =\ 2826\ \text{ft}^2$$

$$\text{Area}\ \times\ \text{Depth}\ \times\ \text{Volume}$$

$$2826\ \text{ft}^2\ \times\ 9\ \text{ft}\ =\ 25{,}434\ \text{(TF Volume)}$$

Key point: To determine the pounds of BOD per 1,000 ft³ in a volume of thousands of cubic feet, we must set up the equation as shown below.

$$\frac{293.6 \text{ lb BOD/day}}{25,434 \text{ ft}^3} \times \frac{1000}{1000}$$

Regrouping the numbers and the units together:

$$\frac{293.6 \text{ lb BOD/day} \times 1000}{25,434 \text{ ft}^3} \times \frac{\text{lb BOD/day}}{1000 \text{ ft}^3} = 11.5 \frac{\text{lb BOD/day}}{1000 \text{ ft}^3}$$

16.5.4 BOD and SS Removed

To calculate the pounds of BOD or suspended solids removed each day, we need to know the milligrams per liter of BOD and SS removed and the plant flow.

Example 16.26

Problem:

If 120 mg/L suspended solids are removed by a trickling filter, how many pounds per day suspended solids are removed when the flow is 4.0 MGD?

Solution:

$$(\text{mg/L}) \, (\text{MGD flow}) \times 8.34 \text{ lb/gal}$$

$$(120 \text{ mg/L}) \, (4.0 \text{ MGD}) \times (8.34 \text{ lb/gal}) = 4003 \text{ lb SS/day}$$

Example 16.27

Problem:

The 3,500,000-gpd influent flow to a trickling filter has a BOD content of 185 mg/L. If the trickling filter effluent has a BOD content of 66 mg/L, how many pounds of BOD are removed daily?

Solution:

$$(\text{mg/L}) \, (\text{MGD flow}) \, (8.34 \text{ lb/gal}) = \text{lb/day removed}$$

$$185 \text{ mg/L} - 66 \text{ mg/L} = 119 \text{ mg/L}$$

$$(119 \text{ mg/L}) \, (3.5 \text{ MGD}) \, (8.34 \text{ lb/gal}) = 3474 \text{ lb/day removed}$$

16.5.5 Recirculation Flow

Recirculation in trickling filters involves the return of filter effluent back to the head of the trickling filter. It can level flow variations and assist in solving operational problems, such as ponding, filter flies, and odors. The operator must check the rate of recirculation to ensure that it is within design specifications. Rates above design specifications indicate hydraulic overloading; rates under them indicate hydraulic underloading. The *trickling filter recirculation ratio* is the ratio of the recirculated trickling filter flow to the primary effluent flow.

The trickling filter recirculation ratio may range from 0.5:1 (0.5) to 5:1 (5). However, the ratio is often 1:1 or 2:1.

$$\text{Recirculation} = \frac{\text{Recirculated Flow, MGD}}{\text{Primary Effluent Flow, MGD}} \qquad (16.20)$$

Example 16.28

Problem:

A treatment plant receives a flow of 3.2 MGD. If the trickling filter effluent is recirculated at the rate of 4.50 MGD, what is the recirculation ratio?

Solution:

$$\text{Recirculation Ratio} = \frac{\text{Recirculated Flow, MGD}}{\text{Primary Effluent Flow, MGD}}$$

$$\frac{4.5 \text{ MGD}}{3.2 \text{ MGD}}$$

$$= 1.4 \text{ Recirculation Ratio}$$

Example 16.29

Problem:

A trickling filter receives a primary effluent flow of 5 MGD. If the recirculated flow is 4.6 MGD, what is the recirculation ratio?

Solution:

$$\text{Recirculation Ratio} = \frac{\text{Recirculated Flow, MGD}}{\text{Primary Effluent Flow, MGD}}$$

$$\frac{4.6 \text{ MGD}}{5 \text{ MGD}}$$

$$= 0.92 \text{ Recirculation Ratio}$$

16.5.6 Trickling Filter Design

In trickling filter design, the parameters used are the hydraulic loading and BOD:

$$\text{Hydraulic Loading} = \frac{Q_o + R}{A} \qquad (16.21)$$

where
Q_o = average wastewater flow rate, million gallons per day
R = recirculated flow = $Q_o \times$ circulation ratio
A = filter area, acres

$$\text{BOD loading} = \frac{8340\,(\text{BODs})(Q_o)}{V} \qquad (16.22)$$

where
BOD_s = settled BOD_5 from primary, milligrams per liter
Q_o = Average wastewater flow rate, million gallons per day
V = filter volume, cubic feet
8340 = conversion of units

16.6 ROTATING BIOLOGICAL CONTACTORS (RBCS)

In essence the rotating biological contactor (RBC) is a variation of the attached growth idea provided by the trickling filter (see Figure 16.5 and Figure 16.6). Still relying on microorganisms that grow on the surface of a medium, the RBC is instead a *fixed film* biological treatment device. The basic biological process, however, is similar to that occurring in trickling filters.

An RBC consists of a series of closely spaced (mounted side by side, circular, plastic [synthetic]) disks, typically about 11.5 ft in diameter. Attached to a rotating horizontal shaft, approximately 40% of each disk is submersed in a tank that contains the wastewater to be treated. As the RBC rotates, the attached biomass film (zoogleal slime) that grows on the surface of the disks moves into and out of the wastewater. While submerged in the wastewater, the microorganisms absorb organics; while they are rotated out of the wastewater, they are supplied with needed oxygen for aerobic decomposition. As the zoogleal slime re-enters the wastewater, excess solids and waste

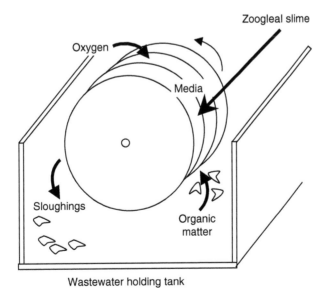

Figure 16.5 Rotating biological contactor (RBC) cross-section and treatment system. (From Spellman, F.R., 1999, *Spellman's Standard Handbook for Wastewater Operators*, Vol. 1, Lancaster, PA: Technomic Publishing Company.)

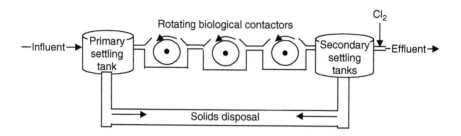

Figure 16.6   Rotating biological contactor (RBC) treatment system. (From Spellman, F.R., 1999, *Spellman's Standard Handbook for Wastewater Operators,* Vol. 1, Lancaster, PA: Technomic Publishing Company.)

products are stripped off the media as sloughings, which are transported with the wastewater flow to a settling tank for removal.

16.6.1 RBC Process Control Calculations

Several process control calculations are useful in the operation of an RBC. These include soluble BOD, total media area, organic loading rate, and hydraulic loading. Settling tank calculations and biosolids pumping calculations may be helpful for evaluation and control of the settling tank following the RBC.

16.6.2 Hydraulic Loading Rate

The manufacturer normally specifies the RBC media surface area, and the hydraulic loading rate is based on the media surface area, usually in square feet. Hydraulic loading is expressed in terms of gallons of flow per day per square foot of media. This calculation can be helpful in evaluating the current operating status of the RBC. Comparison with design specifications can determine if the unit is hydraulically over- or underloaded. Hydraulic loading on an RBC can range from 1 to 3 gpd/ft^2.

Example 16.30

Problem:
    An RBC treats a primary effluent flow rate of 0.244 MGD. What is the hydraulic loading rate in gallons per day per square foot if the media surface area is 92,600 ft^2?

Solution:

$$\frac{244,000 \text{ gpd}}{92,000 \text{ ft}^2} = 2.65 \text{ gpd/ft}^2$$

Example 16.31

Problem:
    An RBC treats a flow of 3.5 MGD. The manufacturer's data indicate a media surface area of 750,000 ft^2. What is the hydraulic loading rate on the RBC?

Solution:

$$\text{Hydraulic Loading Rate } = \frac{\text{Flow, gpd}}{\text{Media Area, ft}^2}$$

$$= \frac{3,500,000 \text{ gpd}}{750,000 \text{ ft}^2} = 4.7 \text{ ft}^2$$

Example 16.32

Problem:

A rotating biological contactor treats a primary effluent flow of 1,350,000 gpd. The manufacturer's data indicate that the media surface area is 600,000 ft². What is the hydraulic loading rate on the filter?

Solution:

$$\text{Hydraulic Loading Rate } = \frac{\text{Flow, gpd}}{\text{Area, ft}^2}$$

$$= \frac{1,350,000 \text{ gpd}}{600,000 \text{ ft}^2} = 2.3 \text{ ft}^2$$

16.6.3 Soluble BOD

The soluble BOD concentration of the RBC influent can be determined experimentally in the laboratory, or it can be estimated using the suspended solids concentration and the "K" factor. This factor is used to approximate the BOD (particulate BOD) contributed by the suspended matter. The K factor must be provided or determined experimentally in the laboratory; for domestic wasters, it is normally in the range of 0.5 to 0.7.

$$\text{Soluble BOD}_5 = \text{Total BOD}_5 - (\text{K Factor} \times \text{Total Suspended Solids}) \qquad (16.23)$$

Example 16.33

Problem:

The suspended solids concentration of a wastewater is 250 mg/L. If the amount of *K*-value at the plant is 0.6, what is the estimated particulate biochemical oxygen demand (BOD) concentration of the wastewater?

Solution:

> *Key point*: The 0.6 *K*-value indicates that about 60% of the suspended solids are organic suspended solids (particulate BOD).

$$(250 \text{ mg/L}) \, (0.6) = 150 \text{ mg/L Particulate BOD}$$

Example 16.34

Problem:

A rotating biological contactor receives a flow of 2.2 MGD with a BOD content of 170 mg/L and suspended solids (SS) concentration of 140 mg/L. If the *K*-value is 0.7, how many pounds of soluble BOD enter the RBC daily?

Solution:

$$\text{Total BOD} = \text{Particulate BOD} + \text{Soluble BOD}$$

$$170 \text{ mg/L} = (140 \text{ mg/L})(0.7) + x \text{ mg/L}$$

$$170 \text{ mg/L} = 98 \text{ mg/L} + x \text{ mg/L}$$

$$170 \text{ mg/L} - 98 \text{ mg/L} = x$$

$$x = 72 \text{ mg/L Soluble BOD}$$

Now, pounds per day of soluble BOD may be determined:

$$(\text{mg/L Soluble BOD})(\text{MGD Flow})(8.34 \text{ lb/gal}) = \text{lb/day}$$

$$(72 \text{ mg/L})(2.2 \text{ MGD})(8.34 \text{ lb/gal}) = 1321 \text{ lb/day Soluble BOD}$$

Example 16.35

Problem:

The wastewater entering a rotating biological contactor has a BOD content of 210 mg/L. The suspended solids content is 240 mg/L. If the *K*-value is 0.5, what is the estimated soluble BOD (milligrams per liter) of the wastewater?

Solution:

$$\text{Total BOD, mg/L} = \text{Particulate BOD, mg/L} + \text{Soluble BOD, mg/L}$$

$$\underset{\text{BOD}}{210 \text{ mg/L}} = \underset{\text{SS}}{(240 \text{ mg/L})(0.5)} + \underset{\text{Sol. BOD}}{x \text{ mg/L}}$$

$$210 \text{ mg/L} = 120 \text{ mg/L} + \underset{\text{Sol. BOD}}{x \text{ mg/L}}$$

$$210 - 120 = x$$

$$\text{Sol. BOD } 90 \text{ mg/L} = x$$

16.6.4 Organic Loading Rate

The organic loading rate can be expressed as total BOD loading in pounds per day per 1000 ft² of media. The actual values can then be compared with plant design specifications to determine the current operating condition of the system.

$$\text{Organic Loading Rate} = \frac{\text{Sol. BOD} \times \text{Flow, MGD} \times 8.34 \text{ lb/gal}}{\text{Media Area, 1000 ft}^2} \qquad (16.24)$$

Example 16.36

Problem:

A rotating biological contactor (RBC) has a media surface area of 500,000 ft² and receives a flow of 1,000,000 gpd. If the soluble BOD concentration of the primary effluent is 160 mg/L, what is the organic loading on the RBC in pounds per day per 1000 ft²?

Solution:

$$\text{Organic Loading Rate} = \frac{\text{Sol. BOD, lb/day}}{\text{Media Area, 1000 ft}^2}$$

$$\frac{(160 \text{ mg/L})(1.0 \text{ MGD}) (8.34 \text{ lb/gal})}{500 \times 1000 \text{ ft}^2}$$

$$\frac{2.7 \text{ lb/day Sol. BOD}}{1000 \text{ ft}^2}$$

Example 16.37

Problem:

The wastewater flow to an RBC is 3,000,000 gpd. The wastewater has a soluble BOD concentration of 120 mg/L. The RBC consists of six shafts (each 110,000 ft²), with two shafts comprising the first stage of the system. What is the organic loading rate in pounds per day per 1000 ft² on the first stage of the system?

Solution:

$$\text{Organic Loading Rate} = \frac{\text{Sol. BOD, lb/day}}{\text{Media Area, 1000 ft}^2}$$

$$\frac{(120 \text{ mg/L}) \times (3.0 \text{ MGD}) (8.34 \text{ lb/gal})}{220 \qquad 1000 \text{ ft}^2}$$

$$= 13.6 \text{ lb Sol. BOD/day/1000 ft}^2$$

16.6.5　Total Media Area

Several process control calculations for the RBC use the total surface area of all the stages within the train. As was the case with the soluble BOD calculation, plant design information or information supplied by the unit manufacturer must provide the individual stage areas (or the total train area) because physical determination of this would be extremely difficult.

$$\text{Total Area} = \text{1st Stage Area} + \text{2nd Stage Area} + \dots + \text{nth Stage Area} \qquad (16.25)$$

16.6.6　Modeling RBC Performance

Although a number of semiempirical formulations have been used, the Schultz–Germain formula for trickling filters is recommended for modeling RBC performance (Spengel and Dzombok, 1992):

$$Se = Sie^{k(V/Q)0.5} \qquad (16.26)$$

where
Se = total BOD of settled effluent, milligrams per liter
Si = total BOD of wastewater applied to filter, milligrams per liter
V = in square meters
Q = in square meters per second

16.6.7　RBC Performance Parameter

The control parameter for RBC performance is soluble BOD (SBOD):

$$\text{SBOD} = \text{TBOD} - \text{Suspended BOD} \qquad (16.27)$$

$$\text{Suspended BOD} = c\,(\text{TSS}) \qquad (16.28)$$

$$\text{SBOD} = \text{TBOD} - c\,(\text{TSS}) \qquad (16.29)$$

where
c = a coefficient
　= 0.5 to 0.7 for domestic wastewater
　= 0.5 for raw domestic wastewater (TSS > TBOD)
　= 0.6 for raw wastewater (TSS ≅ TBOD)
　= 0.6 for primary effluents
　= 0.5 for secondary effluents

Example 16.38

Problem:
　　Average TBOD is 152 mg/L and TSS is 132 mg/L. What is the influent SBOD concentration that can be used for the design of an RBC system? The RBC is used as the secondary treatment unit.

Solution:
　　For the primary effluent (RBC influent)

$$c = 0.6$$

Estimate SBOD concentration of RBC influent using Equation 16.29.

$$SBOD = TBOD - c\,(TSS)$$

$$= 152\ mg/L\ -\ 0.6\ (132\ mg/L)$$

$$= 73\ mg/L$$

16.7 ACTIVATED BIOSOLIDS

The activated biosolids (sludge) process is a man-made process that mimics the natural self-purification process that takes place in streams. In essence, we can state that the activated biosolids treatment process is a "stream in a container."

In wastewater treatment, activated-biosolids processes are used for secondary treatment as well as complete aerobic treatment without primary sedimentation. *Activated biosolids* refers to biological treatment systems that use a suspended growth of organisms to remove BOD and suspended solids. The basic components of an activated biosolids sewage treatment system include an aeration tank and a secondary basin, settling basin, or clarifier. Primary effluent is mixed with settled solids recycled from the secondary clarifier and is then introduced into the aeration tank. Compressed air is injected continuously into the mixture through porous diffusers located at the bottom of the tank, usually along one side.

Wastewater is fed continuously into the aerated tank, where the microorganisms metabolize and biologically flocculate the organics. Microorganisms (activated biosolids) are settled from the aerated mixed liquor under quiescent conditions in the final clarifier and are returned to the aeration tank. Left uncontrolled, the number of organisms would eventually become too great; therefore, some must periodically be removed (waste). A portion of the concentrated solids from the bottom of the settling tank must be removed from the process (waste activated sludge, or WAS). Clear supernatant from the final settling tank is the plant effluent.

16.7.1 Activated Biosolids Process Control Calculations

As with other wastewater treatment unit processes, process control calculations are important tools used to control and optimize process operations. In this section, we review many of the most frequently used activated biosolids process calculations.

16.7.2 Moving Averages

When performing process control calculations, using a 7-day moving average is recommended. The moving average is a mathematical method to level the impact of any single test result. Determine the moving average by adding all of the test results collected during the preceding 7 days and dividing by the number of tests.

$$\text{Moving Average} = \frac{\text{Test 1 + Test 2 + Test 3 + ... Test 6 + Test 7}}{\text{\# of Tests Performed during the Seven Days}} \tag{16.30}$$

Example 16.39

Problem:

Calculate the 7-day moving average for days 7, 8, and 9:

Day	MLSS	Day	MLSS
1	3340	6	2780
2	2480	7	2476
3	2398	8	2756
4	2480	9	2655
5	2558	10	2396

Solution:

1.

$$\text{Moving Average, Day 7} =$$
$$\frac{3340 + 2480 + 2398 + 2480 + 2558 + 2780 + 2476}{7} = 2645$$

2.

$$\text{Moving Average, Day 8} =$$
$$\frac{2480 + 2398 + 2480 + 2558 + 2780 + 2476 + 2756}{7} = 2561$$

3.

$$\text{Moving Average, Day 9} =$$
$$\frac{2398 + 2480 + 2558 + 2780 + 2476 + 2756 + 2655}{7} = 2586$$

16.7.3 BOD or COD Loading

When calculating BOD, COD, or SS loading on an aeration process (or any other treatment process), loading on the process is usually calculated as pounds per day. The following equation is used:

$$\text{BOD, COD, or SS Loading, lb/day} = (\text{mg/L})\,(\text{MGD})\,(8.34\ \text{lb/gal}) \qquad (16.31)$$

Example 16.40

Problem:

The BOD concentration of the wastewater entering an aerator is 210 mg/L. If the flow to the aerator is 1,550,000 gpd, what is the pounds-per-day BOD loading?

Solution:

$$\text{BOD lb/day} = (\text{mg/L})\,(\text{MGD})\,(8.34\ \text{lb/gal})$$

$$= (210\ \text{mg/L})\,(1.55\ \text{MGD})\,(8.34\ \text{lb/gal})$$

$$= 2715\ \text{lb/day}$$

Example 16.41

Problem:
The flow to an aeration tank is 2750 gpm. If the BOD concentration of the wastewater is 140 mg/L, how many pounds of BOD are applied to the aeration tank daily?

Solution:
First, convert the gallons-per-minute flow to gallons-per-day flow:

$$(2750 \text{ gpm}) (1440 \text{ min/day}) = 3,960,000 \text{ gpd}$$

Then calculate pounds per day of BOD:

$$\text{BOD, lb/day} = (\text{BOD, mg/L}) (\text{Flow, MGD}) (8.34 \text{ lb/gal})$$

$$= (140, \text{mg/L}) (3.96 \text{ MGD}) (8.34 \text{ lb/day})$$

$$= 4624 \text{ lb/day}$$

16.7.4 Solids Inventory

In the activated biosolids process, controlling the amount of solids under aeration is important. The suspended solids in an aeration tank are called mixed liquor suspended solids (MLSS). To calculate the pounds of solids in the aeration tank, we need to know the milligrams per liter of MLSS concentration and the aeration tank volume. Then pounds of MLSS can be calculated as follows:

$$\text{lb MLSS} = (\text{MLSS, mg/L}) (\text{MG}) (8.34) \tag{16.32}$$

Example 16.42

Problem:
If the mixed liquor suspended solids concentration is 1200 mg/L and the aeration tank has a volume of 550,000 gal, how many pounds of suspended solids are in the aeration tank?

Solution:

$$\text{Lb} = (\text{mg/L}) (\text{MG Volume}) (8.34 \text{ lb/gal})$$

$$= (1200 \text{ mg/L}) (0.550 \text{ MG}) (8.34 \text{ lb/gal})$$

$$= 5504 \text{ lb MLSS}$$

16.7.5 Food-to-Microorganism Ratio (F/M Ratio)

The food-to-microorganism ratio (F/M ratio) is a process control method/calculation based upon maintaining a specified balance between available food materials (BOD or COD) in the aeration tank influent and the aeration tank mixed liquor volatile suspended solids (MLVSS) concentration.

The chemical oxygen demand (COD) test is sometimes used because the results are available in a relatively short period of time. To calculate the F/M ratio, the following information is required:

- Aeration tank influent flow rate, million gallons per day
- Aeration tank influent BOD or COD, milligrams per liter
- Aeration tank MLVSS, milligrams per liter
- Aeration tank volume, million gallons

$$\text{F/M Ratio} = \frac{\text{Primary Eff. COD/BOD mg/L} \times \text{Flow MGD} \times 8.34 \text{ lb/mg/L/MG}}{\text{MLVSS mg/L} \times \text{Aerator Vol., MG} \times 8.34 \text{ lb/mg/L/MG}} \quad (16.33)$$

Typical F/M ratio for an activated biosolids process is shown in the following table:

Process	Pounds BOD / Pounds MLVSS	Pounds COD / Pounds MLVS
Conventional	0.2–0.4	0.5–1.0
Contact stabilization	0.2–0.6	0.5–1.0
Extended aeration	0.05–0.15	0.2–0.5
Pure oxygen	0.25–1.0	0.5–2.0

Example 16.43

Problem:

The aeration tank influent BOD is 145 mg/L and the aeration tank influent flow rate is 1.6 MGD. What is the F/M ratio if the MLVSS is 2300 mg/L and the aeration tank volume is 1.8 MG?

Solution:

$$\text{F/M ratio} = \frac{145 \text{ mg/L} \times 1.6 \text{ MGD} \times 8.34 \text{ lb/mg/L/MG}}{2300 \text{ mg/L} \times 1.8 \text{ MG} \times 8.34 \text{ lb/mg/L/M}}$$

$$= 0.06 \text{ BOD/lb MLVSS}$$

Key point: If the MLVSS concentration is not available, it can be calculated if the percent of volatile matter (% VM) of the mixed liquor suspended solids (MLSS) is known:

$$\text{MLVSS} = \text{MLSS} \times \% \text{ (decimal) Volatile Matter (VM)} \quad (16.34)$$

Key point: The "F" value in the F/M ratio for computing loading to an activated biosolids process can be BOD or COD. Remember that the reason for biosolids production in the activated biosolids process is to convert BOD to bacteria. One advantage of using COD over BOD for analysis of organic load is that COD is more accurate.

Example 16.44

Problem:

The aeration tank contains 2885 mg/L of MLSS. Lab tests indicate the MLSS is 66% volatile matter. What is the MLVSS concentration in the aeration tank?

Solution:

$$\text{MLVSS, mg/L} = 2885 \text{ mg/L} \times 0.66 = 1904 \text{ mg/L}$$

Required MLVSS Quantity (Pounds)

The pounds of MLVSS required in the aeration tank to achieve the optimum F/M ratio can be determined from the average influent food (BOD or COD) and the desired F/M ratio:

$$\text{MLVSS, lb} = \frac{\text{Primary Effluent BOD or COD} \times \text{Flow, MGD} \times 8.34}{\text{Desired F/M Ratio}} \qquad (16.35)$$

The required pounds of MLVSS determined by this calculation can then be converted to a concentration value by:

$$\text{MLVSS, mg/L} = \frac{\text{Desired MLVSS, lb}}{[\text{Aeration Volume, MG} \times 8.34]} \qquad (16.36)$$

Example 16.45

Problem:

The aeration tank influent flow is 4.0 MGD, and the influent COD is 145 mg/L. The aeration tank volume is 0.65 MG. The desired F/M ratio is 0.3 lb COD/lb MLVSS.

1. How many pounds of MLVSS must be maintained in the aeration tank to achieve the desired F/M ratio?
2. What is the required concentration of MLVSS in the aeration tank?

Solution:

$$\text{MLVSS} = \frac{145 \text{ mg/L} \times 4.0 \text{ MGD} \times 8.34 \text{ lb/gal}}{0.3 \text{ lb COD/lb MLVSS}} = 16{,}124 \text{ lb MLVSS}$$

$$\text{MLVSS, mg/L} = \frac{16{,}124 \text{ lb MLVSS}}{[0.65 \text{ MG} \times 8.34]} = 2974 \text{ mg/L MLVSS}$$

Calculating Waste Rates Using F/M Ratio

Maintaining the desired F/M ratio is accomplished by controlling the MLVSS level in the aeration tank. This may be accomplished by adjustment of return rates; however, the most practical method is by proper control of the waste rate.

$$\text{Waste Vol. Solids, lb/day} = \text{Actual MLVSS, lb} - \text{Desired MLVSS, lb} \qquad (16.37)$$

If the desired MLVSS is greater than the actual MLVSS, wasting is stopped until the desired level is achieved.

Practical considerations demand that the required waste quantity be converted to a required volume to waste ratio per day. This is accomplished by converting the waste pounds to flow rate in million gallons per day or gallons per minute.

$$\text{Waste, MGD} = \frac{\text{Waste Volatile, lb/day}}{[\text{Waste Volatile Conc., mg/L} \times 8.34]} \qquad (16.38)$$

$$\text{Waste, MGD} = \frac{\text{Waste, MGD} \times 1{,}000{,}000 \text{ gpd/MGD}}{1440 \text{ min/day}} \qquad (16.39)$$

Key point: When F/M ratio is used for process control, the volatile content of the waste-activated sludge should be determined.

Example 16.46

Problem:

Given the following information, determine the required waste rate in gallons per minute to maintain an F/M ratio of 0.17-lb COD/lb MLVSS.

Primary effluent COD	140 mg/L
Primary effluent flow	2.2 MGD
MLVSS, mg/L	3549 mg/L
Aeration tank volume	0.75 MG
Waste volatile concentrations	4440 mg/L (volatile solids)

Solution:

$$\text{Actual MLVSS, lb} = 3{,}549 \text{ mg/L} \times 0.75 \text{ MG} \times 8.34 = 22{,}199 \text{ lb}$$

$$\text{Required MLVSS, lb} = \frac{140 \text{ mg/L} \times 2.2 \text{ MGD} \times 8.34}{0.17 \text{ lb COD/lb MLVSS}} = 15{,}110 \text{ lb MLVSS}$$

$$\text{Waste, lb/day} = 22{,}199 \text{ lb} - 15{,}110 \text{ lb} = 7089 \text{ lb}$$

$$\text{Waste, MGD} = \frac{7089 \text{ lb/day}}{4440 \text{ mg/L} \times 8.34} = 0.19 \text{ MGD}$$

$$\text{Waste, gpm} = \frac{0.19 \text{ MGD} \times 1{,}000{,}000 \text{ gpd/MGD}}{1440 \text{ min/day}} = 132 \text{ gpm}$$

16.7.6 Gould Biosolids Age

Biosolids age refers to the average number of days a particle of suspended solids remains under aeration; it is a part of the calculation used to maintain the proper amount of activated biosolids in the aeration tank. This calculation is sometimes referred to as Gould biosolids age, so that it is not confused with similar calculations such as solids retention time, or mean cell residence time.

When considering sludge age, in effect we are asking, "How many days of suspended solids are in the aeration tank?" For example, if 3000 lb SS enter the aeration tank daily and the tank contains 12,000 lb of suspended solids, when 4 days of solids are in the aeration tank, the tank has a sludge age of 4 days.

$$\text{Sludge Age, days} = \frac{\text{SS in Tank, lb}}{\text{SS Added, lb/day}} \qquad (16.40)$$

Example 16.47

Problem:

A total of 2740 lb/day of suspended solids enters an aeration tank in the primary effluent flow. If the aeration tank has a total of 13,800 lb of mixed liquor suspended solids, what is the biosolids age in the aeration tank?

Solution:

$$\text{Sludge Age, day} = \frac{\text{MLSS, lb}}{\text{SS Added, lb/day}}$$

$$= \frac{13,800 \text{ lb}}{2740 \text{ lb/day}}$$

$$= 5.0 \text{ days}$$

16.7.7 Mean Cell Residence Time (MCRT)

Mean cell residence time (MCRT), sometimes called *sludge retention time*, is another process control calculation used for activated biosolids systems. MCRT represents the average length of time an activated biosolids particle remains in the activated biosolids system. It can also be defined as the length of time required at the current removal rate to remove all the solids in the system.

Mean Cell Residence Time, day =

$$\frac{[\text{MLSS mg/L} \times (\text{Aeration Vol.} + \text{Clarifier Vol.}) \times 8.34 \text{ lb/mg/L/MG}]}{[\text{WAS, mg/L} \times (\text{WAS flow} \times 8.34) + (\text{TSS out} \times \text{flow out} \times 8.34)]} \qquad (16.41)$$

Key point: MCRT can be calculated using only the aeration tank solids inventory. When comparing plant operational levels to reference materials, it is necessary to determine the calculation that reference manual uses to obtain its example values. Other methods are available to determine the clarifier solids concentrations. However, the simplest method assumes that the average suspended solids concentration is equal to the aeration tank's solids concentration.

Example 16.48

Problem:
Given the following data, what is the MCRT?

Aerator volume	1,000,000 gal
Final clarifier	600,000 gal
Flow	5.0 MGD
Waste rate	0.085 MGD
MLSS mg/L	2500 mg/L
Waste mg/L	6400 mg/L
Effluent TSS	14 mg/L

Solution:

$$\text{MRCT} = \frac{[2500 \text{ mg/L} \times (1.0 \text{ MG} + 0.60 \text{ MG}) \times 8.34]}{[6400 \text{ mg/L} \times (0.085 \text{ MGD} \times 8.34) + (14 \text{ mg/L} \times 5.0 \text{ MGD} \times 8.34)]} = 6.5 \text{ days}$$

Waste Quantities/Requirements

MCRT for process control requires determination of the optimum range for MCRT values. This is accomplished by comparison of the effluent quality with MCRT values. When the optimum MCRT is established, the quantity of solids to be removed (wasted) is determined by:

Waste, lb/day =

$$\left(\frac{\text{MLSS} \times (\text{Aer., MG} + \text{Clarifier, MG}) \times 8.34}{\text{Desired MCRT}} \right) - [\text{TSS}_{\text{out}} \times \text{Flow} \times 8.34] \qquad (16.42)$$

Example 16.49

$$\frac{3400 \text{ mg/L} \times (1.4 \text{ MG} + 0.50 \text{ MG}) \times 8.34}{8.6 \text{ days}} - [10 \text{ mg/L} \times 5.0 \text{ MGD} \times 8.34]$$

Waste Quality, lb/day = 5848 lb

Waste Rate in Million Gallons per Day

When the quantity of solids to be removed from the system is known, the desired waste rate in million gallons per day can be determined. The unit used to express the rate (million gallons per day; gallons per day; and gallons per minute) is a function of the volume of waste to be removed and the design of the equipment.

$$\text{Waste, MGD} = \frac{\text{Waste lb/day}}{\text{WAS Concentration, mg/L} \times 8.34} \qquad (16.43)$$

$$\text{Waste, gpm} = \frac{\text{Waste MGD} \times 1,000,000 \text{ gpd/MGD}}{1440 \text{ min/day}} \qquad (16.44)$$

Example 16.50

Problem:

Given the following data, determine the required waste rate to maintain an MCRT of 8.8 days:

MLSS, milligrams per liter	2500 mg/L
Aeration volume	1.20 MG
Clarifier volume	0.20 MG
Effluent TSS	11 mg/L
Effluent flow	5.0 MGD
Waste concentration	6000 mg/L

Solution:

$$\text{Waste, lb/day} = \frac{2500 \text{ mg/L} \times (1.20 + 0.20) \times 8.34}{8.8 \text{ days}} - [11 \text{ mg/L} \times 5.0 \text{ MGD} \times 8.34]$$

$$= 3317 \text{ lb/day} - 459 \text{ lb/day}$$

$$= 2858 \text{ lb/day}$$

$$\text{Waste, lb/day} = \frac{2858 \text{ lb/day}}{[6000 \text{ mg/L} \times 8.34]} = 0.057 \text{ MGD}$$

$$\text{Waste, gpm} = \frac{0.057 \text{ MGD} \times 1,000,000 \text{ gpd/MGD}}{1440 \text{ min/day}} = 40 \text{ gpm}$$

16.7.8 Estimating Return Rates from SBV_{60} (SSV_{60})

Many methods are available for estimating the proper return biosolids rate. A simple method described in the *Operation of Wastewater Treatment Plants, Field Study Programs* (1986), developed by the California State University, Sacramento, uses the 60-min percent-settled biosolids (sludge) volume. The percent SBV_{60} test results can provide an approximation of the appropriate return-activated biosolids rate. This calculation assumes that the SBV_{60} results are representative of the actual settling occurring in the clarifier. If this is true, the return rate in percent should be approximately equal to the SBV_{60}. To determine the approximate return rate in million gallons per day, the influent flow rate, current return rate, and SBV_{60} must be known. The results of this calculation can then be adjusted based upon sampling and visual observations to develop the optimum return biosolids rate.

Key point: The percent SBV_{60} must be converted to a decimal percent and total flow rate (wastewater flow and current return rate in million gallons per day must be used).

$$\text{Est. Return Rate, MGD} = \tag{16.45}$$
$$(\text{Influent Flow, MGD} + \text{Current Return Flow, MGD}) \times \%SBV_{60}$$

$$\text{RAS Rate, GPM} = \frac{\text{Return, Biosolids Rate, gpd}}{1440 \text{ min/day}} \tag{16.46}$$

Assume:

- Percent SBV_{60} is representative.
- Return rate in percent equals $\%SBV_{60}$.
- Actual return rate is normally set slightly higher to ensure organisms are returned to the aeration tank as quickly as possible. The rate of return must be adequately controlled to prevent the following:
  - Aeration and settling hydraulic overloads
  - Low MLSS levels in the aerator
  - Organic overloading of aeration
  - Septic return-activated biosolids
  - Solids loss due to excessive biosolids blanket depth

Example 16.51

Problem:

The influent flow rate is 5.0 MGD and the current return-activated sludge flow rate is 1.8 MGD. The SBV_{60} is 37%. Based upon this information, what should be the return biosolids rate in million gallons per day?

Solution:

$$\text{Return, MGD} = (5.0\ \text{MGD} + 1.8\ \text{MGD}) \times 0.37 = 2.5\ \text{MGD}$$

16.7.9 Biosolids (Sludge) Volume Index (BVI)

Biosolids volume index (BVI) is a measure (an indicator) of the settling quality (a quality indicator) of the activated biosolids. As the BVI increases, the biosolids settle more slowly, do not compact as well, and are likely to result in an increase in effluent suspended solids. As the BVI decreases, the biosolids become denser, settling is more rapid, and the biosolids age. BVI is the volume in milliliters occupied by 1 g of activated biosolids. For the settled biosolids volume (milliliters per liter) and the MLSS calculation, milligrams per liter are required. The proper BVI range for any plant must be determined by comparing BVI values with plant effluent quality.

$$\text{Biosolids (Sludge) Volume Index (SBI)} = \frac{\text{SBV, mL/L} \times 1000}{\text{MLSS, mg/L}} \qquad (16.47)$$

Example 16.52

Problem:

The SBV_{30} is 250 mL/L and the MLSS is 2425 mg/L. What is the SBI?

Solution:

$$\text{Biosolids Volume Index (BVI)} = \frac{350\ \text{mL/L} \times 1000}{2425\ \text{mg/L}} = 144$$

BI equals 144. What does this mean? What it means is that the system is operating normally with good settling and low effluent turbidity. How do we know this? We know this because we compare the 144 result with the following parameters to obtain the expected condition (the result).

BVI	Expected condition (indicates)
Less than 100	Old biosolids — possible pin floc Effluent turbidity increasing
100–250	Normal operation — good settling Low effluent turbidity
Greater than 250	Bulking biosolids — poor settling High effluent turbidity

16.7.10 Mass Balance: Settling Tank Suspended Solids

Solids are produced whenever biological processes are used to remove organic matter from wastewater. Mass balance for anaerobic biological process must take into account the solids removed by physical settling processes and the solids produced by biological conversion of soluble organic matter to insoluble suspended matter organisms. Research has shown that the amount of solids produced per pound of BOD removed can be predicted, based upon the type of process used. Although the exact amount of solids produced can vary from plant to plant, research has developed a series of K-factors used to estimate the solids production for plants using a particular treatment process. These average factors provide a simple method to evaluate the effectiveness of a facility's process control program. The mass balance also provides an excellent mechanism to evaluate the validity of process control and effluent monitoring data generated.

16.7.11 Mass Balance Calculation

$$\text{BOD in, lb} = \text{BOD, mg/L} \times \text{Flow, MGD} \times 8.34 \qquad (16.48)$$

$$\text{BOD out, lb} = \text{BOD, mg/L} \times \text{Flow, MGD} \times 8.34$$

$$\text{Solids Produced, lb/day} = [\text{BOD in, lb} - \text{BOD out, lb}] \times K$$

$$\text{TSS out, lb/day} = \text{TSS out, mg/L} \times \text{Flow, MGD} \times 8.34$$

$$\text{Waste, lb/day} = \text{Waste, mg/L} \times \text{Flow, MGD} \times 8.34$$

$$\text{Solids Removed, lb/day} = \text{TSS out, lb/day} + \text{Waste, lb/day}$$

$$\% \text{ Mass Balance} = \frac{(\text{Solids Produced} - \text{Solids Removed}) \times 100}{\text{Solids Produced}}$$

16.7.12 Biosolids Waste Based Upon Mass Balance

$$\text{Waste Rate, MGD} = \frac{\text{Solids Produced, lb/day}}{(\text{Waste Concentration} \times 8.34)} \qquad (16.49)$$

Example 16.53

Problem:

Given the following data, determine the mass balance of the biological process and the appropriate waste rate to maintain current operating conditions.

Process		Extended aeration (no primary)
Influent	Flow	1.1 MGD
	BOD	220 mg/L
	TSS	240 mg/L
Effluent	Flow	1.5 MGD
	BOD	18 mg/L
	TSS	22 mg/L
Waste	Flow	24,000 gpd
	TSS	8710 mg/L

Solution:

$$\text{BOD in} = 220 \text{ mg/L} \times 1.1 \text{ MGD} \times 8.34 = 2018 \text{ lb/day}$$

$$\text{BOD out} = 18 \text{ mg/L} \times 1.1 \text{ MGD} \times 8.34 = 165 \text{ lb/day}$$

$$\text{BOD Removed} = 2018 \text{ lb/day} - 165 \text{ lb/day} = 1853 \text{ lb/day}$$

$$\text{Solids Produced} = 1853 \text{ lb/day} \times 0.65 \text{ lb/lb BOD} = 1204 \text{ lb solids/day}$$

$$\text{Solids Out, lb/day} = 22 \text{ mg/L} \times 1.1 \text{ MGD} \times 8.34 = 202 \text{ lb/day}$$

$$\text{Sludge Out, lb/day} = 8710 \text{ mg/L} \times 0.024 \text{ MGD} \times 8.34 = 1743 \text{ lb/day}$$

$$\text{Solids Removed, lb/day} = (202 \text{ lb/day} + 1743 \text{ lb/day}) = 1945 \text{ lb/day}$$

$$\text{Mass Balance} = \frac{(1204 \text{ lb Solids/day} - 1945 \text{ lb/day}) \times 100}{1204 \text{ lb/day}} = 62\%$$

The mass balance indicates:

- The sampling points, collection methods, and/or laboratory testing procedures are producing nonrepresentative results.
- The process is removing significantly more solids than is required. Additional testing should be performed to isolate the specific cause of the imbalance.

To assist in the evaluation, the waste rate based upon the mass balance information can be calculated.

$$\text{Waste, GPD} = \frac{\text{Solids Produced, lb/day}}{(\text{Waste TSS, mg/L} \times 8.34)} \tag{16.50}$$

$$\text{Waste, GPD} = \frac{1204 \text{ lb/day} \times 1,000,000}{8710 \text{ mg/L} \times 8.34} = 16,575 \text{ gpd}$$

16.7.13 Aeration Tank Design Parameters

The two design parameters of aeration tanks are food-to-microorganism (F/M) ratio and aeration period (similar to detention time). F/M ratio (BOD loading) is expressed as pounds of BOD per day per pound of MLSS:

$$\frac{F}{M} = \frac{133,690 \, (\text{BOD}) \, Q_o}{(\text{MLSS}) \, \cancel{V}} \tag{16.51}$$

where
BOD   = settled BOD from primary tank, milligrams per liter
Q_o = average daily wastewater flow, million gallons per day
MLSS   = mixed liquor suspended solids, milligrams per liter
$\cancel{V}$ = volume of tank, square feet
133,690 = conversion of units

Example 16.54

Problem:
    Using the following given data, design a conventional aeration tank.

    MGD = 1 million gallons per day
    BOD from primary clarifier = 110 mg/L
    MLSS = 2000 mg/L
    Design F/M = 0.5/day
    Design aeration period, t = 6 h

Solution:

$$0.50 = \frac{133,690 \, (110) \, (1)}{(2000) \, \cancel{V}}$$

$$\cancel{V} = 14,706 \text{ ft}^3$$

Aeration tank volume, $\cancel{V} = Qt$

$$= (1 \times 106 \text{ gal/day}) \, (6 \text{ h}) \, [1/7.48 \text{ ft}^3/\text{gal}][1 \text{ day}/24 \text{ h}]$$

$$= 33,422 \text{ ft}^3$$

Assume a depth of 10 ft and a length of twice the width:

$$A = \frac{33,422}{10} = 3342 \text{ ft}^2$$

$$(2w)(w) = 3342$$

$$w = 41 \text{ ft}$$

$$l = 82 \text{ ft}$$

16.7.14 Lawrence and McCarty Design Model

Over the years, numerous design criteria using empirical and rational parameters based on biological kinetic equations have been developed for suspended-growth systems. In practice, the basic Lawrence and McCarty (1970) model is widely used in the industry. We list the Lawrence and McCarty design equations used for sizing suspended-growth systems next.

16.7.14.1 Complete Mix with Recycle

For a complete mix system, the mean hydraulic retention time (HRT) θ for the aeration basin is:

$$\theta = V/Q \tag{16.52}$$

where
θ = hydraulic retention time, days
V = volume of aeration tank, cubic meters
Q = influent wastewater flow, cubic meters per day

The mean cell residence time θ_c (or biosolids age or BRT; i.e., for sludge, SRT) is expressed as:

$$\theta_c = \frac{X}{(\Delta X/\Delta t)} \tag{16.53}$$

$$\theta_c = \frac{VX}{(Q_{wa}X + Q_c X_c)} = \frac{\text{total mass SS in reactor}}{\text{SS wasting rate}} \tag{16.54}$$

where
θ_c = mean cell residence time based on solids in the tank, days
X = concentration of MLVSS maintained in the tank, milligrams per liter
$\Delta X/\Delta t$ = growth of biological sludge over time period Δt, milligrams per (liter days)
Q_{wa} = flow of waste sludge removed from the aeration tank, cubic meters per day
Q_c = flow of treated effluent, cubic meters per day
X_c = microorganism concentration (VSS in effluent, milligrams per liter)

The mean cell residence time for system-drawn sludge from the return line would be:

$$Q_c = \frac{VX}{(Q_{wr}X_r + Q_cX_c)} \tag{16.55}$$

where
Q_{wr} = flow of waste sludge from return sludge line, cubic meters per day
X_r = microorganism concentration in return sludge line, milligrams per liter

Microorganism mass balance. The mass balance for the microorganisms in the entire activated biosolids system is expressed as (Metcalf and Eddy, 1991):

$$V\frac{dX}{dt} = QX_o + V(r_g') - (Q_{wa}X + Q_cX_c) \tag{16.56}$$

where
V = volume of aeration tank, cubic meters
dX/dt = rate of change of microorganisms concentration (VSS), milligrams per
        (liter · cubic meter · day)
Q = flow, cubic meters per day
X_o = microorganisms concentration (VSS) in influent, milligrams per liter
X = microorganisms concentration in tank, milligrams per liter
r_g' = net rate of microorganism growth (VSS), milligrams per (liter · day)

The net rate of bacterial growth is expressed as

$$r_g' = Yr_{su} - K_dX \tag{16.57}$$

where
Y = maximum yield coefficient over finite period of log growth, milligrams per milligram
r_{su} = substrate utilization rate, milligrams per cubic meter
k_d = endogenous decay coefficient, per day

Assuming the cell concentration in the influent is zero and steady-state conditions, Equation 16.58 can be used.

$$\frac{Q_{wa}X + Q_cX_e}{VX} = -Y^r\frac{su}{X} - K_d \tag{16.58}$$

The net specific growth rate can be determined using:

$$\frac{1}{\theta_c} = -Y\frac{r_{su}}{X} - k_d \tag{16.59}$$

The term r_{su} can be computed from:

$$r_{su} = \frac{Q}{V}(S_o - S) = \frac{S_o - S}{\theta} \tag{16.60}$$

where
$S_o - S$ = mass concentration of substrate utilized, milligrams per liter
S_o = substrate concentration in influent, milligrams per liter
S = substrate concentration in effluent, milligrams per liter
θ = hydraulic retention time

16.7.15 Effluent Microorganism and Substrate Concentrations

The mass concentration of microorganisms X in the aeration basin can be computed from the following equation:

$$X = \frac{\theta_c Y (S_o - S)}{\theta (1 + k_d \theta_c)} = \frac{\mu_m (S_o - S)}{k (1 + k_d \theta_c)} \tag{16.61}$$

Aeration basin volume can be computed from the following equation:

$$V = \frac{\theta_c Q Y (S_o - S)}{X (1 + k_d \theta_c)} \tag{16.62}$$

The substrate concentration in effluent S can be determined by the following equation:

$$S = \frac{K_s (1 + \theta_c k_d)}{\theta_c (Yk - k_d) - 1} \tag{16.63}$$

where
S = effluent substrate (soluble BOD) concentration, milligrams per liter
K_s = half-velocity constant, substrate concentration at one half the maximum growth rate, milligrams per liter
k = maximum rate of substrate utilization per unit mass of microorganism per day

Other parameters have been mentioned in previous equations.
    Observed yield in the system can be determined by using the following equation:

$$Y_{obs} = \frac{Y}{1 + Q_{ct}} \tag{16.64}$$

where
Y_{obs} = observed yield in the system with recycle, milligrams per milligram
Q_{ct} = mean of all residence times based on solids in the aeration tank and in the secondary clarifier, days

Other terms have been defined previously.

16.7.15.1 Process Design and Control Relationships

The specific substrate utilization rate (closely related to the F/M ratio widely used in practice) can be computed by:

$$U = \frac{r_{su}}{X} \tag{16.65}$$

$$U = \frac{Q(S_o - S)}{VX} = \frac{S_o - S}{\theta X} \tag{16.66}$$

The net specific growth rate can be computed by:

$$\frac{1}{\theta_c} = YU - k_d \tag{16.67}$$

The flow rate of waste sludge from the sludge return line will be approximately:

$$Q_{wt} = \frac{VX}{\theta_c X_r} \tag{16.68}$$

where X_r = the concentration (in milligrams per liter) of sludge in the sludge return line.

16.7.15.2 Sludge Production

The amount of sludge generated per day can be calculated by:

$$P_x = Y_{obs}Q(S_o - S)(8.34) \tag{16.69}$$

where
P_x = net waste-activated sludge (VSS), kilograms per day or pounds per day
Y_{obs} = observed yield, gallons per gallon or pounds per pound
Q = influent wastewater flow, cubic meters per day or million gallons per day
S_o = influent soluble BOD concentration, milligrams per liter
S = effluent soluble BOD concentration, milligrams per liter
8.34 = conversion factor, (pounds per million gallons):(milligrams per liter)

16.7.15.3 Oxygen Requirements

The theoretical oxygen requirement to remove the carbonaceous organic matter in wastewater for an activated-biosolids process is expressed by Metcalf and Eddy (1991) in SI units and British system:

Mass of O_2/day = total mass of BOD_u used − 1.42 (mass of organisms wasted, p_x)

$$\text{kg } O_2/\text{day} = \frac{Q(S_o - S)}{(1000 \text{ g/kg}) f} - 1.42 P_x \tag{16.70}$$

$$\text{kg } O_2/\text{day} = \frac{Q(S_o - S)}{(1000 \text{ g/kg})} \left(\frac{1}{f} - 1.42 Y_{obs} \right) \tag{16.71}$$

$$\text{lb } O_2/\text{day} = Q(S_o - S) \times 8.34 \left(\frac{1}{f} - 1.42 Y_{obs} \right) \tag{16.72}$$

where

BOD_u = ultimate BOD

P_x = net waste activated sludge (VSS), kilograms per day or pounds per day

Q = influent flow, cubic meters per day or million gallons per day

S_o = influent soluble BOD concentration, milligrams per liter

S = effluent soluble BOD concentration, milligrams per liter

f = conversion factor for converting BOD to BOD_u

Y_{obs} = observed yield, gallons per gallon or pounds per pound

8.34   = conversion factor, pounds per million gallons: (mg/L)

16.8   OXIDATION DITCH DETENTION TIME

Oxidation ditch systems may be used when the treatment of wastewater is amendable to aerobic biological treatment and the plant design capacities generally do not exceed 1.0 MGD. The oxidation ditch is a form of aeration basin in which the wastewater is mixed with returned biosolids; it is essentially a modification of a completely mixed activated biosolids system used to treat wastewater from small communities. An oxidation ditch system can be classified as an extended aeration process and is considered a low loading rate system. This type of treatment facility can remove 90% or more of influent BOD. Oxygen requirements generally depend on the maximum diurnal organic loading, degree of treatment, and suspended solids concentration to be maintained in the aerated channel MLSS. Detention time is the length of time for required wastewater at a given flow rate to pass through a tank. This time is not normally calculated for aeration basins, but it is calculated for oxidation ditches.

> *Key point*: When calculating detention time, the time and volume units used in the equation must be consistent.

$$\text{Detention Time, h} = \frac{\text{Vol. of Oxidation Ditch, gal}}{\text{Flow Rate, gph}} \qquad (16.73)$$

Example 16.55

Problem:

An oxidation ditch has a volume of 160,000 gal. If the flow to the oxidation ditch is 185,000 gpd, what is the detention time in hours?

Solution:

Because detention time is desired in hours, the flow must be expressed as gallons per hour:

$$\frac{185,000 \text{ gpd}}{24 \text{ h/day}} = 7708 \text{ gph}$$

Now calculate detention time:

$$\text{Detention Time, h} = \frac{\text{Vol. of Oxidation Ditch, gal}}{\text{Flow Rate, gph}}$$

$$\frac{160,000 \text{ gallons}}{7708 \text{ gph}}$$

$$= 20.8 \text{ h}$$

16.9 TREATMENT PONDS

The primary goals of wastewater treatment ponds focus on simplicity and flexibility of operation, protection of the water environment, and protection of public health. Ponds are relatively easy to build and manage; they accommodate large fluctuations in flow and can also provide treatment that approaches conventional systems (producing a highly purified effluent) at much lower cost. The cost (the economics) drives many managers to decide on the pond option of treatment.

The actual degree of treatment provided in a pond depends on the types and numbers of ponds used. Ponds can be used as the sole type of treatment, or they can be used in conjunction with other forms of wastewater treatment — that is, other treatment processes followed by a pond or a pond followed by other treatment processes. They can be classified according to their location in the system, by the type of wastes they receive, and by the main biological process occurring in the pond. First, we look at the types of ponds according to their location and the type of wastes they receive: raw sewage stabilization ponds, oxidation ponds, and polishing ponds.

16.9.1 Treatment Pond Parameters

Before we discuss the process control calculations mentioned earlier, we first describe the calculations for determining the area, volume, and flow rate parameters crucial in making treatment pond calculations.

Determining pond area in inches:

$$\text{Area, acres} = \frac{\text{Area, ft}^2}{43,560 \text{ ft}^2/\text{acre}} \qquad (16.74)$$

Determining pond volume in acre-feet:

$$\text{Volume, acre-feet} = \frac{\text{Volume, ft}^3}{43,560 \text{ ft}^2/\text{acre-foot}} \qquad (16.75)$$

Determining flow rate in acre-feet per day:

$$\text{Flow, acre-feet/day} = \text{flow, MGD} \times 3069 \text{ acre-feet/MG} \qquad (16.76)$$

Key point: "Acre-feet" (acre-ft) is a unit that can cause confusion, especially for those not familiar with pond or lagoon operations. One acre-foot is the volume of a box with a 1-acre top and 1 ft of depth; however, the top does not need to be an even number of acres in size to use the acre-feet unit of measurement.

Determining flow rate in acre-inches per day:

$$\text{Flow, acre-inches/day} = \text{flow, MGD} \times 36.8 \text{ acre-inches/MG} \qquad (16.77)$$

16.9.2 Treatment Pond Process Control Calculations

Although there are no recommended process control calculations for treatment ponds, several calculations may be helpful in evaluating process performance or identifying causes of poor performance. These include hydraulic detention time; BOD loading; organic loading rate; BOD removal efficiency; population loading; and hydraulic loading rate. In this section, we provide a few calculations that might be helpful in evaluating pond performance and identifying causes of poor performance, along with other helpful calculations.

16.9.2.1 Hydraulic Detention Time, Days

$$\text{Hydraulic detention time, days} = \frac{\text{Pond volume, acre-ft}}{\text{Influent flow, acre-ft/day}} \qquad (16.78)$$

Key point: Normally, hydraulic detention time ranges from 30 to 120 days for stabilization ponds.

Example 16.56

Problem:

A stabilization pond has a volume of 54.5 acre-ft. What is the detention time in days when the flow is 0.35 MGD?

Solution:

$$\text{Flow, acre-ft/day} = 0.35 \text{ MGD} \times 3.069 \text{ acre-ft/MG}$$

$$= 1.07 \text{ acre-ft/day}$$

$$\text{DT day} = \frac{54.5 \text{ acre/ft}}{1.07 \text{ acre-ft/day}} = 51$$

16.9.2.2 BOD Loading

When calculating BOD loading on a wastewater treatment pond, use the following equation:

$$\text{Lb/day} = (\text{BOD, mg/L}) (\text{flow, MGD}) (8.34 \text{lb/gal}) \qquad (16.79)$$

Example 16.57

Problem:

Calculate the BOD loading (pounds per day) on a pond if the influent flow is 0.3 MGD with a BOD of 200 mg/L.

Solution:

$$\text{Lb/day} = (\text{BOD, mg/L}) (\text{flow, MGD}) (8.34 \text{lb/gal})$$

$$= (200 \text{ mg/L}) (0.3 \text{ MGD}) (8.34 \text{ lb/gal})$$

$$= 500 \text{ lb/day BOD}$$

16.9.2.3 Organic Loading Rate

Organic loading can be expressed as pounds of BOD per acre per day (most common), pounds of BOD per acre-foot per day, or number of people per acre per day.

$$\text{Organic Loading, lb BOD/acre/day} = \frac{\text{BOD, mg/L Influ. Flow, MGD} \times 8.34}{\text{Pond area, acre}} \quad (16.80)$$

Key point: Normal range is 10 to 50 lb BOD per day per acre.

Example 16.58

Problem:
   A wastewater treatment pond has an average width of 370 ft and an average length of 730 ft. The influent flow rate to the pond is 0.10 MGD with a BOD concentration of 165 mg/L. What is the organic loading rate to the pound in pounds per day per acre (lb/day/acre)?

Solution:

$$730 \text{ ft} \times 370 \text{ ft} \times \frac{1 \text{ acre}}{43,560 \text{ ft}^2} = 6.2 \text{ acre}$$

$$0.10 \text{ MGD} \times 165 \text{ mg/L} \times 8.34 \text{ lb/gal} = 138 \text{ lb/day}$$

$$\frac{138 \text{ lb/day}}{6.2 \text{ acre}} = 22.3 \text{ lb/day/acre}$$

16.9.2.4 BOD Removal Efficiency

The efficiency of any treatment process is its effectiveness in removing various constituents from the water or wastewater. BOD removal efficiency is therefore a measure of the effectiveness of the wastewater treatment pond in removing BOD from the wastewater.

$$\% \text{ BOD Removed} = \frac{\text{BOD Removed, mg/L}}{\text{BOD Total, mg/L}} \times 100$$

Example 16.59

Problem:
   The BOD entering a waste treatment pond is 194 mg/L. If the BOD in the pond effluent is 45mg/L, what is BOD removal efficiency of the pond?

Solution:

$$\% \text{ BOD Removed} = \frac{\text{BOD Removed, mg/L}}{\text{BOD Total, mg/L}} \times 100$$

$$= \frac{149 \text{ mg/L}}{194 \text{ mg/L}} \times 100 = 77\%$$

16.9.2.5 Population Loading

$$\text{Pop. loading, people/acre/day} = \frac{\text{BOD, mg/L Infl.flow, MGD} \times 8.34}{\text{Pond area, acre}} \qquad (16.81)$$

16.9.2.6 Hydraulic Loading, Inches/Day (Overflow Rate)

$$\text{Hydraulic Loading, in./day} = \frac{\text{Influent flow, acre-in./day}}{\text{Pond area, acre}} \qquad (16.82)$$

16.9.3 Aerated Ponds

According to Metcalf and Eddy (1991), depending on the hydraulic retention time, the effluent from an aerated pond will contain from one third to one half the concentration of the influent BOD in the form of cell tissue. These solids must be removed by settling before the effluent is discharged. The mathematical relationship for BOD removal in a complete-mix-activated pond is derived from the following equation:

$$QS_o - QS - kSV = 0 \qquad (16.83)$$

Rearranged:

$$\frac{S}{S_o} = \frac{1}{1 + k(V/Q)} = \frac{\text{effluent BOD}}{\text{influent BOD}} \qquad (16.84)$$

$$= \frac{1}{1 + k\theta} \qquad (16.85)$$

where
S = effluent BOD concentration, milligrams per liter
S_o = influent BOD concentration, milligrams per liter
k = overall first-order BOD removal rate, per day = 0.25 to 1.0, based on e
Q = wastewater flow, cubic meters per day or million gallons per day
θ = total hydraulic retention time, days

The resulting temperature in the aerated pond from the influent wastewater temperature, air temperature, surface area, and flow can be computed using the following equation (Mancini and Barnhart, 1968):

$$T_1 - T_w = \frac{(T_w - T_a)fA}{Q} \qquad (16.86)$$

where
T_1 = influent wastewater temperature, degrees Celsius or Fahrenheit
T_w = lagoon water temperature, degrees Celsius or Fahrenheit
T_a = ambient air temperature, degrees Celsius or Fahrenheit

f = proportionality factor = 12×10^{-6} (British system) or 0.5 (for SI units)
A = surface area of lagoon, square meters or square feet
Q = wastewater flow, cubic meters per day or million gallons per day

Using Equation 16.86 rearranged, the pond water temperature is:

$$T_w = \frac{AfT_a + QT_1}{Af + Q} \tag{16.87}$$

16.10 CHEMICAL DOSAGE CALCULATIONS

Note: In Chapter 15 we discussed calculations used in the chlorination processes for treating potable water. Crossover of similar information occurs in this chapter.

16.10.1 Chemical Dosing

Chemicals are used extensively in wastewater treatment (and water treatment) operations. Plant operators add chemicals to various unit processes for slime-growth control; corrosion control; odor control; grease removal; BOD reduction; pH control; biosolids-bulking control; ammonia oxidation; bacterial reduction; and for other reasons.

To apply any chemical dose correctly, making certain dosage calculations is essential. Some of the most frequently used calculations in wastewater/water mathematics are calculations to determine dosage or loading. The general types of milligrams per liter to pounds per day or pound calculations are for chemical dosage; BOD; COD; SS loading/removal; pounds of solids under aeration; and WAS pumping rate. These calculations are usually made using Equation 16.88 or Equation 16.89:

$$\text{(Chemical, mg/L) (MGD flow) (8.34 lb/gal)} = \text{lb/day} \tag{16.88}$$

$$\text{(Chemical, mg/L) (MG volume) (8.34 lb/gal)} = \text{lb} \tag{16.89}$$

Key point: If milligrams per liter concentration represents a concentration in a flow, then million gallons per day flow is used as the second factor. However, if the concentration pertains to a tank or pipeline volume, then million gallons volume is used as the second factor.
Key point: Typically, especially in the past, the expression *parts per million* (ppm) was used as an expression of concentration, because 1 mg/L = 1 ppm. However, current practice is to use milligrams per liter as the preferred expression of concentration.

16.10.2 Chemical Feed Rate

In chemical dosing, a measured amount of chemical is added to the wastewater (or water). The amount of chemical required depends on the type of chemical used, the reason for dosing, and the flow rate being treated. The two expressions most often used to describe the amount of chemical added or required are (1) milligrams per liter (mg/L); and (2) pounds per day (lb/day).

A milligram per liter is a measure of concentration. For example, consider Figure 16.1 and in which it is apparent that the milligrams per liter concentration expresses a ratio of the milligram chemical in each liter of water. As shown, if a concentration of 5 mg/L is desired, then a total of 15 mg chemical would be required to treat 3 L:

$$\frac{5 \text{ mg} \times 3}{L \times 3} = \frac{15 \text{ mg}}{3 \text{ L}}$$

The amount of chemical required therefore depends on two factors:

- Desired concentration (milligrams per liter)
- Amount of wastewater to be treated (normally expressed as million gallons per day)

To convert from milligrams per liter to pounds per day, use Equation 16.88.

Example 16.60

Problem:

Determine the chlorinator setting (pounds per day) needed to treat a flow of 5 MGD with a chemical dose of 3 mg/L.

Solution:

$$\text{Chemical, lb/day} = \text{Chemical, mg/L} \times \text{Flow, MGD} \times 8.34 \text{ lb/gal}$$

$$= 3 \text{ mg/L} \times 5 \text{ MGD} \times 8.34 \text{ lb/gal} = 125 \text{ lb/day}$$

Example 16.61

Problem:

The desired dosage for a dry polymer is 10 mg/L. If the flow to be treated is 2,100,000 gpd, how many pounds per day of polymer are required?

Solution:

$$\text{Polymer, lb/day} = \text{Polymer, mg/L} \times \text{Flow, MGD} \times 8.34 \text{ lb/day}$$

$$= 10 \text{ mg/L Polymer} \times (2.10 \text{ MGD}) (8.34 \text{ lb/day}) = 175 \text{ lb/day Polymer}$$

Key point: To calculate chemical dose for tanks or pipelines, a modified equation must be used. Instead of million gallons per day flow, million gallons volume is used:

$$\text{Lb Chemical} = \text{Chemical, mg/L} \times \text{Tank Volume, MG} \times 8.34 \text{ lb/gal} \qquad (16.90)$$

Example 16.62

Problem:

To neutralize a sour digester, 1 lb of lime is added for every pound of volatile acids in the digester biosolids. If the digester contains 300,000 gal of biosolids with a volatile acid (VA) level of 2200 mg/L, how many pounds of lime should be added?

Solution:

Because the volatile acid concentration is 2200 mg/L, the lime concentration should also be 2200 mg/L:

$$\text{Lb Lime Required} = \text{Lime, mg/L} \times \text{Digester Volume, MG} \times 8.34 \text{ lb/gal}$$

$$= (2200 \text{ mg/L}) (0.30 \text{ MG}) 8.34 \text{ lb/gal}$$

$$= 5504 \text{ lb Lime}$$

16.10.3 Chlorine Dose, Demand, and Residual

Chlorine is a powerful oxidizer commonly used in wastewater and water treatment for disinfection; in wastewater treatment for odor control and bulking control; and in other applications. When chlorine is added to a unit process, obviously, we want to ensure that a measured amount is added.

Chlorine dose depends on two considerations—the chlorine demand and the desired chlorine residual:

$$\text{Chlorine Dose} = \text{Chlorine Demand} + \text{Chlorine Residual} \qquad (16.91)$$

16.10.3.1 Chlorine Dose

In describing the amount of chemical added or required, we use Equation 16.92:

$$\text{lb/day} = \text{Chemical, mg/L} \times \text{MGD} \times 8.34 \text{ lb/day} \qquad (16.92)$$

Example 16.63

Problem:
Determine the chlorinator setting (pounds per day) needed to treat a flow of 8 MGD with a chlorine dose of 6 mg/L.

Solution:

$$(\text{mg/L})(\text{MGD})(8.34) = \text{lb/day}$$

$$(6 \text{ mg/L}) (8 \text{ MGD}) (8.34 \text{ lb/gal} = \text{lb/day}$$

$$= 400 \text{ lb/day}$$

16.10.3.2 Chlorine Demand

Chlorine demand is the amount of chlorine used in reacting with various components of the water — harmful organisms and other organic and inorganic substances, for example. When the chlorine demand has been satisfied, these reactions cease.

Example 16.64

Problem:
The chlorine dosage for a secondary effluent is 6 mg/L. If the chlorine residual after 30 min of contact time is 0.5 mg/L, what is the chlorine demand expressed in milligrams per liter?

Solution:

$$\text{Chlorine Dose} = \text{Chlorine Demand} + \text{Chlorine Residual}$$

$$6 \text{ mg/L} = x \text{ mg/L} + 0.5 \text{ mg/L}$$

$$6 \text{ mg/L} - 0.5 \text{ mg/L} = x \text{ mg/L}$$

$$x = 5.5 \text{ mg/L Chlorine Demand}$$

16.10.3.3   Chlorine Residual

Chlorine residual is the amount of chlorine remaining after the demand has been satisfied.

Example 16.65

Problem:

What should the chlorinator setting be (pounds per day) to treat a flow of 3.9 MGD if the chlorine demand is 8 mg/L and a chlorine residual of 2 mg/L is desired?

Solution:

First calculate the chlorine dosage in milligrams per liter:

$$\text{Chlorine Dose} = \text{Chlorine Demand} + \text{Chlorine Residual}$$

$$= 8 \text{ mg/L} + 2 \text{ mg/L}$$

$$= 10 \text{ mg/L}$$

Then calculate the chlorine dosage (feed rate) in pounds per day:

$$(\text{Chlorine, mg/L}) (\text{MGD flow}) (8.34 \text{ lb/gal}) = \text{lb/day Chlorine}$$

$$(10 \text{ mg/L}) (3.9 \text{ MGD}) (8.34 \text{ lb/gal}) = 325 \text{ lb/day Chlorine}$$

16.10.4 Hypochlorite Dosage

Hypochlorite is less hazardous than chlorine; therefore, it is often used as a substitute chemical for elemental chlorine. Hypochlorite is similar to strong bleach and comes in two forms: dry calcium hypochlorite (often referred to as HTH) and liquid sodium hypochlorite. Calcium hypochlorite contains about 65% available chlorine; sodium hypochlorite contains about 12 to 15% available chlorine (in industrial strengths).

> *Key point*: Because neither type of hypochlorite is 100% pure chlorine, more pounds per day must be fed into the system to obtain the same amount of chlorine for disinfection — an important economical consideration for facilities considering substituting hypochlorite for chlorine. Some

studies indicate that such a switch can increase overall operating expenses by up to three times the cost of using elemental chlorine.

To determine the pounds per day of hypochlorite needed requires a two-step calculation:

Step 1.

$$mg/L \ (MGD) \ (8.34) = lb/day$$

Step 2.

$$\frac{Chorine, \ lb/day}{\dfrac{\% \ available}{100}} = Hypochlorite, \ lb/day \qquad (16.93)$$

Example 16.66

Problem:
A total chlorine dosage of 10 mg/L is required to treat a particular wastewater. If the flow is 1.4 MGD and the hypochlorite has 65% available chlorine, how many pounds per day of hypochlorite are required?

Solution:

Step 1. Calculate the pounds per day of chlorine required using the milligrams-to-liter to pounds-per-day equation:

$$mg/L \ (MGD) \ (8.34) = lb/day$$

$$(10 \ mg/L) \ (1.4 \ MGD) \ (8.34 \ lb/gal) = 117 \ lb/day$$

Step 2. Calculate the pounds per day of hypochlorite required. Because only 65% of the hypochlorite is chlorine, more than 117 lb/day will be required:

$$\frac{117 \ lb/day \ Chlorine}{\dfrac{65 \ \% \ available}{100}} = 180 \ lb/day \ Hypochlorite$$

Example 16.67

Problem:
A wastewater flow of 840,000 gpd requires a chlorine dose of 20 mg/L. If sodium hypochlorite (15% available chlorine) is used, how many pounds per day of sodium hypochlorite are required? How many gallons per day of sodium hypochlorite is this?

Solution:

Step 1. Calculate the pounds per day of chlorine required:

$$\text{mg/L (MGD) (8.34)} = \text{lb/day}$$

$$(20 \text{ mg/L})(0.84 \text{ MGD)} (8.34 \text{ lb/gal}) = 140 \text{ lb/day Chlorine}$$

Step 2. Calculate the pounds per day of sodium hypochlorite:

$$\frac{140 \text{ lb/day Chlorine}}{\dfrac{15 \% \text{ available}}{100}} = 933 \text{ lb/day Hypochlorite}$$

Step 3. Calculate the gallons per day of sodium hypochlorite:

$$= \frac{933 \text{ lb/day}}{8.34 \text{ lb/gal}} = 112 \text{ gal/day Sodium Hypochlorite}$$

Example 16.68

Problem:

How many pounds of chlorine gas are necessary to treat 5,000,000 gal of wastewater at a dosage of 2 mg/L?

Solution:

Step 1. Calculate the pounds of chlorine required.

$$V, 10^6 \text{ gal} = \text{Chlorine Concentration (mg/L)} \times 8.34 = \text{lb Chlorine}$$

Step 2. Substitute

$$5 \times 10^6 \text{ gal} \times 2 \text{ mg/L} \times 8.34 = 83 \text{ lb Chlorine}$$

16.10.5 Chemical Solutions

A *water solution* is a homogeneous liquid consisting of the solvent (the substance that dissolves another substance) and the solute (the substance that dissolves in the solvent). Water is the solvent (see Figure 16.7). The solute (whatever it may be) will dissolve up to a certain point. This is called its *solubility* — that is, the solubility of the solute in the particular solvent (water) at a particular temperature and pressure.

Remember that, in chemical solutions, the substance being dissolved is called the *solute*, and the liquid present in the greatest amount in a solution (and that does the dissolving) is called the *solvent*. We should also be familiar with another term: *concentration* — the amount of solute dissolved in a given amount of solvent. Concentration is measured as:

$$\% \text{ Strength} = \frac{\text{Wt. of solute}}{\text{Wt. of solution}} \times 100 = \frac{\text{Wt. of solute}}{\text{Wt. of solute} + \text{solvent}} \times 100$$

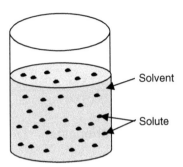

Figure 16.7 Solution with two components: solvent and solute. (From Spellman, F.R., 1999, *Spellman's Standard Handbook for Wastewater Operators*, Vol. 1, Lancaster, PA: Technomic Publishing Company.)

Example 16.69

Problem:

If 30 lb of chemical is added to 400 lb of water, what is the percent strength (by weight) of the solution?

Solution:

$$\% \text{ Strength } = \frac{30 \text{ lb Solute}}{400 \text{ lb Solution}} \times 100 = \frac{30 \text{ lb Solute}}{30 \text{ lb Solute } + \ 400 \text{ lb Water}} \times 100$$

$$= \frac{30 \text{ lb Solute}}{430 \text{ lb Solute/Water}} \times 100$$

$$\% \text{ Strength } = 7.0\%$$

A complete understanding of the dimensional units involved is important to making accurate computations of chemical strength. For example, it is necessay to understand exactly what *milligrams per liter* signifies;

$$\text{Milligrams per Liter (mg/L) } = \frac{\text{Milligrams of Solute}}{\text{Liters of Solution}} \qquad (16.94)$$

Another important dimensional unit commonly used when dealing with chemical solutions is *parts per million* (ppm);

$$\text{Parts per Million (ppm) } = \frac{\text{Parts of Solute}}{\text{Million Parts of Solution}} \qquad (16.95)$$

Key point: "Parts" is usually a weight measurement.

For example:

$$8 \text{ ppm} = \frac{8 \text{ lb solids}}{1{,}000{,}000 \text{ lb solution}}$$

$$8 \text{ ppm} = \frac{8 \text{ mg solids}}{1{,}000{,}000 \text{ mg solution}}$$

16.10.6 Mixing Solutions of Different Strengths

When different percent strength solutions are mixed, we use the following equations, depending upon the complexity of the problem:

$$\% \text{ Strength of mixture} = \frac{\text{Chemical in Mixture, lbs}}{\text{Solution Mixture, lbs}} \times 100 \tag{16.96}$$

$$\% \text{ Strength of mixture} = \frac{\text{lbs Chemical (Sol. 1)} + \text{lbs Chem (Sol. 2)}}{\text{lbs Sol. 1} + \text{lbs Sol. 2}} \times 100 \tag{16.97}$$

$$\% \text{ Strength of Mixture} = \frac{(\text{Sol. 1, lb}) \dfrac{(\% \text{ Strength Sol. 1})}{100} + (\text{Sol. 2}) \dfrac{(\% \text{ Strength Sol. 2})}{100}}{\text{lbs Sol. 1} + \text{lbs Sol. 2}} \tag{16.98}$$

Example 16.70

Problem:

If 25 lb of a 10% strength solution are mixed with 40 lb of a 1% strength solution, what is the percent strength of the solution mixture?

Solution:

$$\% \text{ Strength of Mixture} = \frac{(\text{Sol. 1, lb}) \dfrac{(\% \text{ Strength Sol. 1})}{100} + (\text{Sol. 2}) \dfrac{(\% \text{ Strength Sol. 2})}{100}}{\text{lb Sol. 1} + \text{lb Sol. 2}} \times 100$$

$$= \frac{(25 \text{ lb}) (0.1) + (40 \text{ lb}) (0.01)}{25 \text{ lb} + 40 \text{ lb}} \times 100$$

$$= \frac{2.5 \text{ lb} + 0.4 \text{ lb}}{65 \text{ lb}} \times 100$$

$$= 4.5\%$$

Key point: Percent strength should be expressed in terms of pounds of chemical per pound of solution. For example, when solutions are expressed in terms of gallons, the gallons should be expressed as pounds before continuing with the percent strength calculations.

16.10.7 Solution Mixtures Target Percent Strength

When two different percent strength solutions are mixed to obtain a desired quantity of solution and a target percent strength, we use Equation 16.98 and fill in the given information. Then, we find for the unknown, x.

Example 16.71

Problem:
   What weights of a 3% solution and a 6% solution must be mixed to make 800 lb of a 4% solution?

Solution:

$$\% \text{ Strength of Mixture} = \frac{(\text{Sol. 1, lb})\dfrac{(\% \text{ Strength Sol. 1})}{100} + (\text{Sol. 2})\dfrac{(\% \text{ Strength Sol. 2})}{100}}{\text{lb Sol. 1} + \text{lb Sol. 2}} \times 100$$

$$4 = \frac{(x \text{ lb})(0.03) + (800 - x \text{ lb})(0.06)}{800 \text{ lb}} \times 100$$

$$\frac{(4)}{100} = (800) = 0.03x + 48 - 0.06x$$

$$34 = -0.03x \ 48$$

$$0.03x = 14$$

$$x = 467 \text{ lb of 3\% Solution}$$

Then $800 - 467 = 333$ lb of 6% Solution

16.10.8 Solution Chemical Feeder Setting, GPD

Calculating GPD feeder setting depends on how the solution concentration is expressed: pounds per gallon or percent. If the solution strength is expressed as pounds per gallon, use the following equation:

$$\text{Solution, gpd} = \frac{(\text{Chemical, mg/L})(\text{Flow, MGD})(8.34 \text{ lb/gal})}{\text{lb Chemical Solution}} \qquad (16.99)$$

In water/wastewater operations, a standard, trial-and-error method known as *jar testing* is conducted to determine optimum chemical dosage. This method of testing has been the accepted bench testing procedure for many years. After jar testing results are analyzed to determine the best chemical dosage, the following example problems demonstrate how the actual calculations are made.

Example 16.72

Problem:

Jar tests indicate that the best liquid alum dose for a water is 8 mg/L. The flow to be treated is 1.85 MGD. Determine the gallons per day setting for the liquid alum chemical feeder if the liquid alum contains 5.30 lb of alum per gallon of solution.

Solution:

First, calculate the pounds per day of dry alum required, using the milligrams-per-liter to pounds-per-day equation:

$$\text{lb/day} = (\text{Dose, mg/L}) (\text{Flow, MGD}) (8.34 \text{ lb/gal})$$

$$= (8 \text{ mg/L}) (1.85 \text{ MGD}) (8.34 \text{ lb/gal})$$

$$= 123 \text{ lb/day dry alum}$$

Then, calculate gallons per day of solution required.

$$\text{Alum Solution, gpd} = \frac{123 \text{ lb/day Alum}}{5.30 \text{ lb Alum/gal solution}}$$

$$\text{Feeder Setting} = 23 \text{ gpd Alum Solution}$$

If the solution strength is expressed as a percent, we use the following equation:

$$(\text{Chem., mg/L}) (\text{Flow Treated, MGD}) (8.34 \text{ lb/gal}) =$$
$$(\text{Sol., mg/L}) (\text{Sol. Flow, MGD}) (8.34 \text{ lb/gal})$$

(16.100)

Example 16.73

Problem:

The flow to a plant is 3.40 MGD. Jar testing indicates that the optimum alum dose is 10 mg/L. What should the gallons per day setting be for the solution feeder if the alum solution is a 52% solution?

Solution:

A solution concentration of 52% is equivalent to 520,000 mg/L:

$$\text{Desired Dose, lb/day} = \text{Actual Dose, lb/day}$$

$$(\text{Chem., mg/L}) (\text{Flow Treated, MGD}) (8.34 \text{ lb/gal}) = (\text{Sol., mg/L}) (\text{Sol. Flow, MGD}) (8.34 \text{ lb/gal})$$

$$(10 \text{ mg/L}) (3.40 \text{ MGD}) (8.34 \text{ lb/gal}) = (520,000 \text{ mg/L}) (x \text{ MGD}) (8.34 \text{ lb/gal})$$

$$x = \frac{(10)\,(3.40)\,(8.34)}{(520,000)\,(8.34)}$$

$$x = 0.0000653 \text{ MGD}$$

This can be expressed as gallons per day of flow:

$$0.0000653 \text{ MGD} = 65.3 \text{ gpd flow}$$

16.10.9 Chemical Feed Pump — Percent Stroke Setting

Chemical feed pumps are generally positive displacement pumps (also called "piston" pumps). This type of pump displaces, or pushes out, a volume of chemical equal to the volume of the piston. The length of the piston, called the stroke, can be lengthened or shortened to increase or decrease the amount of chemical delivered by the pump. In calculating percent stroke setting, use the following equation:

$$\% \text{ Stroke Setting} = \frac{\text{Required Feed, gpd}}{\text{Maximum Feed, gpd}} \qquad (16.101)$$

Example 16.74

Problem:
    The required chemical pumping rate has been calculated at 8 gpm. If the maximum pumping rate is 90 gpm, what should the percent stroke setting be?

Solution:
    The percent stroke setting is based on the ratio of the gallons per minute required to the total possible gallons per minute:

$$\% \text{ Stroke Setting} = \frac{\text{Required Feed, gpd}}{\text{Maximum Feed, gpd}} \times 100$$

$$= \frac{8 \text{ gpm}}{90 \text{ gpm}} \times 100$$

$$= 8.9\%$$

16.10.10 Chemical Solution Feeder Setting, Milliliters per Minute

Some chemical solution feeders dispense chemical as milliliters per minute (mL/min). To calculate the milliliters per minute of solution required, use the following equation:

$$\text{Solution, mL/min} = \frac{(\text{gpd})\,(3785 \text{ mL/gal})}{1440 \text{ min/day}} \qquad (16.102)$$

Example 16.75

Problem:

The desired solution feed rate was calculated at 7 gpd. What is this feed rate expressed as milliliters per minute?

Solution:

Because the gallons-per-day flow has already been determined, the milliliters per minute flow rate can be calculated directly:

$$\text{Feed Rate mL/min} = \frac{(\text{gpd})\,(3785 \text{ mL/gal})}{1440 \text{ min/day}}$$

$$= \frac{(7 \text{ gpd})\,(3785 \text{ mL/gal})}{1440 \text{ min/day}}$$

$$= 18 \text{ mL/min Feed Rate}$$

16.10.11 Chemical Feed Calibration

Routinely, to ensure accuracy, we need to compare the actual chemical feed rate with the feed rate indicated by the instrumentation. To accomplish this, we use calibration calculations.

To calculate the actual chemical feed rate for a dry chemical feed, place a container under the feeder, weigh the container when it is empty, and then weigh it again after a specified length of time, such as 30 min. Actual chemical feed rate can then be determined as:

$$\text{Chemical Feed Rate, lb/min} = \frac{\text{Chemical Applied, lb}}{\text{Length of Application, min}} \qquad (16.103)$$

Example 16.76

Problem:

Calculate the actual chemical feed rate, pounds per day, if a container is placed under a chemical feeder and a total of 2.2 lb is collected during a 30-min period.

Solution:

First, calculate the pounds per minute feed rate:

$$\text{Chemical Feed Rate, lb/min} = \frac{\text{Chemical Applied, lb}}{\text{Length of Application, min}}$$

$$= \frac{2.2 \text{ lb}}{30 \text{ min}}$$

$$= 0.07 \text{ lb/min Feed Rate}$$

Then, calculate the pounds per day feed rate:

$$\text{Chemical Feed Rate, lb/day} = (0.07 \text{ lb/min}) (1440 \text{ min/day})$$

$$= 101 \text{ lb/day Feed Rate}$$

Example 16.77

Problem:

A chemical feeder must be calibrated. The container to be used to collect the chemical is weighed (0.35 lb) and placed under the chemical feeder. After 30 min, the weight of the container and chemical is measured at 2.2 lb. Based on this test, what is the actual chemical feed rate, in pounds per day?

Solution:

First, calculate the pounds per minute feed rate:

Key point: The chemical applied is the weight of the container and chemical minus the weight of the empty container.

$$\text{Chemical Feed Rate, lb/min} = \frac{\text{Chemical Applied, lb}}{\text{Length of Application, min}}$$

$$= \frac{2.2 \text{ lb} - 0.35 \text{ lb}}{30 \text{ min}}$$

$$= \frac{1.85 \text{ lb}}{30 \text{ min}}$$

$$= 0.062 \text{ lb/min Feed Rate}$$

Then calculate the pounds per day feed rate:

$$(0.062 \text{ lb/min}) (1440 \text{ min/day}) = 89 \text{ lb/day Feed Rate}$$

When the chemical feeder is for a solution, the calibration calculation is slightly more difficult than that for a dry chemical feeder. As with other calibration calculations, the actual chemical feed rate is determined and then compared with the feed rate indicated by the instrumentation. Use these calculations for solution feeder calibration:

$$\text{Flow Rate, gpd} = \frac{(\text{mL/min}) (1440 \text{ min/day})}{3785 \text{ mL/gal}} = \text{gpd} \qquad (16.104)$$

Then, calculate chemical dosage, pounds per day:

$$\text{Chemical, lb/day} = (\text{Chemical, mg/L}) (\text{Flow, MGD}) (8.34 \text{ lb/day}) \qquad (16.105)$$

Example 16.78

Problem:

A calibration test is conducted for a solution chemical feeder. During 5 min, the solution feeder delivers a total of 700 mL. The polymer solution is a 1.3% solution. What is the pounds-per-day feed rate? (Assume the polymer solution weighs 8.34 lb/gal.)

Solution:

The milliliters-per-minute flow rate is calculated as:

$$\frac{700 \text{ mL}}{5 \text{ min}} = 140 \text{ mL/min}$$

Then convert milliliters-per-minute flow rate to gallons-per-day flow rate:

$$\frac{(140 \text{ mL/min}) (1440 \text{ min/day})}{3785 \text{ mL/gal}} = 53 \text{ gpd flow rate}$$

and calculate pounds-per-day fee rate:

$$\text{Chemical, lb/day} = (\text{Chemical, mg/L}) (\text{Flow, MGD}) (8.34 \text{ lb/day})$$

$$(13,000 \text{ mg/L}) (0.000053 \text{ MGD}) (8.34 \text{ lb/day}) = 5.7 \text{ lb/day polymer}$$

Actual pumping rates can be determined by calculating the volume pumped during a specified time frame. For example, if 120 gal are pumped during a 15-min test, the average pumping rate during the test is 8 gpm. The gallons pumped can be determined by measuring the drop in tank level during the timed test:

$$\text{Flow, gpm} = \frac{\text{Volume Pumped, gal}}{\text{Duration of Test, min}} \tag{16.106}$$

Then the actual flow rate (gallons per minute) is calculated using

$$\text{Flow, gpm} = \frac{(0.785) (D^2) (\text{Drop in Level, ft}) (7.48 \text{ gal/ft}^3)}{\text{Duration of Test, min}} \tag{16.107}$$

Example 16.79

Problem:

A pumping rate calibration test is conducted for a 5-min period. The liquid level in the 4-ft diameter solution tank is measured before and after the test. If the level drops 0.4 ft during the 5-min test, what is the pumping rate in gallons per minute?

Solution:

$$Flow,\ gpm\ =\ \frac{(0.785)\ (D^2)\ (Drop\ in\ Level,\ ft)\ (7.48\ gal/ft^3)}{Duration\ of\ Test,\ min}$$

$$=\ \frac{(0.785)\ (D^2)\ (4\ ft)\ (4\ ft)\ (0.4\ ft)\ (7.48\ gal/ft^3)}{5\ min}$$

Pumping Rate = 7.5 gpm

16.10.12 Average Use Calculations

During a typical shift, operators log in or record several parameter readings. The data collected are important in monitoring plant operation — in providing information on how to best optimize plant or unit process operation. One of the important parameters monitored each shift or each day is the actual use of chemicals. From the recorded chemical use data, expected chemical use can be forecast. These data are also important for inventory control; determination can be made when additional chemical supplies will be required.

In determining average chemical use, we first must determine the average daily chemical use:

$$Average\ Use,\ lb/day\ =\ \frac{Total\ Chemical\ Used,\ lb}{Number\ of\ Days} \qquad (16.108)$$

or

$$Average\ Use,\ gpd\ =\ \frac{Total\ Chemical\ Used,\ gal}{Number\ of\ Days} \qquad (16.109)$$

Then calculate day's supply in inventory:

$$Days\ Supply\ in\ Inventory\ =\ \frac{Total\ Chemical\ in\ Inventory,\ lb}{Average\ Use,\ lb/day} \qquad (16.110)$$

or

$$Days\ Supply\ in\ Inventory\ =\ \frac{Total\ Chemical\ in\ Inventory,\ gal}{Average\ Use,\ gpd} \qquad (16.111)$$

Example 16.80

Problem:

The chemical amount used for each day during a week is given in the following table. Based on these data, what was the average pounds-per-day chemical use during the week?

Day	Amount (lb/day)
Monday	92
Tuesday	94
Wednesday	92
Thursday	88
Friday	96
Saturday	92
Sunday	88

Solution:

$$\text{Average Use, lb/day} = \frac{\text{Total Chemical Used, lb}}{\text{Number of Days}}$$

$$= \frac{642 \text{ lb}}{7 \text{ days}}$$

$$\text{Average Use} = 91.7 \text{ lb/day}$$

Example 16.81

Problem:

The average chemical use at a plant is 83 lb/day. If the chemical inventory in stock is 2600 lb, how many days' supply is this?

Solution:

$$\text{Days Supply in Inventory} = \frac{\text{Total Chemical in Inventory, lb}}{\text{Average Use, lb/day}}$$

$$= \frac{2600 \text{ lb in Inventory}}{83 \text{ lb/day Average Use}}$$

$$= 31.3 \text{ days supply}$$

16.11 BIOSOLIDS PRODUCTION AND PUMPING CALCULATIONS

16.11.1 Process Residuals

The wastewater unit treatment processes remove solids and biochemical oxygen demand from the waste stream before the liquid effluent is discharged to its receiving waters. What remains to be disposed is a mixture of solids and wastes called *process residuals* — more commonly referred to as biosolids (or sludge).

> *Key point*: Sludge is the commonly accepted name for wastewater residual solids. However, if wastewater sludge is used for beneficial reuse (as a soil amendment or fertilizer), it is commonly called biosolids. We choose to refer to process residuals as biosolids in this text.

The most costly and complex aspect of wastewater treatment can be the collection, processing, and disposal of biosolids because the quantity of biosolids produced may be as high as 2% of the original volume of wastewater, depending somewhat on the treatment process used. Biosolids can be as much as 97% water content and the costs of disposal are related to the volume of biosolids processed; thus, one of the primary purposes or goals of biosolids treatment (along with stabilizing it so that it is no longer objectionable or environmentally damaging) is to separate as much of the water from the solids as possible.

16.11.2 Primary and Secondary Solids Production Calculations

We point out that when making calculations pertaining to solids and biosolids, the term "solids" refers to *dry solids* and the term "biosolids" refers to the *solids and water*. The solids produced during primary treatment depend on the solids that settle in or are removed by the primary clarifier. In making primary clarifier solids production calculations, we use the milligrams-per-liter to pounds-per-day equation:

$$\text{Susp. Solids (SS) Removed, lb/day } =$$

$$\text{(SS Removed, mg/L) (Flow, MGD) (8.34 lb/gal)}$$

(16.112)

16.11.3 Primary Clarifier Solids Production Calculations

Example 16.82

Problem:
  A primary clarifier receives a flow of 1.80 MGD with suspended solids concentrations of 340 mg/L. If the clarifier effluent has a suspended solids concentration of 180 mg/L, how many pounds of solids are generated daily?

Solution:

$$\text{Susp. Solids (SS) Removed, lb/day } = \text{(SS Removed, mg/L) (Flow, MGD) (8.34 lb/gal)}$$

$$= \text{(160 mg/L) (1.80 MGD) (8.34 lb/gal)}$$

$$\text{Solids } = 2402 \text{ lb/day}$$

Example 16.83

Problem:
  The suspended solids content of the primary influent is 350 mg/L and the primary influent is 202 mg/L. How many pounds of solids are produced during a day on which the flow is 4,150,000 gpd?

Solution:

$$\text{SS, lb/day Removed } = \text{(SS Removed, mg/L) (Flow, MGD) (8.34 lb/gal)}$$

$$= \text{(148 mg/L) (4.15 MGD) (8.34 lb/gal)}$$

$$\text{Solids Removed } = 5122 \text{ lb/day}$$

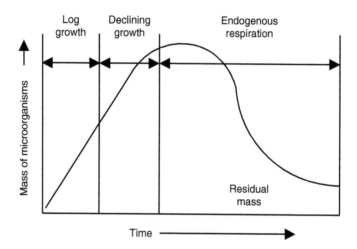

Figure 16.8   Bacteria growth curve. (From Spellman, F.R., 1999, *Spellman's Standard Handbook for Wastewater Operators*, Vol. 1, Lancaster, PA: Technomic Publishing Company.)

16.11.4 Secondary Clarifier Solids Production Calculation

Solids produced during secondary treatment depend on many factors, including the amount of organic matter removed by the system and the growth rate of bacteria (see Figure 16.8). Because precise calculations of biosolids production are complex, we provide a rough estimate method of solids production that uses an estimated growth rate (unknown) value: the BOD-removed, pounds-per-day equation:

$$\text{BOD Removed, lb/day} = (\text{BOD Removed, mg/L})\,(\text{Flow, MGD})\,(8.34 \text{ lb/day}) \quad (16.113)$$

Example 16.84

Problem:

   The 1.5-MGD influent to the secondary system has a BOD concentration of 174 mg/L. The secondary effluent contains 22 mg/L BOD. If the bacteria growth rate (unknown *x*-value) for this plant is 0.40 lb SS per pound of BOD removed, how many pounds of dry biosolids solids are produced each day by the secondary system?

Solution:

$$\text{BOD Removed, lb/day} = (\text{BOD, mg/L})\,(\text{Flow, MGD})\,(8.34 \text{ lb/day})$$

$$= (152 \text{ mg/L})\,(1.5 \text{ MGD})\,8.34 \text{ lb/gal}$$

$$= 1902 \text{ lb/day}$$

Then use the unknown *x*-value to determine pounds per day of solids produced.

$$\frac{0.44 \text{ lb SS Produced}}{1 \text{ lb BOD Removed}} = \frac{x \text{ lb SS Produced}}{1902 \text{ lb/day BOD Removed}}$$

$$= \frac{(0.44)(1902)}{1} = x$$

837 lb/day Solids Produced = x

Key point: Typically, for every pound of food consumed (BOD removed) by the bacteria, between 0.3 and 0.7 lb of new bacteria cells are produced; these are solids that must be removed from the system.

16.11.5 Percent Solids

Biosolids are composed of water and solids. The vast majority of biosolids are water — usually in the range of 93 to 97%. To determine the solids content of biosolids, a sample is dried overnight in an oven at 103 to 105°F. The solids that remain after drying represent the total solids content of the biosolids. Solids content may be expressed as a percent or as milligrams per liter. Either of two equations is used to calculate percent solids:

$$\% \text{ Solids} = \frac{\text{Total Solids, g (grams)}}{\text{Biosolids Sample, g}} \times 100 \qquad (16.114)$$

$$\% \text{ Solids} = \frac{\text{Solids, lb/day}}{\text{Biosolids, lb/day}} \times 100 \qquad (16.115)$$

Example 16.85

Problem:

The total weight of a biosolids sample (sample only, not the dish) is 22 g. If the weight of the solids after drying is 0.77 g, what is the percent total solids of the biosolids?

Solution:

$$\% \text{ Solids} = \frac{\text{Total Solids, g}}{\text{Biosolids Sample, g}} \times 100$$

$$= \frac{0.77 \text{ g}}{22 \text{ g}} \times 100$$

$$= 3.5\%$$

16.11.6 Biosolids Pumping

While on shift, wastewater operators are often required to make various process control calculations. One important calculation (covered in this subsection) involves biosolids pumping.

16.11.7 Estimating Daily Biosolids Production

The calculation for estimation of the required biosolids-pumping rate provides a method to establish an initial pumping rate or to evaluate the adequacy of the current withdrawal rate:

$$\text{Est. pump rate} = \frac{(\text{Influ. TSS Conc.} - \text{Effluent TSS Conc.}) \times \text{Flow} \times 8.34}{\% \text{ Solids in Sludge} \times 8.34 \times 1440 \text{ min/day}} \quad (16.116)$$

Example 16.86

Problem:

The biosolids withdrawn from the primary settling tank contain 1.4% solids. The unit influent contains 285 mg/L TSS, and the effluent contains 140 mg/L TSS. If the influent flow rate is 5.55 MGD, what is the estimated biosolids withdrawal rate in gallons per minute (assuming the pump operates continuously)?

Solution:

$$\text{Biosolids Rate, gpm} = \frac{(285 \text{ mg/L} - 140 \text{ mg/L} \times 5.55 \times 8.34}{0.014 \times 8.34 \times 1440 \text{ min/day}} = 40 \text{ gpm}$$

16.11.8 Biosolids Production (Pounds per Million Gallons)

A common method of expressing biosolids production is in pounds of biosolids per million gallons of wastewater treated:

$$\text{Biosolids, lb/MG} = \frac{\text{Total Biosolids Production, lb}}{\text{Total Wastewater Flow, MG}} \quad (16.117)$$

Example 16.87

Problem:

Records show that a plant has produced 85,000 gal of biosolids during the past 30 days. The average daily flow for this period was 1.2 MGD. What was the plant's biosolids production in pounds per million gallons?

Solution:

$$\text{Biosolids, lb/MG} = \frac{85,000 \text{ gal} \times 8.34 \text{ lb/gal}}{1.2 \text{ MGD} \times 30 \text{ days}} = 19,692 \text{ lb/MG}$$

16.11.9 Biosolids Production (Wet Tons per Year)

Biosolids production can also be expressed in terms of the amount of biosolids (water and solids) produced per year. This is normally expressed in wet tons per year:

$$\text{Biosolids, wet tons/yr} \; \frac{\begin{array}{c}\text{Biosolids Prod., lb/MG} \times \text{Ave. Daily Flow,} \\ \text{MGD} \times 365 \text{ day/yr}\end{array}}{2000 \text{ lb/ton}} = 40 \text{ gpm} \quad (16.118)$$

Example 16.88

Problem:
   A plant is currently producing biosolids at the rate of 16,500 lb/MG. The current average daily wastewater flow rate is 1.5 MGD. What is the total amount of biosolids produced per year in wet tons per year?

Solution:

$$\text{Biosolids, Wet Tons/yr} = \frac{16{,}500 \text{ lb/MG} \times 1.5 \text{ MGD} \times 365 \text{ days/yr}}{2000 \text{ lb/ton}}$$

$$= 4517 \text{ Wet Tons/yr}$$

16.11.10 Biosolids Pumping Time

The biosolids pumping time is the total time the pump operates during a 24-h period, expressed in minutes.

$$\text{Pump Operating Time} = \text{Time/Cycle, min} \times \text{Frequency, cycles/day} \qquad (16.119)$$

Note: The following information is used for Example 16.89 through Example 16.94:

Frequency	24 times per day
Pump rate	120 gpm
Solids	3.70%
Volatile matter	66%

Example 16.89

Problem:
   What is the pump operating time?

Solution:

$$\text{Pump Operating Time} = 15 \text{ min/h} \times 24 \text{ (cycles)/day} = 360 \text{ min/day}$$

Biosolids pumped per day in gallons:

$$\text{Biosolids, gpd} = \text{Operating Time, min/day} \times \text{Pump Rate, gpm} \qquad (16.120)$$

Example 16.90

Problem:
   What is the amount of biosolids pumped per day in gallons?

Solution:

$$\text{Biosolids, gpd} = 360 \text{ min/day} \times 120 \text{ gpm} = 43{,}200 \text{ gpd}$$

Biosolids pumped per day in pounds:

$$\text{Sludge, lb/day} = \text{Gallons of Biosolids Pumped} \times 8.34 \text{ lb/gal} \qquad (16.121)$$

Example 16.91

Problem:
   What is the amount of biosolids pumped per day in pounds?

Solution:

$$\text{Biosolids, lb/day} = 43{,}200 \text{ gal/day} \times 8.34 \text{ lb/gal} = 360{,}300 \text{ lb/day}$$

Biosolids pumped per day in pounds:

$$\text{Solids Pumped, lb/day} = \text{Biosolids Pumped, lb/day} \times \% \text{ Solids}$$

Example 16.92

Problem:
   What are the solids pumped per day?

Solution:

$$\text{Solids Pumped lb/day} = 360{,}300 \text{ lb/day} \times 0.0370 = 13{,}331 \text{ lb/day}$$

Volatile matter pumped per day in pounds:

$$\text{Vol. Matter (lb/day)} = \text{Solids Pumped, lb/day} \times \% \text{ Volatile Matter} \qquad (16.122)$$

Example 16.93

Problem:
   What is the volatile matter in pounds per day?

Solution:

$$\text{Volatile Matter, lb/day} = 13{,}331 \text{ lb/day} \times 0.66 = 8798 \text{ lb/day}$$

Pounds of solids/pounds of volatile solids per day:
   If we wish to calculate the pounds of solids or the pounds of volatile solids removed per day, the individual equations demonstrated earlier can be combined into single calculations:

$$\text{Solids, lb/day} =$$

$$\text{Pump Time, min/cyc} \times \text{Freq., cyc/day} \times \text{Rate, gpm} \times 8.34 \text{ lb/gal} \times \text{solids} \qquad (16.123)$$

$$\text{Vol. Mat., lb/day} =$$

$$\text{Time, min/cyc} \times \text{Freq., cyc/day} \times \text{Rate, gpm} \times 8.34 \times \% \text{ Solids} \times \% \text{ VM} \tag{16.124}$$

Example 16.94

$$\text{Solids, lb/day} = 15 \text{ min/cyc} \times 24 \text{ cyc/day} \times 120 \text{ gpm} \times 8.34 \times 0.0370$$

$$= 13,331 \text{ lb/day}$$

$$\text{VM, lb/day} = 15 \text{ min/cyc} \times 24 \text{ cyc/day} \times 120 \text{ gpm} \times 8.34 \times 0.0370 \times .66$$

$$= 8,798 \text{ lb/day}$$

16.12 BIOSOLIDS THICKENING

16.12.1 Thickening

Biosolids thickening (or concentration) is a unit process used to increase the solids content of the biosolids by removing a portion of the liquid fraction. In other words, biosolids thickening is all about volume reduction. By increasing the solids content, more economical treatment of the biosolids can be effected. Biosolids thickening processes include:

- Gravity thickeners
- Flotation thickeners
- Solids concentrators

Biosolids thickening calculations are based on the concept that the solids in the primary or secondary biosolids are equal to the solids in the thickened biosolids. The solids are the same. Primarily, water has been removed to thicken the biosolids, resulting in higher percent solids. In unthickened biosolids, the solids might represent 1 to 4% of the total pounds of biosolids. When some of the water is removed, those same-amount solids might represent 5 to 7% of the total pounds of biosolids.

Key point: The key to biosolids thickening calculations is that solids remain constant.

16.12.2 Gravity/Dissolved Air Flotation Thickener Calculations

Biosolids thickening calculations are based on the concept that the solids in the primary or secondary biosolids are equal to the solids in the thickened biosolids. Assuming a negligible amount of solids is lost in the thickener overflow, the solids are the same. Note that the water is removed to thicken the biosolids and results in higher percent solids.

16.12.2.1 Estimating Daily Sludge Production

The calculation for estimating the required biosolids-pumping rate provides a method to establish an initial pumping rate or to evaluate the adequacy of the current pump rate:

$$\text{Est. Pump Rate} = \frac{(\text{Influent TSS Conc.} - \text{Eff. TSS Conc.}) \times \text{Flow} \times 8.34}{\% \text{ Solids in Biosolids} \times 8.34 \times 1440 \text{ min/day}} \qquad (16.125)$$

Example 16.95

Problem:

The biosolids withdrawn from the primary settling tank contain 1.5% solids. The unit influent contains 280 mg/L TSS, and the effluent contains 141 mg/L TSS. If the influent flow rate is 5.55 MGD, what is the estimated biosolids withdrawal rate in gallons per minute (assuming the pump operates continuously)?

Solution:

$$\text{Biosolids Withdrawal Rate, gpm} = \frac{(280 \text{ mg/L} - 141 \text{ mg/L}) \times 5.55 \text{ MGD} \times 8.34}{0.015 \times 8.34 \times 1440 \text{ min/day}}$$

$$= 36 \text{ gpm}$$

16.12.2.2 Surface Loading Rate, Gallons per Day per Square Foot

Surface loading rate (surface settling rate) is hydraulic loading — the amount of biosolids applied per square foot of gravity thickener.

$$\text{Surface Loading, gal/day/ft}^2 = \frac{\text{Biosolids Applied to the Thickener, gpd}}{\text{Thickener Area, ft}^2} \qquad (16.126)$$

Example 16.96

Problem:

A 70-ft diameter gravity thickener receives 32,000 gpd of biosolids. What is the surface loading in gallons per square foot per day?

Solution:

$$\text{Surface Loading} = \frac{32,000 \text{ gpd}}{0.785 \times 70 \text{ ft} \times 70 \text{ ft}} = 8.32 \text{ gpd/ft}^2$$

16.12.2.3 Solids Loading Rate, Pounds per Day per Square Foot

The solids loading rate is the pounds of solids per day applied to 1 ft² of tank surface area. The calculation uses the surface area of the bottom of the tank. It assumes that the floor of the tank is flat and has the same dimensions as the surface:

$$\text{Surface Loading, lb/day/ft}^2 =$$

$$\frac{\% \text{ Biosolids Solids} \times \text{Biosolids Flow, gpd} \times 8.34 \text{ lb/gal}}{\text{Thickener Area, ft}^2} \qquad (16.127)$$

Example 16.97

Problem:
    The thickener influent contains 1.6% solids. The influent flow rate is 39,000 gpd. The thickener is 50 ft in diameter and 10 ft deep. What is the solid loading in pounds per day?

Solution:

$$\text{Surface Loading, lb/day/ft}^2 = \frac{0.016 \times 39,000 \text{ gpd} \times 8.34 \text{ lb/gal}}{0.785 \times 50 \text{ ft} \times 50 \text{ ft}} = 2.7 \text{ lb/ft}^2$$

16.12.3 Concentration Factor (*Cf*)

The concentration factor (*CF*) represents the increase in concentration resulting from the thickener. It is a means of determining the effectiveness of the gravity thickening process.

$$CF = \frac{\text{Thickened Biosolids Concentration, \%}}{\text{Influent Biosolids Concentration, \%}} \qquad (16.128)$$

Example 16.98

Problem:
    The influent biosolids contain 3.5% solids. The thickened biosolids–solids concentration is 7.7%. What is the concentration factor?

Solution:

$$CF = \frac{7.7\%}{3.5\%} = 2.2$$

16.12.4 Air-to-Solids Ratio

Air–solids ratio is the ratio between the pounds of solids entering the thickener and the pounds of air applied:

$$\text{Air:Solids Ratio} = \frac{\text{Air Flow ft}^3/\text{min} \times 0.0785 \text{ lb/ft}^3}{\text{Biosolids Flow, gpm} \times \% \text{ Solids} \times 8.34 \text{ lb/gal}} \qquad (16.129)$$

Example 16.99

Problem:
    The biosolids pumped to the thickener are 0.85% solids. The airflow is 13 ft³/min. What is the air-to-solids ratio if the current biosolids flow rate entering the unit is 50 gpm?

Solution:

$$\text{Air:Solids Ratio} = \frac{13 \text{ cfm} \times 0.075 \text{ lb/ft}^3}{50 \text{ gpm} \times 0.0085 \times 8.34 \text{ lb/gal}} = 0.28$$

16.12.5 Recycle Flow in Percent

The amount of recycle flow is expressed as a percent:

$$\text{Recycle \%} = \frac{\text{Recycle Flow Rate, gpm} \times 100}{\text{Sludge Flow, gpm}} = 175\% \tag{16.130}$$

Example 16.100

Problem:

The sludge flow to the thickener is 80 gpm. The recycle flow rate is 140 gpm. What is the percent of recycle?

Solution:

$$\text{\% Recycle} = \frac{140 \text{ gpm} \times 100}{80 \text{ gpm}} = 175\%$$

16.12.6 Centrifuge Thickening Calculations

A centrifuge exerts a force on the biosolids thousands of times greater than gravity. Sometimes polymer is added to the influent of the centrifuge to help thicken the solids. The two most important factors that affect the centrifuge are the volume of the biosolids put into the unit (gallons per minute) and the pounds of solids put in. The water that is removed is called *centrate*.

Normally, hydraulic loading is measured as flow rate per unit of area. However, because of the variety of sizes and designs, hydraulic loading to centrifuges does not include area considerations and is expressed only as gallons per hour. The equations to be used if the flow rate to the centrifuge is given as gallons per day or gallons per minute are:

$$\text{Hydraulic Loading, gph} = \frac{\text{Flow, gpd}}{24 \text{ h/day}} \tag{16.131}$$

$$\text{Hydraulic Loading, gpm} = \frac{(\text{gpm flow})(60 \text{ min})}{\text{h}} \tag{16.132}$$

Example 16.101

Problem:

A centrifuge receives a waste-activated biosolids flow of 40 gpm. What is the hydraulic loading on the unit in gallons per hour?

Solution:

$$\text{Hydraulic Loading, gph} = \frac{(\text{gpm flow})(60 \text{ min})}{\text{h}}$$

$$= \frac{(40 \text{ gpm})(60 \text{ min})}{\text{h}}$$

$$= 2400 \text{ gph}$$

Example 16.102

Problem:
A centrifuge receives 48,600 gal of biosolids daily. The biosolids concentration before thickening is 0.9%. How many pounds of solids are received each day?

Solution:

$$\frac{48,600 \text{ gal}}{\text{day}} \times \frac{8.34 \text{ lb}}{\text{gal}} \times \frac{0.9}{100} = 3648 \text{ lb/day}$$

16.13 STABILIZATION

16.13.1 Biosolids Digestion

A major problem in designing wastewater treatment plants is the disposal of biosolids into the environment without causing damage or nuisance. It is even more difficult to dispose of untreated biosolids; they must be stabilized to minimize disposal problems. In most cases, the term *stabilization* is considered synonymous with digestion.

> *Key point*: The stabilization of organic matter is accomplished biologically using a variety of organisms. The microorganisms convert the colloidal and dissolved organic matter into various gases and into protoplasm. Because protoplasm has a specific gravity slightly higher than that of water, it can be removed from the treated liquid by gravity.

Biosolids digestion is a process in which biochemical decomposition of the organic solids occurs; in the decomposition process, the organics are converted into simpler and more stable substances. Digestion also reduces the total mass or weight of biosolids solids, destroys pathogens, and makes drying or dewatering the biosolids easier. Well-digested biosolids have the appearance and characteristics of a rich potting soil.

Biosolids may be digested under aerobic or anaerobic conditions. Most large municipal wastewater treatment plants use anaerobic digestion. Aerobic digestion finds application primarily in small, package-activated biosolids treatment systems.

16.13.2 Aerobic Digestion Process Control Calculations

The purpose of aerobic digestion is to stabilize organic matter, reduce volume, and eliminate pathogenic organisms. Aerobic digestion is similar to the activated biosolids process. Biosolids are aerated for 20 days or more and volatile solids are reduced by biological activity.

16.13.2.1 *Volatile Solids Loading, Pounds per Square Foot per Day*

Volatile solids (organic matter) loading for the aerobic digester is expressed in pounds of volatile solids entering the digester per day per cubic foot of digester capacity.

$$\text{Volatile Solids Loading, lb/day/ft}^3 = \frac{\text{Volatile Solids Added, lb/day}}{\text{Digester Volume, ft}^3} \quad (16.133)$$

Example 16.103

Problem:

An aerobic digester is 20 ft in diameter and has an operating depth of 20 ft. The biosolids added to the digester daily contain 1500 lb of volatile solids. What is the volatile solids loading in pounds per day per cubic foot?

Solution:

$$\text{Volatile Solids Loading, lb/day/ft}^3 = \frac{1500 \text{ lb/day}}{0.785 \times 20 \text{ ft} \times 20 \text{ ft} \times 20 \text{ ft}} = 0.24 \text{ lb/day/ft}^3$$

16.13.2.2 Digestion Time, Days

The theoretical time that biosolids remain in the aerobic digester is:

$$\text{Digestion Time, Days} = \frac{\text{Digester Volume, gallons}}{\text{Biosolids Added, gpd}} \qquad (16.134)$$

Example 16.104

Problem:

The digester volume is 240,000 gal. Biosolids are added to the digester at the rate of 15,000 gpd. What is the digestion time in days?

Solution:

$$\text{Digestion Time, Days} = \frac{240,000 \text{ gal}}{15,000 \text{ gpd}} = 16 \text{ days}$$

16.13.2.3 pH Adjustment

In many instances, the pH of the aerobic digester falls below the levels required for good biological activity. When this occurs, the operator must perform a laboratory test to determine the amount of alkalinity required to raise the pH to the desired level. The results of the lab test must then be converted to the actual quantity required by the digester:

$$\text{Digestion Time, Days} = \frac{\text{Chem. Used in Lab Test, mg} \times \text{Dig. Vol.} \times 3.785}{\text{Sample Vol., L} \times 454 \text{ g/lb} \times 1000 \text{ mg/g}} \qquad (16.135)$$

Example 16.105

Problem:

The pH of a 1-L sample of the aerobic digester contents will be increased to pH 7.1 by 240 mg of lime. The digester volume is 240,000 gal. How many pounds of lime are required to increase the digester pH to 7.3?

Solution:

$$\text{Chemical Required, lb} = \frac{240 \text{ mg} \times 240{,}000 \text{ gal} \times 3.785 \text{ L/gal}}{1\text{L} \times 454 \text{ g/lb} \times 1{,}000 \text{ mg/g}} = 480 \text{ lb}$$

16.13.3 Aerobic Tank Volume

The aerobic tank volume can be computed in situations in which no significant nitrification will occur by the following equation (WPCF, 1985):

$$V = \frac{Q_i(X_i + YS_i)}{X(K_dP_v + 1/\theta_C)} \tag{16.136}$$

where
V = volume of aerobic digester, cubic feet
Q_i = influent average flow rate to digester, cubic feet per day
X_i = influent suspended solids concentration, milligrams per liter
Y = fraction of the influent BOD consisting of raw primary sludge, in decimals
S_i = influent BOD, milligrams per liter
X = digester suspended solids concentration, milligrams per liter
K_d = reaction-rate constant, days^{-1}
P_v = volatile fraction of digester suspended solids, in decimals
θ = solids retention time, days

Example 16.106

Problem:
    The pH of an aerobic digester has declined to 6.1. How much sodium hydroxide must be added to raise the pH to 7.0? The volume of the digester is 370 m³. Results from jar tests show that 34 mg of caustic soda will raise the pH to 7.0 in a 2-L jar.

Solution:

$$\text{NaOH required per m}^3 = 34 \text{ mg/2L} = 17 \text{ mg/L}$$

$$= 17 \text{ g/m}^3$$

$$\text{NaOH to be added} = 17 \text{ g/m}^3 \times 370 \text{ m}^3$$

$$= 6290 \text{ g}$$

$$= 6.3 \text{ kg} = 13.9 \text{ lb}$$

16.13.4 Anaerobic Digestion Process Control Calculations

The purpose of anaerobic digestion is the same that of aerobic digestion: to stabilize organic matter, reduce volume, and eliminate pathogenic organisms. Equipment used in anaerobic digestion includes an anaerobic digester of the floating or fixed cover type. These include biosolids pumps for biosolids addition and withdrawal, as well as heating equipment such as heat exchangers, heaters and pumps, and mixing equipment for recirculation. Typical ancillaries include gas storage, cleaning equipment, and safety equipment such as vacuum relief and pressure relief devices, flame traps, and explosion-proof electrical equipment.

In the anaerobic process, biosolids enter the sealed digester where organic matter decomposes anaerobically. Anaerobic digestion is a two-stage process:

1. Sugars, starches, and carbohydrates are converted to volatile acids, carbon dioxide, and hydrogen sulfide.
2. Volatile acids are converted to methane gas.

We cover key anaerobic digestion process control calculations in the following subsections.

16.13.4.1 Required Seed Volume in Gallons

$$\text{Seed Volume (Gallons)} = \text{Digester Volume, gal} \times \% \text{ Seed} \qquad (16.137)$$

Example 16.107

Problem:

The new digester requires as seed 25% to achieve normal operation within the allotted time. If the digester volume is 280,000 gal, how many gallons of seed material are required?

Solution:

$$\text{Seed Volume} = 280{,}000 \times 0.25 = 70{,}000 \text{ gal}$$

16.13.4.2 Volatile Acids-to-Alkalinity Ratio

The volatile acids–alkalinity ratio can be used to control an anaerobic digester:

$$\text{Ratio} = \frac{\text{Volatile Acids Concentration}}{\text{Alkalinity Concentration}} \qquad (16.138)$$

Example 16.108

Problem:

The digester contains 240 mg/L volatile acids and 1840 mg/L alkalinity. What is the volatile acids–alkalinity ratio?

Solution:

$$\text{Ratio} = \frac{240 \text{ mg/L}}{1840 \text{ mg/L}} = 0.13$$

Key point: Increases in the ratio normally indicate a potential change in the operating condition of the digester.

16.13.4.3 Biosolids Retention Time

The length of time the biosolids remain in the digester is:

$$BRT = \frac{\text{Digester Volume in Gallons}}{\text{Biosolids Volume added per day, gpd}}$$

Example 16.109

Problem:
Biosolids are added to a 520,000-gal digester at the rate of 12,600 gal/day. What is the biosolids retention time?

Solution:

$$BRT = \frac{520,000 \text{ gal}}{12,600 \text{ gpd}} = 41.3 \text{ days}$$

16.13.4.4 Estimated Gas Production (Cubic Feet per Day)

The rate of gas production is normally expressed as the volume of gas (cubic feet) produced per pound of volatile matter destroyed. The total cubic feet of gas that a digester will produce per day can be calculated by:

$$\text{Gas production, ft}^3/\text{day} =$$
$$\text{Vol. Matter In, lb/day} \times \% \text{ Vol. Mat. Reduction} \times \text{Prod. Rate ft}^3/\text{lb} \qquad (16.140)$$

Key point: Multiplying the volatile matter added to the digester per day by the percent of volatile matter reduction (in decimal percent) gives the amount of volatile matter destroyed by the digestion process per day.

Example 16.110

Problem:
The digester reduces 11,500 lb of volatile matter per day. Currently, the volatile matter reduction achieved by the digester is 55%. The rate of gas production is 11.2 ft³ of gas per pound of volatile matter destroyed.

Solution:

$$\text{Gas Prod.} = 11,500 \text{ lb/day} \times 0.55 \times 11.2 \text{ ft}^3/\text{lb} = 70,840 \text{ ft}^3/\text{day}$$

16.13.4.5 Volatile Matter Reduction (Percent)

Because of the changes occurring during biosolids digestion, the calculation used to determine percent of volatile matter reduction is more complicated:

$$\% \text{ Red. } = \frac{(\% \text{ Vol. Matter}_{in} - \% \text{ Vol. Matter}_{out}) \times 100}{[\% \text{ Vol. Matter}_{in} - (\% \text{ Vol. Matter}_{in} \times \% \text{ Vol. Matter}_{out})]} \qquad (16.141)$$

Example 16.111

Problem:

Using the digester data provided here, determine the percent of volatile matter reduction for the digester: raw biosolids volatile matter, 71%, and digested biosolids volatile matter, 54%.

Solution:

$$\% \text{ Volatile Matter Reduction } = \frac{0.71 - 0.54}{[0.71 - (0.71 \times 0.54)]} = 52\%$$

16.13.4.6   Percent Moisture Reduction in Digested Biosolids

$$\% \text{ Moisture Reduction } = \frac{(\% \text{ Moisture}_{in} - \% \text{ Moisture}_{out}) \times 100}{[\% \text{ Moisture}_{in} - (\% \text{ Moisture}_{in} \times \% \text{ Moisture}_{out})]} \qquad (16.142)$$

Key point: Percent of moisture = 100% minus percent of solids.

Example 16.112

Problem:

Using the digester data provided in the following table, determine the percent of moisture reduction and percent of volatile matter reduction for the digester.

	% Solids	% Moisture
Raw biosolids	9	91 (100 − 9)
Digested biosolids	15	85 (100 − 15)

Solution:

$$\% \text{ Moisture Reduction } = \frac{(0.91 - 0.85) \times 100}{[0.91 - (0.91 \times 0.85)]} = 44\%$$

16.13.4.7   Gas Production

In measuring the performance of a digester, gas production is one of the most important parameters. Typically, gas production ranges from 800 to 1125 L of digester gas per kilogram of volatile solids destroyed. Gas produced from a properly operated digester contains approximately 68% methane and 32% carbon dioxide. If carbon dioxide exceeds 35%, the digestion system is operating incorrectly. The quantity of methane gas produced can be calculated by these equations, in SI and British units, respectively, derived by McCarty (1964):

$$V = 350[Q(S_o - S) / (1000) - 1.42P_x] \qquad (16.143)$$

$$V = 5.62[Q(S_o - S)8.34 - 1.42P_x] \qquad (16.144)$$

where
V = volume of methane produced at standard conditions (0°C, 32°F and 1 atm), liters per
                day or cubic feet per day
350, 5.62 = theoretical conversion factor for the amount of methane produced per kilogram (pound)
                of ultimate BOD oxidized, 350 L/kg or 5.62 ft³/lb
1000   = 1000 g/kg
Q = flow rate, cubic meters per or million gallons per day
S_o = influent ultimate BOD, milligrams per liter
S = effluent ultimate BOD, milligrams per liter
8.34   = conversion factor, pounds/(million gallons per day) (milligrams per liter)
P_x = net mass of cell tissue produced, kilograms per day or pounds per day

For a complete-mix, high-rate, two-stage anaerobic digester (without recycle), the mass of
biological solids synthesized daily, P_x, can be estimated by the following equations (in SI and
British system units, respectively):

$$P_x = \frac{Y[Q(S_o - S)]}{1 + k_d \theta_c} \qquad (16.145)$$

$$P_x = \frac{Y[Q(S_o - S)8.34]}{1 + k_d \theta_c} \qquad (16.146)$$

where
Y = yield coefficient, kilograms per kilogram or pounds per pound
k_d = endogenous coefficient, per day
θ_c = mean cell residence time, days

Other terms have been defined previously.

Example 16.113

Problem:
Determine the amount of methane generated per kilogram of ultimate BOD stabilized. Use
Glucose, $C_6H_{12}O_6$, as BOD.

Given:
Molecular weight of glucose: 180
Molecular weight of methane and carbon dioxide: 48
48/180 = 0.267
Oxidation of methane and carbon dioxide and water = 1.07 kg

Solution:

Step 1. Calculate the rate of the amount of methane generated per kilogram of BOD converted.

$$\frac{0.267}{1.07} = \frac{0.25}{1.0}$$

Thus, 0.25 kg of methane is produced by each kilogram of BOD stabilized.

Step 2. Calculate the volume equivalent of 0.25 kg of methane at the standard conditions (0°C and 1 atm).

$$\text{Volume} = (0.25 \times 1000 \text{ g})(1 \text{ mol}/16 \text{ g})(22.4 \text{ l/mol})$$

$$= 350 \text{ liters}$$

16.14 BIOSOLIDS DEWATERING AND DISPOSAL

16.14.1 Biosolids Dewatering

The process of removing enough water from liquid biosolids to change its consistency to that of a damp solid is called *biosolids dewatering*. Although the process is also called *biosolids drying,* the "dry" or dewatered biosolids may still contain a significant amount of water, often as much as 70%. At moisture contents of 70% or less, however, the biosolids no longer behave as a liquid and can be handled manually or mechanically.

Several methods are available to dewater biosolids. The particular types of dewatering techniques/devices used best describe the actual processes used to remove water from biosolids and change their form from a liquid to a damp solid. The commonly used techniques/devices include:

- Filter presses
- Vacuum filtration
- Sand drying beds

Key point: Centrifugation is also used in the dewatering process. However, in this text we concentrate on the unit processes traditionally used for biosolids dewatering.

Note that an ideal dewatering operation would capture all of the biosolids at minimum cost and the resultant dry biosolids solids or cake would be capable of being handled without causing unnecessary problems. Process reliability, ease of operation, and compatibility with the plant environment would also be optimized.

16.14.2 Pressure Filtration Calculations

In pressure filtration, the liquid is forced through the filter media by a positive pressure. Several types of presses are available, but the most commonly used types are plate and frame presses and belt presses.

16.14.3 Plate and Frame Press

The plate and frame press consists of vertical plates held in a frame and pressed together between a fixed and moving end. A cloth filter medium is mounted on the face of each individual plate. The press is closed, and biosolids are pumped into the press at pressures up to 225 psi and passed through feed holes in the trays along the length of the press. Filter presses usually require a precoat material, such as incinerator ash or diatomaceous earth, to aid in solids retention on the cloth and to allow easier release of the cake.

Performance factors for plate and frame presses include feed biosolids characteristics, type and amount of chemical conditioning, operating pressures, and type and amount of precoat. Filter press calculations typically used in wastewater solids handling operations include solids loading rate; net filter yield; hydraulic loading rate; biosolids feed rate; solids loading rate; flocculant feed rate; flocculant dosage; total suspended solids; and percent recovery.

16.14.3.1 Solids Loading Rate

The solids loading rate is a measure of the pounds per hour of solids applied per square foot of plate area, as shown in Equation 16.147:

$$\text{Sol. Loading Rate, lb/h/ft}^2 = \frac{(\text{Biosolids, gph}) (8.34, \text{lb/gal}) (\% \text{ Sol.}/100)}{\text{Plate Area, ft}^2} \qquad (16.147)$$

Example 16.114

Problem:
A filter press used to dewater digested primary biosolids receives a flow of 710 gal during a 2-h period. The biosolids have a solids content of 3.3%. If the plate surface area is 120 ft², what is the solids loading rate in pounds per hour per square foot?
The flow rate is given as gallons per 2 h. First, express this flow rate as gallons per hour: 710 gal/2 h = 355 gal/h.

Solution:

$$\text{Sol. Loading Rate, lb/h/ft}^2 = \frac{(\text{Biosolids, gph}) (8.34 \text{ lb/gal}) \dfrac{(\% \text{ Sol.}/100)}{100}}{\text{Plate Area, ft}^2}$$

$$= \frac{(b355 \text{ gph}) (8.34 \text{ lb/gal}) \dfrac{(3.3)}{100}}{120 \text{ ft}^2}$$

$$= 0.81 \text{ lb/h/ft}^2$$

Key point: The solids loading rate measures the pounds per hour of solids applied to each square foot of plate surface area. However, this does not reflect the time when biosolids feed to the press is stopped.

16.14.3.2 Net Filter Yield

Operated in the batch mode, biosolids are fed to the plate and frame filter press until the space between the plates is completely filled with solids. The biosolids flow to the press is then stopped and the plates are separated, allowing the biosolids cake to fall into a hopper or conveyor below. The *net filter yield,* measured in pounds per hour per square foot, reflects the run time, as well as the down time of the plate and frame filter press. To calculate the net filter yield, simply multiply the solids loading rate (in pounds per hour per square foot) by the ratio of filter run time to total cycle time as:

$$\text{N.F.Y.} = \frac{(\text{Biosolids, gph}) (8.34 \text{ lb/gal}) (\% \text{ Sol}/100)}{\text{Plate Area, ft}^2} \frac{\text{Filter Run Time}}{\text{Total Cycle Time}} \qquad (16.148)$$

Example 16.115

Problem:

A plate and frame filter press receives a flow of 660 gal of biosolids during a 2-h period. The solids concentration of the biosolids is 3.3% and the surface area of the plate is 110 ft². If the down time for biosolids cake discharge is 20 min, what is the net filter yield in pounds per hour per square foot?

Solution:

First, calculate solids loading rate; then multiply that number by the corrected time factor:

$$\text{Sol. Loading Rate} = \frac{(\text{Biosolids, gph}) \, (8.34 \text{ lb/gal}) \, (\% \text{ Sol.}/100)}{\text{Plate Area, ft}^2}$$

$$= \frac{(330 \text{ gph}) \, (8.34 \text{ lb/gal}) \, (3.3/100)}{100 \text{ ft}^2}$$

$$= 0.83 \text{ lb/h/ft}^2$$

Next, calculate net filter yield, using the corrected time factor:

$$\text{Net Filter Yield, lb/h/ft}^2 = \frac{(0.83 \text{ lb/h/ft}^2) \, (2 \text{ h})}{2.33 \text{ h}}$$

$$= 0.71 \text{ lb/h/ft}^2$$

16.14.4 Belt Filter Press

The belt filter press consists of two porous belts. The biosolids are sandwiched between the two porous belts (see Figure 16.9). The belts are pulled tightly together as they are passed around a series of rollers to squeeze water out of the biosolids. Polymer is added to the biosolids just before they get to the unit. The biosolids are then distributed across one of the belts to allow for some of the water to drain by gravity. The belts are then put together with the biosolids between them.

16.14.4.1 Hydraulic Loading Rate

Hydraulic loading for belt filters is a measure of gallons per minute of flow per foot or belt width.

$$\text{Hydraulic Loading Rate, gpm/ft} = \frac{\text{Flow, gpm}}{\text{Belt Width, ft}} \qquad (16.149)$$

Example 16.116

Problem:

A 6-ft wide belt press receives a flow of 110 gpm of primary biosolids. What is the hydraulic loading rate in gallons per minute per foot?

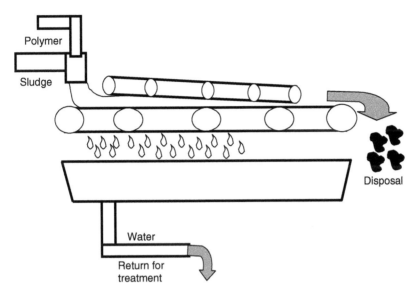

Figure 16.9 Belt filter press. (From Spellman, F.R., 1999, *Spellman's Standard Handbook for Wastewater Operators,* Vol. 1, Lancaster, PA: Technomic Publishing Company.)

Solution:

$$\text{Hydraulic Loading Rate, gpm/ft} = \frac{\text{Flow, gpm}}{\text{Belt Width, ft}}$$

$$= \frac{110 \text{ gpm}}{6 \text{ ft}}$$

$$= 18.3 \text{ gpm/ft}$$

Example 16.117

Problem:
A belt filter press 5 ft wide receives a primary biosolids flow of 150 gpm. What is the hydraulic loading rate in gallons per minute per square foot?

Solution:

$$\text{Hydraulic Loading Rate, gpm/ft} = \frac{\text{Flow, gpm}}{\text{Belt Width, ft}}$$

$$= \frac{150 \text{ gpm}}{5 \text{ ft}}$$

$$= 30 \text{ gpm/ft}$$

16.14.4.2 Biosolids Feed Rate

The biosolids feed rate to the belt filter press depends on several factors, including the biosolids pounds per day that must be dewatered; the maximum solids feed rate in pounds per hour that will produce an acceptable cake dryness; and the number of hours per day the belt press is in operation. The equation used in calculating biosolids feed rate is:

$$\text{Biosolids Feed Rate, lb/h} = \frac{\text{Biosolids to be dewatered, lb/day}}{\text{Operating Time, h/day}} \qquad (16.150)$$

Example 16.118

Problem:

   The amount of biosolids to be dewatered by the belt filter press is 20,600 lb/day. If the belt filter press is to be operated 10 h each day, what should the biosolids feed rate be to the press in pounds per hour?

Solution:

$$\text{Biosolids Feed Rate, lb/h} = \frac{\text{Biosolids to be dewatered, lb/day}}{\text{Operating Time, h/day}}$$

$$= \frac{20,600 \text{ lb/day}}{10 \text{ h/day}}$$

$$= 2060 \text{ lb/h}$$

16.14.5 Solids Loading Rate

The solids loading rate may be expressed as pounds per hour or as tons per hour. In either case, the calculation is based on biosolids flow (or feed) to the belt press and percent of milligrams per liter concentration of total suspended solids (TSS) in the biosolids. The equation used in calculating solids loading rate is:

$$\text{Sol. Load. Rate, lb/h} = (\text{Feed, gpm}) (60 \text{ min/h}) (8.34 \text{ lb/gal}) (\% \text{ TSS}/100) \qquad (16.151)$$

Example 16.119

Problem:

   The biosolids feed to a belt filter press is 120 gpm. If the total suspended solids concentration of the feed is 4%, what is the solids loading rate, in pounds per hour?

Solution:

$$\text{Sol. Load Rate, lb/h} = (\text{Feed, gpm}) (60 \text{ min/h}) (8.34 \text{ lb/gal}) (\% \text{ TSS}/100)$$

$$= (120 \text{ gpm}) (60 \text{ min/h}) (8.34 \text{ lb/gal}) (4/100)$$

$$= 2402 \text{ lb/h}$$

16.14.6 Flocculant Feed Rate

The flocculant feed rate may be calculated like all other milligrams-per-liter to pounds-per-day calculations:

$$\text{Flocculant Feed, lb/h} = \frac{(\text{Floc., mg/L)(Feed Rate, MGD)(8.34 lb/gal)}}{24 \text{ h/day}} \qquad (16.152)$$

Example 16.120

Problem:
The flocculent concentration for a belt filter press is 1% (10,000 mg/L). If the flocculent feed rate is 3 gpm, what is the flocculent feed rate in pounds per hour?

Solution:
First, calculate pounds per day of flocculent using the milligrams-per-liter to pounds-per-day calculation. Note that the gallons-per-minute feed flow must be expressed as million-gallons-per-day feed flow:

$$= \frac{(3 \text{ gpm)(1440 min/day)}}{1,000,000} = 0.00432 \text{ MGD}$$

$$\text{Flocculant Feed, lb/day} = (\text{mg/L Floc)(Feed Rate, MGD)(8.34 lb/gal)}$$

$$= (10,000 \text{ mg/L)(0.00432 MGD)(8.34 lb/gal)}$$

$$= 360 \text{ lb/day}$$

Then, convert pounds per day of flocculent to pounds per hour:

$$= \frac{360 \text{ lb/day}}{24 \text{ h/day}} = 15 \text{ lb/h}$$

16.14.7 Flocculant Dosage

Once the solids loading rate (tons per hour) and flocculant feed rate (pounds per hour) have been calculated, the flocculant dose in pounds per ton can be determined. The equation used to determine flocculant dosage is

$$\text{Flocculant Dosage, lb/ton} = \frac{\text{Flocculant, lb/h}}{\text{Solids Treated, ton/h}} \qquad (16.153)$$

Example 16.121

Problem:
A belt filter has solids loading rate of 3100 lb/h and a flocculant feed rate of 12 lb/h. Calculate the flocculant dose in pounds per ton of solids treated.

Solution:

First, convert pounds per hour of solids loading to tons per hour of solids loading:

$$\frac{3100 \text{ lb/h}}{2000 \text{ lb/ton}} = 1.55 \text{ ton/h}$$

Now calculate pounds of flocculant per ton of solids treated:

$$\text{Flocculant Dosage, lb/ton} = \frac{\text{Flocculant, lb/h}}{\text{Solids Treated, ton/h}}$$

$$\frac{12 \text{ lb/h}}{1.55 \text{ ton/h}}$$

$$= 7.8 \text{ lb/ton}$$

16.14.8 Total Suspended Solids

The feed biosolids solids comprise two types of solids: suspended solids and dissolved solids. Suspended solids that will not pass through a glass fiber filter pad can be further classified as total suspended solids (TSS), volatile suspended solids, and/or fixed suspended solids. They can also be separated into three components based on settling characteristics: settleable solids, floatable solids, and colloidal solids. Total suspended solids in wastewater are normally in the range of 100 to 350 mg/L. Dissolved solids that will pass through a glass fiber filter pad can also be classified as total dissolved solids (TDS), volatile dissolved solids, and fixed dissolved solids. Total dissolved solids are normally in the range of 250 to 850 mg/L.

Two lab tests can be used to estimate the TSS concentration of the feed biosolids concentration of the feed biosolids to the filter press: the total residue test, which measures suspended and dissolved solids concentrations, and the total filterable residue test, which measures only the dissolved solids concentration. Subtracting the total filterable residue from the total residue yields the total nonfilterable residue (TSS), as shown in Equation 16.154:

$$\text{Total Res., mg/L} - \text{Total Filterable Residue, mg/L} =$$

$$\text{Total Non-Filterable Residue, mg/L}$$

(16.154)

Example 16.122

Lab tests indicate that the total residue portion of a feed biosolids sample is 22,000 mg/L. The total filterable residue is 720 mg/L. On this basis, what is the estimated total suspended solids concentration of the biosolids sample?

$$\text{Total Res., mg/L} - \text{Total Filterable Residue, mg/L} = \text{Total Non-Filterable Residue, mg/L}$$

$$22{,}000 \text{ mg/L} - 720 \text{ mg/L} = 21{,}280 \text{ mg/L Total SS}$$

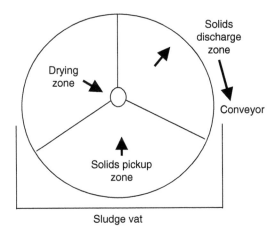

Figure 16.10   Vacuum filter. (From Spellman, F.R., 1999, *Spellman's Standard Handbook for Wastewater Operators,* Vol. 1, Lancaster, PA: Technomic Publishing Company.)

16.14.9 Rotary Vacuum Filter Dewatering Calculations

The rotary vacuum filter (see Figure 16.10) is a device used to separate solid material from liquid. The vacuum filter consists of a large drum with large holes in it covered with a filter cloth. The drum is partially submerged and rotated through a vat of conditioned biosolids. The rotary vacuum filter is capable of excellent solids capture and high-quality supernatant/filtrate; solids concentrations of 15 to 40% can be achieved.

16.14.9.1 Filter Loading

The filter loading for vacuum filters is a measure of pounds per hour of solids applied per square foot of drum surface area. The equation used in this calculation is:

$$\text{Filter Loading, lb/h/ft}^2 = \frac{\text{Solids to Filter, lb/h}}{\text{Surface Area, ft}^2} \tag{16.155}$$

Example 16.123

Problem:

Digested biosolids are applied to a vacuum filter at a rate of 70 gpm, with a solids concentration of 3%. If the vacuum filter has a surface area of 300 ft², what is the filter loading in pounds per hour per square foot?

Solution:

$$\text{Filter Loading, lb/h/ft}^2 = \frac{(\text{Biosolids, gpm}) (60 \text{ min/h}) (8.34 \text{ lb/gal}) (\% \text{ Sol.}/100)}{\text{Surface Area, ft}^2}$$

$$= \frac{(70 \text{ gpm}) (60 \text{ min/h}) (8.34 \text{ lb/gal}) (3/100)}{300 \text{ ft}^2}$$

$$= 3.5 \text{ lb/h/ft}^2$$

16.14.10 Filter Yield

One of the most common measures of vacuum filter performance is filter yield, the pounds per hour of dry solids in the dewatered biosolids (cake) discharged per square foot of filter area. It can be calculated using Equation 16.156:

$$\text{Filter Yield, lb/h/ft}^2 \ = \ \frac{(\text{Wet Cake Flow, lb/h})\dfrac{(\% \text{ Solids in Cake})}{100}}{\text{Filter Area, ft}^2} \tag{16.156}$$

Example 16.124

Problem:

The wet cake flow from a vacuum filter is 9000 lb/h. If the filter area is 300 ft² and the percent solids in the cake is 25%, what is the filter yield in pounds per hour per square foot?

Solution:

$$\text{Filter Yield, lb/h/ft}^2 \ = \ \frac{(\text{Wet Cake Flow, lb/h})\dfrac{(\% \text{ Solids in Cake})}{100}}{\text{Filter Area, ft}^2}$$

$$= \ \frac{(9000 \text{ lb/h})\dfrac{(25)}{100}}{300 \text{ ft}^2}$$

$$= \ 7.5 \text{ lb/h ft}^2$$

16.14.11 Vacuum Filter Operating Time

The vacuum filter operating time required to process given pounds per day of solids can be calculated using Equation 16.156. The vacuum filter operating time, of course, is the unknown factor, designated by x.

Example 16.125

Problem:

A total of 4000 lb/day of primary biosolids solids are to be processed by a vacuum filter. The vacuum filter yield is 2.2 lb/h/ft². The solids recovery is 95%. If the area of the filter is 210 ft², how many hours per day must the vacuum filter remain in operation to process these solids?

Solution:

$$\text{Filter Yield, lb/h/ft}^2 \ = \ \frac{\dfrac{\text{Sol. To Filter, lb/day}}{\text{Filter Oper., lb/day}}}{\text{Filter Area, ft}^2} \ \frac{(\% \text{ Recovery})}{100}$$

$$2.2 \text{ lb/h/ft}^2 = \frac{\dfrac{4000 \text{ lb/day}}{x \text{ h/day Oper.}}}{210 \text{ ft}^2} \frac{(95)}{100}$$

$$2.2 \text{ lb/h/ft}^2 = \frac{(4000 \text{ lb/day})}{x \text{ h/day}} \frac{(1)}{210 \text{ ft}^2} \frac{(95)}{100}$$

$$x = \frac{(4000)(1)(95)}{(2.2)(210)(100)}$$

$$x = 8.2 \text{ h/day}$$

16.14.12 Percent Solids Recovery

The function of the vacuum filtration process is to separate the solids from the liquids in the biosolids processed. Therefore, the percent of feed solids "recovered" (sometimes referred to as the percent solids "capture") is a measure of the efficiency of the process. Equation 16.157 is used to determine percent solids recovery:

$$\% \text{ Sol. Rec.} = \frac{(\text{Wet Cake Flow, lb/h}) \dfrac{(\% \text{ Sol. in Cake})}{100}}{(\text{Biosolids Feed, lb/h}) \dfrac{(\% \text{ Sol. in Feed})}{100}} \times 100 \qquad (16.157)$$

Example 16.126

Problem:
   The biosolids feed to a vacuum is 3400 lb/day, with a solids content of 5.1%. If the wet cake flow is 600 lb/h with 25% solids content, what is the percent solids recovery?

Solution:

$$\% \text{ Sol. Rec.} = \frac{(\text{Wet Cake Flow, lb/h}) \dfrac{(\% \text{ Sol. in Cake})}{100}}{(\text{Biosolids Feed, lb/h}) \dfrac{(\% \text{ Sol. in Feed})}{100}} \times 100$$

$$\% \text{ Sol. Rec.} = \frac{(600 \text{ lb/h}) \dfrac{(25)}{100}}{(3400 \text{ lb/h}) \dfrac{(5.1)}{100}} \times 100$$

$$= \frac{150 \text{ lb/h}}{173 \text{ lb/h}} \times 100$$

$$\% \text{ Sol. Rec.} = 87\%$$

16.14.13 Sand Drying Beds

Drying beds are generally used for dewatering well-digested biosolids. Biosolids drying beds consist of a perforated or open joint drainage system in a support media, usually gravel or wire mesh. Drying beds are usually separated into workable sections by wood, concrete, or other materials and may be enclosed or opened to the weather. They may rely entirely on natural drainage and evaporation processes or may use a vacuum to assist the operation.

The oldest biosolids dewatering technique, sand drying beds consist of 6 to 12 in. of coarse sand underlain by layers of graded gravel ranging from 0.133 to 0.25 in. at the top and 0.75 to 1.5 in. at the bottom. The total gravel thickness is typically about 1 ft. Graded natural earth (4 to 6 in.) usually makes up the bottom, with a web of drain tile placed on 20- to 30-ft centers. Sidewalls and partitions between bed sections are usually of wooden planks or concrete and extend about 14 in. above the sand surface.

16.14.14 Sand Drying Beds Process Control Calculations

Typically, three calculations are used to monitor sand drying bed performance: total biosolids applied, solids loading rate, and biosolids withdrawal to drying beds.

16.14.14.1 Total Biosolids Applied

The total gallons of biosolids applied to sand drying beds may be calculated using the dimensions of the bed and depth of biosolids applied, as shown by Equation 16.158:

$$\text{Volume, gal} = (\text{length, ft}) (\text{width, ft}) (\text{depth, ft}) (7.48, \text{gal/ft}^3) \qquad (16.158)$$

Example 16.127

Problem:

A drying bed is 220 ft long and 20 ft wide. If biosolids are applied to a depth of 4 in., how many gallons of biosolids are applied to the drying bed?

Solution:

$$\text{Volume, gal} = (\text{l}) (\text{w}) (\text{d}) (7.48, \text{gal/ft}^3)$$

$$= (220 \text{ ft}) (20 \text{ ft}) (0.33 \text{ ft}) (7.48 \text{ gal/ft}^3)$$

$$= 10,861 \text{ gal}$$

16.14.14.2 Solids Loading Rate

The biosolids loading rate may be expressed as pounds per year per square foot. The loading rate is dependent on biosolids applied per applications; pounds; percent solids concentration; cycle length; and square feet of sand bed area. The equation for biosolids loading rate is:

$$\text{Sol. Load. Rate, lb/yr/ft}^2 = \dfrac{\dfrac{\text{lb Biosolids Applied}}{\text{Days of Application}} (365 \text{ day/yr}) \dfrac{(\% \text{ Solids})}{100}}{(\text{length, ft}) (\text{width, ft})} \qquad (16.159)$$

Example 16.128

Problem:

A biosolids bed is 210 ft long and 25 ft wide. A total of 172,500 lb of biosolids is applied during each application to the sand drying bed. The biosolids have a solids content of 5%. If the drying and removal cycle requires 21 days, what is the solids loading rate in pounds per year per square foot?

Solution:

$$\text{Sol. Load. Rate, lb/yr/ft}^2 = \frac{\dfrac{\text{lb Biosolids Applied}}{\text{Days of Application}}(365 \text{ day/yr})\dfrac{(\% \text{ Solids})}{100}}{\text{Bed area, ft}^2}$$

$$= \frac{\dfrac{172,500 \text{ lb}}{21 \text{ Days}}(365 \text{ day/yr})\dfrac{(5)}{100}}{(210 \text{ ft})(25 \text{ ft})}$$

$$= 37.5 \text{ lb/yr/ft}^2$$

16.14.14.3 Biosolids Withdrawal to Drying Beds

Pumping digested biosolids to drying beds, thus making the dried biosolids useful as a soil conditioner, is one method among many for dewatering biosolids. Depending upon the climate of a region, the drying bed depth may range from 8 to 18 in. Therefore, the area covered by these drying beds may be substantial. For this reason, the use of drying beds is more common for smaller plants than for larger ones.

When calculating biosolids withdrawal to drying beds, use:

$$\text{Biosolids Withdrawn, ft}^3 = (0.785)(D^2)(\text{Drawdown, ft}) \qquad (16.160)$$

Example 16.129

Problem:

Biosolids are withdrawn from a digester with a diameter of 40 ft. If the biosolids are drawn down 2 ft, how many cubic feet are sent to the drying beds?

Solution:

$$\text{Biosolids Withdrawn, ft}^3 = (0.785)(D^2)(\text{ft drop})$$

$$= (0.785)(40 \text{ ft})(40 \text{ ft})(2 \text{ ft})$$

$$= 2512 \text{ ft}^3 \text{ withdrawn}$$

16.14.15 Biosolids Disposal

In the disposal of biosolids, land application, in one form or another, has become not only necessary (because of the banning of ocean dumping in the U.S. in 1992 and the shortage of landfill space since then) but also quite popular as a beneficial reuse practice. *Beneficial reuse* means that the biosolids are disposed of in an environmentally sound manner by recycling nutrients and soil conditions. Biosolids are being applied throughout the U.S. to agricultural and forest lands.

For use in land applications, the biosolids must meet certain conditions. They must comply with state and federal biosolids management/disposal regulations and must also be free of materials dangerous to human health (toxicities and pathogenic organisms, for example) and/or dangerous to the environment (toxicity, pesticides, and heavy metals, for example). Biosolids are applied to land by direct injection, by application and incorporation (plowing in), or by composting.

16.15 LAND APPLICATION CALCULATIONS

Land application of biosolids requires precise control to avoid problems. Use of process control calculations is part of overall process control. Calculations include determining disposal cost; plant available nitrogen (PAN); application rate (dry tons and wet tons per acre); metals loading rates; maximum allowable applications based upon metals loading; and site life based on metals loading.

16.15.1 Disposal Cost

The cost of disposal of biosolids can be determined by:

$$\text{Cost} = \text{Wet Tons Biosolids Produced/Year} \times \% \text{ Solids} \times \text{Cost/dry ton} \qquad (16.161)$$

Example 16.130

Problem:

The treatment system produces 1925 wet tons of biosolids for disposal each year. The biosolids are 18% solids. A contractor disposes of the biosolids for $28.00 per dry ton. What is the annual cost for biosolids disposal?

Solution:

$$\text{Cost} = 1925 \text{ wet tons/year} \times 0.18 \times \$28.00/\text{dry ton} = \$9702$$

16.15.2 Plant Available Nitrogen (PAN)

One factor considered when applying biosolids to land is the amount of nitrogen in the biosolids available to the plants grown on the site. This includes ammonia nitrogen and organic nitrogen. The organic nitrogen must be mineralized for plant consumption; only a portion of the organic nitrogen is mineralized per year. The mineralization factor (f_1) is assumed at 0.20. The amount of ammonia nitrogen available is directly related to the time elapsed between applying the biosolids and incorporating (plowing) the biosolids into the soil. We provide volatilization rates based upon the following example:

$$\text{PAN, lb/dry ton} =$$

$$[(\text{Or. Nit., mg/kg} \times f_1) + (\text{Amm. Nit., mg/kg} \times V_1)] \times 0.002 \text{ lb/dry ton} \qquad (16.162)$$

where:
f_1 = mineral rate for organic nitrogen (assume 0.20)
V_1 = volatilization rate ammonia nitrogen
V_1 = 1.00 if biosolids are injected
V_1 = 0.85 if biosolids are plowed in within 24 h
V_1 = 0.70 if biosolids are plowed in within 7 days

Example 16.131

Problem:
    The biosolids contain 21,000 mg/kg of organic nitrogen and 10,500 mg/kg of ammonia nitrogen and are incorporated into the soil within 24 h after application. What is the PAN per dry ton of solids?

Solution:

$$\text{PAN, lb/dry ton} = [(21,000 \text{ mg/kg} \times 0.20) + (10,500 \times 0.85)] \times 0.002$$

$$= 26.3 \text{ lb PAN/dry ton}$$

16.15.3 Application Rate Based on Crop Nitrogen Requirement

In most cases, the application rate of domestic biosolids to crop lands is controlled by the amount of nitrogen the crop requires. The biosolids application rate based upon the nitrogen requirement is determined by:

1. Using an agriculture handbook to determine the nitrogen requirement of the crop to be grown
2. Determining the amount of biosolids in dry tons required to provide this much nitrogen

$$\text{Dry ton/acre} = \frac{\text{Plant Nitrogen Requirement, lb/acre}}{\text{Plant Available Nitrogen, lb/dry ton}} \qquad (16.163)$$

Example 16.132

Problem:
    The crop to be planted on the land application site requires 150 lb nitrogen per acre. What is the required biosolids application rate if the PAN of the biosolids is 30 lb/dry ton?

Solution:

$$\text{Dry ton/acre} = \frac{150 \text{ lb nitrogen nitrogen/acre}}{30 \text{ lb/dry ton}} = 5 \text{ dry ton/acre}$$

16.15.4 Metals Loading

When biosolids are applied to land, metals concentrations are closely monitored and their loading on land application sites calculated:

$$\text{Loading, lb/acre} = \text{Metal Conc., mg/kg} \times 0.002 \text{ lb/dry ton} \times \text{Appl. Rate, dry ton/acre} \quad (16.164)$$

Example 16.133

Problem:

The biosolids contain 14 mg/kg of lead. Biosolids are currently applied to the site at a rate of 11 dry tons per acre. What is the metals loading rate for lead in pounds per acre?

Solution:

$$\text{Loading, lb/acre} = 14 \text{ mg/kg} \times 0.002 \text{ lb/dry ton} \times 11 \text{ dry ton} = 0.31 \text{ lb/acre}$$

16.15.5 Maximum Allowable Applications Based upon Metals Loading

If metals are present, they may limit the total number of applications that a site can receive. Metals loadings are normally expressed in terms of the maximum total amount of metal that can be applied to a site during its use:

$$\text{Applications} = \frac{\text{Max. Allowable Cumulative Load for the Metal, lb/ac}}{\text{Metal Loading, lb/acre/application}} \quad (16.165)$$

Example 16.134

Problem:

The maximum allowable cumulative lead loading is 48.0 lb/acre. Based upon the current loading of 0.35 lb/acre, how many applications of biosolids can be made to this site?

Solution:

$$\text{Applications} = \frac{48.0 \text{ lb/acre}}{0.35 \text{ lb/acre}} = 137 \text{ applications}$$

16.15.6 Site Life Based on Metals Loading

The maximum number of applications based upon metals loading and the number of applications per year can be used to determine the maximum site life:

$$\text{Applications} = \frac{\text{Maximum Allowable Applications}}{\text{Number of Applications Planned/Year}} \quad (16.166)$$

Example 16.135

Problem:

Biosolids are currently applied to a site twice annually. Based upon the lead content of the biosolids, the maximum number of applications is determined at 135 applications. Based upon the lead loading and the applications rate, how many years can this site be used?

Solution:

$$\text{Site Life} = \frac{135 \text{ applications}}{2 \text{ applications/yr}} = 68 \text{ yr} \qquad (16.167)$$

> *Key point*: When more than one metal is present, the calculations must be performed for each metal. The site life is the lowest value generated by these calculations.

16.16 BIOSOLIDS TO COMPOST

The purpose of composting biosolids is to stabilize the organic matter, reduce volume, eliminate pathogenic organisms, and produce a product that can be used as a soil amendment or conditioner. Composting is a biological process in which dewatered solids are usually mixed with a bulking agent (hardwood chips, for example) and stored until biological stabilization occurs. The composting mixture is ventilated during storage to provide sufficient oxygen for oxidation and to prevent odors. After the solids are stabilized, they are separated from the bulking agent. The composted solids are then stored for curing and later applied to farmlands or used in other beneficial ways. Expected performance of the composting operation for percent volatile matter reduction and percent moisture reduction ranges from 40 to 60%. Performance factors related to biosolids composting include moisture content; temperature; pH; nutrient availability; and aeration.

The biosolids must contain sufficient moisture to support the biological activity. If the moisture level is too low (40% or less), biological activity will be reduced or stopped. If the moisture level exceeds approximately 60%, it prevents sufficient airflow through the mixture.

The composting process operates best when temperatures are maintained within an operating range of 130 to 140°F; biological activities provide enough heat to increase the temperature well above this range. Forced air ventilation or mixing is used to remove heat and maintain the desired operating temperature range. The temperature of the composting solids, when maintained at the required levels, is sufficient to remove pathogenic organisms.

The influent pH can affect the performance of the process if it is extreme (less than 6.0 or greater than 11.0). The pH during composting may have some impact on the biological activity, but does not appear to be a major factor. Composted biosolids generally have a pH in the range of 6.8 to 7.5.

The critical nutrient in the composting process is nitrogen. The process works best when the ratio of nitrogen to carbon is in the range of 26 to 30 carbon to one nitrogen. Above this ratio, composting is slowed. Below this ratio, the nitrogen content of the final product may be less attractive as compost.

Aeration is essential to provide oxygen to the process and to control the temperature. In forced air processes, some means of odor control should be included in the design of the aeration system.

16.16.1 Composting Calculations

Pertinent composting process control calculations include determining percent of moisture of compost mixture and compost site capacity.

16.16.1.1 Blending Dewatered Biosolids with Composted Biosolids

Blending composted material with dewatered biosolids is similar to blending two different percent solids biosolids. The percent solids (or percent moisture) content of the mixture will always fall somewhere between the percent solids (or percent moisture) concentrations of the two materials being mixed. Equation 16.168 is used to determine percent moisture of mixture:

$$\% \text{ Moist. of Mixture } =$$

$$\frac{(\text{Biosolids, lb/day})\dfrac{(\% \text{ Moist.})}{100} + (\text{Compost, lb/day})\dfrac{(\% \text{ Moist.})}{100}}{(\text{Biosolids, lb/day}) \quad + \quad (\text{Compost, lb/day})} \times 100 \qquad (16.168)$$

Example 16.136

Problem:

If 5000 lb/day of dewatered biosolids is mixed with 2000 lb/day of compost, what is the percent moisture of the blend? The dewatered biosolids have a solids content of 25% (75% moisture) and the compost has 30% moisture content.

Solution:

$$\% \text{ Moist. of Mixture } = \frac{(\text{Biosolids, lb/day})\dfrac{(\% \text{ Moist.})}{100} + (\text{Compost, lb/day})\dfrac{(\% \text{ Moist.})}{100}}{(\text{Biosolids, lb/day}) \quad + \quad (\text{Compost, lb/day})} \times 100$$

$$= \frac{(5000 \text{ lb/day})\dfrac{(75)}{100} + (2000 \text{ lb/day})\dfrac{(30)}{100}}{(5000 \text{ lb/day}) \quad + \quad (2000 \text{ lb/day})} \times 100$$

$$= \frac{3750 \text{ lb/day } + 600 \text{ lb/day}}{7000 \text{ lb/day}}$$

$$= 62\%$$

16.16.1.2 Compost Site Capacity Calculation

An important consideration in compost operation is the solids processing capability (fill time), pounds per day or pounds per week. Equation 16.169 is used to calculate site capacity:

$$\text{Fill Time, days } = \frac{\text{Total Available Capacity, yd}^3}{\dfrac{\text{Wet Compost, lb/day}}{\text{Compost Bulk Density, lb/yd}^3}} \qquad (16.169)$$

Example 16.137

Problem:

A composting facility has an available capacity of 7600 yd³. If the composting cycle is 21 days, how many pounds per day of wet compost can be processed by this facility? Assume a compost bulk density of 900 lb/yd³.

Solution:

$$\text{Fill Time, days} = \frac{\dfrac{\text{Total Available Capacity, yd}^3}{\text{Wet Compost, lb/day}}}{\text{Compost Bulk Density, lb/yd}^3}$$

$$21 \text{ days} = \frac{\dfrac{7600 \text{ yd}^3}{x \text{ lb/day}}}{900 \text{ lb/yd}^3}$$

$$21 \text{ days} = \frac{(7600 \text{ yd}^3)\,(900 \text{ lb/yd}^3)}{x \text{ lb/day}}$$

$$x \text{ lb/day} = \frac{(7600 \text{ yd}^3)\,(900 \text{ lbs/yd}^3)}{21 \text{ days}}$$

$$x = 325,714 \text{ lb/day}$$

16.17 WASTEWATER LAB CALCULATIONS

16.17.1 The Wastewater Lab

Wastewater treatment plants are sized to meet the need (hopefully for the present and the future). No matter what the size of the treatment plant is, some space or area within the plant is designated as the "lab" area (ranging from closet-size to fully equipped and staffed environmental laboratories). Wastewater laboratories usually perform a number of different tests and provide operators with the information necessary to operate the treatment facility. Laboratory testing usually includes pH, COD, total phosphorus, fecal coliform count, and BOD (seeded) test. The standard reference for performing wastewater testing is *Standard Methods for the Examination of Water and Wastewater*.

In this subsection, we focus on wastewater lab tests that involve various calculations. Specifically, we focus on calculations used to determine proportioning factor for composite sampling; BOD; molarity and moles; normality; settleability; settleable solids; biosolids total; fixed and volatile solids; suspended solids and volatile suspended solids; and biosolids volume index and biosolids density index.

16.17.2 Composite Sampling Calculation (Proportioning Factor)

In preparing oven-baked food, a cook pays close attention in setting the correct oven temperature. Usually, the cook sets the temperature at the correct setting and then moves on to some other chore; the oven thermostat makes sure that the oven-baked food is cooked at the correct temperature. Unlike the cook, in wastewater treatment plant operations, the operator does not have the luxury of setting a plant parameter and then walking off and forgetting about it. To optimize plant operations, various adjustments to unit processes must be made on an on-going basis.

The operator makes unit process adjustments based on local knowledge (experience) and on lab test results. However, before lab tests can be performed, samples must be taken. Two basic types of samples are in common use: grab samples and composite samples. The type of sample taken depends on the specific test, the reason the sample is being collected, and the requirements in the plant discharge permit.

A *grab sample* is a discrete sample collected at one time and one location. It is primarily used for any parameter whose concentration can change quickly (dissolved oxygen, pH, temperature, and total chlorine residual, for example) and is representative only of the conditions at the time of collection. A *composite sample* consists of a series of individual grab samples taken at specified time intervals and in proportion to flow. The individual grab samples are mixed together in proportion to the flow rate at the time the sample was collected to form the composite sample. The composite sample represents the character of the wastewater over a period of time.

16.17.3 Composite Sampling Procedure and Calculation

Because knowledge of the procedure used in processing composite samples is important (a basic requirement) to the wastewater operator, in this subsection, we cover the actual procedure used.

Procedure:

1. Determine the total amount of sample required for all tests to be performed on the composite sample.
2. Determine the treatment system's average daily flow.
   Key point: Average daily flow can be determined by using several months of data; this provides a more representative value.
3. Calculate a proportioning factor:

$$\text{Prop. Factor (PF)} = \frac{\text{Total Sample Volume Required, mm}}{\text{\# of Samples to be Calculated} \times \text{Av. Daily Flow, MGD}} \quad (16.170)$$

   Key point: Round the proportioning factor to the nearest 50 units (50, 100, 150, etc.) to simplify calculation of the sample volume.
4. Collect the individual samples in accordance with the schedule (once per hour, once per 15 min, etc.).
5. Determine flow rate at the time the sample was collected.
6. Calculate the specific amount to add to the composite container:

$$\text{Required Volume, mL} = \text{Flow}^{T} \times \text{PF} \quad (16.171)$$

   where T = time sample was collected.

7. Mix the individual sample thoroughly, measure the required volume, and add to composite storage container.
8. Refrigerate the composite sample throughout the collection period.

Example 16.138

Problem:

The effluent testing requires 3645 mL of sample. The average daily flow is 4.05 MGD. Using the flows given in the following table, calculate the amount of sample to be added at each of the times shown:

Time	Flow, MGD
8 A.M.	3.88
9 A.M.	4.10
10 A.M.	5.05
11 A.M.	5.25
12 Noon	3.80
1 P.M.	3.65
2 P.M.	3.20
3 P.M.	3.45
4 P.M.	4.10

Solution:

$$\text{Proportioning Factor (PF)} = \frac{3650 \text{ mL}}{9 \text{ Samples} \times 4.05 \text{ MGD}}$$

$$= 100$$

$$\text{Volume}_{8 \text{ AM}} = 3.88 \times 100 = 388 \ (400) \text{ mL}$$
$$\text{Volume}_{9 \text{ AM}} = 4.10 \times 100 = 410 \ (400) \text{ mL}$$
$$\text{Volume}_{10 \text{ AM}} = 5.05 \times 100 = 505 \ (500) \text{ mL}$$
$$\text{Volume}_{11 \text{ AM}} = 5.25 \times 100 = 525 \ (550) \text{ mL}$$
$$\text{Volume}_{12 \text{ N}} = 3.80 \times 100 = 380 \ (400) \text{ mL}$$
$$\text{Volume}_{1 \text{ PM}} = 3.65 \times 100 = 365 \ (350) \text{ mL}$$
$$\text{Volume}_{2 \text{ PM}} = 3.20 \times 100 = 320 \ (300) \text{ mL}$$
$$\text{Volume}_{3 \text{ PM}} = 3.45 \times 100 = 345 \ (350) \text{ mL}$$
$$\text{Volume}_{4 \text{ PM}} = 4.10 \times 100 = 410 \ (400) \text{ mL}$$

16.17.4 Biochemical Oxygen Demand (BOD) Calculations

BOD_5 measures the amount of organic matter that can be biologically oxidized under controlled conditions (5 days at 20°C in the dark). Several criteria are used in selecting which BOD_5 dilutions should be used for calculating test results. Consult a laboratory testing reference manual (such as *Standard Methods*) for this information. Two basic calculations are used for BOD_5. The first is used for unseeded samples and the second must be used whenever BOD_5 samples are seeded. We introduce both methods and provide examples next.

16.17.4.1 BOD₅ (Unseeded)

$$BOD_5 \text{ (Unseeded)} = \frac{(DO_{start}, \text{ mg/L} - DO_{final}, \text{ mg/L}) \times 300 \text{ mL}}{\text{Sample Volume, mL}} \qquad (16.172)$$

Example 16.139

Problem:

The BOD_5 test is completed. Bottle 1 of the test had dissolved oxygen (DO) of 7.1 mg/L at the start of the test. After 5 days, bottle 1 had a DO of 2.9 mg/L. Bottle 1 contained 120 mg/L of sample. Determine 30 D_5 (unseeded).

Solution:

$$BOD_5 \text{ (Unseeded)} = \frac{(7.1 \text{mg/L} - 2.9 \text{ mg/L}) \times 300 \text{ mL}}{120 \text{ mL}} = 10.5 \text{ mg/L}$$

16.17.4.2 BOD_5 (Seeded)

If the BOD_5 sample has been exposed to conditions that could reduce the number of healthy, active organisms, the sample must be seeded with organisms. Seeding requires using a correction factor to remove the BOD_5 contribution of the seed material:

$$\text{Seed Correction} = \frac{\text{Seed Material } BOD_5 \times \text{ Seed in Dilution, mL}}{300 \text{ mL}} \tag{16.173}$$

$$BOD_5 \text{ (Seeded)} = \frac{[(DO_{start}, \text{ mg/L} - DO_{final}, \text{ mg/L}) - \text{ Seed Corr.}] \times 300}{\text{Sample Volume, mL}} \tag{16.174}$$

Example 16.140

Problem:

Using the data provided in the following table, determine the BOD_5:

BOD_5 of seed material		90 mg/L
Dilution #1	Milliliters of seed material	3 mL
	Milliliters of sample	100 mL
	Start DO	7.6 mg/L
	Final DO	2.7 mg/L

Solution:

$$\text{Seed Correction} = \frac{90 \text{ mg/L} \times 3 \text{ mL}}{300 \text{ mL}} = 0.90 \text{ mg/L}$$

$$BOD_5 \text{ (Seeded)} = \frac{[(7.6 \text{ mg/L} - 2.7 \text{ mg/L}) - 0.90] \times 300}{100 \text{ mL}} = 12 \text{ mg/L}$$

16.17.5 BOD 7-Day Moving Average

Because the BOD characteristic of wastewater varies from day to day, even hour to hour, operational control of the treatment system is most often accomplished based on trends in data rather than individual data points. The BOD 7-day moving average is a calculation of the BOD trend.

Key point: The 7-day moving average is called that because a new average is calculated each day, adding the new day's value and the six previous days' values:

$$7\text{-day Average BOD} = \frac{\underset{\text{Day 1}}{\text{BOD}} + \underset{\text{Day 2}}{\text{BOD}} + \underset{\text{Day 3}}{\text{BOD}} + \underset{\text{Day 4}}{\text{BOD}} + \underset{\text{Day 5}}{\text{BOD}} + \underset{\text{Day 6}}{\text{BOD}} + \underset{\text{Day 7}}{\text{BOD}}}{7} \quad (16.175)$$

Example 16.141

Problem:

Given the following primary effluent BOD test results, calculate the 7-day average.

Date	Milligrams per liter
June 1	200
June 2	210
June 3	204
June 4	205
June 5	222
June 6	214
June 7	218

Solution:

$$7\text{-day average BOD} = \frac{200 + 210 + 204 + 205 + 222 + 214 + 218}{7}$$

$$= 210 \text{ mg/L}$$

16.17.6 Moles and Molarity

Chemists have defined a very useful unit called the *mole*. Moles and molarity, a concentration term based on the mole, have many important applications in water/wastewater operations. A mole is defined as a gram molecular weight; that is, the molecular weight expressed as grams. For example, a mole of water is 18 g of water and a mole of glucose is 180 g of glucose. A mole of any compound always contains the same number of molecules. The number of molecules in a mole is called Avogadro's number and has a value of 6.022×10^{23}.

> *Interesting point*: How big is Avogadro's number? An Avogadro's number of soft drink cans would cover the surface of the Earth to a depth of over 200 miles.
> *Key point*: Molecular weight is the weight of one molecule; it is calculated by adding the weights of all the atoms present in one molecule. The units are atomic mass units (amu). The molecular weight is the weight of one molecule in daltons. The reason all moles have the same number of molecules is because the value of the mole is proportional to the molecular weight.

16.17.6.1 Moles

A mole is a quantity of a compound equal in weight to its formula weight. For example, the formula weight for water (H_2O; see Figure 16.11) can be determined using the periodic table of elements:

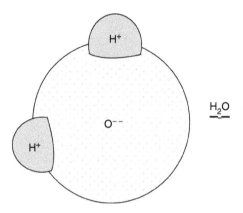

Figure 16.11 A molecule of water. (From Spellman, F.R., 1999, *Spellman's Standard Handbook for Wastewater Operators,* Vol. 1, Lancaster, PA: Technomic Publishing Company.)

$$
\begin{aligned}
\text{Hydrogen } (1.008) \times 2 &= 2.016 \\
\text{Oxygen} &= 16.000 \\
\hline
\text{Formula weight of } H_2O &= 18.016
\end{aligned}
$$

Because the formula weight of water is 18.016, a mole is 18.016 units of weight. A *gram-mole* is 18.016 g of water. A *pound-mole* is 18.016 lb of water. For our purposes in this text, the term "mole" means "gram-mole." The equation used in determining moles is:

$$
\text{Moles} = \frac{\text{Grams of Chemical}}{\text{Formula Wt of Chemical}} \tag{16.176}
$$

Example 16.142

Problem:
The atomic weight of a certain chemical is 66. If 35 g of the chemical are used in making up a 1-L solution, how many moles are used?

Solution:

$$
\text{Moles} = \frac{\text{Grams of Chemical}}{\text{Formula Wt of Chemical}}
$$

$$
= \frac{66 \text{ g}}{35 \text{ g/mol}}
$$

$$
= 1.9 \text{ mol}
$$

The *molarity* of a solution is calculated by taking the moles of solute and dividing by the liters of solution:

$$\text{Molarity} = \frac{\text{Moles of solute}}{\text{Liters of solution}} \qquad (16.177)$$

Example 16.143

Problem:

What is the molarity of 2 mol of solute dissolved in 1 L of solvent?

Solution:

$$\text{Molarity} = \frac{2 \text{ mol}}{1 \text{ L}} = 2\text{M}$$

Key point: Measurement in moles is a measurement of the *amount* of a substance. Measurement in molarity is a measurement of the *concentration* of a substance — the amount (moles) per unit volume (liters).

16.17.6.2 Normality

The molarity of a solution refers to its concentration (the solute dissolved in the solution). The normality of a solution refers to the number of equivalents of solute per liter of solution. Defining the chemical equivalent depends on the substance or type of chemical reaction under consideration. Because the concept of equivalents is based on the "reacting power" of an element or compound, it follows that a specific number of equivalents of one substance will react with the same number of equivalents of another substance. When the concept of equivalents is taken into consideration, chemicals are less likely to be wasted as excess amounts.

Keeping in mind that normality is a measure of the reacting power of a solution (1 equivalent of a substance reacts with 1 equivalent of another substance), we use the following equation to determine normality.

$$\text{Normality} = \frac{\text{No. of Equivalents of Solute}}{\text{Liters of Solution}} \qquad (16.178)$$

Example 16.144

Problem:

If 2.0 equivalents of a chemical are dissolved in 1.5 L of solution, what is the normality of the solution?

Solution:

$$\text{Normality} = \frac{\text{No. of Equivalents of Solute}}{\text{Liters of Solution}}$$

$$\text{Normality} = \frac{2.0 \text{ Equivalents}}{1.5 \text{ L}} = 1.33 \text{ N}$$

Example 16.145

Problem:

An 800-mL solution contains 1.6 equivalents of a chemical. What is the normality of the solution?

Solution:

First, convert 800 mL to liters:

$$= \frac{800 \text{ mL}}{1000 \text{ mL}} = 0.8 \text{ L}$$

Then, calculate the normality of the solution:

$$\text{Normality} = \frac{\text{No. of Equivalents of Solute}}{\text{Liters of Solution}}$$

$$= \frac{1.6 \text{ Equivalents}}{0.8 \text{ Liters}} = 2 \text{ N}$$

16.17.7 Settleability (Activated Biosolids Solids)

The settleability test is a test of the quality of the activated biosolids solids or activated sludge solids (mixed liquor suspended solids). Settled biosolids volume (SBV) or settled sludge volume (SSV) is determined at specified times during sample testing. Observations of 30 and 60 min are used for control. Subscripts (SBV_{30} or SSV_{30} and SBV_{60} or SSV_{60}) indicate settling time.

A sample of activated biosolids is taken from the aeration tank, poured into a 2000-mL graduated cylinder, and allowed to settle for 30 or 60 min. The settling characteristics of the biosolids in the graduated cylinder give a general indication of the settling of the MLSS in the final clarifier. From the settleability test, the percent of settleable solids can be calculated using:

$$\% \text{ Settleable Solids} = \frac{\text{mL Settled Solids}}{2000 \text{ mL Sample}} \times 100 \qquad (16.179)$$

Example 16.146

Problem:

The settleability test is conducted on a sample of MLSS. What is the percent of settleable solids if 420 mL settle in the 2000-mL graduate?

Solution:

$$\% \text{ Settleable Solids} = \frac{\text{mL Settled Solids}}{2000 \text{ mL Sample}} \times 100$$

$$= \frac{420 \text{ mL}}{2000 \text{ mL}} \times 100$$

$$= 21\%$$

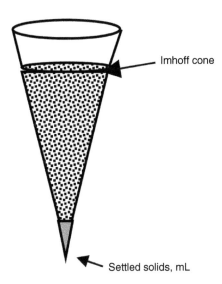

Figure 16.12   One-liter Imhoff cone. (From Spellman, F.R., 1999, *Spellman's Standard Handbook for Waste-water Operators,* Vol. 1, Lancaster, PA: Technomic Publishing Company.)

Example 16.147

Problem:

A 2000-mL sample of activated biosolids is tested for settleability. If the settled solids are measured as 410 mL, what is the percent of settled solids?

Solution:

$$\% \text{ Settleable Solids } = \frac{\text{mL Settled Solids}}{2000 \text{ mL Sample}} \times 100$$

$$= \frac{410 \text{ mL}}{2000 \text{ mL}} \times 100$$

$$= 20.5\%$$

16.17.8 Settleable Solids

The settleable solids test is an easy, quantitative method to measure sediment found in wastewater. An Imhoff cone (a plastic or glass 1-L cone; see Figure 16.12) is filled with 1 L of sample wastewater, stirred, and allowed to settle for 60 min. The settleable solids test, unlike the settleability test, is conducted on samples from sedimentation tank or clarifier influent and effluent to determine percent of removal of settleable solids. The percent of settleable solids is determined by:

$$\% \text{ Settleable Solids Removed } = \frac{\text{Set. Solids Removed, mL/L}}{\text{Set. Solids in Influent, mL/L}} \times 100 \qquad (16.180)$$

Example 16.148

Problem:

Calculate the percent removal of settleable solids if the settleable solids of the sedimentation tank influent are 15 mL/L and the settleable solids of the effluent are 0.4 mL/L.

Solution:

First, subtract 0.4 mL/L from 15.0, which equals 14.6 mL/L removed settleable solids. Next, insert parameters into Equation 16.180:

$$\% \text{ Set. Sol. Removed } = \frac{14.6 \text{ mL/L}}{15.0 \text{ mL/L}} \times 100$$

$$= 97\%$$

Example 16.149

Problem:

Calculate the percent removal of settleable solids if the settleable solids of the sedimentation tank influent are 13 mL/L and the settleable solids of the effluent are 0.5 mL/L.

Solution:

First, subtract 0.5 mL/L from 13 mL/L, which equals 12.5 mL/L of removed settleable solids:

$$\% \text{ Set. Sol. Removed } = \frac{12.5 \text{ mL/L}}{13.0 \text{ mL/L}} \times 100$$

$$= 96\%$$

16.17.9 Biosolids Total Solids, Fixed Solids, and Volatile Solids

Wastewater consists of water and solids (see Figure 16.13). The total solids may be further classified as *volatile* (organics) or *fixed* (inorganics). Normally, total solids and volatile solids are expressed as percents; suspended solids are generally expressed as milligrams per liter.

In calculating percents or miligrams per liter of concentrations, certain concepts must be understood:

- Total solids — the residue left in the vessel after evaporation of liquid from a sample and subsequent drying in an oven at 103 to 105°C.
- Fixed solids — the residue left in the vessel after a sample is ignited (heated to dryness at 550°C).
- Volatile solids — the weight loss after a sample is ignited (heated to dryness at 550°C).

Determinations of fixed and volatile solids do not distinguish precisely between inorganic and organic matter because the loss on ignition is not confined to organic matter; it includes losses due to decomposition or volatilization of some mineral salts.

> *Key point*: When the word *biosolids* is used, it may be understood to mean a semiliquid mass composed of solids and water. The term *solids*, however, is used to mean dry solids after the evaporation of water.

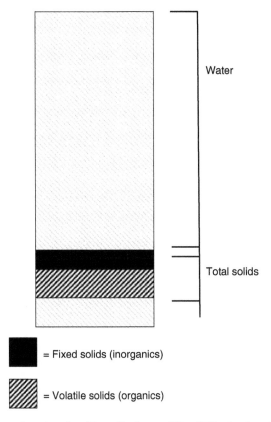

Figure 16.13 Composition of wastewater. (From Spellman, F.R., 1999, *Spellman's Standard Handbook for Wastewater Operators,* Vol. 1, Lancaster, PA: Technomic Publishing Company.)

Percents of total solids and volatile solids are calculated as:

$$\% \text{ Total Solids } = \frac{\text{Total Solids Weight}}{\text{Biosolids Sample Weight}} \times 100 \qquad (16.181)$$

$$\% \text{ Volatile Solids } = \frac{\text{Volatile Solids Weight}}{\text{Total Solids Weight}} \times 100 \qquad (16.182)$$

Example 16.150

Problem:

Given the information below, determine the percent of solids in the sample and the percent of volatile solids in the biosolids sample:

	Biosolids Sample	After Drying	After Burning (Ash)
Weight of Sample & Dish	73.43 g	24.88	22.98
Weight of Dish (tare weight)	22.28 g	22.28	22.28

Solution:

Step 1. To calculate the percent of total solids, the grams of total solids (solids after drying) and grams of biosolids sample must be determined:

<table>
<tr><td align="center">Total Solids</td><td align="center">Biosolids Sample</td></tr>
<tr><td align="center">24.88 g Total Solids and Dish</td><td align="center">73.43 g Biosolids and Dish</td></tr>
<tr><td align="center">−22.28 g Weight of Dish</td><td align="center">−22.28 g Weight of Dish</td></tr>
<tr><td align="center">2.60 g Total Solids</td><td align="center">51.15 g Biosolids</td></tr>
</table>

$$\% \text{ Total Solids} = \frac{\text{Wt. of Total Solids}}{\text{Wt. of Biosolids Sample}} \times 100$$

$$= \frac{2.60 \text{ g}}{51.15 \text{ g}} \times 100$$

$$= 5\% \text{ Total Solids}$$

Step 2. To calculate the percent of volatile solids, the grams of total solids and grams of volatile solids must be determined. Because percent of total solids has already been calculated in Step 1, only volatile solids must be calculated:

<table>
<tr><td align="center">Volatile Solids</td></tr>
<tr><td align="center">24.88 g Sample and Dish before Burning</td></tr>
<tr><td align="center">−22.98 g Sample and Dish after Burning</td></tr>
<tr><td align="center">1.90 g Solids Lost in Burning</td></tr>
</table>

$$\% \text{ Volatile Solids} = \frac{\text{Weight of Volatile Solids}}{\text{Weight of Total Sample}} \times 100$$

$$= \frac{1.90 \text{ g}}{2.60 \text{ g}} \times 100$$

$$= 73\% \text{ Volatile Solids}$$

16.17.10 Wastewater Suspended Solids and Volatile Suspended Solids

Total suspended solids (TSS) are the amount of filterable solids in a wastewater sample. Samples are filtered through a glass fiber filter. The filters are dried and weighed to determine the amount of total suspended solids in milligrams per liter of sample. Volatile suspended solids (VSS) are solids lost on ignition (heating to 500°C). They are useful because they give a rough approximation of the amount of organic matter present in the solid fraction of wastewater, activated biosolids, and industrial wastes. With the exception of the required drying time, the suspended solids and volatile suspended solids tests of wastewater are similar to those of the total and volatile solids performed for biosolids (described earlier).

Key point: The total and volatile solids of biosolids are generally expressed as percents, by weight. The biosolids samples are 100 mL and are unfiltered.

Calculations of suspended solids and volatile suspended solids are demonstrated in the following example.

Example 16.151

Problem:

Given the following information regarding a primary effluent sample, calculate the milligrams per liter of suspended solids and the percent of volatile suspended solids of the sample:

	After Drying (before Burning)	After Burning (Ash)
Weight of Sample and Dish	24.6268 g	24.6232 g
Weight of Dish (tare wt.)	24.6222 g	24.6222 g

Sample Volume = 50 mL

Solution:

Step 1. To calculate the milligrams of suspended solids per liter of sample, it is necessary first to determine grams of suspended solids:

$$
\begin{array}{ll}
24.6268 \text{ g} & \text{Dish and Suspended Solids} \\
- 24.6222 \text{ g} & \text{Dish} \\
\hline
00.0046 \text{ g} & \text{Suspended Solids}
\end{array}
$$

Next, calculate milligrams per liter of suspended solids (using a multiplication factor of 20 [this number will vary with sample volume] to make the denominator equal to 1 L (1000 mL):

$$
= \frac{0.0046 \text{ g SS}}{50 \text{ mL}} \times \frac{1000 \text{ mg}}{1 \text{ g}} \times \frac{20}{20} = \frac{92 \text{ mg}}{1000 \text{ mL}} = 92 \text{ mg/L SS}
$$

Step 2. To calculate percent of volatile suspended solids, it is necessary to know the weight of total suspended solids (calculated in Step 1) and volatile suspended solids:

$$
\begin{array}{l}
24.6268 \text{ g Dish and SS before Burning} \\
- 24.6232 \text{ g Dish and SS after Burning} \\
\hline
0.0036 \text{ g Solids Lost in Burning}
\end{array}
$$

$$
\% V_{SS} = \frac{\text{Wt. of Volatile Solids}}{\text{Wt. of Suspended Solids}} \times 100
$$

$$\% \text{ VSS} = \frac{0.0036 \text{ g VSS}}{0.0046 \text{ g}} \times 100$$

$$= 78\% \text{ VSS}$$

16.17.11 Biosolids Volume Index (BVI) and Biosolids Density Index (BDI)

Two variables are used to measure the settling characteristics of activated biosolids and to determine the return biosolids pumping rate. These are the volume of the biosolids (BVI) and the density of the biosolids (BDI) indices:

$$\text{BVI} = \frac{\% \text{ MLSS volume after 30 min settling}}{\% \text{ MLSS mg/L MLSS}} = \text{mL settled biosolids} \times 1000 \quad (16.183)$$

$$\text{BDI} = \frac{\text{MLSS (\%)}}{\% \text{ volume MLSS after 30 min settling}} \times 100 \quad (16.184)$$

These indices relate the weight of biosolids to the volume that the biosolids occupy. They show how well the liquids–solids separation part of the activated biosolids system is performing its function on the biological floc that has been produced and is to be settled out and returned to the aeration tanks or wasted. The better the liquid–solids separation is, the smaller the volume occupied by the settled biosolids and the lower the pumping rate required to keep the solids in circulation are.

Example 16.152

Problem:
The settleability test indicates that after 30 min, 220 mL of biosolids settle in the 1-L graduated cylinder. If the MLSS concentration in the aeration tank is 2400 mg/L, what is the biosolids volume?

Solution:

$$\text{BVI} = \frac{\text{Volume (determined by settleability test)}}{\text{Density (determined by the MLSS conc.)}} \quad (16.185)$$

$$\text{BVI} = \frac{220 \text{ mL/L}}{2400 \text{ mg/L}}$$

$$= \frac{220 \text{ mL/L}}{2400 \text{ mg/L}} (\text{convert milligrams to grams}) \frac{220 \text{ mL}}{2.4 \text{ g}}$$

$$= 92$$

The biosolids density index (BDI) is also a method of measuring the settling quality of activated biosolids; however, like the BVI parameter, it may or may not provide a true picture of the quality of the biosolids in question, unless compared with other relevant process parameters. It differs from

BVI in that the higher the BDI value is, the better the settling quality of the aerated mixed liquor is. Similarly, the lower the BDI is, the poorer the settling quality of the mixed liquor is.

BDI is the concentration in percent solids that the activated biosolids will assume after settling for 30 min. BDI will range from 2.00 to 1.33, and biosolids with values of one or more are generally considered to have good settling characteristics. In making the calculation of BDI, we simply invert the numerators and denominators and multiply by 100.

Example 16.153

Problem:

The MLSS concentration in the aeration tank is 2500 mg/L. If the activated biosolids settleability test indicates 225 mL settled in the 1-L graduated cylinder, what is the BDI?

Solution:

$$\text{BDI} = \frac{\text{Density (determined by the MLSS concentration)}}{\text{Volume (Determined by the settleability test)}} \times 100 \qquad (16.186)$$

$$\text{BDI} = \frac{2500 \text{ mg}}{225 \text{ mL}} \times 100 \text{(convert milligrams to grams)} = \frac{2.5 \text{ g}}{225 \text{ mL}} \times 100$$

$$= 1.11$$

REFERENCES

American Public Health Association. (1995). *Standard Methods for the Examination of Water and Wastewater,* 19th ed. Washington, D.C.: American Public Health Association.

Camp, T.R. (1946). Grit chamber design. *Sewage Works J.*, 14, 368–389.

Lawrence, A.W. and McCarty, P.L. (1970). Unified basis for biological treatment design and operation. *J. Sanitary Eng. Div. Proc. ASCE.* 96 (SA3), 757–888.

Mancini, J.L. and Barnhart, E.L. (1968). Industrial waste treatment in aerated lagoons. in Gloyna, E.R. and Eckenfelder, W.W. Jr., (Eds.), *Advances in Water Quality Improvement*. Austin: University of Texas Press.

McCarty, P.L. (1964). Anaerobic waste treatment fundamentals. *Public Works,* 95(9), 107–112.

Metcalf & Eddy, Inc. (1991). *Wastewater Engineering Treatment, Disposal, and Reuse*. New York: McGraw-Hill.

Operation of Wastewater Treatment Plants, Field Study Programs. (1986). Sacramento: California State University.

Russo, R. (2001). *Empire Falls*. New York: Vintage Books: Random House.

Spellman, F.R. (1999). *Spellman's Standard Handbook for Wastewater Operators*, Vol. 1. Lancaster, PA: Technomic Publishing Company

Spengel, D.B. and Dzombak, D.A. (1992). Biokinetic modeling and scale-up considerations for biological contractors, *Water Environ. Res.*, 64(3), 223–234.

WPCF (Water Pollution Control Federation) (1985). Sludge stabilization. Manual of Practice FD-9. Alexandria, Virginia: Water Pollution Control Federation.

PART VI

Math Concepts: Stormwater Engineering

CHAPTER 17

Stormwater Engineering Calculations

"Come Watson, come! The game is afoot!" (Doyle, 1930). Wayne County has operated an illicit Connection and Discharge Elimination Program for over 15 years. Its staff has gained valuable investigative expertise by experimenting with many different methods, committing lots of trial and error, and having a little bit of luck. Investigating for illicit discharges in the field is very similar to Holmes and Watson solving a case — it requires a mix of science, detection, deduction, and persistence.

Dean Tuomari and Susan Thompson, 2003

17.1 INTRODUCTION

For the environmental engineer involved with stormwater compliance programs, March 10, 2003, was a very significant date — the municipality deadline for compliance with new National Pollutant Discharge Elimination System (NPDES) permit applications for previously exempt municipal separate storm sewer systems (MS4s). The affected MS4s include federal- and state-regulated operations serving fewer than 100,000 people for areas that include military installations, prisons, hospitals, universities, and others. These operations are now required (since March 10, 2003) to comply with the Storm Water Phase II Rule, published December 8, 1999. State regulators may also subject certain other entities to regulations, such as municipally owned industrial sources, construction sites that disturb less than 1 acre, and other sources that contribute to a significant degradation of water quality.

To comply with the new stormwater regulations, environmental engineers must design stormwater discharge control systems. In the design phase, several mathematical computations are made to ensure that the finished stormwater discharge control system meets regulatory requirements.

This chapter provides guidelines for performing various engineering calculations associated with the design of stormwater management facilities, including extended-detention and retention basins and multistage outlet structures. Prerequisite to using these calculations is determining the hydrologic characteristic of the contributing watershed in the form of the peak discharge (in cubic feet per second) or a runoff hydrograph, depending on the hydrologic and hydraulic routing methods. Thus, before discussing the various math computations used in engineering a stormwater discharge system, we begin by defining general stormwater terms and acronyms, and discuss hydrologic methods.

Note: Much of the information contained in this chapter is adapted from Spellman and Drinan (2003) or excerpted from Federal and State Regulations, Soil Conservation Service (SCS) Technical Release Nos. 20 and 55 (TR-20 and TR-55), *Virginia Stormwater Management Handbook* (1999). The stormwater terms and acronyms that follow are from *Virginia Stormwater Management Handbook* (1999).

17.2   STORMWATER TERMS AND ACRONYMS

Antiseep collar — a device constructed around a pipe or other conduit, then placed into a dam, levee, or dike for the purpose of reducing seepage losses and piping failures along the conduit it surrounds.

Antivortex device — a device placed at the entrance of a pipe conduit structure to help prevent swirling action and cavitation from reducing the flow capacity of the conduit system.

Aquatic bench — A 10- to 15-ft wide bench around the inside perimeter of a permanent pool that ranges in depth from 0 to 12 in. Vegetated with emergent plants, the bench augments pollutant removal, provides habitat, protects the shoreline from the effects of water fluctuations, and enhances safety.

Aquifer — a porous, water-bearing geologic formation generally restricted to materials capable of yielding an appreciable supply of water.

Atmospheric deposition — the process by which atmospheric pollutants reach the land surface, as dry deposition or as dissolved or particulate matter contained in precipitation.

Average land cover condition — the percentage of impervious cover considered to generate an equivalent amount of phosphorus as the total combined land uses within the watershed.

Bankfull flow — condition in which flow fills a stream channel to the top of bank, at a point where the water begins to overflow onto a floodplain.

Base flow — discharge of water independent of surface runoff conditions, usually a function of groundwater levels.

Basin — a facility designed to impound stormwater runoff.

Best management practice (BMP) — structural or nonstructural practice designed to minimize the impacts of changes in land use on surface and groundwater systems. Structural BMP refers to basins or facilities engineered for the purpose of reducing the pollutant load in stormwater runoff, including bioretention and constructed stormwater wetlands. Nonstructural BMP refers to land use or development practices determined effective in minimizing the impact on receiving stream systems, including preservation of open space and stream buffers, and disconnection of impervious surfaces.

Biochemical oxygen demand (BOD) — an indirect measure of the concentration of biologically degradable material present in organic wastes. BOD usually reflects the amount of oxygen consumed in 5 days by biological processes breaking down organic waste.

Biological processes — a pollutant removal pathway in which microbes break down organic pollutants and transform nutrients.

Bioretention basin — water quality BMP engineered to filter the water quality volume through an engineered planting bed, consisting of a vegetated surface layer (vegetation, mulch, ground cover), planting soil, and sand bed (optional), and into the *in-situ* material; also called rain gardens.

Bioretention filter — a bioretention basin with the addition of a sand layer and collector pipe system beneath the planting bed.

COE — United States Army Corps of Engineers

Catch basin — an inlet chamber, usually built at the curb line of a street or low area for collection of surface runoff and admission into a sewer or subdrain. These structures commonly have a sediment sump at the base (below the sewer or subdrain discharge elevation) designed to retain solids below the point of overflow.

Channel stabilization — the introduction of natural or manmade materials placed within a channel to prevent or minimize the erosion of the channel bed and/or banks.

Check dam — a small dam constructed in a channel for the purpose of decreasing the flow velocity, minimizing channel scour, and promoting deposition of sediment. Check dams are a component of grassed swale BMPs.

Chemical oxygen demand (COD) — a measure of the oxygen required to oxidize all compounds, organic and inorganic, in water.

Chute — a high-velocity open channel for conveying water to a lower level without erosion.

Compaction — the process by which soil grains are rearranged to decrease void space and bring them in closer contact with one another, thereby reducing the permeability and increasing the soil's unit weight, and shear and bearing strength.

Constructed stormwater wetlands — areas intentionally designed and created to emulate the water quality improvement function of wetlands for the primary purpose of removing pollutants from stormwater.

Contour — a line representing a specific elevation on the land surface or a map.

Cradle — a structure, usually of concrete, shaped to fit around the bottom and sides of a conduit to support the conduit, increase its strength, and, in dams, to fill all voids between the underside of the conduit and soil.

Crest — the top of a dam, dike, spillway, or weir, frequently restricted to the overflow portion.

Curve number (CN) — a numerical representation of a given area's hydrologic soil group, plant cover, impervious cover, interception, and surface storage derived in accordance with Natural Resource Conservation Service methods. This number is used to convert rainfall depth into runoff volume; sometimes referred to as *runoff curve number.*

Cut — a reference to an area or material that has been excavated in the process of a grading operation.

Design storm — a selected rainfall hyetograph of specified amount, intensity, duration, and frequency used as a basis for design.

Detention basin — a stormwater management facility that temporarily impounds runoff and discharges it through a hydraulic outlet structure to a downstream conveyance system. Although a certain amount of outflow may also occur via infiltration through the surrounding soil, such amounts are negligible when compared to the outlet structure discharge rates and therefore are not considered in the facility's design. An extended detention basin impounds runoff only temporarily; it is normally dry during nonrainfall periods.

Disturbed area — an area in which the natural vegetative soil cover or existing surface treatment has been removed or altered and therefore is susceptible to erosion.

Diversion — a channel or dike constructed to direct water to areas where it can be used, treated, or disposed of safely.

Drainage basin — an area of land that contributes stormwater runoff to a designated point; also called a drainage area or, on a larger scale, a watershed.

Drop structure — a man-made device constructed to transition water to a lower elevation.

Duration — the length of time over which precipitation occurs.

Embankment — a man-made deposit of soil, rock, or other material used to form an impoundment.

Energy dissipator — a device used to reduce the velocity or turbulence of flowing water.

Erosion — the wearing away of the land surface by running water, wind, ice, or other geological agent.

  Accelerated erosion — erosion in excess of what is presumed or estimated to be naturally occurring levels and which is a direct result of human activities.

  Gully erosion — erosion process whereby water accumulates in narrow channels and removes the soil to depths ranging from a few inches to 1 or 2 ft to as much as 75 to 100 ft.

  Rill erosion — erosion process in which numerous small channels only several inches deep are formed.

  Sheet erosion — spattering of small soil particles caused by the impact of raindrops on wet soils. The loosened and spattered particles may subsequently be removed by surface runoff.

Extended detention basin — a stormwater management facility that temporarily impounds runoff and discharges it through a hydraulic outlet structure over a specified period of time to a downstream conveyance system for the purpose of water quality enhancement or stream channel erosion control. Although a certain amount of outflow may also occur via infiltration through the surrounding soil, such amounts are negligible when compared to outlet structure discharge rates and therefore are not considered in the facility's design. Because an extended detention basin impounds runoff only temporarily, it is normally dry during nonrainfall periods.

Extended detention basin — enhanced — an extended detention basin modified to increase pollutant removal by providing a shallow marsh in the lower stage of the basin.

Exfiltration — the downward movement of runoff through the bottom of a stormwater facility and into the soil.

Filter bed — the section of a constructed filtration device that houses the filtering media.

Filter strip — an area of vegetation, usually adjacent to a developed area, constructed to remove sediment, organic matter, and other pollutants from sheet flow runoff.

First flush — the first portion of runoff resulting from a rainfall event, usually defined as a depth in inches, considered to contain the highest pollutant concentration.

Floodplain — for a given flood event, that area of land adjoining a continuous water course that has been covered temporarily by water.

Flow splitter — an engineered hydraulic structure designed to divert a portion of storm flow to a BMP located out of the primary channel, to direct stormwater to a parallel pipe system, or to bypass a portion of baseflow around a BMP.

Forebay — storage space, commonly referred to as a sediment forebay, located near a stormwater BMP inlet that serves to trap incoming coarse sediments before they accumulate in the main treatment area.

Freeboard — the vertical distance between the surface elevation of the design high water and the top of a dam, levee, or diversion ridge.

Frequency (design storm frequency) — the recurrence interval of storm events having the same duration and volume. The frequency of a specified design storm can be expressed in terms of exceedance probability or return period.

    Exceedance probability — the probability that an event having a specified volume and duration will be exceeded in one time period, usually assumed to be 1 year. If a storm has a 1% chance of occurring in any given year, then it has an exceedance probability of 0.01.

    Return period — the average length of time between events having the same volume and duration. If a storm has a 1% chance of occurring in any given year, then it has a return period of 100 years.

GIS — geographic information system. A method of overlaying spatial land and land use data of different kinds. The data are referenced to a set of geographical coordinates and encoded in a computer software system. GIS is used by many localities to map utilities and sewer lines and to delineate zoning areas.

Gabion — A flexible woven wire basket composed of rectangular cells filled with large cobbles or riprap. Gabions may be assembled into many types of structures, including revetments, retaining walls, channel liners, drop structures, diversions, check dams, and groins.

Grassed swale — an earthen conveyance system that is broad and shallow, with check dams, vegetated with erosion resistant and flood-tolerant grasses. Grassed swales are engineered to remove pollutants from stormwater runoff by filtration through grass and infiltration into the soil.

HEC-1 — hydraulic engineering circular-1; a rainfall-runoff event simulation computer model sponsored by the U.S. Corps of Engineers.

Head — the height of water above any plane or object of reference; also used to express kinetic or potential energy, measured in feet, possessed by each unit weight of a liquid.

Hydric soil — a soil that is saturated, flooded, or ponded long enough during the growing season to develop anaerobic conditions in the upper part.

Hydrodynamic structure — an engineered flow-through structure that uses gravitational settling to separate sediments and oils from stormwater runoff.

Hydrograph — a plot showing the rate of discharge, depth, or velocity of flow vs. time for a given point on a stream or drainage system.

Hydrologic cycle — a continuous process by which water is cycled from the oceans to the atmosphere to the land and back to the oceans.

Hydrologic soil group (HSG) — SCS classification system of soils based on the permeability and infiltration rates of the soils. "A" type soils are primarily sandy with a high permeability, while "D" type soils are primarily clayey with low permeability.

Hyetograph — a graph of the time distribution of rainfall over a watershed.

Impervious cover — a surface composed of any material that significantly impedes or prevents natural infiltration of water into soil. Impervious surfaces include but are not limited to roofs, buildings, streets, parking areas, and any concrete, asphalt, or compacted gravel surface.

Impoundment — an artificial collection or storage of water, including reservoirs, pits, dugouts, and sumps.

Industrial stormwater permit — NPDES permit issued to a commercial industry for regulating the pollutant levels associated with industrial stormwater discharges. The permit may specify on-site pollution control strategies.

Infiltration facility — a stormwater management facility that temporarily impounds runoff and discharges it via infiltration through the surrounding soil. Although an infiltration facility may also be equipped with an outlet structure to discharge impounded runoff, such discharge is normally reserved for overflow and other emergency conditions. Because an infiltration facility impounds runoff only temporarily, it is normally dry during nonrainfall periods. Infiltration trenches, infiltration dry wells, and porous pavement are considered infiltration facilities.

Initial abstraction — the maximum amount of rainfall that can be absorbed under specific conditions without producing runoff; also called *initial losses*.

Intensity — the depth of rainfall divided by duration.

Invert — the lowest flow line elevation in any component of a conveyance system, including storm sewers, channels, and weirs.

Kjeldahl nitrogen (TKN) — a measure of the ammonia and organic nitrogen present in a water sample.

Lag time — the interval between the center of mass of the storm precipitation and the peak flow of the resultant runoff.

Low-impact development (LID) — hydrologically functional site design with pollution prevention measures to reduce impacts and compensate for development impacts on hydrology and water quality.

Manning's formula — equation used to predict the velocity of water flow in an open channel or pipeline.

Micropool — a smaller permanent pool incorporated into the design of larger stormwater ponds to avoid resuspension of particles, provide varying depth zones, and minimize impacts to adjacent natural features.

Modified rational method — a variation of the rational method used to calculate the critical storage volume whereby the storm duration can vary and does not necessarily equal the time of concentration.

Nonpoint source pollution — contaminants whose sources cannot be pinpointed that include sediment; nitrogen and phosphorous; hydrocarbons; heavy metals; and toxins, which are washed from the land surface in a diffuse manner by stormwater runoff.

Normal depth — depth of flow in an open conduit during uniform flow for the given conditions.

Off-line — stormwater management system designed to manage a portion of the stormwater diverted from a stream or storm drain. A flow splitter is typically used to divert the desired portion of the flow.

On-line — stormwater management system designed to manage stormwater in its original stream or drainage channel.

Peak discharge — the maximum rate of flow associated with a given rainfall event or channel.

Percolation rate — the velocity at which water moves through saturated granular material.

Point source — any discernible, confined, and discrete conveyance (including but not limited to any pipe, ditch, channel, tunnel, conduit, well, container, concentrated animal feeding operation, or landfill leachate collection system) from which pollutants may be discharged. This term does not include return flows from irrigated agriculture or agricultural storm water runoff.

Porosity — the ratio of pore or open space volume to total solids volume.

Principal spillway — the primary spillway or conduit for the discharge of water from an impoundment facility; generally constructed of permanent material and designed to regulate the rate of discharge.

Rational method — means of computing peak storm drainage flow rates based on average percent imperviousness of the site, mean rainfall intensity, and drainage area.

Recharge — replenishment of groundwater reservoirs by infiltration and transmission of water through permeable soils.

Redevelopment — any construction of, alteration of, or improvement to existing development.

Retention — permanent storage of stormwater.

Retention basin — a stormwater management facility, which includes a permanent impoundment or normal pool of water for the purpose of enhancing water quality, and therefore is normally wet, even during nonrainfall periods. Storm runoff inflows may be temporarily stored above this permanent impoundment for the purpose of reducing flooding or stream channel erosion.

Riprap — broken rock, cobbles, or boulders placed on earth surfaces (such as the face of a dam or the bank of a stream) for protection against erosive forces such as flow velocity and waves.

Riser — a vertical structure that extends from the bottom of an impoundment facility and houses the control devices (weirs/orifices) to achieve the desired rates of discharge for specific designs.

Roughness coefficient — a factor in velocity and discharge formulas representing the effect of channel roughness on energy losses in flowing water. Manning's "*n*" is a commonly used roughness coefficient.

Routing — a method of measuring the inflow and outflow from an impoundment structure while considering the change in storage volume over time.

Runoff — the portion of precipitation, snow melt, or irrigation water that runs off the land into surface waters.

Runoff coefficient — the fraction of total rainfall that appears as runoff; represented as *C* in the rational method formula.

SCS — Soil Conservation Service (now called Natural Resource Conservation Service, NRCS), a branch of the U.S. Department of Agriculture.

Safety bench — a flat area above the permanent pool and surrounding a stormwater pond designed to provide a separation to adjacent slopes. See also *bench*.

Sand filter — a contained bed of sand that acts to filter the first flush of runoff. The runoff is then collected beneath the sand bed and conveyed to an adequate discharge point or infiltrated into the *in-situ* soils.

Sediment forebay — a settling basin or plunge pool constructed at the incoming discharge points of a stormwater facility.

Soil test — chemical analysis of soil to determine the need for fertilizers or amendments for the species of plant being grown.

Stage — water surface elevation above any chosen datum.

Storm sewer — a system of pipes, separate from sanitary sewers, that only carries runoff from buildings and land surfaces.

Stormwater filtering (or filtration) — a pollutant removal method for stormwater runoff in which stormwater is passed through filter media such as sand, peat, grass, compost, or other materials to strain or filter pollutants out of the stormwater.

Stormwater hot spot — an area where the land use or activities are considered to generate runoff with concentrations of pollutants in excess of those typically found in stormwater.

Stream buffers — the zones of variable width located along both sides of a stream and designed to provide a protective natural area along a stream corridor.

Surcharge — flow condition occurring in closed conduits when the hydraulic grade line is above the crown of the sewer. This condition usually results in localized flooding or stormwater flowing out the top of inlet structures and manholes.

SWMM (storm water management model) — Rainfall-runoff event simulation model sponsored by the USEPA.

Technical release no. 20 (TR-20) — Project Formulation Hydrology; SCS watershed hydrology computer model used to compute runoff volumes and route storm events through stream valleys and/or impoundments.

Technical release no. 55 (TR-55) — Urban Hydrology for Small Watersheds; SCS watershed hydrology computation model used to calculate runoff volumes and provide a simplified routing for storm events through stream valleys and/or ponds.

Time of concentration — the time required for water to flow from the hydrologic most distant point (in time of flow) of the drainage area to the point of analysis (outlet). This time varies, generally depending on the slope and character of the surfaces.

Trash rack — a structural device used to prevent debris from entering a spillway or other hydraulic structure.

Travel time — the time required for water to flow from the outlet of a drainage sub-basin to the outlet of the entire drainage basin being analyzed. Travel time is normally concentrated flow through an open or closed channel.

Ultimate condition — full watershed build-out based on existing zoning.

Ultra–urban — densely developed urban areas in which little pervious surface exists.

Urban runoff — stormwater from city streets and adjacent domestic or commercial properties that carries nonpoint source pollutants of various kinds into the sewer systems and receiving waters.

Water quality window — the volume equal to the first ½ in. of runoff, multiplied by the impervious surface of the land development project.

Water surface profile — longitudinal profile assumed by the surface of a stream flowing in an open channel; hydraulic grade line.

Water table — upper surface of the free groundwater in a zone of saturation.

Watershed — a defined land area drained by a river, stream, or drainage way, or by a system of connecting rivers, streams, or drainage ways. In a watershed, all surface water within the area flows through a single outlet.

Wet weather flow — combination of dry weather flows and stormwater runoff.

Wetted perimeter — the length of the wetted surface of a natural or man-made channel.

17.3 HYDROLOGIC METHODS

Hydrology is the study of the properties, distribution, and effects of water on the Earth's surface, as well as in the soils, underlying rocks, and atmosphere. The hydrologic cycle (see Figure 17.1) is the closed loop through which water travels as it moves from one phase or surface to another. Water lost from the Earth's surface to the atmosphere by evaporation from the surface of lakes, rivers, and oceans or through the transpiration of plants forms clouds that condense to deposit moisture on the land and sea. A drop of water may travel thousands of miles between the time it evaporates and the time it falls to Earth again as rain, sleet, or snow. The water that collects on land flows to the ocean in streams and rivers or seeps into the earth, joining groundwater. Even groundwater eventually flows toward the ocean for recycling. When humans intervene in the natural

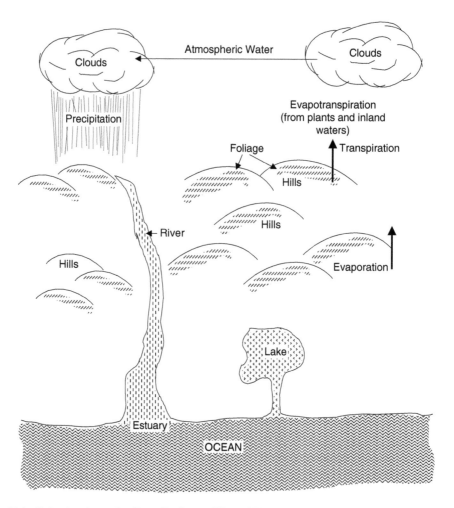

Figure 17.1   Natural water cycle. (From Spellman, F.R. and Drinan, J., 2000, *The Drinking Water Handbook.*
Lancaster, PA: Technomic Publishing Company.)

water cycle, they generate artificial water cycles or *urban* water cycles (local subsystems of the
water cycle, or integrated water cycles; see Figure 17.2) (Spellman and Drinan, 2000).

The hydrologic cycle is complex, and to simulate just a small portion of it (such as the
relationship between precipitation and surface runoff) can be an inexact science. Many variables
and dynamic relationships must be accounted for and, in most cases, reduced to basic assumptions.
However, these simplifications and assumptions make possible developing solutions to the flooding,
erosion, and water quality impacts associated with changes in land cover and hydrologic characteristics.

Proposed engineering solutions typically involve identifying a storm frequency as a benchmark
for controlling these impacts. The 2-, 10-, and 100-year frequency storms have traditionally been
used for hydrological modeling, followed by an engineered solution designed to offset increased
peak flow rates. The hydraulic calculations inherent in this process are dependent upon the engineer's ability to predict the amount of rainfall and its intensity. Recognizing that the frequency of
a specific rainfall depth or duration is developed from statistical analysis of historical rainfall data,
the engineer cannot presume to predict the characteristics of a future storm event accurately.

This section provides guidance for preparing acceptable calculations for various elements of
the hydrologic and hydraulic analysis of a watershed.

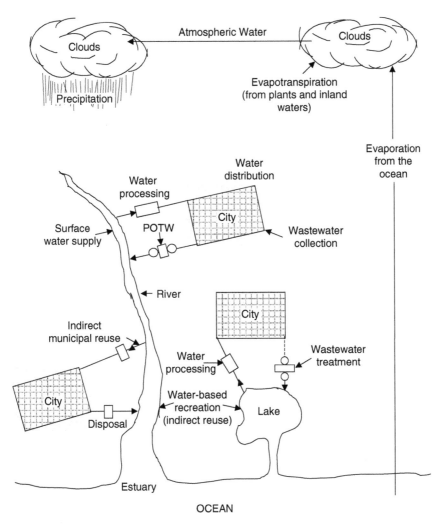

Figure 17.2   Urban water cycle. (From Spellman, F.R. and Drinan, J., 2000, *The Drinking Water Handbook.* Lancaster, PA: Technomic Publishing Company.)

17.3.1 Precipitation

Precipitation is a random event that cannot be predicted from historical data. However, any given precipitation event has several distinct and independent characteristics that can be quantified:

- *Duration* — the length of time over which precipitation occurs (hours)
- *Depth* — the amount of precipitation occurring throughout the storm duration (inches)
- *Frequency* — the recurrence interval of events with the same duration and volume
- *Intensity* — the depth divided by the duration (inches per hour)

A specified amount of rainfall may occur from many different combinations of intensities and durations (see Table 17.1). Note that the peak intensity of runoff associated with each combination varies widely. Storm events with the same intensity may have significantly different volumes and durations if the specified storm frequency (2, 10, or 100 years) is different (see, for example, Table 17.2). That some regulatory criterion specifies the volume (or intensity) and the duration for a specified frequency design storm becomes critical.

Table 17.1 Variations of Duration and Intensity for a Given Volume

Duration (h)	Intensity (in./h)	Volume (in.)
0.5	3.0	1.5
1.0	1.5	1.5
1.5	1.0	1.5
6.0	0.25	1.5

Source: *Virginia Stormwater Management Handbook*, 1999, Virginia Department of Conservation and Recreation, Division of Soil and Water Conservation.

Table 17.2 Variations of Volume, Duration, and Return Frequency for a Given Intensity

Duration (h)	Volume (in.)	Intensity (in./h)	Frequency (yr)
1.0	1.5	1.5	2
2.0	3.0	1.5	10
3.0	4.5	1.5	100

Source: *Virginia Stormwater Management Handbook*, 1999, Virginia Department of Conservation and Recreation, Division of Soil and Water Conservation.

Although specifying one combination of volume and duration may limit the analysis with regard to the critical variables for any given watershed (erosion, flooding, water, quality), such simplified parameters do allow us to establish a baseline from which to work. (This analysis supports the SCS 24-h design storm because an entire range of storm intensities is incorporated into the rainfall distribution.) Localities may choose to establish criteria based on specific watershed and receiving channel conditions that then dictate the appropriate design.

17.3.1.1 Frequency

The frequency of a specified design storm can be expressed in terms of exceedance probability or return period. *Exceedance probability* is the probability that an event with a specified volume and duration will be exceeded in one time period, usually assumed to be 1 year. *Return period* is the average length of time between events with the same volume and duration.

If a storm of a specified duration and volume has a 1% chance of occurring in any given year, it then has an exceedance probability of 0.01 and a return period of 100 years. The return period concept is often misunderstood in that it implies that a 100-year flood will occur only once in a 100-year period. This will not always hold true because storm events cannot be predicted deterministically. Because these events are random, the exceedance probability indicates that a finite probability exists (0.01 for this example) that the 100-year storm may occur in any given year or consecutive years, regardless of the historic occurrence of that storm event.

17.3.1.2 Intensity–Duration–Frequency (I–D–F) Curves

To establish the importance of the relationship between average intensity, duration, and frequency, the U.S. Weather Bureau compiled intensity–duration–frequency (I–D–F) curves based on historic rainfall data for most localities across the country. The rational method uses the I–D–F curves directly, while SCS methods generalize the rainfall data taken from the I–D–F curves and create rainfall distributions for various regions of the country.

Debate is occurring concerning which combinations of storm durations and intensities are appropriate to use in a hydrologic analysis for a typical urban development. Working within the limitations of the methodology as described later in this section, it is possible to model small drainage areas (1 to 20 acres) in an urban setting accurately using SCS or rational methods. The belief that the short, very intense storm generates the greatest need for stormwater management often leads engineers to use the rational method for stormwater management design because this method is based on short duration storms. However, the SCS 24-h storm method is also appropriate for short duration storms because it includes short storm intensities within the 24-h distribution.

17.3.1.3 SCS 24-H Storm Distribution

The SCS 24-h storm distribution curve was derived from the National Weather Bureau's Rainfall Frequency Atlases of compiled data for areas less than 400 mi², for durations up to 24 h, and for frequencies from 1 to 100 years. Data analysis resulted in four regional distributions: Type I and IA for use in Hawaii, Alaska, and the coastal side of the Sierra Nevada and Cascade Mountains in California, Washington, and Oregon; Type II distribution for most of the remainder of the U.S.; and Type III for the Gulf of Mexico and Atlantic coastal areas. The Type III distribution represents the potential impact of tropical storms, which can produce large 24-h rainfall amounts.

> *Note:* For a more detailed description of the development of dimensionless rainfall distributions, refer to the USDA Soil Conservation Service's *National Engineering Handbook*, Section 4 (U.S. SCS, 1956).

The SCS 24-h storm distributions are based on the generalized rainfall depth–duration–frequency relationships collected for rainfall events lasting from 30 min up to 24 h. The next largest 30-min incremental depth occurs just after the maximum depth; the third largest rainfall depth occurs just prior to the maximum depth, etc. This continues with each decreasing 30-min incremental depth until the smaller increments fall at the beginning and end of the 24-h rainfall (see Figure 17.3). Note that this process includes all of the critical storm intensities within the 24-h

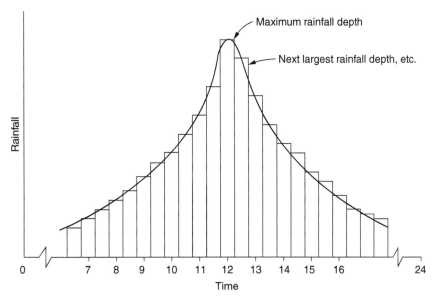

Figure 17.3 Typical 24-h rainfall distribution. (USDA SCS, 1956.)

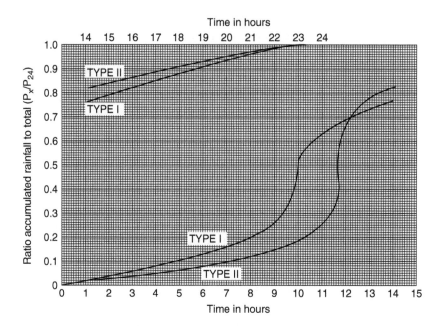

TYPE 1 - Coastal side of Sierra Nevada and Cascade Mountains in California, Oregon
and Washington; the Hawaiian Islands and Alaska

TYPE II - Remaining United States, Puerto Rico, and Virgin Islands

Figure 17.4   SCS 24-h rainfall distribution. (USDA SCS, 1986.)

distributions. The SCS 24-h storm distributions are, therefore, appropriate for rainfall and runoff modeling for small and large watersheds for the entire range of rainfall depths.

One of the stated disadvantages of using the SCS TR-55 method for hydrologic modeling is its restriction to the use of the 24-h storm. The following discussion, taken from Appendix B of the TR-55 manual (U.S. SCS, 1986) addresses this limitation:

> To avoid the use of a different set of rainfall intensities for each drainage area's size, a set of synthetic rainfall distributions having "nested" rainfall intensities was developed. The set "maximizes" the rainfall intensities by incorporating selected short-duration intensities within those needed for larger durations at the same probability level.

> For the size of the drainage areas for which SCS usually provides assistance, a storm period of 24 hours was chosen for the synthetic rainfall distributions. The 24-hour storm, while longer than that needed to determine peaks for these drainage areas, is appropriate for determining runoff volumes. Therefore, a single storm duration and associated synthetic rainfall distribution can be used to represent not only the peak discharges but also the runoff volumes for a range of drainage area sizes.

Figure 17.4 shows the SCS 24-h rainfall distribution, a graph of the fraction of total rainfall at any given time, t. Note that the peak intensity for the Type II distribution occurs between time $t =$ 11.5 h and $t = 12.5$ h.

17.3.1.4   Synthetic Storms

The alternative to a given rainfall "distribution" is to input a custom design storm into the model. This can be compiled from data gathered from a single rainfall event in a particular area or a synthetic storm created to test the response characteristics of a watershed under specific rainfall

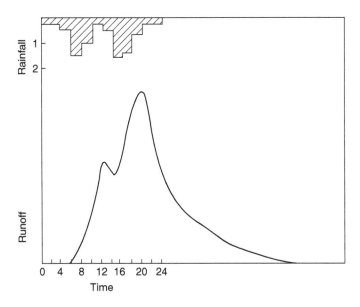

Figure 17.5 Rainfall hyetograph and associated runoff hydrograph. (From *Virginia Stormwater Management Handbook*, 1999, Virginia Department of Conservation and Recreation, Division of Soil and Water Conservation.)

conditions. Note, however, that a single historic design storm of known frequency is inadequate for such design work. A more accurate modeling method is to synthesize data from the longest possible grouping of rainfall data and derive a frequency relationship as described with the I–D–F curves.

17.3.1.5 Single Event vs. Continuous Simulation Computer Models

The fundamental requirement of a stormwater management plan is a quantitative analysis of the watershed hydrology, hydraulics, and water quality, with consideration for associated facility costs. Computers have greatly reduced the time required to complete such an analysis and have also greatly simplified the statistical analysis of compiled rainfall data. In general, hydrologic computer models fit into two main categories: single-event computer models and continuous-simulation models.

Single-event computer models require a minimum of one design-storm hyetograph as input. A *hyetograph* is a graph of rainfall intensity on the vertical axis vs. time on the horizontal axis (see Figure 17.5). A hyetograph shows the volume of precipitation at any given time as the area beneath the curve and the time variation of the intensity. The hyetograph can be a *synthetic* hyetograph or an *historic storm* hyetograph. When a frequency or recurrence interval is specified for the input hyetograph, the modeling inputs are set so that the resulting output runoff has the same recurrence interval. (This is one of the general assumptions made for most single-event models.)

Continuous simulation models incorporate the entire meteorological record of a watershed as their input, which may consist of decades of precipitation data. The computer model processes the data, producing a continuous runoff hydrograph. The continuous hydrograph output can be analyzed using basic statistical analysis techniques to provide discharge–frequency relationships, volume–frequency relationships, flow–duration relationships, and so forth. The extent to which the output hydrograph may be analyzed depends upon the input data available. The principal advantage of the continuous simulation model is that it eliminates the need to choose a design storm, instead providing long-term response data for a watershed that can then be statistically analyzed for the desired frequency storm.

Computer advances have greatly reduced the analysis time and related expenses associated with continuous models. We can expect that future models, which combine some features of continuous modeling with the ease of single-event modeling, will offer quick and more accurate analysis procedures. The hydrologic methods discussed in this text are limited to single-event methodologies based on historic data. Further information regarding the derivation of the I–D–F curves and the SCS 24-h rainfall distribution can be found in the *National Engineering Handbook*, Section 4 (U.S. SCS, 1985).

17.4 RUNOFF HYDROGRAPHS

A runoff hydrograph is a graphical plot of the runoff or discharge from a watershed with respect to time. Runoff occurring in a watershed flows downstream in various patterns influenced by many factors, including the amount and distribution of the rainfall, rate of snowmelt, stream channel hydraulics, infiltration capacity of the watershed, and others more difficult to define. No two flood hydrographs are alike.

Empirical relationships, however, have been developed from which complex hydrographs can be derived. The critical element of the analysis, as with any hydrologic analysis, is the accurate description of the watershed's rainfall-runoff relationship, flow paths, and flow times. From these data, runoff hydrographs can be generated. Some types of hydrographs used for modeling include:

- *Natural hydrographs* — obtained directly from the flow records of a gauged stream
- *Synthetic hydrographs* — obtained by using watershed parameters and storm characteristics to simulate a natural hydrograph
- *Unit hydrographs* — natural or synthetic hydrographs adjusted to represent 1 inch of direct runoff
- *Dimensionless unit hydrographs* — designed to represent many unit hydrographs by using the time to peak rates as basic units and plotting the hydrographs in ratios of these units

17.5 RUNOFF AND PEAK DISCHARGE

Despite its simplification of the complex rainfall–runoff process, the practice of estimating runoff as a fixed percentage of rainfall has been used in the design of storm drainage systems for many years. This method can be accurate when drainage areas are subdivided into homogeneous units and when the designer has enough data and experience to use the appropriate factors.

For watersheds or drainage areas comprising primarily pervious cover (including open space, woods, lawns, or agricultural land uses) the rainfall–runoff analysis becomes much more complex. Soil conditions and types of vegetation are two of the variables that play a larger role in determining the amount of rainfall that becomes runoff. In addition, other types of flow have a larger effect on stream flow (and measured hydrograph) when the watershed is less urbanized. These factors include:

- *Surface runoff*, which occurs only when the rainfall rate is greater than the infiltration rate and the total volume of rainfall exceeds the interceptions, infiltration, and surface detention capacity of the watershed. The runoff flows on the land surface collect in the stream network.
- *Subsurface flow*, which occurs when infiltrated rainfall meets an underground zone of low transmission and travels above the zone to the soil surface to appear as a seep or spring.
- *Base flow*, which occurs when a fairly steady flow into a stream channel occurs from natural storage. The flow comes from lakes or swamps or from an aquifer replenished by infiltrated rainfall or surface runoff.

In watershed hydrology, dealing separately with base flow and combining all other types of flow into direct runoff is customary. Depending upon the requirements of the study, the designer can calculate the *peak flow rate* in cubic feet per second (cfs) of the direct runoff from the watershed or determine the *runoff hydrograph* for the direct runoff from the watershed. A hydrograph shows the volume of runoff as the area beneath the curve, and the time variation of the discharge rate.

If the purpose of a hydrologic study is to measure the impact of various developments on the drainage network within a watershed or to design flood control structures, a hydrograph is needed. If the purpose of a study is to design a roadway culvert or other simple drainage improvement, only the peak rate of flow is needed. Therefore, the purpose of a given study dictates the most suitable methodology. Note that the rational method and TR-55 graphical peak discharge method do not generate runoff hydrographs. The TR-55 tabular method and the modified rational method do generate runoff hydrographs.

17.6 CALCULATION METHODS

Because different tasks related to stormwater design require different types of input and generate different types of results, environmental engineers responsible for stormwater design should be familiar with the different methods for calculating runoff from a watershed. The methods covered here are the rational method, modified rational method, and SCS methods' TR-55, Urban Hydrology for Small Watersheds (U.S. SCS, 1986): Graphical Peak Discharge and Tabular Hydrograph Methods. Note that many computer programs are available that develop these methodologies using the rainfall–runoff relationship described previously. Many of these programs also "route" the runoff hydrograph through a stormwater management facility, calculating the peak rate of discharge and a discharge hydrograph.

Examples provided later use SCS TR-20 Project Formulation, Hydrology (U.S. SCS, 1982). Other readily available computer programs also utilize SCS methods. The computer model accuracy is based upon the accuracy of the input, typically generated through the rational or SCS methodologies covered here. Again, engineers should be familiar with all of the methods covered because any one may be appropriate for the specific site or the watershed being modeled.

> *Note*: All the methods presented next make assumptions and have limitations on accuracy. However, when these methods are used correctly, they will provide reasonable estimates of the peak rate of runoff from a drainage area or watershed.
>
> *Important point*: For small storm events (<2 in. rainfall), TR-55 tends to underestimate the runoff, although it provides fairly accurate estimates for larger storm events. Similarly, the rational formula is fairly accurate on smaller homogeneous watersheds, while tending to lose accuracy in the larger, more complex watersheds. The following discussion provides further explanation of these methods, including assumptions, limitations, and information needed for the analysis.

17.6.1 The Rational Method

The rational method was devised for determining the peak discharges from drainage areas. Though frequently criticized for its simple approach, its simplicity has made the rational method one of the most widely used techniques today. The rational formula estimates the peak rate of runoff at any location in a drainage area as a function of the runoff coefficient, mean rainfall intensity, and drainage area. The rational formula is expressed as:

$$Q = CIA \qquad (17.1)$$

where
Q = maximum rate of runoff, cubic feet per second
C = dimensionless runoff coefficient, dependent upon land use

I = design rainfall intensity, in inches per hour, for a duration equal to the time of concentration of the watershed

A = drainage area, in acres

17.6.1.1 Assumptions

The rational method is based on the following assumptions:

- Under steady rainfall intensity, the maximum discharge will occur at the watershed outlet at the time when the entire area above the outlet is contributing runoff. This "time" is commonly known as the *time of concentration*, t_c, and is defined as the time required for runoff to travel from the most hydrologically distant point in the watershed to the outlet. The assumption of steady rainfall dictates that even during longer events (when factors such as increasing soil saturation are ignored), the maximum discharge occurs when the entire watershed is contributing to the peak flow, at time $t = t_c$. The time of concentration is equal to the minimum duration of peak rainfall. The time of concentration reflects the minimum time required for the entire watershed to contribute to the peak discharge. The rational method assumes that all types of discharge do not increase as a result of soil saturation, decreased conveyance time, or other factors. Therefore, the time of concentration is not necessarily intended to be a measure of the actual storm duration, but simply the critical time period used to determine the average rainfall intensity from the I–D–F curves.
- The frequency or return period of the computed peak discharge is the same as the frequency or return period of rainfall intensity (design storm) for the given time of concentration. Frequencies of peak discharges depend not only on the frequency of rainfall intensity, but also on the response characteristics of the watershed. For small and mostly impervious areas, rainfall frequency is the dominant factor because response characteristics are relatively constant. However, for larger watersheds, the response characteristics have a much greater impact on the frequency of the peak discharge because of drainage structures, restrictions within the watershed, and initial rainfall losses from interception and depression storage.
- The fraction of rainfall that becomes runoff is independent of rainfall intensity or volume. This assumption is reasonable for impervious areas (streets, rooftops, and parking lots). For pervious areas, the fraction of rainfall that becomes runoff varies with rainfall intensity, and the runoff will increase. This fraction is represented by the dimensionless runoff coefficient, C. Therefore, the accuracy of the rational method is dependent on the careful selection of a coefficient appropriate for the storm, soil, and land use conditions. We can easily see why the rational method becomes more accurate as the percentage of impervious cover in the drainage area approaches 100%.
- The peak rate of runoff is sufficient information for the design of stormwater detention and retention facilities.

17.6.1.2 Limitations

Because of the preceding assumptions, the rational method should only be used when the following criteria are met:

- The given watershed has a time of concentration, t_c, of less than 20 min.
- The drainage area is less than 20 acres.

For larger watersheds, attenuation of peak flows through the drainage network begins to be a factor in determining peak discharge. Although ways to adjust runoff coefficients (C factors) to account for attenuation or routing effects are possible, using a hydrograph method or computer simulation for these more complex situations produces more accurate and useful results.

Similarly, the presence of bridges, culverts, or storm sewers may act as restrictions that ultimately have an impact on the peak rate of discharge from the watershed. The peak discharge upstream of the restriction can be derived using a simple calculation procedure such as the rational method; however, a detailed storage routing procedure that considers the storage volume above the restriction should be used to determine the discharge downstream of the restriction accurately.

17.6.1.3 Design Parameters

The following is a brief summary of the design parameters used in the rational method.

Time of Concentration, t_c

The most consistent source of error in the use of the rational method is oversimplifying the time of concentration calculation procedure. Because the origin of the rational method is rooted in the design of culverts and conveyance systems, the main components of the time of concentration are *inlet time* (or overland flow) and *pipe* or *channel flow time*. The inlet overland flow time is defined as the time required for runoff to flow overland from the furthest point in the drainage area over the surface to the inlet or culvert. The pipe or channel flow time is defined as the time required for the runoff to flow through the conveyance system to the design point. When an inlet time of less than 5 min is encountered, the time is rounded up to 5 min; this time is then used to determine the rainfall intensity, *I*, for that inlet.

Variations in the time of concentration can affect the calculated peak discharge. When the procedure for calculating the time of concentration is oversimplified, the rational method's accuracy is greatly compromised. To prevent this oversimplification, a more rigorous procedure for determining the time of concentration should be used, such as the one presented in Chapter 15, Section 4 of SCS *National Engineering Handbook* (U.S. SCS, 1985).

Many procedures are available for estimating the time of concentration. Some were developed with a specific type or size watershed in mind, while others were based on studies of a specific watershed. The selection of any given procedure should include a comparison of the hydrologic and hydraulic characteristics used in the formation of the procedure vs. the characteristics of the watershed under study. The engineer should be aware that if two or more methods of determining time of concentration are applied to a given watershed, a wide range in results occurs. The SCS method is recommended because it provides a means of estimating overland sheet flow time and shallow concentrated flow time as a function of readily available parameters, such as land slope and land surface conditions. Regardless of which method is used, the result should be reasonable when compared to an average flow time over the total length of the watershed.

Rainfall Intensity, I

The rainfall intensity, *I*, is the average rainfall rate, in inches per hour, for a storm duration equal to the time of concentration for a selected return period (for example, 1, 2, 10, or 25 years). Once a particular return period has been selected and the time of concentration has been determined for the drainage area, the rainfall intensity can be read from the appropriate rainfall I–D–F curve for the geographic area in which the drainage area is located. These charts were developed from data furnished by the National Weather Service for regions of the U.S.

Runoff Coefficients, C

The runoff coefficients for different land uses within a watershed are used to generate a single, weighted coefficient that represents the relationship between rainfall and runoff for that watershed. Recommended values are found in Table 17.3. In an attempt to make the rational method more accurate, adjustments to the runoff coefficients were made in order to represent more accurately the integrated effects of drainage basin parameters: *land use, soil type,* and *average land slope.* Table 17.3 provides recommended coefficients based on urban land use only.

Table 17.3   Rational Equation Runoff Coefficients

Land use	"*C*" value
Business, industrial and commercial	0.90
Apartments	0.75
Schools	0.60
Residential — lots of 10,000 sq ft	0.50
— lots of 12,000 sq ft	0.45
— lots of 17,000 sq ft	0.45
— lots of ½ acre or more	0.40
Parks, cemeteries, and unimproved areas	0.34
Paved and roof areas	0.90
Cultivated areas	0.60
Pasture	0.45
Forest	0.30
Steep grass slopes (2:1)	0.70
Shoulder and ditch areas	0.50
Lawns	0.20

Source: USDOT, Federal Highway Administration, 2001,
Urban Drainage Design Manual. Washington, D.C.:
Department of Transportation.

A good understanding of these parameters is essential in choosing an appropriate coefficient. As the slope of a drainage basin increases, runoff velocities increase for sheet flow and shallow concentrated flow. As the velocity increases, the ability of the surface soil to absorb the runoff decreases. This decrease in infiltration results in an increase in runoff. In this case, the designer should select a higher runoff coefficient to reflect the increase from slope.

Soil properties influence the relationship between runoff and rainfall even further because soils have differing rates of infiltration. Historically, the rational method was used primarily for the design of storm sewers and culverts in urbanizing areas; soil characteristics were not considered, especially when the watershed was largely impervious. In such cases, a conservative design simply meant a larger pipe and less headwater. For stormwater management purposes, however, the existing condition (prior to development, usually with large amounts of pervious surfaces) often dictates the allowable postdevelopment release rate and therefore must be accurately modeled.

Soil properties can change throughout the construction process because of compaction, cut, and fill operations. If these changes are not reflected in the runoff coefficient, the accuracy of the model decreases. Some localities arbitrarily require an adjustment in the runoff coefficient for pervious surfaces because of the effects of construction on soil infiltration capacities.

Adjustment for Infrequent Storms

The rational method has undergone further adjustment to account for infrequent, higher intensity storms. This adjustment is in the form of a frequency factor, C_f, which accounts for the reduced impact of infiltration and other effects on the amount of runoff during larger storms. With this adjustment, the rational formula is expressed as:

$$Q = CC_f IA \tag{17.2}$$

where C_f = the values listed in Table 17.4. The product of $C_f \times C$ should not exceed 1.0.

Table 17.4   Rational Equation Frequency Factors

C_f	Storm return frequency
1.0	10 yr or less
1.1	25 yr
1.2	50 yr
1.25	100 yr

17.6.2 Modified Rational Method

The modified rational method is a variation of the rational method developed mainly for the sizing of detention facilities in urban areas. The modified rational method is applied similarly to the rational method except that it uses a fixed rainfall duration. The selected rainfall duration depends on the requirements of the user. For example, when sizing a detention basin, the designer might perform an iterative calculation to determine the rainfall duration that produces the maximum storage volume requirement.

17.6.2.1 Assumptions

The modified rational method is based on the following assumptions:

- All of the assumptions used with the rational method apply. The most significant difference is that the time of concentration for the modified rational method is equal to the rainfall intensity-averaging period, rather than the actual storm duration. This assumption means that any rainfall, or any runoff generated by the rainfall, that occurs before or after the rainfall averaging period is unaccounted for. Thus, when used as a basin sizing procedure, the modified rational method may seriously underestimate the required storage volume (Walesh, 1989).
- The runoff hydrograph for a watershed can be approximated as triangular or trapezoidal in shape. This assumption implies a linear relationship between peak discharge and time for any and all watersheds.

17.6.2.2 Limitations

All of the limitations listed for the rational method apply to the modified rational method. The key difference is the assumed shape of the resulting runoff hydrograph. The rational method produces a triangular shaped hydrograph that, when modified, can generate triangular or trapezoidal hydrographs for a given watershed (see Figure 17.6).

17.6.2.3 Design Parameters

The equation $Q = C I A$ (the rational equation) is used to calculate the peak discharge for all three hydrographs shown in Figure 17.6. Notice that the only difference between the rational method and the modified rational method is the incorporation of the storm duration, d, into the modified rational method to generate a volume of runoff in addition to the peak discharge.

The rational method generates the peak discharge that occurs when the entire watershed is contributing to the peak (at a time $t = t_c$), and ignores the effects of a storm that lasts longer than time t. The modified rational method, however, considers storms with a longer duration than the watershed t_c, which may have a smaller or larger peak rate or discharge but will produce a greater volume of runoff (area under the hydrograph) associated with the longer duration of rainfall. Figure 17.7 shows a family of hydrographs representing storms of different durations. The storm duration

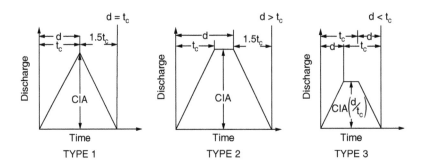

Type 1 — Storm duration, d, = time of concentration, t_c.
Type 2 — Storm duration, d, > the time of concentration, t_c.
Type 3 — Storm duration, d, < time of concentration, t_c.

Figure 17.6 Modified rational method runoff hydrographs. (Adapted from Walesh, S.G., 1989, *Urban Surface Water Treatment*, New York: John Wiley & Sons, Inc.)

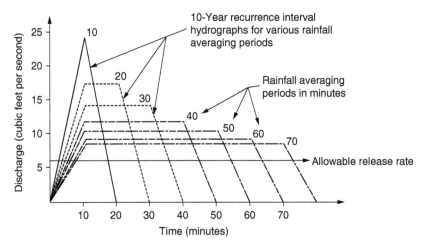

Figure 17.7 Modified rational method family of runoff hydrographs. (From *Virginia Stormwater Management Handbook*, 1999, Virginia Department of Conservation and Recreation, Division of Soil and Water Conservation.)

that generates the greatest volume of runoff may not necessarily produce the greatest peak rate of discharge.

Note that the duration of the receding limb of the hydrograph is set to equal the time of concentration, t_c, or $1.5 \times t_c$. The direct solution (discussed later) uses $1.5\ t_c$ as the receding limb, justified because it is more representative of actual storm and runoff dynamics. (It is also more similar to the SCS unit hydrograph in which the receding limb extends longer than the risking limb.) Using $1.5 \times t_c$ in the direct solution methodology provides for a more conservative design and is used in this text.

The modified rational method allows the designer to analyze several different storm durations to determine the one that requires the greatest storage volume with respect to that allowable release rate. This storm duration is referred to as the critical storm duration and is used as a basin-sizing tool. We discuss the technique in more detail later.

17.6.3 SCS Methods — TR-55 Estimating Runoff

The U.S. Soil Conservation Service (SCS) published the second edition of *Urban Hydrology for Small Watershed*, technical release number 55 (TR-55) in June 1986. The techniques outlined in TR-55 require the same basic data as the rational method: drainage area, time of concentration, land use, and rainfall. The SCS approach, however, is more sophisticated in that it allows the designer to manipulate the time distribution of the rainfall, the initial rainfall losses to interception and depression storage, and the moisture condition of the soils prior to the storm. The procedures developed by SCS are based on a dimensionless rainfall distribution curve for a 24-h storm.

TR-55 presents two general methods for estimating peak discharges from urban watersheds: the graphical method and the tabular method. The *graphical method* is limited to watersheds whose runoff characteristics are fairly uniform and whose soils, land use, and ground cover can be represented by a single runoff curve number (CN). The graphical method provides a peak discharge only and is not applicable for situations in which a hydrograph is required.

The *tabular method* is a more complete approach and can be used to develop a hydrograph at any point in a watershed. For large areas, dividing the area into subwatersheds may be necessary in order to account for major land use changes, analyze specific study points within subwatersheds, or locate stormwater drainage facilities and assess their effects on peak flows. The tabular method can generate a hydrograph for each subwatershed for the same storm event. The hydrographs can then be routed through the watershed and combined to produce a partial composite hydrograph at the selected study point. The tabular method is particularly useful in evaluating the effects of altered land use in a specific area within a given watershed.

Prior to using the graphical or tabular method, the designer must determine the volume of runoff resulting from a given depth of precipitation and the time of concentration, t_c, for the watershed being analyzed. In this section, we briefly discuss the methods for determining these values. However, the reader is strongly encouraged to obtain a copy of the TR-55 manual from the SCS to gain more insight into the procedures and limitations.

The SCS runoff CN method is used to estimate runoff. This method is described in detail in the SCS *National Engineering Handbook*, Section 4 (U.S. SCS, 1985). The runoff equation (found in TR-55 and discussed later in this section) provides a relationship between runoff and rainfall as a function of the CN. The CN is a measure of the land's ability to infiltrate or otherwise detain rainfall, with the excess becoming runoff. The CN is a function of land cover (woods, pasture, agricultural use, percent impervious surface, for example), hydrologic condition, and soils.

17.6.3.1 Limitations

- TR-55 has simplified the relationship between rainfall and runoff by reducing all of the initial losses before runoff begins, or initial abstractions, to the term I_a, and approximating the soil and cover conditions using the variable S, potential maximum retention. The terms I_a and S are functions of the runoff curve number. Runoff curve numbers describe average conditions that are useful for design purposes. If the purpose of the hydrologic study is to model an historical storm event, average conditions may not be appropriate.
- The designer should understand the assumption reflected in the initial abstraction term, I_a. This term represents interception; initial infiltration; surface depression storage; evapotranspiration; and other watershed factors, and is generalized as a function of the runoff curve number based on data from agricultural watersheds. This can be especially important in an urban application because the combination of impervious area with pervious area can imply a significant initial loss that may not take place. On the other hand, the combination of impervious and pervious area can underestimate initial losses if the urban area has significant surface depression storage. (To use a relationship other than the one established in TR-55, the designer must redevelop the runoff equation by using the original rainfall–runoff data to establish new curve number relationships for each cover and hydrologic soil group — a large data collection and analysis effort.)

- Runoff from snowmelt or frozen ground cannot be estimated using these procedures.
- The runoff curve number method is less accurate when the runoff is less than 0.5 in. As a check, use another procedure to determine runoff.
- The SCS runoff procedures apply only to surface runoff and do not consider subsurface flow or high groundwater.
- Manning's kinetic solution (discussed later) should not be used to calculate the time of concentration for sheet flow longer than 300 ft. This limitation affects the time of concentration calculations. Note that many jurisdictions consider 150 ft to be the maximum length of sheet flow before shallow concentrated flow develops.
- The minimum t_c used in TR-55 is 0.1 h.

17.6.3.2 Information Needed

Generally, a good understanding of the physical characteristics of the watershed is needed to solve the runoff equation and determine the time of concentration. Some features (including topography and channel geometry) can be obtained from topographic maps such as the USGS 1 in. = 2000 ft quadrangle maps. Various sources of information may be accurate enough for a watershed study; however, the accuracy of the study is directly related to the accuracy and level of detail of the base information. Ideally, a site investigation and filed survey should be conducted to verify specific features such as channel geometry and material, culvert sizes, drainage divides, and ground cover. Depending on the size and scope of the study, however, a site investigation may not be economically feasible.

The data needed to solve the runoff equation and determine the watershed time of concentration, t_c, are listed below. We discuss these items in more detail later.

- Soil information (to determine the hydrologic soil group)
- Ground cover type (impervious, woods, grass)
- Treatment (cultivated or agricultural land)
- Hydrologic conditions (for design purposes, the hydrologic condition should be considered "good" for the predeveloped condition)
- Urban impervious area modifications (connected, unconnected)
- Topography detailed enough to identify divides, t_c and T_t flow paths and channel geometry, and surface condition (roughness coefficient) accurately

17.6.3.3 Design Parameters

Soils

In hydrograph applications, runoff is often referred to as *rainfall excess* or *effective rainfall* and is defined as the amount of rainfall that exceeds the land's capability to infiltrate or otherwise retain the rainwater. The soil type or classification, the land use and land treatment, and the hydrologic condition of the cover are the watershed factors with the most significant impact on estimating the volume of rainfall excess or runoff.

Hydrologic soil group classification. SCS has developed a soil classification system that consists of four groups, identified as *A*, *B*, *C*, and *D*. Soils are classified into one of these categories based upon their minimum infiltration rate. By using information obtained from local SCS offices, soil and water conservation district offices, or from SCS soil surveys (published for many counties across the country), one can identify the soils in a given area. Preliminary soil identification is especially useful for watershed analysis and planning in general. When a stormwater management plant is prepared for a specific site, soil borings should be taken to verify the hydrologic soil classification. Soil characteristics associated with each hydrologic soil group are generally described as:

- *Group A*: soils with low runoff potential because of high infiltration rates, even when thoroughly wetted. These soils consist primarily of deep, well-drained to excessively drained sands and gravels with high water transmission rates (0.30 in./h). Group A soils include sand, loamy sand, and sandy loam.
- *Group B*: soils with moderately low runoff potential because of moderate infiltration rates when thoroughly wetted. These soils consist primarily of moderately deep to deep and moderately well-drained to well-drained soils. Group B soils have moderate water transmission rates (0.15 to 0.30 in./h) and include silt loam and loam.
- *Group C*: soils with moderately high runoff potential because of slow infiltration rates when thoroughly wetted. These soils typically have a layer near the surface that affects the downward movement of water or soils. Group C soils have low water transmission rates (0.05 to 0.15 in./h) and include sandy clay loam.
- *Group D*: soils with high runoff potential because of very slow infiltration rates. These soils consist primarily of clays with high swelling potential; soils with permanently high water tables; soils with a claypan or clay layer at or near the surface; and shallow soils over nearly impervious parent material. Group D soils have very low water transmission rates (0 0.05 in./h) and include clay loam; silty-clay loam; sandy clay; silty clay; and clay.

Any disturbance of a soil profile can significantly alter the soil's infiltration characteristics. With urbanization, the hydrologic soil group for a given area can change because of soil mixing; introduction of fill material from other areas; removal of material during mass grading operations; or compaction from construction equipment. A layer of topsoil may typically be saved and replaced after the earthwork is completed, but the native underlying soils have been dramatically altered. Therefore, any disturbed soil should be classified by its physical characteristics as given for each soil group.

Some jurisdictions require all site developments to be analyzed using an HSG classification that is one category below the actual predeveloped HSG. For example, a site with a predeveloped HSG classification of *B*, as determined from the soil survey, would be analyzed in its developed state using an HSG classification of *C*.

Hydrologic Condition

Hydrologic condition represents the effects of cover type and treatment on infiltration and runoff; it is generally estimated from the density of plant and residue cover across the drainage area. Good hydrologic condition indicates that the cover has a low runoff potential and poor hydrologic condition indicates a high runoff potential. This condition is used in describing nonurbanized lands (woods, meadow, brush, agricultural land) and open spaces associated with urbanized areas (lawns, parks, golf courses, and cemeteries). Treatment is a cover type modifier to describe the management of cultivated agricultural lands. Table 17.5a and Table 17.5b provide an excerpt from Table 2-2 in TR-55 that shows the treatment and hydrologic condition for various land uses.

When a watershed is analyzed to determine the impact of proposed development, many stormwater management regulations require the designer to consider all existing or undeveloped land to be in hydrologically good condition. This results in lower existing condition peak runoff rates, which in turn results in greater postdevelopment peak control. In most cases, undeveloped land is in good hydrologic condition unless it has been altered in some way. Because the goal of most stormwater programs is to reduce the peak flows from developed or altered areas to their predeveloped or prealtered rates, this is a reasonable approach. In addition, this approach eliminates any inconsistencies in judging the condition of undeveloped land or open space.

Runoff Curve Number (RCN) Determination

The soil group classification, cover type, and the hydrologic condition are used to determine the runoff curve number, *RCN*. The RCN indicates the runoff potential of an area when the ground is

Table 17.5a   Runoff Curve Numbers for Urban Areas[a]

Cover description		Curve numbers for hydrologic soil group:			
Cover type and hydrologic condition	Average percent impervious area	A	B	C	D
Fully developed urban areas (vegetation established):					
Open space (lawns, parks, golf courses, cemeteries, etc.)[b]					
Good condition (grass cover >75%)		39	61	74	80
Impervious areas:					
Paved parking lots, roofs, driveways, etc. (excluding right-of-way), streets and roads:		98	98	98	98
Paved; curbs and storm sewers (excluding right-of-way		98	98	98	98
Paved; open ditches (including right-of-way)		83	89	92	93
Gravel (including right-of-way)		76	85	89	91
Dirt (including right-of-way)		72	82	87	89
Urban districts:					
Commercial and business	85%	89	92	94	95
Industrial	72%	81	88	91	93
Residential districts by average lot size:					
1/8 acre or less (town houses)	65%	77	85	90	92
1/4 acre	38%	61	75	83	87
1/3 acre	30%	57	72	81	86
1/2 acre	25%	54	70	80	85
1 acre	20%	51	68	79	84
2 acres	12%	46	65	77	82
Developing urban areas:					
Newly graded areas (pervious areas only, no vegetation)[c]		77	86	91	94
Idle lands (CNs are determined using cover types similar to those in TR-55 Table 2-2c)					

[a] Refer to TR-55 for additional cover types and general assumptions and limitations.
[b] For specific footnotes, see TR-55 Table 2-2a.

Note: Average runoff condition and $I_a = 0.25$.
Source: Adapted from TR-55 Table 2-2a — Runoff Curve Numbers for Urban Areas.

Table 17.5b   Runoff Curve Numbers for Other Agricultural Areas[a]

Cover type	Hydrologic conditions	A	B	C	D
Pasture, grassland, or range — continuous forage for grazing[b]	Good	39	61	74	80
Meadow — continuous grass, protected from grazing and generally mowed for hay	—	30	58	71	78
Brush — brush–weed–grass mixture with brush the major element[b]	Good	[b]30	48	65	73
Woods–grass combination (orchard or tree farm)[b]	Good	32	58	72	79
Woods[b]	Good	[b]30	55	70	77
Farmsteads — buildings, lanes, driveways, and surrounding lots	—	59	74	82	86

[a] Refer to TR-55 for additional cover types and general assumptions and limitations.
[b] For specific footnotes, see TR-55 Table 2-2b.
Note: Average runoff condition and $I_a = 0.25$.

Source: Adapted from TR-55 Table 2-2b — Runoff Curve Numbers for Other Agricultural Lands.

not frozen. Table 17.5a and Table 17.5b, excerpted from TR-55 (which gives a more complete range of data), provide the RCNs for various land use types and soil groups.

Several factors should be considered when choosing an RCN for a given land use. First, the designer should realize that the curve numbers in Table 17.5a, Table 17.5b, and TR-55 are for the *average antecedent runoff* or *moisture condition, ARC.* The ARC is the index of runoff potential before a storm event. It can have a major impact on the relationship between rainfall and runoff for a watershed. Average ARC runoff curve numbers can be converted to dry or wet values; the average antecedent runoff condition is recommended for design purposes. Environmental engineers must consider the list of assumptions made in developing the runoff curve numbers as provided in Table 17.5a and Table 17.5b and in TR-55. We outline some of these assumptions next.

> *Note*: The decision to use "wet" or "dry" antecedent runoff conditions should be based on thorough fieldwork, such as carefully monitored rain gauge data.

RCN Determination Assumptions (TR-55)

- The urban curve number for such land uses as residential, commercial, and industrial is computed with the percentage of imperviousness as shown. A composite curve number should be recomputed using the actual percentage of imperviousness if it differs from the value shown.
- Impervious areas are directly connected to the drainage system.
- Impervious areas have a runoff curve number of 98.
- Pervious areas are considered equivalent to open space in good hydraulic condition.

> *Note*: These assumptions, as well as others, are footnoted in TR-55, Table 2-2. TR-55 provides a graphical solution for modification of the given RCNs if any of these assumptions does not hold true.

The engineer should become familiar with the definition of connected vs. unconnected impervious areas, along with the graphical solutions and the impact that their use can have on the resulting RNC. After some experience in using this section of TR-55, the designer will be able to make field evaluations of the various criteria used in the determination of the RCN for a given site. In addition, the designer will need to determine if the watershed contains sufficient diversity in land use to justify dividing it into several subwatersheds. If a watershed or drainage area cannot be adequately described by one weighted curve number, the designer must divide the watershed into subareas and analyze each one individually, generate individual hydrographs, and add those hydrographs together to determine the composite peak discharge for the entire watershed.

Figure 17.8 shows the decision-making process for analyzing a drainage area. The flow chart can be used to select the appropriate tables or figures in TR-55 from which to choose the runoff curve numbers. Worksheet 2 in TR-55 is then used to compute the weighted curve number for the area or subarea.

The Runoff Equation — The SCS runoff equation is used to solve for runoff as a function of the initial abstraction, I_a, and the potential maximum retention, S, of a watershed, both of which are functions of the RCN. This equation attempts to quantify all the losses before runoff begins, including infiltration, evaporation, depression storage, and water intercepted by vegetation.

TR-55 provides a graphical solution for the runoff equation in Chapter 2 of TR-55: Estimating Runoff. Both the equation and graphical solution solve for depth of runoff expected from a watershed or subwatershed of a specified RCN, for any given frequency storm. Additional information can be found in the *National Engineering Handbook*, Section 4 (U.S. SCS, 1985). By providing the basic relationship between rainfall and runoff, these procedures are the basis for any hydrological study based on SCS methodology. Therefore, the designer must conduct a thorough

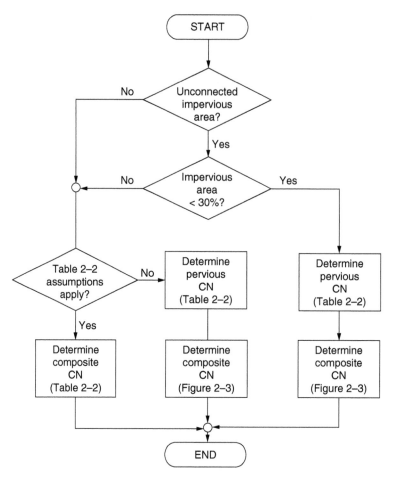

Figure 17.8 Runoff curve number selection flowchart. (From U.S. Soil Conservation Service, 1986, Technical Release No. 55.)

site visit and consider the entire site's features and characteristics (including soil types and hydrologic condition) when analyzing a watershed or drainage area.

Time of Concentration and Travel Time — The time of concentration, t_c, is the length of time required for a drop of water to travel from the most hydraulically distant point in the watershed or subwatershed to the point of analysis. The travel time, T_t, is the time it takes that same drop of water to travel from the study point at the bottom of the subwatershed to the study point at the bottom of the whole watershed. The travel time, T_t, is descriptive of the subwatershed by providing its location relative to the study point of the entire watershed.

Similar to the rational method, time of concentration, t_c, plays an important role in developing the peak discharge for a watershed. Urbanization usually decreases t_c, which results in an increase in peak discharge. For this reason, to model the watershed accurately, engineers must be aware of any conditions that may act to decrease the flow time, such as channelization and channel improvements. They must also be aware of conditions within the watershed that may actually lengthen the flow time, such as surface ponding above undersized conveyance systems and culverts.

- *Heterogeneous watersheds.* A heterogeneous watershed is one with two or more hydrologically defined drainage areas of differing land uses, hydrologic conditions, times of concentration, or other runoff characteristics contributing to the study point.

- *Flow segments.* The time of concentration is the sum of the time increments for each flow segment present in the t_c flow path, such as overland or sheet flow, shallow concentrated flow, and channel flow. These flow types are influenced by surface roughness, channel slope, flow patterns, and slope.
  - *Overland (sheet) flow* is shallow flow over plane surfaces. For the purposes of determining time of concentration, overland flow usually exists in the upper reaches of the hydraulic flow path. TR-55 utilizes Manning's kinematic solution to compute t_c for overland sheet flow. The roughness coefficient is the primary culprit in the misapplication of the kinematic t_c equation. Care should be taken to identify the surface conditions for overland flow accurately. Table 17.6a in this text and Table 3-1 in TR-55 provide selected coefficients for various surface conditions. Refer to TR-55 for the use of Manning's kinematic equation.
  - *Shallow concentrated flow* usually begins where overland flow converges to form small rills or gullies. This flow can exist in small man-made drainage ditches (paved and unpaved) and in curbs and gutters. TR-55 provides a graphical solution for shallow concentrated flow. The input information needed to solve for this flow segment is the land slope and the surface condition (paved or unpaved).
  - *Channel flow* occurs where flow converges in gullies, ditches, or swales and in natural or man-made water conveyances (including storm drainage pipes). This flow is assumed to exist in perennial streams or wherever well-defined channel cross-sections are found. The Manning equation is used for open channel flow and pipe flow and usually assumes full flow or bank-full velocity. Manning coefficients can be found in Table 17.6b through Table 17.6d for open channel flow (natural and man-made channels) and closed channel flow. Coefficients can also be obtained from standard textbooks, including *Open Channel Hydraulics* (Chow, 1959) or *Handbook of Hydraulics* (King and Brater, 1976).

Table 17.6a   Roughness Coefficient "*n*" for the Manning Equation — Sheet Flow

Surface description	"*n*" Value[a]
Smooth surfaces (concrete, asphalt, gravel, or bare soil)	0.011
Fallow (no residue)	0.05
Cultivated soils:	
Residue cover <20%	0.06
Residue cover >20%	0.17
Grass:	
Short grass prairie	0.15
Dense grasses[b]	0.24
Bermuda grass	0.41
Range (natural)	0.13
Woods[c]:	
Light underbrush	0.40
Dense underbrush	0.80

[a] The "*n*" values are composite of information compiled by Engman (1986).
[b] Includes species such as weeping lovegrass, bluegrass, buffalo grass, blue grama grass, and native grass mixtures.
[c] When selecting it, consider cover to a height of about 0.1 ft. This is the only part of the plant cover that will obstruct sheet flow.

Source: U.S. Soil Conservation Service (SCS), 1986, Urban hydrology for small watersheds. Technical Release No. 55.

Table 17.6b   Roughness Coefficient "n" for the Manning Equation —
Pipe Flow

Material	"n" Value range	
	From	To
Coated cast-iron	0.010	0.014
Uncoated cast-iron	0.011	0.015
Vitrified sewer pipe	0.010	0.017
Concrete pipe	0.010	0.017
Common clay drainage tile	0.011	0.017
Corrugated metal ($2\frac{2}{3} \times \frac{1}{2}$)	0.023	0.026
Corrugated metal (3×1 and 6×1)	0.026	0.029
Corrugated metal (6×2 structural plate)	0.030	0.033

Source: Adapted from King, H.W. and Brater, E.F., 1976, *Handbook of Hydraulics*, 6th ed., New York: McGraw-Hill.

Table 17.6c   Roughness Coefficient "n" for the Manning
Equation — Constructed Channels

Lining material	"n" Value range	
	From	To
Concrete lined	0.012	0.016
Cement rubble	0.017	0.025
Earth, straight and uniform	0.017	0.022
Rock cuts, smooth and uniform	0.025	0.033
Rock cuts, jagged and irregular	0.035	0.045
Winding, sluggish canals	0.022	0.027
Dredged earth channels	0.025	0.030
Canals with rough stony beds	0.025	0.035

Weed on earth banks:

Earth bottom, rubble sides	0.028	0.033

Small grass channels:

Long grass — 13 in.	0.042
Short grass — 3 in.	0.034

Source: Adapted from King, H.W. and Brater, E.F., 1976, *Handbook of Hydraulics*, 6th ed., New York: McGraw-Hill.

Table 17.6d   Roughness Coefficient "n" for the Manning Equation — Natural Stream Channels

	Channel lining	"n" Value range	
		From	To
1.	Clean, straight bank, full stage, no rifts or deep pools	0.025	0.030
2.	Same as #1, but some weeds and stones	0.030	0.035
3.	Winding, some pools and shoals, clean	0.033	0.040
4.	Same as #3, lower stages, more ineffective slope and sections	0.040	0.050
5.	Same as #3, some weeds and stones	0.035	0.045
6.	Same as #4, stony sections	0.045	0.055
7.	Sluggish river reaches, rather weedy with very deep pools	0.050	0.070
8.	Very weedy reaches	0.075	0.125

Source: Adapted from King, H.W. and Brater, E.F., 1976, *Handbook of Hydraulics*, 6th ed., New York: McGraw-Hill.

17.6.4 TR-55 Graphical Peak Discharge Method

The graphical peak discharge method was developed from hydrograph analyses using TR-20, Computer Program for Project Formulation — Hydrology (U.S. SCS, 1982). The graphical method develops the peak discharge in cubic feet per second for a given watershed.

17.6.4.1 Limitations

Engineers should be aware of several limitations before using the TR-55 graphical method:

- The watershed studied must be hydrologically homogeneous (the land use, soils, and cover distributed uniformly throughout the watershed and described by one curve number).
- The watershed may have only one main stream or flow path. If more than one is present, they must have nearly equal times of concentration so that one t_c represents the entire watershed.
- The analysis of the watershed cannot be part of a larger watershed study, which would require adding hydrographs because the graphical method does not generate a hydrograph.
- For the same reason, the graphical method should not be used if a runoff hydrograph is to be routed through a control structure.
- When the initial abstraction (rainfall ratio, I_a/P) falls outside the range of the unit peak discharge curves (0.1 to 0.5), the limiting value of the curve must be used.

The reader is encouraged to review the TR-55 manual to become familiar with these and other limitations associated with the graphical method.

The graphical method can be used as a planning tool to determine the impact of development or land use changes within a watershed, or to anticipate or predict the need for stormwater management facilities or conveyance improvements. Sometimes, the graphical method can be used in conjunction with the TR-55 short-cut method for estimating storage volume required for post-developed peak discharge control. This short-cut method is found in Chapter 6 of TR-55. However, note that a more sophisticated computer model such as TR-20 or HEC-1 or even TR-55 tabular hydrograph method should be used for complex urbanizing watersheds.

17.6.4.2 Information Needed

The following parameters are needed to compute the peak discharge of a watershed using the TR-55 graphical peak discharge method:

- The drainage area, in square miles
- Time of concentration, t_c, in hours
- Weighted runoff curve number, RCN
- Rainfall amount, P, for specified design storm, in inches
- Total runoff, Q, in inches
- Initial abstraction, I_a, for each subarea
- Ratio of I_a/P for each subarea;
- Rainfall distribution (Type I, IA, II, or III)

17.6.4.3 Design Parameters

The TR-55 peak discharge equation is:

$$q_p = q_u A_m Q F_p \qquad (17.3)$$

where

q_p = peak discharge, cubic feet per second
q_u = unit peak discharge, cubic feet per second per square mile per inch (csm/in)
A_m = drainage area, square miles
Q = runoff, inches
F_p = pond and swamp adjustment factor

All the required information has been determined earlier, except for the unit peak discharge, q_u, and the pond and swamp adjustment factor, F_p.

The unit peak discharge, q_u, is a function of the initial abstraction, I_a, precipitation, P, and the time of concentration, t_c, and can be determined from the unit peak discharge curves in TR-55. The unit peak discharge is expressed in cubic feet per second per square mile per inch of runoff. Initial abstraction is a measure of all the losses that occur before runoff begins, including infiltration, evaporation, depression storage, and water intercepted by vegetation, and can be calculated from empirical equations or Table 4-1 in TR-55. The pond and swamp adjustment factor is an adjustment in the peak discharge to account for pond and swamp areas if they are spread throughout the watershed and are not considered in the t_c computation. Refer to TR-55 for more information on pond and swamp adjustment factors.

The unit peak discharge, q_u, is obtained by using t_c and the I_a/P ratio with Exhibits 4-I, 4-IA, 4-II, or 4-III (depending on the rainfall distribution type) in TR-55. As the fifth limitation discussed earlier indicates, the ratio of I_a/P must fall between 0.1 and 0.5. The engineer must use the limiting value on the curves when the computed value is not within this range. The unit peak discharge is determined from these curves and entered into the preceding equation to calculate the peak discharge.

17.6.5 TR-55 Tabular Hydrograph Method

The tabular hydrograph method can be used to analyze large heterogeneous watersheds. This method can develop partial composite flood hydrographs at any point in a watershed by dividing the watershed into homogeneous subareas. The method is especially applicable for estimating the effects of land use change in a portion of a watershed.

The tabular hydrograph method provides a tool to analyze several subwatersheds efficiently in order to verify the combined impact at a downstream study point. It is especially useful to verify the timing of peak discharges. Sometimes, the use of detention in a lower subwatershed may actually increase the combined peak discharge at the study point. This procedure allows a quick check to verify the timing of the peak flows and to decide if a more detailed study is necessary.

17.6.5.1 Limitations

Some of the basic limitations of which the engineer should be aware before using the TR-55 tabular method include:

- The travel time, T_t, must be less than 3 h (largest T_t in TR-55, Exhibit 5).
- The time of concentration, t_c, must be less than 2 h (largest t_c in TR-55, Exhibit 5).
- The acreage of the individual subwatersheds should not differ by a factor of 5 or more.

When these limitations cannot be met, the engineer should use the TR-20 computer program or other available computer models that provide more accurate and detailed results. The reader is encouraged to review the TR-55 manual to become familiar with these and other limitations associated with the tabular method.

17.6.5.2 Information Needed

The following parameters are needed to compute the peak discharge of a watershed using the TR-55 tabular method:

- Subdivision of the watershed into relatively homogeneous areas
- The drainage area of each subarea, in square miles
- Time of concentration, t_c, for each subarea in hours
- Travel time, T_t, for each routing reach, in hours
- Weighted runoff curve number, RCN, for each subarea
- Rainfall amount, P, in inches, for each specified design storm
- Total runoff, Q, in inches for each subarea
- Initial abstraction, I_a, for each subarea
- Ratio of I_a/P for each subarea
- Rainfall distribution (I, IA, II, or III)

17.6.5.3 Design Parameters

The use of the tabular method requires that the engineer determine the travel time through the entire watershed. Because the entire watershed is divided into smaller subwatersheds that must be related to one another and to the whole watershed with respect to time, the result is that the time of peak discharge is known for any one subwatershed relative to any other subwatershed or for the entire watershed. Travel time, T_t, represents the time for flow to travel from the study point at the bottom of a subwatershed to the bottom of the entire watershed. This information must be compiled for each subwatershed.

> *Note*: The data for up to 10 subwatersheds can be compiled on one TR-55 worksheet (TR-55 worksheets 5a and 5b).

To obtain the peak discharge using the graphical method, the unit peak discharge is read off of a curve. However, the tabular method provides this information in the form of a table of values, found in TR-55, Exhibit 5. These tables are arranged by rainfall type (I, IA, II, and III), I_a/P, t_c, and T_t. In most cases, the actual values for these variables (other than the rainfall type) will be different from the values shown in the table. Therefore, a system of rounding these values has been established in the TR-55 manual. The I_a/P term is simply rounded to the nearest table value. The t_c and T_t values are rounded together in a procedure outlined on pages 5-2 and 5-3 of the TR-55 manual. The accuracy of the computed peak discharge and time of peak discharge is highly dependent on the proper use of these procedures.

The following equation, along with the information compiled on TR-55 worksheet 5b, is then used to determine the flow at any time:

$$q = q_t A_m Q \qquad (17.4)$$

where
q = hydrograph coordinate in cubic feet per second, at hydrograph time t
q_t = tabular hydrograph unit discharge at hydrograph time t from TR-55 Exhibit 5, csm per inch
A_m = drainage area of individual subarea, square miles
Q = runoff, inches

The product $A_m Q$ is multiplied by each table value in the appropriate unit hydrograph in TR-55 Exhibit 5, (each subwatershed may use a different unit hydrograph) to generate the actual hydrograph for the subwatershed. This hydrograph is tabulated on TR-55 worksheet 5b, then added together with the hydrographs from the other subwatersheds, taking care to use the same time increment of each subwatershed. The result is a composite hydrograph at the bottom of the worksheet for the entire watershed.

> *Note*: The preceding discussion on the tabular method is taken from TR-55 and is *not* complete. The engineer should obtain a copy of TR-55 and learn the procedures and limitations as outlined in that document. Examples and worksheets are provided in TR-55 that lead the reader through the procedures for each chapter.

17.7   GENERAL STORMWATER ENGINEERING CALCULATIONS

This section provides guidelines for performing various engineering calculations associated with the design of stormwater management facilities, including extended-detention and retention basins, and multistage outlet structures.

17.7.1   Detention, Extended-Detention, and Retention Basin Design Calculations

In general, basin stormwater management regulations require that stormwater management basins be designed to control water quantity (for flood control and channel erosion control) and to enhance (or treat) water quality. The type of basin selected (extended detention, retention, and infiltration) and the relationships among its design components (design inflow, storage volume, and outflow) dictate the size of the basin and serve as the basis for its hydraulic design. Some design component parameters (design storm return frequency and allowable discharge rates) may be specified by the local regulatory authority, based upon the specific needs of certain watersheds or stream channels within that locality. Occasionally, as in stream channel erosion control, the engineer must document and analyze the specific needs of the downstream channel and establish the design parameters.

The design inflow is the peak flow or the runoff hydrograph from the developed watershed. This inflow becomes the input data for the basin sizing calculations, often called *routings*. Various routing methods are available. Note that the format of the hydrologic input data is usually dictated by the chosen routing method. (The methods discussed in this text require the use of a peak discharge or an actual runoff hydrograph.) Generally, larger and more complex projects require a detailed analysis, which includes a runoff hydrograph. Preliminary studies and small projects may be designed using simpler, short-cut techniques that only require a peak discharge. For all projects, the designer must document the hydrologic conditions to support the inflow portion of the hydraulic relationship.

Manipulation of the site grades and strategic placement of permanent features like buildings and parking lots can usually accomplish achieving adequate storage volume within a basin. Sometimes, the site topography and available outfall location dictate the location of a stormwater facility.

17.7.2   Allowable Release Rates

The allowable release rates for a stormwater facility depend on the proposed function of that facility, such as flood control, channel erosion control, or water quality enhancement. For example, a basin used for water quality enhancement is designed to detain the water quality volume and slowly release it over a specified amount of time. This water quality volume is the first flush of runoff, which is considered to contain the largest concentration of pollutants (Schueler, 1987). In contrast, a basin used for flood or channel erosion control is designed to detain and release runoff from a given storm event at a predetermined maximum release rate. This release rate may vary from one watershed to another, based on predeveloped conditions.

Through stormwater management and erosion control ordinances, localities have traditionally set the allowable release rates for given frequency storm events to equal the watershed's predeveloped rates. This technique has become a convenient and consistent mechanism for establishing the design parameters for a stormwater management facility, particularly as it relates to flood control or stream channel erosion control.

Depending on location, the allowable release rate for controlling stream channel erosion or flooding may be established by ordinance using the state's minimum criteria, or by analyzing specific downstream topographic, geographic, or geologic conditions to select alternate criteria. Obviously, the engineer should be aware of the local requirements before beginning the design. The design examples and calculations in this text use minimum requirements for illustrative purposes.

17.7.3 Storage Volume Requirements Estimates

Stormwater management facilities are designed using a trial and error process. The engineer does many iterative routings to select a minimum facility size with the proper outlet controls. Each iterative routing requires that the facility size (stage–storage relationship) and the outlet configuration (stage–discharge relationship) be evaluated for performance against the watershed requirements. A graphical evaluation of the inflow hydrograph vs. an approximation of the outflow-rating curve provides the engineer with an estimate of the required storage volume. Starting with this assumed required volume, the number of iterations is reduced.

The graphical hydrograph analysis requires that the evaluation of the watershed's hydrology produce a runoff hydrograph for the appropriate design storms. Generally, local stormwater management regulations allow the use of SCS methods or the modified rational method (critical storm duration approach) for analysis. Many techniques are available to generate the resulting runoff hydrographs based on these methods. The engineer holds the responsibility for being familiar with the limitations and assumptions of the methods as they apply to generating hydrographs.

Graphical procedures can be time consuming, especially when dealing with multiple storms, and are therefore not practical when designing a detention facility for small site development. Shortcut procedures have been developed to allow the engineer to approximate the storage volume requirements. Such methods include TR-55: Storage Volume for Detention Basins, Section 5-4.2, and Critical Storm Duration — Modified Rational Method — Direct Solution, Section 5-4.4, which can be used as planning tools. Final design should be refined using a more accurate hydrograph routing procedure. Sometimes, these short-cut methods may be used for final design, but they must be used with caution because they only approximate the required storage volume.

Note that the TR-55 tabular hydrograph method does not produce a full hydrograph. The tabular method generates only the portion of the hydrograph that contains the peak discharge and some of the time steps just before and just after the peak. The missing values must be extrapolated, thus potentially reducing the accuracy of the hydrograph analysis. If SCS methods are used, a full hydrograph should be generated using one of the available computer programs. The analysis can only be as accurate as the hydrograph used.

17.7.4 Graphical Hydrograph Analysis — SCS Methods

The following analysis presents a graphical hydrograph analysis that results in the approximation of the required storage volume for a proposed stormwater management basin. We present the procedure to illustrate this technique. See Table 17.7 for a summary of the hydrology. The TR-20 computer-generated hydrograph is used for this example. The allowable discharge from the proposed basin has been established by ordinance (based on predeveloped watershed discharge).

Table 17.7　Hydrologic Summary, SCS Methods

Condition	DA	RCN	t_c	Q_2	Q_{10}
TR-55 graphical peak discharge					
PRE-DEV	25 acre	64	0.87 h	8.5 cfs[a]	26.8 cfs[a]
POST-DEV	25 acre	75	0.35 h	29.9 cfs	70.6 cfs
TR-20 computer run					
PRE-DEV	25 acre	64	0.87 h	8.0 cfs[a]	25.5 cfs[a]
POST-DEV	25 acre	75	0.35 h	25.9 cfs	61.1 cfs

[a] Allowable release rate.

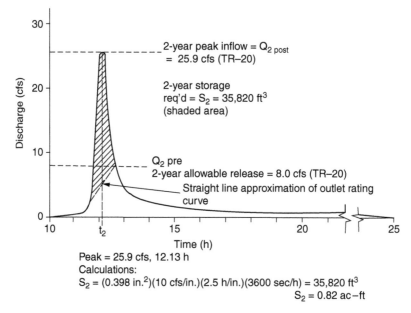

Figure 17.9　SCS runnoff hydrograph, 2-year postdeveloped. (From U.S. Soil Conservation Service, 1986, Technical Release No. 55 and U.S. Soil Conservation Services, 1982, Technical Release No. 20.)

17.7.4.1　Procedure

The pre- and postdeveloped hydrology (which includes the predeveloped peak rate of runoff (allowable release rate) and the postdeveloped runoff hydrograph (inflow hydrograph)) is required for hydrograph analysis (see Table 17.7; see Figure 17.9 for the 2-year developed inflow hydrograph and Figure 17.10 for the 10-year developed inflow hydrograph)

1. Commencing with the plot of the 2-year developed inflow hydrograph (discharge vs. time), the 2-year allowable release rate, $Q_2 = 8$ cfs, is plotted as a horizontal line starting at time $t = 0$ and continuing to the point where it intersects the falling limb of the hydrograph.
2. A diagonal line is then drawn from the beginning of the inflow hydrograph to the intersection point described previously. This line represents the hypothetical rating curve of the control structure and approximates the rising limb of the outflow hydrograph for the 2-year storm.
3. The storage volume is then approximated by calculating the area under the inflow hydrograph, less the area under the rising limb of the outflow hydrograph (shown as the shaded area in Figure 17.9). The storage volume required for the 2-year storm can be estimated by measuring the shaded area with a planimeter to approximate S_2.

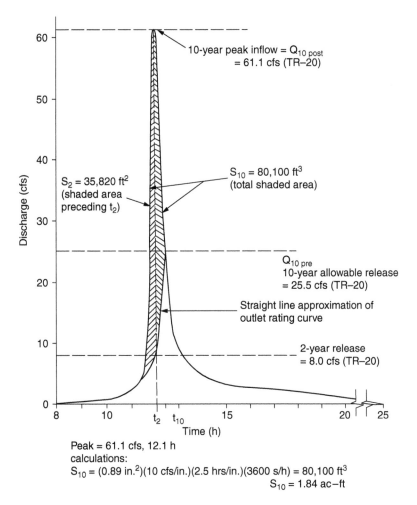

Peak = 61.1 cfs, 12.1 h
calculations:
S_{10} = (0.89 in.^{2})(10 cfs/in.)(2.5 hrs/in.)(3600 s/h) = 80,100 ft^3
S_{10} = 1.84 ac–ft

Figure 17.10 SCS runoff hydrograph, 10-year postdeveloped. (From U.S. Soil Conservation Service, 1986, Technical Release No. 55 and U.S. Soil Conservation Services, 1982, Technical Release No. 20.)

The vertical scale of a hydrograph is in cubic feet per second and the horizontal scale is in hours. Therefore, the area, as measured in square inches is multiplied by scale conversion factors of cubic feet per second per inch, hours per inch, and 3600 sec/h, to yield an area in cubic feet. The conversion is:

$$S^2 = (0.398 \text{ in}^2)(10 \text{ cfs/in.})(2.5 \text{ h/in.})(3,600 \text{ sec/h}) = 35,820 \text{ ft}^3 = 0.82 \text{ acre-ft}$$

1. On a plot of the 10-year inflow hydrograph, the 10-year allowable release rate, Q_{10}, is plotted as a horizontal line extending from time zero to the point where it intersects the falling limb of the hydrograph.
2. By trial and error, the time t_2 at which the S_2 volume occurs while maintaining the 2-year release is determined by planimeter. The shaded area to the left of t_2 on Figure 17.10 represents this. From the intersection point of t_2 and the 2-year allowable release rate, Q_2, a line is drawn to connect to the intersection point of the 10-year allowable release rate and the falling limb of the hydrograph. This intersection point is t_{10}, and the connecting line is a straight-line approximation of the outlet-rating curve.

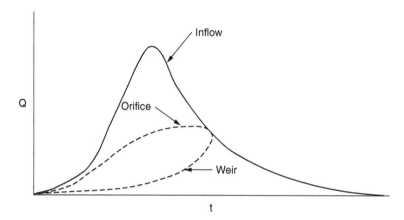

Q

Inflow

Orifice

Weir

t

Figure 17.11   Typical outlet rating curves for orifice and weir outlet devices. (From U.S. Soil Conservation Service, 1986, Technical Release No. 55 and U.S. Soil Conservation Services, 1982, Technical Release No. 20.)

3. The area under the inflow hydrograph from time t_2 to time t_{10}, less the area under the rising limb of the hypothetical rating curve, represents the additional volume (shaded area to the right of t_2 on Figure 17.10) needed to meet the 10-year storm storage requirements.
4. The total storage volume required, S_{10}, can be computed by adding this additional storage volume to S_2. The total shaded area under the hydrograph represents this.

$$S_{10} \;=\; (0.89 \text{ in.}^2)(10 \text{ cfs/in.})(2.5 \text{ h/in.})(3,600 \text{ sec/h}) \;=\; 80,100 \text{ ft}^3 \;=\; 1.84 \text{ acre-ft}$$

These steps may be repeated if storage of the 100-year storm (or any other design frequency storm) is required by ordinance of downstream conditions.

In summary, the total volume of storage required is the area *under* the runoff hydrograph curve and *above* the basin outflow curve. Note that the outflow-rating curve is approximated as a straight line. The actual shape of the outflow-rating curve depends on the type of outlet device used. Figure 17.11 shows the typical shapes of outlet rating curves for orifice and weir outlet structures. The straight-line approximation is reasonable for an orifice outlet structure. However, this approximation will likely underestimate the storage volume required when a weir outlet structure is used. Depending on the complexity of the design and the need for an exact engineered solution, the use of a more rigorous sizing technique, such as a storage indication routing, may be necessary.

17.7.5   TR-55: Storage Volume for Detention Basins (Short-Cut Method)

The TR-55 Storage Volume for Detention Basins, or TR-55 short-cut procedure, provides results similar to the graphical analysis. This method is based on average storage and routing effects for many structures. TR-55 can be used for single-stage or multistage outflow devices. The only constraints are that (1) each stage requires a design storm and a computation of the storage required for it; and (2) the discharge of the upper stages includes the discharge of the lower stages. Refer to TR-55 for more detailed discussions and limitations.

17.7.5.1   *Information Needed*

To calculate the required storage volume using TR-55, the pre- and postdeveloped hydrology per SCS methods is needed. This includes the watershed's predeveloped peak rate of discharge, or allowable release rate, Q_o; the watershed's postdeveloped peak rate of discharge, or inflow, Q_i, for

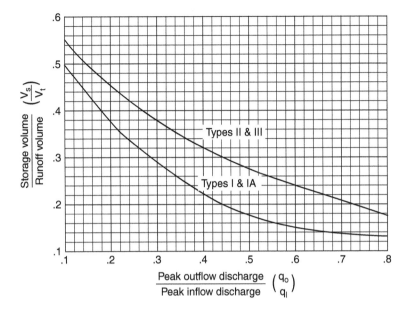

Figure 17.12   Approximate detention basin routing for rainfall types I, IA, II, and III. (From U.S. Soil Conservation Service, 1986, Technical Release No. 55, Figure 6.1.)

the appropriate design storms; and the watershed's postdeveloped runoff, Q, in inches. (Note that this method does not require a hydrograph.) Once these parameters are known, the TR-55 manual can be used to approximate the storage volume required for each design storm. The following procedure summarizes the TR-55 short-cut method using the 25-acre watershed.

Example 17.1

Procedure

Step 1. Determine the peak-developed inflow, Q_i, and the allowable release rate, Q_o, from the hydrology for the appropriate design storm. Use the 2-year storm flow rates given below, based on TR-55, graphical peak discharge:

$$Q_{o2} = 8.5 \text{ cfs}; Q_{i2} = 29.9 \text{ cfs}$$

Using the ratio of the allowable release rate, Q_o, to the peak developed inflow, Q_i, or Q_o/Q_i, for the appropriate design storm, use Figure 17.12 (or Figure 6-1 in TR-55) to obtain the ratio of storage volume, V_r, V_s/V_r.

$$Q_{o2} / Q_{i2} = 8.5/29.9 = 0.28$$

From Figure 17.12 or TR-55 Figure 6.1:

$$V_{s2} / V_{r2} = .39$$

Step 2. Determine the runoff volume, V_r, in acre-feet, from the TR-55 worksheets for the appropriate design storm.

Table 17.8   Storage Volume Requirements

Method	2-yr Storage required	10-yr Storage required
Graphical hydrograph analysis	0.82 acre-ft	1.84 acre-ft
TR-55 shortcut method	1.05 acre-ft	1.96 acre-ft

$$V_r = Q A_m \; 53.33$$

where:

Q = runoff, inches, from TR-55 worksheet 2 = 1.30 in.

A_m = drainage area, in square miles (25 acre/640 acre/mi^2 = 0.039 mi^2)

$$V_r = 1.30(0.39) \; 53.33 = 2.70 \text{ acre-ft}$$

Step 3. Multiply the V_s/V_r ratio from Step 1 by the runoff volume, V_r, from Step 2 to determine the volume of storage required, V_s, in acre-feet:

$$\left(\frac{V_s}{V_r}\right) V_r = V_s = (0.39)(2.70 \text{ acre-ft}) = 1.05 \text{ acre-ft}$$

Step 4. Repeat these steps for each additional design storm as required to determine the approximate storage requirements. We present the 10-year storm storage requirements here:

  a. $Q_o = 26.8$ cfs
  b. $Q_i = 70.6$ cfs

$$Q_o/Q_i = 26.8/70.6 = 0.38$$

   From Figure 17.12 or TR-55 Figure 6-1: $V_s/V_r = 0.33$

  c. $V_s = (V_s/V_r)V_r = (0.33)5.93$ acre-ft = 1.96 acre-ft

   This volume represents the total storage required for the 2-year storm and the 10-year storm.

Step 5. Note that the volume from Step 4 may need to be increased if additional storage is required for water quality purposes or channel erosion control.

The design presented here should be used with TR-55 worksheet 6a. The worksheet includes an area to plot the *stage–storage curve*, from which actual elevations corresponding to the required storage volumes can be derived. Table 17.8 provides a summary of the required storage volumes using the graphical SCS hydrograph analysis and the TR-55 short-cut method.

17.7.6 Graphical Hydrograph Analysis, Modified Rational Method Critical Storm Duration

The modified rational method uses the *critical storm duration* — the storm duration that generates the greatest volume of runoff and therefore requires the most storage — to calculate the maximum storage volume for a detention facility. In contrast, the rational method produces a triangular runoff hydrograph that gives the peak inflow at time $= t_c$ and falls to zero flow at time $= 2.5t_c$. In theory, this hydrograph represents a storm whose duration equals the time of concentration, t_c, resulting in the greatest peak discharge for the given return frequency storm. The volume of runoff, however, is of greater consequence in sizing a detention facility. A storm whose duration is longer than the

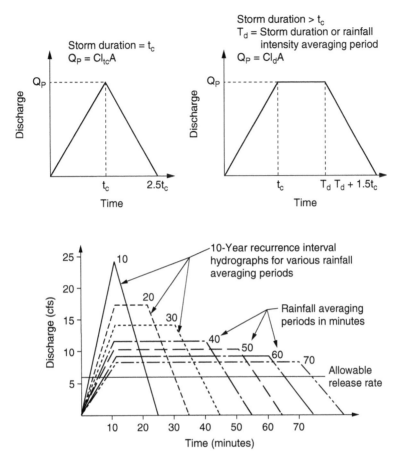

Figure 17.13 Modified rational method hydrographs. (From *Virginia Stormwater Management Handbook*, 1999, Virginia Department of Conservation and Recreation, Division of Soil and Water Conservation.)

t_c may not produce as large a peak rate of runoff, but it may generate a greater volume of runoff. By using the modified rational method, the designer can evaluate several different storm durations to verify which one requires the greatest volume of storage with respect to the allowable release rate. The basin must be designed to detain the maximum storage volume.

The first step in determining the critical storm duration is to use the postdeveloped time of concentration, t_c, to generate a postdeveloped runoff hydrograph. Rainfall intensity averaging periods, T_d, representing time periods incrementally longer than the t_c, are then used to generate a "family" of runoff hydrographs for the same drainage area. These hydrographs are trapezoidal, with the peak discharges, Q_i, based upon the intensity, I, of the averaging period, T_d. Figure 17.13 shows the construction of a typical triangular and trapezoidal hydrograph using the modified rational method and a family of trapezoidal hydrographs representing storms of different durations.

Note that the duration of the receding limb of the trapezoidal hydrograph (Figure 17.13) is set to equal 1.5 times the time of concentration, t_c. Also, the total hydrograph duration is 2.5 t_c vs. 2 t_c. This longer duration is considered more representative of actual storm and runoff dynamics and is more analogous to the SCS unit hydrograph where the receding limb extends longer than the rising limb.

The modified rational method assumes that the rainfall intensity averaging period is equal to the actual storm duration. This means that the rainfall and runoff that occur before and after the rainfall-averaging period are not accounted for. Therefore, the modified rational method may underestimate the required storage volume for any given storm event.

Table 17.9 Hydrologic Summary, Rational Method

Condition	DA	C	T_2	Q_2	Q_{10}
		Rational method			
Predeveloped	25 acre	0.38	0.87 h, 52 min	17 cfs	24 cfs
Postdeveloped	25 acre	0.59	0.35 h, 21 min	49 cfs	65 cfs

The rainfall intensity averaging periods are chosen arbitrarily. However, the designer should select periods for which the corresponding I–D–F curves are available (10 min, 20 min, 30 min and so on). The shortest period selected should be the time of concentration, t_c. A straight line starting at $Q = 0$ and $t = 0$ and intercepting the inflow hydrograph on the receding limb at the allowable release rate, Q_o, represents the outflow rating curve. The time-averaging period hydrograph that represents the greatest required storage volume is the one with the largest area between the inflow hydrograph and outflow-rating curve. This determination is made by a graphical analysis of the hydrographs.

The next procedure presents a graphical analysis very similar to the one described earlier. Note that the rational and modified rational methods should normally be used in homogeneous drainage areas of less than 20 acres, with a t_c of less than 20 min. Although the watershed in our example has a drainage area of 25 acres and a t_c of greater than 20 min, we use it here for illustrative purposes. Note that the pre- and postdeveloped peak discharges are much greater than those calculated using the SCS method applied to the same watershed. This difference may be the result of the large acreage and t_c values. A summary of the hydrology is found in Table 17.9. Note that the t_c calculations were performed using the more rigorous SCS TR-55 method.

17.7.6.1 Information Needed

The modified rational method-critical storm duration approach is very similar to SCS methods because it requires pre- and postdeveloped hydrology in the form of a predeveloped peak rate of runoff (allowable release rate) and a postdeveloped runoff hydrograph (inflow hydrograph), as developed using the rational method.

Procedure

(See Figure 17.14 and Figure 17.15.)

Step 1. Plot the 2-year developed condition inflow hydrograph (triangular) based on the developed condition, t_c.

Step 2. Plot a family of hydrographs, with the time averaging period, T_d, of each hydrograph increasing incrementally from 21 min (developed condition t_c) to 60 min, as shown in Figure 17.14. Note that the first hydrograph is a Type 1 modified rational method triangular hydrograph, where the storm duration, d, or T_d, is equal to the time of concentration, t_c. The remaining hydrographs are trapezoidal, or Type 2 hydrographs. The peak discharge for each hydrograph is calculated using the rational equation, $Q = CIA$, where the intensity, I, from the I–D–F curve is determined using the rainfall intensity averaging period as the storm duration.

Step 3. Superimpose the outflow-rating curve on each inflow hydrograph. The area between the two curves then represents the storage volume required (see Figure 17.14). Similar cautions as those described in the SCS method regarding the straight-line approximation of the outlet discharge curve apply here as well. The actual shape of the outflow curve depends on the type of outlet device.

Step 4. Compute and tabulate the required storage volume for each of the selected durations or time averaging periods, T_d.

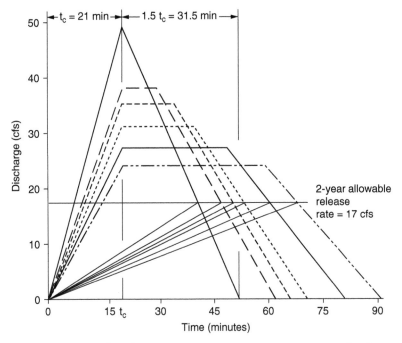

Conversion: in.^{3} to ft^3 = 10 cfs/in. × 0.25 h/in. × 3600 sec/h = 9000 ft^3/in.^{2}

Rainfall avg. period (T)	Intensity I(in./h)	Q peak (cfs)	Area (in.^{2})	Storage volume (ft^3)
21 min	3.3	49	5.62	50,580
30 min	2.6	38	5.51	49,590
35 min	2.4	35	5.53	49,770
40 min	2.1	31	5.29	47,610
50 min	1.8	27	5.25	47,250
60 min	1.6	24	5.24	47,160

Figure 17.14 Modified rational method runoff hydrograph, 2-year postdeveloped condition. (From *Virginia Stormwater Management Handbook*, 1999, Virginia Department of Conservation and Recreation, Division of Soil and Water Conservation.)

The storm duration that requires the maximum storage is the *critical storm* and is used for the sizing of the basin. (Storm duration equal to the t_c produces the largest storage volume required for the 2-year storm presented here.)

Step 5. Repeat Step 1 through Step 4 for the analysis of the 10-year storage area requirements. (Figure 17.15 represents this procedure repeated for the 10-year design storm.)

17.7.7 Modified Rational Method, Critical Storm Duration — Direct Solution

A direct solution to the modified rational method, critical storm duration was developed to eliminate the time-intensive iterative process of generating multiple hydrographs. This direct solution takes into account storm duration and allows the engineer to solve for the time at which the storage volume curve has a slope equal to zero, which corresponds to maximum storage. The basis derivation of this method is provided next, followed by the procedure as applied to our examples.

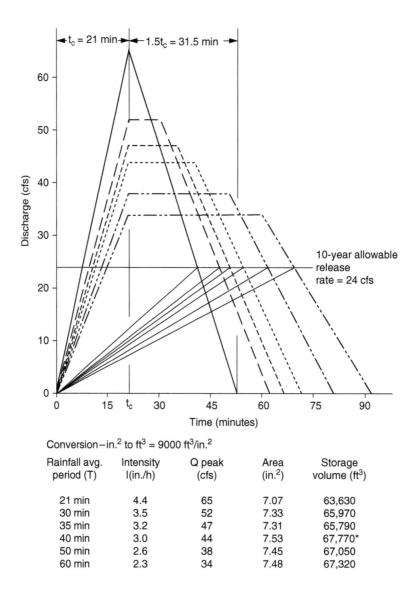

Conversion – in.² to ft³ = 9000 ft³/in.²

Rainfall avg. period (T)	Intensity I(in./h)	Q peak (cfs)	Area (in.²)	Storage volume (ft³)
21 min	4.4	65	7.07	63,630
30 min	3.5	52	7.33	65,970
35 min	3.2	47	7.31	65,790
40 min	3.0	44	7.53	67,770*
50 min	2.6	38	7.45	67,050
60 min	2.3	34	7.48	67,320

Figure 17.15 Modified rational method runoff hydrograph, 10-year postdeveloped condition. (From *Virginia Stormwater Management Handbook*, 1999, Virginia Department of Conservation and Recreation, Division of Soil and Water Conservation.)

17.7.7.1 *Storage Volume*

The runoff hydrograph developed with the modified rational method, critical storm duration will be triangular or trapezoidal in shape. The outflow hydrograph of the basin is approximated by a straight line starting at 0 cfs at the time $t = 0$ and intercepting the receding leg of the runoff hydrograph at the allowable discharge, q_o.

> *Note*: The straight-line representation of the outflow hydrograph is a conservative approximation of the shape of the outflow hydrograph for an orifice control release structure. This method should be used with caution when designing a weir control release structure.

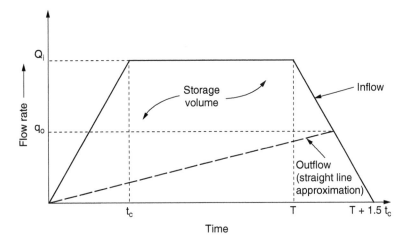

Figure 17.16 Trapezoidal hydrograph storage volume estimate. (From *Virginia Stormwater Management Handbook*, 1999, Virginia Department of Conservation and Recreation, Division of Soil and Water Conservation.)

The area between the inflow hydrograph and the outflow hydrograph in Figure 17.16 represents the required storage volume. This area — the trapezoidal hydrograph storage volume — can be approximated using the following equation:

$$V = \left[Q_i T_d + \frac{Q_i T_c}{4} - \frac{q_o T_d}{2} - \frac{3 q_o T_c}{4} \right] 60 \qquad (17.5)$$

where
V = required storage volume, cubic feet
Q_i = inflow peak discharge, cubic feet per second, for the critical storm duration, T_d
T_c = postdeveloped time of concentration, minutes
q_o = allowable peak outflow, cubic feet per second
T_d = critical storm duration, minutes

Regulatory ordinance or downstream conditions establish the allowable peak outflow. The critical storm duration, T_d, is an unknown and must be determined to solve for the intensity, I, and, ultimately, to calculate the peak inflow, Q_i. Therefore, a relationship between rainfall intensity, I, and the critical storm duration, T_d, must be established.

17.7.7.2 Rainfall Intensity

The rainfall intensity as taken from the I–D–F curves is dependent on the time of concentration, t_c, of a given watershed. Setting the storm duration, T_d, equal to the time of concentration, t_c, provides the maximum peak discharge, but does not necessarily generate the maximum volume of discharge. Because this maximum volume of runoff is of interest and the storm duration is unknown, the rainfall intensity, I, must be represented as a function of time, frequency, and location. The relationship is expressed by the modified rational method intensity (I) equation as:

$$I = \frac{a}{b + T_d} \qquad (17.6)$$

Table 17.10 Rainfall Constants for Virginia

| | Duration — 5 min to 2 h | | |
Station	Rainfall frequency	Constant A	Constant B
1	2	117.7	19.1
	5	168.6	23.8
	10	197.8	25.2
2	2	118.8	17.2
	5	158.9	20.6
	10	189.8	22.6
3	2	130.3	18.5
	5	166.9	20.9
	10	189.2	22.1
4	2	126.3	17.2
	5	173.8	22.7
	10	201.0	23.9
5	2	143.2	21.0
	5	173.9	22.7
	10	203.9	24.8

Source: The preceding constants are based on linear regression analyses of the frequency intensity-duration curves contained in the Virginia Department of Transportation (VDOT) Drainage Manual. (Adapted from DCR Course "C" Training Notebook.)

where

I = rainfall intensity, inches per hour

T_d = rainfall duration or rainfall intensity averaging period, minutes

a and b = rainfall constants developed for storms of various recurrence intervals and various geographic locations (see Table 17.10)

The rainfall constants, a and b, were developed from linear regression analyses of the I–D–F curves and can be generated for any area where such curves are available. The limitations associated with the I–D–F curves, such as duration and return frequency, also limit development of the constants. Table 17.10 provides rainfall constants for various regions in Virginia. Substituting Equation 17.6 into the rational equation results in the rearranged rational equation, Equation 17.7:

$$Q = C\left(\frac{a}{b+T_d}\right)A \qquad (17.7)$$

where

Q = peak rate of discharge, cubic per second

a and b = rainfall constants developed for storms of various recurrence intervals and various geographic locations (see Table 17.10)

T_d = critical storm duration, minutes

C = runoff coefficient

A = drainage area, acres

Substituting this relationship for Q, Equation 17.5 then becomes:

$$V = \left[\left[C\left(\frac{a}{b+T_d}\right)A\right]T_d + \frac{\left[C\left(\frac{a}{b+T_d}\right)A\right]t_c}{4} - \frac{q_oT_d}{2} - \frac{3q_oT_c}{4}\right]60 \qquad (17.8)$$

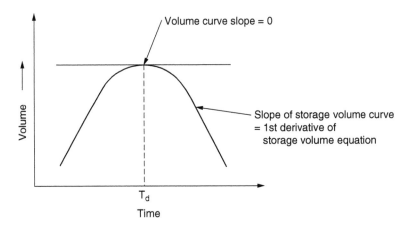

Figure 17.17   Storage volume vs. time curve. (From *Virginia Stormwater Management Handbook*, 1999, Virginia Department of Conservation and Recreation, Division of Soil and Water Conservation.)

17.7.7.3 Maximum Storage Volume

The first derivative of storage volume with respect to time (Equation 17.8) is an equation that represents the slope of the storage volume curve plotted vs. time. When this equation is set to equal zero and solved for T_d, at which the slope of the storage volume curve is zero or at a maximum (see Figure 17.17), Equation 17.9 presents the first derivative of the storage volume equation with respect to time and can be solved for critical storm duration, T_d.

$$T_d = \sqrt{\frac{2CAa(b - T_c/4)}{q_o}} - b \qquad (17.9)$$

where
T_d = critical storm duration, minutes
C = runoff coefficient
A = drainage area, acres
a and b = rainfall constants developed for storms of various recurrence intervals and various geographic locations (see Table 17.10)
t_c = time concentration, minutes
q_o = allowable peak outflow, cubic feet per second

Equation 17.9 is solved for T_d. Then, T_d is substituted into Equation 17.7 to solve for Q_i and, when Q_i is known, the outlet structure and the stormwater facility can be sized. This method provides a direct solution to the graphical analysis of the family of hydrographs and is quicker to use. The following procedure illustrates this method.

17.7.7.4 Information Needed

The modified rational method–direct solution is similar to the previous methods because it requires determination of the pre- and postdeveloped hydrology, resulting in a predeveloped peak rate of runoff (allowable release rate) and a postdeveloped runoff hydrograph. Table 17.9 provides a summary of the hydrology. The rainfall constants a and b for the watershed are determined from Table 17.10.

Procedure

Step 1. Determine the 2-year critical storm duration by solving Equation 17.9:

$$T_{d2} = \sqrt{\frac{2CAa(b - T_c/4)}{q_{o2}}} - b$$

Given:
T_{d2} = 2-year critical storm duration, min
C = developed condition runoff coefficient = 0.59
A = drainage area = 25.0 acres
T_c = postdeveloped time of concentration = 21 min
q_{o2} = allowable peak outflow = 17 cfs (predeveloped peak rate of discharge)
a_2 = 2-year rainfall constant = 130.3
b_2 = 2-year rainfall constant = 18.5

$$T_{d2} = \sqrt{\frac{2(.59)(25.0)(130.3)(18.5 - 21/4)}{17}} - 18.5$$

$$\sqrt{2995.0} - 18.5$$

$$T_{d2} = 36.2 \text{ min.}$$

Step 2. Solve for 2-year critical storm duration intensity, I_2, using Equation 17.6 and the 2-year critical storm duration, T_{d2}:

$$I_2 = \frac{a}{b + T_{d2}}$$

where:
T_{d2} = critical storm duration = 36.2 min
a = 2-year rainfall constant = 130.3
b = 2-year rainfall constant = 18.5

$$I_2 = \frac{130.3}{18.5 + 36.2} = 2.38 \text{ in./h}$$

Step 3. Determine the 2-year peak inflow, Q_{i2}, using the rational equation and the critical storm duration intensity, I_2:

$$Q_{i2} = CI_2A$$

where:

Q_{i2} = 2-year peak inflow, cubic feet per second
C = developed condition runoff coefficient = 0.59
I_2 = critical storm intensity = 2.38 in./h
A = drainage area = 25 acres

$$Q_{i2} = (0.59)(2.38)(25)$$

$$Q_{i2} = 35.1 \text{ cfs}$$

Step 4. Determine the 2-year required storage volume for the 2-year critical storm duration, T_{d2}, using Equation 17.5.

$$V_2 = \left[Q_{i2}T_{d2} + \frac{Q_{i2}T_c}{4} - \frac{q_{o2}T_{d2}}{2} - \frac{3q_{o2}T_c}{4} \right]60$$

where
V_2 = 2-year required storage, cubic feet
Q_{i2} = 2-year peak inflow for critical storm = 35.1 cfs
C = developed runoff coefficient = 0.59
A = area = 25.0 acres
T_{d2} = critical storm duration = 36.2 min
t_c = developed condition time of concentration = 21 min
q_{o2} = 2-year allowable peak outflow = 17 cfs

$$V_2 = \left[(35.1)(36.2) + \frac{(35.1)(21)}{4} - \frac{(17)(36.2)}{2} - \frac{3(17)(21)}{4} \right]60$$

$$V_2 = 52,764 \text{ ft}^3 = 1.21 \text{ acre-ft}$$

Repeat Step 2 through Step 4 for the 10-year storm:
Step 5. Determine the 10-year critical storm duration T_{d10}, using Equation 17.9:

Given:
T_{d10} = 10-year critical storm duration, minutes
C = developed condition runoff coefficient = 0.59
A = drainage area = 25 acres
t_c = post-developed time of concentration = 21 min
q_{o10} = 24 cfs
a_{10} = 189.2
b_{10} = 22.1

$$T_{d10} = \sqrt{\frac{2(.59)(25.0)(189.2)(22.1 - 21/4)}{24}} - 22.1$$

$$T_{d10} = \sqrt{3918.6} - 22.1$$

$$T_{d10} = 40.5 \text{ min.}$$

Step 6. Solve for the 10-year critical storm duration intensity, I_{10}, using Equation 17.6 and the 10-year critical storm duration, T_{d10}.

$$I_{10} = \frac{a}{b + T_{d10}}$$

$$I_{10} = \frac{189.2}{22.1 + 40.5} = 3.02 \text{ in./h}$$

Step 7. Determine the 10-year peak inflow, Q_{i10}, using the rational equation and the critical storm duration intensity I_{10}:

Given:
Q_{i10} = 10-year peak inflow
C = developed condition runoff coefficient = 0.59
I_{10} = critical storm intensity = 3.02 in./h
A = drainage area = 25.0 acres

$$Q_{i10} = CI_{10}A$$

$$Q_{i10} = 44.5 \text{ cfs}$$

Step 8. Determine the required 10-year storage volume for the 10-year critical storm duration, T_{d10}, using Equation 17.5:

Given:
V_{10} = required storage, cubic feet
Q_{i10} = 44.5 cfs
C = 0.59
A = 25.0 acres
T_{d10} = 40.5 min
t_c = 21 min
q_{o10} = 24 cfs

$$V_{10} = \left[(44.5)(40.5) + \frac{(44.5)(21)}{4} - \frac{(24)(40.5)}{2} - \frac{3(24)(21)}{4} \right] 60$$

$$V_{10} = 70{,}308 \text{ ft}^3 = 1.61 \text{ acre-ft}$$

V_2 and V_{10} represent the total storage volume required for the 2- and 10-year storms, respectively. Table 17.11 provides a summary of the four different sizing procedures used in this section. The engineer should choose one of these methods, based on the complexity and size of the watershed and the chosen hydrologic method. Using the stage–storage curve, a multistage riser structure can then be designed to control the appropriate storms and, if required, the water quality volume.

Table 17.11 Summary of Results: Storage Volume Requirement Estimates

Method	2-yr Storage required	10-yr Storage required
Graphical hydrograph analysis	0.82 acre-ft	1.84 acre-ft
TR-55 shortcut method	1.05 acre-ft	1.96 acre-ft
Modified rational method	1.16 acre-ft	1.56 acre-ft
Modified rational method — critical	1.21 acre-ft	1.61 acre-ft
Storm duration — direct solution	T_d = 36.2 min.	T_d = 40.5 min.

17.7.8 Stage–Storage Curve

By using one of the preceding methods for determining storage volume requirements, the engineer now has sufficient information to place and grade the proposed stormwater facility. Remember that this is a preliminary sizing that must be refined during the actual design. Trial and error can achieve the approximate required volume by designing the basin to fit the site geometry and topography. The storage volume can be computed by planimetering the contours and creating a stage–storage curve.

17.7.8.1 Storage Volume Calculations

For retention/detention basins with vertical sides (tanks and vaults, for example), the storage volume is simply the bottom surface area times the height. For basins with graded (2H:1V, 3H:1V, etc.) side slopes or an irregular shape, the stored volume can be computed by the following procedure. (Figure 17.18 represents the stage–storage computation worksheet completed for our example. Note that other methods for computing basin volumes, such as the conic method for reservoir volumes, are available that are not presented here.)

Procedure

Step 1. Planimeter or otherwise compute the area enclosed by each contour and enter the measured value into column 1 and column 2 of Figure 17.18. The invert of the lowest control orifice represents zero storage, which corresponds to the bottom of the facility for extended-detention or detention facilities, or the permanent pool elevation for retention basins.

Step 2. Convert the planimetered area (often in square inches) to units of square feet in column 3 of Figure 17.18.

Step 3. Calculate the average area between each contour. The average area between two contours is computed by adding the area planimetered for the first elevation, column 3, to the area planimetered for the second elevation, also column 3, and then dividing their sum by 2. This average is then written in column 4 of Figure 17.18. From this figure:

$$\text{Average area, elevation 81 to 82: } \frac{0+1800}{2} = 900\,\text{ft}^2$$

$$\text{Average area, elevation 82 to 84: } \frac{1800+3240}{2} = 2520\,\text{ft}^2$$

$$\text{Average area, elevation 84 to 86: } \frac{3240+5175}{2} = 4207\,\text{ft}^2$$

This procedure is repeated to calculate the average area found between any two consecutive contours.

PROJECT: _____EXAMPLE 1_____   SHEET ____ OF _____

COUNTY: _____   COMPUTED BY: _____   DATE: _____

DESCRIPTION: _____

ATTACH COPY OF TOPO: SCALE- 1" = 30 ft.

1	2	3	4	5	6	7	8
ELEV.	AREA (in.²)	AREA (ft²)	AVG. AREA (ft²)	INTERVAL	VOL. (ft³)	TOTAL VOLUME (ft³)	TOTAL VOLUME (ac.ft.)
81	0	0				0	0
			900	1	900		
82	2.0	1800				900	.02
			2520	2	5040		
84	3.6	3240				5940	.14
			4207	2	8414		
86	5.75	5175				14354	.33
			7614	2	15228		
88	11.17	10053				29582	.68
			12991	2	25982		
90	17.7	15930				55564	1.28
			20700	2	41400		
92	28.3	25470				96964	2.23
			31102	1	31102		
93	40.8	36734				128066	2.94
			38105	1	38105		
94	43.9	39476				166171	3.81

Figure 17.18 Stage-storage computation worksheet. (From *Virginia Stormwater Management Handbook*, 1999, Virginia Department of Conservation and Recreation, Division of Soil and Water Conservation.)

Step 4. Calculate the volume between each contour by multiplying the average area from Step 3 (column 4) by the contour interval and placing the product in column 6. From Figure 17.18:
• Contour interval between 81 and 82 = 1 ft. × 900 ft² = 900 ft³
• Contour interval between 82 and 84 = 2 ft. × 2520 ft² = 5040 ft³
Repeat this procedure for each measured contour interval.
Step 5. Sum the volume for each contour interval in column 7, using Figure 17.18. This is simply the sum of the volumes computed in the previous step:
• Contour 81, volume = 0
• Contour 82, volume = 0 + 900 = 900 ft³
• Contour 84, volume = 900 + 5040 = 5940 ft³
• Contour 86, volume = 5940 + 8414 = 14,354 ft³
• Column 8 allows for the volume to be tabulated in units of acre-feet: ft³ + 43,560 ft²/acre.
Repeat this procedure for each measured contour interval.
Step 6. Plot the stage–storage curve with *stage* on the *y*-axis vs. *storage* on the *x*-axis. Figure 17.19 represents the stage–storage curve for our example in units of feet (stage) vs. acre-feet (storage).

The stage–storage curve allows the engineer to estimate the design high water elevation for each of the design storms if the required storage volume has been determined. This allows for a preliminary design of the riser orifice sizes and their configuration.

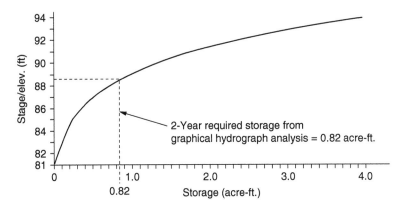

Figure 17.19 Stage vs. storage curve. (From *Virginia Stormwater Management Handbook*, 1999, Virginia Department of Conservation and Recreation, Division of Soil and Water Conservation.)

17.7.9 Water Quality and Channel Erosion Control Volume Calculations

Various local stormwater management regulations (readers should check with local requirements) require that the first flush of runoff (the water quality volume) be treated to enhance water quality. The water quality volume, V_{wq} (the first 0.5 in. of runoff from the impervious area of development) must be treated using one best management practice (BMP) or a combination of BMPs, depending on the total size of the contributing watershed, the amount of impervious area, and site conditions.

Calculate the water quality volume:

$$V_{wq} = \text{Impervious area (ft}^2) \times (1/2 \text{ in.})/ (12 \text{ in./ft})\qquad(17.10)$$

$$V_{wq} \text{ (acre-ft)} = V_{wq} \text{ (ft}^3)/43{,}560 \text{ ft}^2/\text{acre}\qquad(17.11)$$

The water quality volume for a wet BMP may be dependent on the specific design criteria for that BMP, based on the watershed's imperviousness or the desired pollutant removal efficiency (using performance-based or technology-based criteria, respectively). This discussion focuses on the calculations associated with the control of the water quality volume in extended-detention and retention basins.

Various local stormwater management regulations allow for the control of downstream channel erosion by detaining a specified volume of runoff for a period of time. Specifically, 24-h extended detention of the runoff from the 1-year frequency storm is proposed as an alternate criterion to the 2-year peak rate reduction. The channel erosion control volume (V_{ce}) is calculated by first determining the depth of runoff (in inches) based on the fraction of rainfall to runoff (runoff curve number) and then multiplying the runoff depth by the drainage area to be controlled.

17.7.9.1 Retention Basins — Water Quality Volume

The permanent pool feature of a retention basin allows for settling of particulate pollutants, such as sediment and other pollutants that adsorb to these particulates. Therefore, the volume of the pool must be large enough and properly configured to prevent short-circuiting. (Short-circuiting results when runoff enters the pool and exits without sufficient time for the settling process to occur.)

The permanent pool (or "dead" storage volume) of a retention facility is a function of the water quality volume. For example, a permanent pool sized to contain four times the water quality volume provides greater pollutant removal capacity than a permanent pool sized to contain two times the water quality volume. Our example analyzes a 25-acre watershed. The water quality volume and permanent pool volume calculations for a retention basin serving this watershed are calculated next.

Procedure

Step 1. Calculate the water quality volume, V_{wq}, for the given watershed. Assume the commercial/industrial development disturbs 11.9 acres, with 9.28 acres (404,236 ft²) of impervious cover after development.

$$V_{wq} = 404,236 \text{ ft}^2 \times \text{ ½ in./12 in./ft } = 16,843 \text{ ft}^3 = 16,843 \text{ ft}^3/43,560 \text{ ft}^2/\text{acre}$$

$$V_{wq} = 0.38 \text{ acre-ft}$$

Step 2. Size the permanent pool based on the desired pollutant removal efficiency of the drainage area impervious cover. The pool volume is sized based upon the desired pollutant removal efficiency. The permanent pool must be sized for $4 \times V_{wq}$ for a pollutant removal efficiency of 65%.

$$\text{Permanent Pool Volume} = V_{wq} \times 4.0$$

$$\text{Permanent Pool Volume} = 0.38 \text{ acre-ft} \times 4.0 = 1.52 \text{ acre-ft}$$

17.7.9.2 *Extended-Detention Basins — Water Quality Volume and Orifice Design*

A water quality extended-detention basin treats the water quality volume by detaining it and releasing it over a specified length of time. In theory, extended detention of the water quality volume allows the particulate pollutants to settle out of the first flush or runoff, functioning similarly to a permanent pool. Various local stormwater management regulations pertaining to water quality specify a 30-h drawdown time for the water quality volume. This is a *brim drawdown* time, beginning at the time of peak storage of the water quality volume — the time required for the entire calculated volume to drain out of the basin. This assumes that the brim volume is present in the basin prior to any discharge. In reality, however, water is flowing out of the basin prior to the full or brim volume being reached. Therefore, the extended detention orifice can be sized using either of the following methods:

- Use the maximum hydraulic head associated with the water quality volume (V_{wq}) and calculate the orifice size needed to achieve the required drawdown time, routing the water quality volume through the basin to verify the actual storage volume used and the drawdown time.
- Approximate the orifice size using the average hydraulic head associated with the water quality volume (V_{wq}) and the required drawdown time.

The two methods for calculating the required size of the extended detention orifice allow for a quick and conservative design (Method 2) and a similarly quick estimation with a routing to verify the performance of the design (Method 1).

Method 1, which uses the maximum hydraulic head and maximum discharge in the calculation, results in a slightly larger orifice than the same procedure using the average hydraulic head (Method 2). The routing allows the designer to verify the performance of the calculated orifice size. As a result of the routing effect, the actual basin storage volume used to achieve the drawdown time will be less than the computed brim drawdown volume. Note that the routing of the extended detention of the runoff from storms larger than the water quality storm (such as the 1-year frequency storm for channel erosion control) will result in proportionately larger reduction in the actual storage volume needed to achieve the required extended detention.

The procedure used to size an extended detention orifice includes the first steps of the design of a multistage riser for a basin controlling water quality and/or channel erosion, and peak discharge. These steps are repeated for sizing the 2- and 10-year release openings. Other design storms may be used as required by ordinance or downstream conditions.

Method 1: Water quality orifice design using maximum hydraulic head and routing of the water quality volume. A water quality extended-detention basin sized for two times the water quality volume is used here to illustrate the sizing procedure for an extended-detention orifice.

Procedure

Step 1. Calculate the water quality volume, V_{wq}, required for treatment. Assume:

$$V_{wq} = 404,236 \text{ ft}^2 \times \frac{1}{2} \text{ in./12 in./ft} = 16,843 \text{ ft}^3$$

$$V_{wq} = 16,843 \text{ ft}^3/43,560 \text{ ft}^2/\text{acre} = 0.38 \text{ acre-ft}$$

For extended-detention basins,

$$2 \times V_{wq} = 2(0.38 \text{ acre-ft}) = 0.76 \text{ acre-ft} = 33,106 \text{ ft}^3$$

Step 2. Determine the maximum hydraulic head, h_{max}, corresponding to the required water quality volume. Assume from our example stage vs. storage curve (Figure 17.19) that 0.76 acre-ft occurs at elevation 88 ft (approximate). Therefore, $h_{max} = 88 - 81 = 7.0$ ft.

Step 3. Determine the maximum discharge, Q_{max}, resulting from the 30-h drawdown requirement. The maximum discharge is calculated by dividing the required volume, in cubic feet, by the required time, in seconds, to find the average discharge, and then multiplying by 2 to determine the maximum discharge. Assume:

$$Q_{avg} = \frac{33,106 \text{ ft}^3}{(30 \text{ h}) (3,600 \text{ sec/h})} = 0.30 \text{ cfs}$$

$$Q_{max} = 2 \times 0.30 \text{ cfs} = 0.60 \text{ cfs}$$

Step 4. Determine the required orifice diameter by rearranging orifice equation 17.12 to solve for the orifice area, in square feet, and then diameter, in feet. Insert the values Q_{max} and h_{max} into rearranged orifice Equation 17.13, solve for the orifice area, and then solve for the orifice diameter.

$$Q = Ca\sqrt{2gh} \tag{17.12}$$

$$a = \frac{Q}{C\sqrt{2gh}} \tag{17.13}$$

where:

Q = discharge, cubic feet per second
C = dimensionless coefficient = 0.6
a = area of the orifice, square feet
g = gravitational acceleration, 32.2 ft/sec^2
h = head, feet

Assume:
    For orifice area:

$$a = \frac{0.6}{0.6\sqrt{(2)(32.2)(7.0)}}$$

    For orifice diameter:

$$a = 0.047 \text{ ft}^2 = \pi d^2 = \pi d^2/4$$

$$d = \sqrt{\frac{4a}{\pi}} = \sqrt{\frac{4(0.047 \text{ ft}^2)}{\pi}}$$

$$d = \text{orifice diameter} = 0.245 \text{ ft} = 2.94 \text{ in.}$$

Use a 3-in. diameter water quality orifice.

Routing the water quality volume (V_{wq}) of 0.76 acre-ft, occurring at elevation 88 ft through a 3-in. water quality orifice, will allow the engineer to verify the drawdown line, as well as the maximum elevation of 88 ft.

Route the Water Quality Volume

This calculation gives engineers the inflow–storage–outflow relationship to verify the actual storage volume needed for the extended detention of the water quality volume. The routing procedure takes into account the discharge that occurs before maximum or brim storage of the water quality volume, as opposed to the brim drawdown described in Method 2. The routing procedure is simply a more accurate analysis of the storage volume used while water is flowing into and out of the basin; therefore, the actual volume of the basin used will be less than the volume as defined by the regulation. This procedure is useful if the site to be developed is tight and the area needed for the stormwater basin must be "squeezed" as much as possible.

The routing effect of water entering and discharging from the basin simultaneously also results in the actual drawdown time being less than the calculated 30 h. Use best judgment to determine whether the orifice size should be reduced to achieve the required 30 h or the actual time achieved will provide adequate pollutant removal.

Note: The designer will notice a significant reduction in the actual storage volume used when routing the extended detention of the runoff from the 1-year frequency storm (channel erosion control).

Routing the water quality volume depends on the ability to work backwards from the design runoff volume of 0.5 in. to find the rainfall amount. Using SCS methods, the rainfall needed to generate 0.5 in. of runoff from an impervious surface (RCN = 98) is 0.7 in. The SCS design storm is the Type II, 24-h storm. Therefore, the water quality storm using SCS methods is defined as the SCS Type II, 24-h storm, with a rainfall depth = 0.7 in.

The rational method does not provide a design storm from a specified rainfall depth. Its rainfall depth depends on the storm duration (watershed t_c) and the storm return frequency. Because the water quality storm varies with runoff amount rather than the design storm return frequency, an input runoff hydrograph representing the water quality volume cannot be generated using rational method parameters. Therefore, Method 1, routing of the water quality volume, must use SCS methods.

Continuing with our example, the procedure is:

Procedure (cont.)

Step 5. Calculate a stage–discharge relationship using orifice Equation 17.12 and the orifice size determined in Step 4. Using the 3-in. diameter orifice, the calculation is:

$$Q = Ca\sqrt{2gh}$$

$$Q = 0.6(0.047)\sqrt{(2)(32.2)(h)}$$

$$Q = 0.22\sqrt{h}$$

where h = water surface elevation minus the orifice's centerline elevation, in feet.

> *Note*: If the orifice size is small relative to the anticipated head, h, values of h may be defined as the water surface elevation minus the invert of the orifice elevation.

Step 6. Complete a stage–discharge table for the range of elevations in the basin (see Table 17.12).

Step 7. Determine the time of concentration for the impervious area. From our example, the developed time of concentration, t_c = 0.46 h. The impervious area time of concentration, t_{cimp} = 0.09 h, or 5.4 min.

Step 8. Using t_{cimp}, the stage–discharge relationship, the stage–storage relationship, and the impervious acreage (RCN = 98), route the water quality storm through the basin. The water quality storm for this calculation is the SCS Type 2, 24-h storm, rainfall depth = 0.7 in. (Note that the rainfall depth is established as the amount of rainfall required to generate 0.5 in. of runoff from the impervious area.) The water quality volume may be routed using a variety of computer programs such as TR-20 HEC-1 or other storage indication routing programs.

Table 17.12 Stage–Discharge Table: Water Quality Orifice Design

Elevation	h (ft)	Q (cfs)
81	0	0
82	1	0.2
83	2	0.3
84	3	0.4
85	4	0.4
86	5	0.5
87	6	0.5
88	7	0.6

Step 9. Evaluate the discharge hydrograph to verify that the drawdown time from maximum storage
to zero discharge is at least 30 h. (Note that the maximum storage corresponds to the maximum
rate of discharge on the discharge hydrograph.) The routing of the water quality volume using TR-
20 results in a maximum storage elevation of 85.69 ft vs. the approximated 88.0 ft. The brim
drawdown time is 17.5 h (peak discharge occurs at 12.5 h and .01 discharge occurs at 30 h). For
this example, the orifice size may be reduced to provide a more reasonable drawdown time and
another routing performed to find the new water quality volume elevation.

Method 2: Water quality orifice design using average hydraulic head and average discharge.
For the previous example, Method 2 resulted in a 2.5-in. orifice (vs. a 3.0-in. orifice), and the design
extended-detention water surface elevation is set at 88 ft (vs. 85.69 ft). (Note that Trial 2 of the
method as noted earlier might result in a design water surface elevation closer to 88 ft.) If the basin
must control additional storms, such as the 2- and/or 10-year storms, the additional storage volume
would be "stacked" just above the water quality volume. The invert for the 2-year control, for
example, would be set at 88.1 ft.

17.7.9.3 Extended-Detention Basins — Channel Erosion Control Volume and Orifice Design

Extended detention of a specified volume of stormwater runoff can also be incorporated into a
basin design to protect downstream channels from erosion. Virginia's stormwater management
regulations, for example, recommend 24-h extended detention of the runoff from the 1-year frequency storm as an alternative to the 2-year peak rate reduction. The discussion presented here is
for the design of a channel erosion control extended-detention orifice.

The design of a channel erosion control extended-detention orifice is similar to the design of
the water quality orifice in that two methods can be employed:

- Use the maximum hydraulic head associated with the specified channel erosion control (V_{ce}) storage
  volume and calculate the orifice size needed to achieve the required drawdown time, routing the
  1-year storm through the basin to verify the storage volume and the draw downtime.
- Approximate the orifice size using the average hydraulic head associated with the channel erosion
  control volume (V_{ce}) and drawdown time.

The routing procedure takes into account the discharge that occurs before maximum or brim
storage of the channel erosion control volume (V_{ce}), providing a more accurate accounting of the
storage volume used while water is flowing into and out of the basin. This results in less storage
volume associated with the maximum hydraulic head. The actual storage volume needed for
extended detention of the runoff generated by the 1-year frequency storm will be approximately
60% of the calculated volume (V_{ce}) of runoff for curve numbers between 75 and 95 and time of
concentration between 0.1 and 1 h.

The following procedure illustrates the design of the extended-detention orifice for channel
erosion control.

Procedure

Step 1. Calculate the channel erosion control volume, V_{ce}. Determine the rainfall amount (inches) of
the 1-year frequency storm for the local area where the project is located. With the rainfall amount
and the runoff curve number (RCN), determine the corresponding runoff depth using the runoff
equation (Chapter 4: Hydrologic Methods [SCS TR-55] or the Rainfall – Runoff Depth Charts.

Given:
1-year rainfall = 2.7 in., RCN = 75
1-year frequency depth of runoff = 0.8 in.; therefore,

$$V_{ce} = 25 \text{ acre} \times 0.8 \text{ in.} \times 1/12 \text{ in.} = 1.66 \text{ acre-ft}$$

To account for the routing effect, reduce the channel erosion control volume:

$$V_{ce} = (0.6)(1.66 \text{ acre-ft}) = 1.0 \text{ acre-ft} = 43,560 \text{ ft}^3$$

Step 2. Determine the average hydraulic head, h_{ave}, corresponding to the required channel erosion control volume.

$$h_{avg} = (89 - 81)/2 = 4.0 \text{ ft}$$

Step 3. Determine the average discharge, Q_{avg}, resulting from the 24-h drawdown requirement. Calculate the average discharge by dividing the required volume, in cubic feet, by the required time, in seconds, to find the average discharge.

$$Q_{avg} = \frac{43,560 \text{ ft}^3}{(24 \text{ h}) (3,600 \text{ sec/h})} . = 0.5 \text{ cfs}$$

Step 4. Determine the required orifice diameter by rearranging orifice Equation 17.12 to solve for the orifice area, in square feet, and then diameter, in feet. Insert the values for Q_{avg} and h_{avg} into the rearranged orifice equation to solve for the orifice area, and then solve for the orifice diameter.

$$Q = Ca\sqrt{2gh}$$

$$a = \frac{Q}{Ca\sqrt{2gh}}$$

where:
Q = discharge, cubic feet per second
C = dimensionless coefficient = 0.6
a = area of the orifice, square feet
g = gravitational acceleration, 32.2 ft/sec²
h = head, feet

For orifice area:

$$a = \frac{0.5}{0.6\sqrt{(2)(32.2)(4.0)}}$$

$$a = 0.052 \text{ ft}^2 = \pi r^2 = \pi r^2/4$$

For orifice diameter:

$$d = \sqrt{\frac{4a}{\pi}} = \sqrt{\frac{4(0.052 \text{ ft}^2)}{\pi}}$$

$$d = \text{orifice diameter} = 0.257 \text{ ft} = 3.09 \text{ in}$$

Use 3.0-in. diameter channel erosion extended-detention orifice.

Method 1 results in a 3.7-in. diameter orifice and a routed water surface elevation of 88.69 ft. Additional storms may be "stacked" just above this volume if additional controls are desired.

17.7.10 Multistage Riser Design

For the drop inlet structure, a principal spillway system that controls the rate of discharge from a stormwater facility will often use a multistage riser — a structure that incorporates separate openings or devices at different elevations to control the rate of discharge from a stormwater basin during multiple design storms. Permanent multistage risers are typically constructed of concrete to help increase their life expectancy; they can be precast or cast in place. The geometry of risers varies from basin to basin. The engineer can be creative in finding ways to provide the most economical and hydraulically efficient riser design possible.

In stormwater management basin design, the multistage riser is of utmost importance because it controls the design water surface elevations. In designing the multistage riser, many iterative routings are usually required to arrive at a minimum structure size and storage volume that provides proper control. Each iterative routing requires that the facility's size (stage–storage curve) and outlet shape (state–discharge table or rating curve) be designed and tested for performance. Prior to final design, approximating the required storage volume and outlet shape is helpful, using one of the "short-cut" methods. In doing this, the number of iterations may be reduced. The following procedures outline methods for approximating and then completing the design of a riser structure.

17.7.10.1 Information Needed

- The hydrology for the watershed or drainage area to be controlled
- The allowable release rates for the facility, as established by ordinance or downstream conditions

The design procedure provided here incorporates the traditional 2- and 10-year design storms and the predeveloped hydrology establishes the allowable discharge rates of the developed watershed. Note that any design storm (1- or 5-year, for example) could be substituted into this design procedure, as required.

Procedure

Step 1. Determine water quality or extended detention requirements. Calculate the water quality volume and decide what method (extended detention or retention) will be used to treat it. Also calculate the channel erosion control volume for extended detention, if required.
- *Water quality extended-detention basin*: the water quality volume must be detained and released over 30 h. The established pollutant removal efficiency is based on a 30-h drawdown.
- *Water quality retention basin*: the site impervious cover or the desired pollutant removal efficiency establishes the volume of the permanent pool.
- *Channel erosion control extended-detention basin*: the channel erosion control volume must be detained and released over 24 h.

Step 2. Compute allowable release rates. Compute the pre- and postdeveloped hydrology for the watershed. Sometimes the predeveloped hydrology will establish the allowable release rate from the basin. Other times, the release rate is established by downstream conditions. In either case, the postdeveloped hydrology provides the peak inflow into the basin, as a peak rate (cubic feet per second) or a runoff hydrograph.

Step 3. Estimate the required storage volume. Estimate the storage volume required using one of the short-cut volume estimate methods described earlier. The information required includes the developed condition peak rate of runoff or runoff hydrograph and the allowable release rates for each of the appropriate design storms.

Step 4. Grade the basin; create stage–storage curve. After considering the site geometry and topography, select a location for the proposed stormwater management basin. By trial and error, size the basin to hold the approximate required storage volume. Ensure that the storage volume is measured from the lowest stage outlet. Remember that this is a preliminary sizing that must be fine-tuned during the final stage.

Step 5a. Design water quality orifice (extended detention). The procedure for sizing the water quality orifice for an extended-detention basin was covered earlier. Using Method 1 or Method 2, the engineer establishes the size of the water quality or stream channel erosion control orifice and the design maximum water surface elevation. The lowest stage outlet of an extended-detention basin is the invert of the extended-detention (or water quality) orifice, which corresponds to zero storage.

Step 5b. Set permanent pool volume (retention). In a retention pond, the permanent pool volume from Step 1 establishes the lowest stage outlet for the riser structure (not including a pond drain, if provided). The permanent pool elevation, therefore, corresponds to zero storage for the design of the "dry" storage volume stacked on top of the permanent pool.

Step 6. Size 2-year control orifice. (The 2-year storm is used here to show the design procedure. Other design storms or release requirements can be substituted into the procedure.) Knowing the 2-year storm storage requirement from design Step 3 and the water quality volume from design Step 1, the engineer can create a preliminary design for the 2-year release opening in the multistage riser. To complete the design, some iterations may be required to meet the allowable release rate performance criteria. This procedure is very similar to the water quality orifice sizing calculations:

1. Approximate the 2-year maximum head, h_{2max}. Establish the approximate elevation of the 2-year maximum water surface elevation using the stage storage curve and the preliminary size calculations. Subtract the water quality volume elevation from the approximate 2-year maximum water surface elevation to find the 2-year maximum head. If no water quality requirements are needed, use the elevation of the basin bottom or invert.

2. Determine the maximum allowable 2-year discharge rate, from Step 2.

3. Calculate the size of the 2-year control release orifice using rearranged orifice Equation 17.13 and solve for the area, a, in square feet. The engineer may choose to use any one of a variety of orifice shapes or geometrics. Regardless of the selection, the orifice will initially act as a weir until the top of the orifice is submerged. Therefore, the discharges for the first stages of flow are calculated using the weir equation:

$$Q_w = C_w L h^{1.5} \qquad (17.14)$$

where

Q_w = weir flow discharge, cubic feet per second

C_w = dimensionless weir flow coefficient, typically equal to 3.1 for sharp crested weirs (see Table 17.13)

L = length of weir crest, feet

H = head, feet, measured from the water surface elevation to the crest of the weir

Flow through the rectangular opening will transition from weir flow to orifice flow once the water surface has risen above the top of the opening. This orifice flow is expressed by the orifice equation. The area, a, of a rectangular orifice is written as $a = L \times H$.

where

L = length of opening, feet

H = height of opening, feet

Figure 17.20 shows a rectangular orifice acting as a weir at the lower stages and as an orifice after the water surface rises to height H, the height of the opening.

4. Develop the stage–storage–discharge relationship for the 2-year storm. Calculate the discharge using the orifice equation and, if a rectangular opening is used, the weir equation as needed for each elevation specified on the stage-storage curve. Record the discharge on a stage–storage–discharge worksheet.

Table 17.13 Weir Flow Coefficients

Weir flow coefficients, C		
Measured head, h, (ft)	Breadth of weir crest (ft)		

Measured head, h, (ft)	0.50	0.75	1.00
0.2	2.80	2.75	2.69
0.4	2.92	2.80	2.72
0.6	3.08	2.89	2.75
0.8	3.30	3.04	2.85
1.0	3.32	3.14	2.98
1.2	3.32	3.20	3.08
1.4	3.32	3.26	3.20
1.6	3.32	3.29	3.28
1.8	3.32	3.32	3.31
2.0	3.32	3.32	3.30
3.0	3.32	3.32	3.32
4.0	3.32	3.32	3.32
5.0	3.32	3.32	3.32

Source: Adapted from King, H.W. and Brater, E.R., 1976, *Handbook of Hydraulics*, 6th ed. New York: McGraw-Hill.

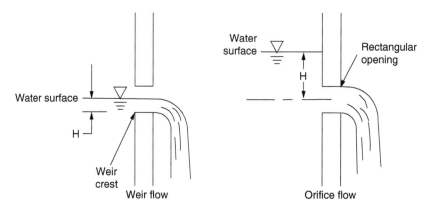

Figure 17.20   Weir and orifice flow. (From *Virginia Stormwater Management Handbook*, 1999, Virginia Department of Conservation and Recreation, Division of Soil and Water Conservation.)

Step 7. Check performance of 2-year opening. Note that this step may not necessary if the design is to be completed using one of the shortcut routing procedures where the water surface elevations are established by the required storage volume and not by an actual routing.
- Check the performance of the 2-year control opening by (1) reservoir routing the 2-year storm through the basin using an acceptable reservoir routing computer program; or (2) doing the long-hand calculations outlined in Section 17.7.8. Verify that the 2-year release rate is less than or equal to the allowable release rate. If not, reduce the size of the opening or provide additional storage and repeat Step 6.

This procedure presents just one of many riser configurations. Engineers may choose to use any type of opening geometry for controlling the design storms and, with experience, may come to recognize the most efficient way to configure the riser. Note that if a weir is chosen for 2-year storm control, engineers may substitute the appropriate values for the 2-year storm and use the procedures outlined here for the 10-year storm.

Step 8. Size 10-year control opening. The design of the 10-year storm control opening is similar to the procedure used in sizing the 2-year control opening:
  1. From the routing results, identify the exact 2-year water surface elevation.

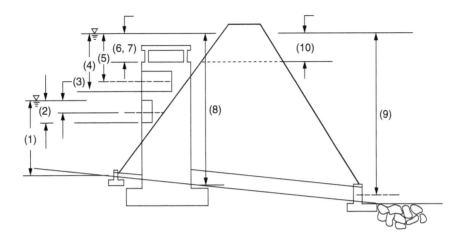

1. WQ orifice: H measured from water surface elevation (WSE) to centerline of pipe, or orifice (orifice flow)
2. 2-Year control: Weir flow – H measured from WSE to invert of 2-year control weir
3. 2-Year control: Orifice flow – H measured from WSE to centerline of opening (submerged)
4. 10-Year control: Weir flow – H measured from WSE to crest of opening
5. 10-Year control: Orifice flow – H measured from WSE to centerline of opening
6. Riser structure: Weir flow – H measured from WSE to crest of riser top (if open)
7. Riser structure: Orifice flow – H measured from WSE to crest of riser top, acting as horizontal orifice
8. Barrel flow: Inlet control – H measured from WSE to upstream invert of outlet barrel
9. Barrel flow: Outlet control – H measured from WSE to centerline of outlet of barrel or tailwater—whichever is higher
10. Emergency spillway: H measured from WSE to crest of emergency spillway

Figure 17.21 Typical hydraulic head values — multistage rise. (From *Virginia Stormwater Management Handbook*, 1999, Virginia Department of Conservation and Recreation, Division of Soil and Water Conservation.)

2. Set the invert of the 10-year control just above the 2-year design water surface elevation and determine the corresponding storage volume from the stage–storage curve. Add this elevation, storage and 2-year discharge to the stage–storage–discharge worksheet.

Note: The 10-year control invert may be set at a small distance (such as 0.1 ft minimum) above the 2-year maximum water surface elevation. If the 2-year orifice is also used for the 10-year control, the head is measured from the maximum water surface elevation to the centerline of the 2-year orifice (see Figure 17.21).

3. Establish the approximate 10-year maximum water surface elevation using the stage-storage curve and the preliminary sizing calculations. Subtract the invert elevation of the 10-year control (from Step 2) from the approximate 10-year maximum water surface elevation to find the 10-year maximum head, h_{10max}.
4. Determine the maximum allowable 10-year discharge rate, $Q_{10alloawable}$, from Step 2.
5. Calculate the required size of the 10-year release opening. The engineer may choose between a circular and rectangular orifice, or a weir. If a weir is chosen, the weir flow equation can be rearranged to solve for L as follows.

$$Q_w = C_w L h^{1.5} \tag{17.15}$$

$$L = Q_{10allowable}/C_w h^{1.5} \qquad (17.16)$$

where:

L = length of weir required, feet.
C_w = dimensionless weir flow coefficient; see Table 17.13
$Q_{10allowable}$ = 10-year allowable riser weir discharge, cubic feet per second
h = hydraulic head; water surface elevation minus the weir crest elevation

6. Develop the stage–storage–discharge relationship for the 10-year storm. Calculate the discharge for each elevation specified on the stage–storage curve, and record the discharge on a stage–storage–discharge worksheet.

Note: Any weir length lost to the trash rack or debris catcher must be accounted for.

Step 9. Check performance of 10-year opening. Note that this step may not be necessary if the design is completed using one of the short-cut routing procedures in which the water surface elevations are established by the required volume and not by an actual routing. Check the performance of the 10-year control opening by (1) reservoir routing the 2-year and 10-year storms through the basin using an acceptable reservoir routing computer program; or (2) doing the long-hand calculations outline in Section 17.7.12. Verify that the 10-year release rate is less than or equal to the allowable release rate. If not, reduce the size of the opening, and/or provide additional storage and repeat Step 8.

Step 10. Perform hydraulic analysis. At this point, several iterations may be required to calibrate and optimize the hydraulics of the riser and barrel system. Drop inlet spillways should be designed so that full flow is established in the outlet conduit and riser at the lowest head over the riser crest as is practical. Also, the structure should operate without excessive surging, noise, vibration, or vortex action at any stage. This requires the riser to have a larger cross-sectional area than the outlet conduit.

As the water passes over the rim of the riser, the riser acts as a weir (Figure 17.22a); this discharge is described as *riser weir flow control*. However, when the water surface reaches a certain height over the trim of the riser, the riser will begin to act as a submerged orifice (Figure 17.22b); such discharge is called *riser orifice flow control*. The engineer must compute the elevation at which this transition from riser weir flow control to riser orifice flow control takes place. (This transition usually occurs during high hydraulic head conditions, such as between the 10- and 100-year design high water elevations.) Note that in Figure 17.22a and Figure 17.22b, the riser crest controls the flow, not the barrel. Thus, either condition can be described as riser flow control. Figure 17.22c and Figure 17.22d illustrate *barrel flow control*, which occurs when the barrel controls the flow at the upstream entrance to the barrel (*barrel inlet flow control*, Figure 17.22c), or along the barrel length (*barrel pipe flow control*, Figure 17.22d).

Barrel flow control conditions, illustrated in Figure 17.22c and Figure 17.22d, are desirable because they reduce or even eliminate cavitation forces or surging and vibration in the riser and barrel system. Cavitation forces may also cause vibrations that can damage the riser (especially corrugated metal risers) and the connection between the riser and barrel. This connection may crack and lose its watertight seal. Additionally, if a concrete riser is excessively tall, with a minimum amount of the riser secured in the embankment, cavitation forces may cause the riser to rock on its foundation, risking possible structural failure.

Surging, vibrations, and other cavitation forces result when the riser is restricting flow to the barrel so that the former is flowing full and the latter is not. This condition occurs when the flow through the riser structure transitions from riser weir flow control to riser orifice flow control before the barrel controls. Therefore, the barrel and riser system should be designed so that as the storm continues and the hydraulic head on the riser increases, the barrel controls the flow before the riser transitions from riser weir flow control to riser orifice flow control. This can be accomplished by checking the flow rates for the riser weir, riser orifice, barrel inlet, and outlet flow control at each stage of discharge. The lowest discharge for any given stage will be the controlling flow.

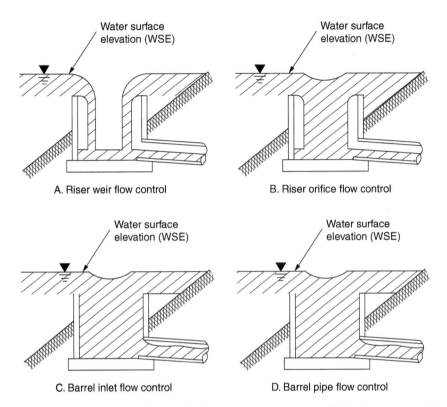

A. Riser weir flow control

B. Riser orifice flow control

C. Barrel inlet flow control

D. Barrel pipe flow control

Figure 17.22   Riser flow diagrams. (From U.S. Soil Conservation Service, 1984, *Engineering Field Manual*, Chapter 6, U.S. Department of Agriculture.

Use the following procedures for designing and checking riser and barrel system hydraulics:

a. *Riser flow control.* During the design of the control orifices and riser weir, the geometry of the riser is established. Subsequently, the riser must be checked to determine the stage at which it transitions from riser weir to riser orifice flow control. The riser weir controls the flow initially; then, as the water rises, the top of the riser acts as a submerged horizontal orifice. Thus, the flow transitions from riser weir flow control to riser orifice flow control as the water in the basin rises. The flow capacity of the riser weir is determined using weir Equation 17.14, and the flow capacity of the riser orifice is determined using orifice Equation 17.13 for each elevation. The smaller of the two flows for any given elevation is the controlling flow.

1. Calculate the flow in cubic feet per second over the riser weir, using standard weir Equation 17.14 for each elevation specified on the stage–storage–discharge worksheet. Record the flows on the worksheet. The weir length, L, is the circumference or length of the riser structure, measured at the crest, less any support posts or trash rack. The head is measured from the water surface elevation to the crest of the riser structure (refer to Figure 17.21).

2. Calculate the flow in cubic feet per second through the riser structure, using standard orifice Equation 17.13 for each elevation specified on the stage–storage–discharge worksheet. Record the flows on the worksheet. The orifice flow area, a, is measured from the inside dimensions of the riser structure. The head is measured from the water surface elevation to the elevation of the orifice centerline or, because the orifice is horizontal, to the elevation of the riser crest.

3. Compare the riser weir flow discharges to the riser orifice flow discharges. The smaller of the two discharges is the controlling flow for any given stage.

b. *Barrel flow control.* Two types of barrel flow exist: (1) barrel flow with inlet control (Figure 17.22c); and (2) barrel flow with outlet, or pipe flow control (Figure 17.22d). For both types, different factors and formulae are used to compute the hydraulic capacity of the barrel. During barrel inlet flow control, the diameter of the barrel, amount of head acting on the barrel, and

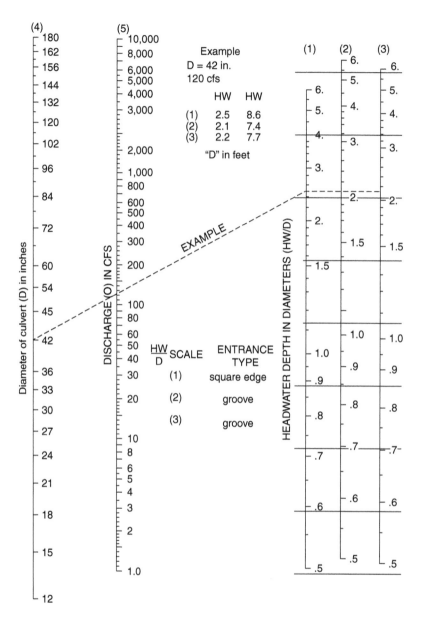

Figure 17.23   Headwater depth for concrete pipe culverts with inlet control. (U.S. Bureau of Public Roads, 1995.)

barrel entrance shape play a part in controlling the flow. Consideration is given to the length, slope, and roughness of the barrel, and the elevation of the tailwater, if any, in the outlet channel for barrel outlet, or pipe flow, control.

1. *Barrel inlet flow control.* Barrel inlet flow control means that the capacity of the barrel is controlled at the barrel entrance by the depth of headwater and the barrel entrance, which acts as a submerged orifice. The flow through the barrel entrance can be calculated using orifice Equation 17.14, or by simply using the pipe flow nomograph shown in Figure 17.23. This nomograph provides stage–discharge relationships for concrete culverts of various sizes. (Additional nomographs for other pipe materials and geometrics are available; refer to the U.S. Bureau of Public Roads (BPR) Hydraulic Engineering Circular (H.E.C.) 5.) The *headwater*, or depth of ponding, is the vertical distance measured from the water surface elevation to the invert at the entrance to the barrel. Refer to Figure 17.25 for ratios of headwater to pipe diameter, or HW/D. This nomograph, based on the orifice equation,

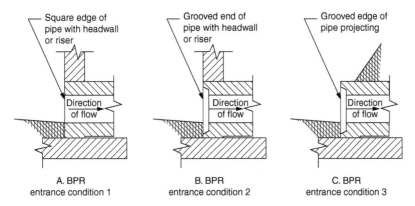

Figure 17.24 Headwater depth entrance conditions. (From *Virginia Stormwater Management Handbook*, 1999, Virginia Department of Conservation and Recreation, Division of Soil and Water Conservation.)

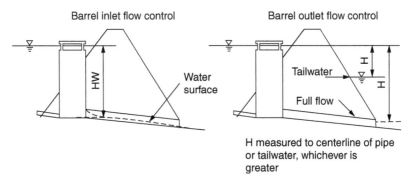

Figure 17.25 Hydraulic head values — barrel flow. (From *Virginia Stormwater Management Handbook*, 1999, Virginia Department of Conservation and Recreation, Division of Soil and Water Conservation.)

provides flow rates for three possible hydraulic entrance shapes (see Figure 17.24). During barrel inlet flow control, neither the barrel's length nor its outlet conditions are factors in determining the barrel's capacity. Note that when the HW/D design values exceed the chart values, the engineer may use the orifice equation to solve for flow rate.

Note: The inlet control nomographs are not truly representative of barrel inlet flow. These nomographs should be used carefully and with the understanding that they were developed to predict flow through highway culverts operating under inlet control. However, depending on the size relationship between the riser and outlet conduit, the inlet control nomograph may provide a reasonable estimate.

The following procedure outlines the steps used to calculate the discharge during barrel inlet flow control conditions:

a. Determine the entrance condition of the barrel.

b. Determine the headwater to pipe diameter ratio (HW/D) for each elevation specified on the stage–storage–discharge worksheet. Headwater is measured from the water surface elevation to the upstream invert of the barrel (see Figure 17.21 and Figure 17.25).

c. Determine the discharge, Q, in cubic feet per second, using the inlet control nomograph for circular concrete pipe presented in Figure 17.23 (for BPR H.E.C. 5 pipe flow nomographs for other pipe materials) or the orifice equation (for HW/D values that exceed the range of the nomographs) for each elevation specified on the stage–storage–discharge worksheet. Enter the values on the worksheet.

2. *Barrel outlet flow control.* Barrels flowing under outlet or pipe flow control experience full flow for all or part of the barrel length (see Figure 17.22d). The general pipe flow equation is derived using the Bernoulli and continuity principles and is simplified to:

$$Q = a = \sqrt{\frac{2gh}{1 + K_m + K_p L}} \qquad (17.17)$$

where:
Q = discharge, cubic feet per second
a = flow area of the barrel, square feet
g = acceleration due to gravity, feet per square second
h = elevation head differential, feet (see Figure 17.25)
K_m = coefficient of minor losses: $K_e + K_b$
K_e = entrance loss coefficient (see Table 17.12)
K_b = vend loss coefficient, typically = 0.5 for riser and barrel system
K_p = coefficient of pipe friction (see Table 17.14)
l = length of the barrel, feet

This equation is derived and further explained in the U.S. SCS's *Engineering Field Manual*, Chapter 3 (1984). The following procedure outlines the steps to check for barrel outlet control:

1. Determine the discharge for each elevation specified in the stage–storage–discharge table using general pipe flow Equation 17.17.
2. Record the discharge on the stage–storage–discharge worksheet.
3. Compare the barrel inlet flow control discharges with the barrel outlet flow control discharges. The smaller of the two discharges is the controlling flow for any given stage.

Step 11. Size 100-year release opening or emergency spillway. All stormwater impoundment structures should have a vegetated emergency spillway, if possible. This provides a degree of safety to prevent overtopping of the embankment if the principal spillway should become clogged or otherwise inoperative. If an emergency spillway is not practical because of site constraints, the 100-year storm must be routed through the riser and barrel system.

100-Year release opening. The design procedure for sizing the 100-year release opening is the same as that of the 10-year design, except that the 100-year storm values are used instead of the 10-year values.

Emergency spillway. An emergency spillway is a broad, crested weir. It can act as a control structure by restricting the release of flow or can be used to pass 100-year storm flow safely with a minimum of storage. The impact of the 100-year storm on the required storage is lessened by using an emergency spillway because of the spillway's ability to pass significant volumes of flow with little head. If an emergency spillway is not used, additional storage may be needed because the riser and barrel will usually pass only a small portion of the 100-year inflow. This remains true unless the riser and barrel are sized for the 100-year storm, in which case they will be oversized for the 2- and 10-year storms. Use the following procedure to design an emergency spillway that will safely pass or control the rate of discharge from the 100-year storm:

1. Identify the 10-year maximum water surface elevation based on the routing from Step 9. This elevation is used to establish the elevation of the 100-year release structure.
2. Determine the storage volume that corresponds to the 100-year control elevation from the stage–storage curve. Add this elevation, storage, and appropriate storm discharges to the stage–storage–discharge worksheet.
3. See the invert of the emergency spillway at the 10-year high water elevation.
4. Determine the 100-year developed inflow from the hydrology.

Note: A minimum distance of 0.1 ft is recommended between the 10-year high water mark and the invert of the emergency spillway.

Table 17.14 Headloss Coefficient, K_p, for Circular and Square Conduits Flowing Full

$$K_p = \frac{5087n^2}{d_1}$$

Pipe diam. (in.)	Flow area (ft²)	Manning's coefficient of roughness "n"															
		0.010	0.011	0.012	0.013	0.014	0.015	0.016	0.017	0.018	0.019	0.020	0.021	0.022	0.023	0.024	0.025
6	0.196	0.0467	0.0565	0.0672	0.0789	0.0914	0.1050	0.1194	0.1340	0.151	0.168	0.187	0.206	0.226	0.247	0.269	0.292
8	0.349	0.0318	0.0385	00.0458	0.0537	0.0623	0.0715	0.0814	0.0919	0.1030	0.1148	0.1272	0.140	0.154	0.168	0.183	0.199
10	0.545	0.0236	0.0286	0.0340	0.0399	0.0463	0.0531	0.0604	0.0682	0.0765	0.0852	0.0944	0.1041	0.1143	0.1249	0.136	0.148
12	0.785	0.0165	0.0224	0.0217	0.0313	0.0365	0.0417	0.0474	0.0535	0.0600	0.668	0.0741	0.0817	0.0896	0.980	0.1067	0.1157
14	1.069	0.0151	0.0182	0.0217	0.0255	0.0235	0.0339	0.0386	0.0436	0.0488	0.0544	0.0603	0.0665	0.0730	0.0798	0.0865	0.0942
15	1.23	0.0138	0.0166	0.0198	0.0232	0.0270	0.0309	0.0352	0.0397	0.0446	0.0496	0.0550	0.0606	0.0666	0.0727	0.0792	0.0859
16	1.40	0.0126	0.0153	0.0182	0.0213	0.0247	0.0284	0.0323	0.0365	0.0409	0.0455	0.0505	0.0556	0.0611	0.0667	0.0727	0.0798
18	1.77	0.01078	0.0130	0.0155	0.0182	0.0214	0.0243	0.0275	0.0312	0.0349	0.0389	0.0431	0.0476	0.0522	0.0570	0.0621	0.0674
21	2.41	0.00878	0.01062	0.0129	0.0148	0.0172	0.0198	0.0225	0.0252	0.0284	0.0317	0.0351	0.0387	0.0425	0.0464	0.0506	0.0549
24	3.14	0.00735	0.00889	0.01058	0.0124	0.0144	0.0165	0.0188	0.0212	0.0238	0.0265	0.0294	0.0324	0.0356	0.0398	0.0423	0.0459
27	3.98	0.00628	0.00760	0.00904	0.01481	0.0123	0.0141	0.0161	0.0181	0.0205	0.0227	0.0251	0.0277	0.0304	0.0352	0.0364	0.0395
30	4.91	0.00546	0.00660	0.00786	0.00322	0.01070	0.0228	0.0140	0.0158	0.0177	0.0187	0.0218	0.0241	0.0264	0.0285	0.0314	0.0341
36	7.07	0.00428	0.00518	0.00616	0.00723	0.00839	0.00963	0.01086	0.0124	0.0139	0.0154	0.0171	0.0189	0.0207	0.0226	0.0286	0.0267
42	9.62	0.00348	0.00422	0.00502	0.00589	0.00685	0.00782	0.00892	0.01007	0.01129	0.0126	0.0139	0.0154	0.0169	0.0184	0.0201	0.0218
48	12.57	0.00292	0.00353	0.00820	0.00485	0.00572	0.00656	0.00747	0.00443	0.00945	0.01058	0.01166	0.0129	0.0141	0.0154	0.0168	0.0182
54	15.50	0.00219	0.00302	0.00339	0.00421	0.00486	0.00561	0.00638	0.00720	0.00808	0.00900	0.00997	0.01099	0.0121	0.0132	0.0144	0.0156
60	19.63	0.00217	0.00262	0.00312	0.00386	0.00228	0.00487	0.00534	0.00626	0.00702	0.00782	0.00864	0.00855	0.01046	0.0115	0.0125	0.0135

Source: From Virginia Stormwater Management Handbook, 1999, Virginia Department of Conservation and Recreation, Division of Soil and Water Conservation.

Table 17.15   Pipe Entrance Loss Coefficients, K_e

Type of structure and design of entrance	Coefficient, K_e
Pipe, concrete	
Projecting from fill, socket end (groove end)	0.2
Projecting from fill, square cut end	0.5
Headwall or headwall and wingwalls	0.2
Socket end of pipe — groove end	0.5
— Square end	0.5
— Rounded (radius = 1/12D)	0.2
Mitered to conform to fill slope	0.7
End-section conforming to fill slope	0.5
Pipe, or pipe-arch, corrugated metal	
Projecting from fill (no headwall)	0.9
Headwall or headwall and wingwalls square end	0.5
Mitered to conform to fill slope	0.7
End-section conforming to fill slope	0.5

Note: "End-section conforming to fill slope," made of metal or concrete, is the section commonly available from manufacturers. Based on limited hydraulic tests, it appears to be equivalent in operation to a headwall in inlet or outlet control.

Source: Federal Highway Administration, Bureau of Public Roads.

5. Using the design procedure provided in Section 17.7.11, determine the required bottom width of the spillway, the length of the spillway level section, and the depth of flow through the spillway that adequately passes the 100-year storm within the available free board. The minimum free board required is 1 ft from the 100-year water surface elevation to the settled top of the embankment.

6. Develop the stage–storage–discharge relationship for the 100-year storm. Calculate the discharge for each elevation specified on the stage–storage curve and record the discharge on the stage–storage–discharge worksheet. If a release rate is specified, the TR-55 short-cut method can be used to calculate the approximate storage volume requirement. If a fixed storage volume is available, the same short-cut method can be used to decide what the discharge must be to ensure that the available storage is not exceeded. Refer to TR-55.

Step 12. Calculate total discharge and check performance of 100-year control opening.

1. Calculate total discharge. The stage–storage–discharge table is now complete and the total discharge from the riser and barrel system and emergency spillway can be determined. The designer should verify that the barrel flow controls before the riser transitions from riser weir flow control to riser orifice flow control. At some point, the combined flows from the water quality orifice, the 2-year opening, the 10-year opening, and the riser will exceed the capacity of the barrel. At this water surface elevation and discharge, the system transitions from riser flow control to barrel flow control. The total discharge for each elevation is simply the sum of the flows through the control orifices of the riser, or the controlling flow through the barrel and riser, whichever is less.

2. Check the performance of the 100-year control by (1) reservoir routing 2-, 10-, and 100-year storms through the basin using an acceptable reservoir routing computer program; or (2) doing the long-hand calculations outlined in Section 17.7.12. Verify that the design storm release rates are less than or equal to the allowable release rates and that the 100-year design high water is:

   • At least 2 ft lower than the settled top of embankment elevation if an emergency spillway is not used, or

   • At least 1 ft lower than the settled top of the embankment if an emergency spillway is used

The designer should also verify that the release rates for each design storm are not too low, which would result in more storage provided than is required.

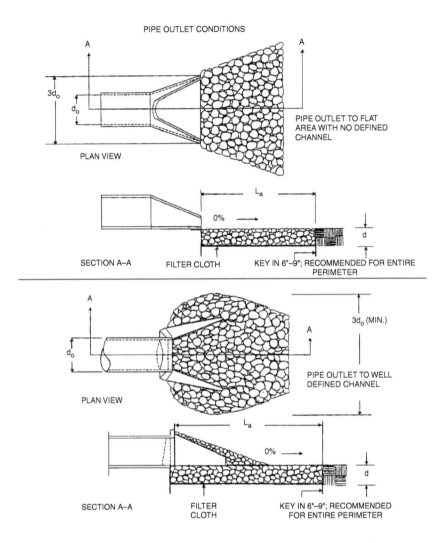

PIPE OUTLET CONDITIONS

PLAN VIEW

$3d_o$

d_o

PIPE OUTLET TO FLAT
AREA WITH NO DEFINED
CHANNEL

L_a

0%

SECTION A–A   FILTER CLOTH   KEY IN 6"–9"; RECOMMENDED FOR ENTIRE
PERIMETER

d

$3d_o$ (MIN.)

d_o

PIPE OUTLET TO WELL
DEFINED CHANNEL

PLAN VIEW

L_a

0%

SECTION A–A   FILTER
CLOTH   KEY IN 6"–9"; RECOMMENDED
FOR ENTIRE PERIMETER

d

NOTES: 1. APRON LINING MAY BE RIPRAP, GROUTED RIPRAP, GABION BASKET, OR CONCRETE.
2. L_a IS THE LENGTH OF THE RIPRAP APRON.
3. d = 1.5 TIMES THE MAXIMUM STONE DIAMETER, BUT NOT LESS THAN 6 IN.

Figure 17.26 Outlet protection detail. (From *Virginia Stormwater Management Handbook*, 1999, Virginia Department of Conservation and Recreation, Division of Soil and Water Conservation.)

Step 13. Design outlet protection. With the total discharge known for the full range of design storms, adequate outlet protection can now be designed. Protection is necessary to prevent scouring at the outlet and to help reduce the potential for downstream erosion by reducing the velocity and energy of the concentrated discharge. The most common form of outlet protection is a riprap-lined apron, constructed at zero grade for a specified distance, as determined by the outlet flow rate and tailwater elevation. The design procedure follows. (*Note*: This procedure is for riprap outlet protection at the downstream end of an embankment conduit. It does not apply to continuous rock linings of channels or streams. Refer to Figure 17.26.)

1. Determine the tailwater depth for the appropriate design storm, immediately below the discharge pipe. Typically, the discharge pipe from a stormwater management facility is sized to carry the allowable discharge from the 10-year frequency design storm. Manning's equation can be used to find the water surface elevation in the receiving channel for the 10-year storm, which represents the tailwater elevation. If the tailwater depth is less than half the outlet pipe diameter, it is called a *minimum tailwater condition*. If the tailwater depth is greater than half the outlet

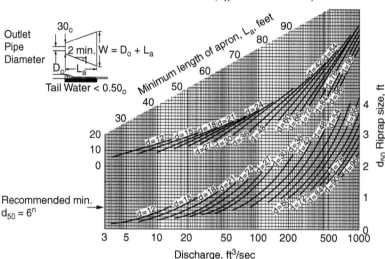

Figure 17.27   Minimum tailwater condition. (From *Virginia Stormwater Management Handbook*, 1999, Virginia Department of Conservation and Recreation, Division of Soil and Water Conservation.)

pipe diameter, it is called a *maximum tailwater condition*. Stormwater basins that discharge onto flat areas with no defined channel may be assumed to have a minimum tailwater condition.

Note: Outflows from stormwater management facilities must be discharged to an adequate channel. Basins discharging onto a flat area with no defined channel will usually require a channel that can convey the design flows.

2. Determine the required riprap size, D_{50}, and apron length, L_a. Enter the appropriate figure — Figure 17.27: minimum tailwater condition or Figure 17.28: maximum tailwater condition — with the design discharge of the pipe spillway to read the required apron length, L_s. (The apron length should not be less than 10 ft.)
3. Determine the required riprap apron width, W. When the pipe discharges into a well-defined channel, the apron will extend across the channel bottom and up the channel banks to an elevation 1 ft above the maximum tailwater depth or the top of bank, whichever is less. If the pipe discharges onto a flat area with no defined channel, the width of the apron will be determined as follows:
    a. The upstream end of the apron, next to the pipe, will be three times wider than the diameter of the outlet pipe.
    b. For a minimum tailwater condition, the width of the apron's downstream end will equal the pipe diameter plus the length of the apron.
    c. For a maximum tailwater condition, the width of the apron's downstream end will equal the pipe diameter plus 0.4 times the length of the apron.

Using the same figure as in the preceding step, determine the D_{50} riprap size and select the appropriate class of riprap (see Table 17.16). Values falling between the table values should be rounded up to the next class size.

4. Determine the required depth of the riprap blanket. The depth of the riprap blanket is approximated as: $2.25 \times D_{50}$.

Additional design considerations and specifications can be found in Minimum Standard 3.02, Principal Spillway and Std. and Spec. 3.18 and 3.19 of the 1992 edition of the *Virginia Erosion and Sediment Control Handbook*.

Step 14. Perform buoyancy calculation. The design of a multistage riser structure must include a buoyancy analysis for the riser and footing. When the ground is saturated and runoff is at an

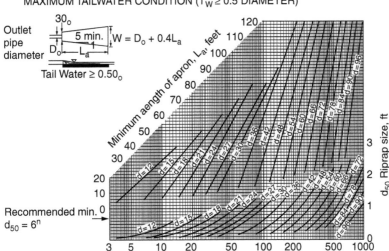

Figure 17.28 Maximum tailwater condition. (From *Virginia Stormwater Management Handbook*, 1999, Virginia Department of Conservation and Recreation, Division of Soil and Water Conservation.)

TABLE 17.16 Graded Riprap Design Values

Riprap class	D_{15} Weight (lb)	Mean D_{15} spherical diameter (ft)	Mean D_{50} spherical diameter (ft)
Class AI	25	0.7	0.9
Class I	50	0.8	1.1
Class II	150	1.3	1.6
Class III	500	1.9	2.2
Type I	1500	2.6	2.8
Type I	6000	4.0	4.5

Source: Virginia Department of Transportation.

elevation higher than the footing of the riser structure, the structure acts like a vessel. During this time, the riser is subject to uplifting, buoyant forces relative in strength to the volume of water displaced. Flotation occurs when the weight of the structure is less than or equal to the buoyant force exerted by the water. Flotation forces on the riser can lead to failure of the connection between the riser structure and barrel, and any other rigid connections. Eventually, this can also lead to the failure of the embankment.

A buoyancy calculation is the summation of all forces acting on the riser. The upward force is the weight of the water, or 62.4 lb/ft^3. The downward force includes the weight of the riser structure, any components (such as trash racks), and the weight of the soil above the footing. Note that conventional reinforced concrete weighs about 150 lb/ft^3 and the unit weight of soil is approximately 120 lb/ft^3. The weight of components such as trash racks, antivortex devices, hoods, etc. is specific to each structure and, depending upon the design, may or may not be significant in comparison to the other forces. If an extended base footing is used below the ground surface to support the control structure, the weight of the soil above the footing may also be a significant force.

The outlet pipe is excluded from the buoyancy analysis for the control structure. However, the barrel should be analyzed to ensure that it is not subject to flotation. The method used to attach the control structure to the outlet pipe is considered to have no bearing on the potential for these components to float.

The following procedure compares the upward force (buoyant force) to the downward force (structure weight). To maintain adequate stability, the downward force should be a minimum of 1.25 times the upward force.

1. Determine the buoyant force. The buoyant force is the total volume of the riser structure and base, using outside dimensions (the total volume displacement of the riser structure) multiplied by the unit weight of water (62.4 lb/ft^3).

2. Determine the downward or resisting force. The downward force is the total volume of the riser walls below the crest, including any top slab, footing, etc., less the openings for any pipe connections, multiplied by the unit weight of reinforced concrete (150 lb/ft^3).

3. Determine if the downward force is greater than the buoyant force by a factor of 1.25 or more. If the downward force is not greater than the buoyant force by this factor, additional weight must be added to the structure. Sinking the riser footing more deeply into the ground and adding concrete to the base can do this. Note that this will also increase the buoyant force; however, because the unit weight of concrete is more than twice that of water, the net result will be an increase in the downward force. The downward and buoyant forces should be adjusted accordingly and this step repeated.

Step 15. Provide seepage control. Seepage control should be provided for the pipe through the embankment. The two most common devices for controlling seepage are (1) filter and drainage diaphragms; and (2) antiseep collars. Note that filter and drainage diaphragms are preferred over antiseep collars for controlling seepage along pipe conduits.

- Filter and drainage diaphragms. The design of filter and drainage diaphragms depends on the foundation and embankment soils and is outside the scope of this text. When filter and drainage diaphragms are warranted, a registered professional engineer should supervise their design and construction. Design criteria and construction procedures for filter and drainage diaphragms can be found in the following references:
  - USDA SCS TR-60
  - USDA SCS Soil Mechanics Note No. 1: Guide for Determining the Gradation of Sand and Gravel Filters (includes design procedures and examples)
  - USDA SCS Soil Mechanics Note No. 3: Soil Mechanics Consideration for Embankment Drains (includes design procedures and examples)
  - U.S. Department of the Interior ACER Technical Memorandum No. 9: Guidelines for Controlling Seepage along Conduits Through Embankments
- Antiseep collars. The Bureau of Reclamation, the U.S. Army Corps of Engineers, and the Soil Conservation Service no longer recommend the use of antiseep collars. In 1987, the Bureau of Reclamation issued Technical Memorandum No. 9, which states: "When a conduit is selected for a waterway through an earth or rockfill embankment, cutoff [antiseep] collars will *not* be selected as the seepage control measure."

Alternative measures to antiseep collars include *graded filters* (or *filter diaphragms*) and *drainage blankets*. These devices not only are less complicated and more cost effective to construct than cutoff collars, but also allow easier placement of the embankment fill. Despite these restrictions, antiseep collars may be appropriate for certain situations. We provide a design procedure next.

1. Determine the length of the barrel within the saturated zone using Equation 17.18:

$$L_s = Y(Z + 4)\left(1 + \frac{S}{0.25 - S}\right) \tag{17.18}$$

where:

L_s = length of the barrel in the saturated zone, feet
Y = the depth of water at the principal spillway crest (10-year frequency storm water surface elevation), feet
Z = slope of the upstream face of the embankment, in A ft horizontal to 1 ft vertical (Z ft H:1V)
S = slope of the barrel, feet per foot

The length of pipe within the saturated zone can also be determined graphically on a *scale*

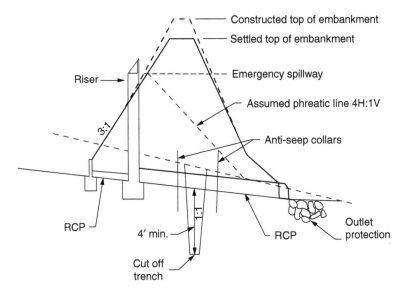

Figure 17.29 Phreatic time graphical determination. (From *Virginia Stormwater Management Handbook*, 1999, Virginia Department of Conservation and Recreation, Division of Soil and Water Conservation.)

profile of the embankment and barrel. The saturated zone of the embankment can be approximated as: starting at a point where the 10-year storm water surface elevation intersects the embankment slope, extend a line at a 4H: 1V slope downward until it intersects the barrel. The area under this line represents the theoretical zone of saturation (see Figure 17.29).

2. Determine the length required by multiplying 15% times the seepage length: 0.15 L_s. The increase in seepage length represents the total collar projection. One or multiple collars can provide for this.

3. Choose a collar size that is at least 4 ft larger than the barrel diameter (2 ft above and 2 ft below the barrel). For example, a 7-ft square collar would be selected for a 36-in. diameter barrel.

4. Determine the collar projection by subtracting the pipe diameter from the collar size.

5. Determine the number of collars required. The number of collars is found by dividing the seepage length increase found in Step 2 by the collar projection from Step 4. To reduce the number of collars required, the collar size can be increased. Alternatively, providing more collars can decrease the collar size.

Summary of Multistage Riser Design Procedure

Step 1. Determine water quality volume requirements.
  a. Extended detention
  b. Retention
Step 2. Compute allowable release rates.
Step 3. Estimate required storage volume.
Step 4. Grade the basin; create stage–storage curve.
Step 5a. Design water quality orifice (extended detention).
Step 5b. Set permanent pool volume (retention).
Step 6. Size 2-year control orifice.
Step 7. Check performance of 2-year opening.
Step 8. Size 10-year control opening.
Step 9. Check performance of 10-year opening.
Step 10. Perform hydraulic analysis.
  a. Riser flow control

   b. Barrel flow control
    1. Barrel inlet flow control
    2. Barrel outlet flow control
Step 11. Size 100-year release opening or emergency spillway.
Step 12. Calculate total discharge and check performance of 100-year control opening.
Step 13. Design outlet protection.
Step 14. Perform buoyancy calculation.
Step 15. Provide seepage control.

17.7.11 Emergency Spillway Design

A vegetated emergency spillway is designed to convey a predetermined design flood volume without excessive velocities and without overtopping the embankment.

We present two design methods here. The first (Procedure 1) is a conservative design procedure also found in the 1992 edition of the *Virginia Erosion and Sediment Control Handbook*, Std. and Speck. 3.14. This procedure is acceptable for typical stormwater management basins. Procedure 2 uses the roughness or retardance and durability of the vegetation and soils within the vegetated spillway. This second design is appropriate for larger or regional stormwater facilities in which construction inspection and permanent maintenance are more readily enforced. These larger facilities typically control relatively large watersheds and are located so that the stability of the emergency spillway is essential to safeguard downstream features.

The following design procedures establish a stage–discharge relationship (H_p vs. Q) for a vegetated emergency spillway serving a stormwater management basin (see Figure 17.30). The information required for these designs includes the determination of the hydrology for the watershed draining to the basin. Any of the methods may be used. The design should include calculations for the allowable release rate from the basin if the spillway is to be used to control a design frequency storm. Otherwise, the design peak flow rate should be calculated based on the spillway design flood or downstream conditions.

> *Note*: In general, a vegetated emergency spillway should not be used as an outlet for any storm with less than the 100-year frequency, unless it is armored with a nonerodible material. The engineer must consider the depth of the riprap blanket when riprap is used to armor the spillway. As noted previously, Class I riprap would require a blanket thickness or stone depth of 30 in., which may add considerable height to the embankment.
>
> *Note*: The design maximum water surface elevations for the emergency spillway should be at least 1 ft lower than the settled top of the embankment.

Procedure 1

1. Determine the design peak rate of inflow from the spillway design flood into the basin, using the developed condition hydrology, or determine the allowable design peak release rate, Q, from the basin, based on downstream conditions or watershed requirements.
2. Estimate the maximum water surface elevation and calculate the maximum flow through the riser and barrel system at this elevation (refer to the stage–storage discharge table). Subtract this flow volume from the design peak rate of inflow to determine the desired maximum spillway design discharge.
3. Determine the crest elevation of the emergency spillway. This is usually a small increment (0.1 ft) above the design high water elevation of the next smaller storm, typically the 10-year frequency storm.
4. Enter Table 17.17 with the maximum H_p value (maximum design water surface elevation from Step 2, less the crest elevation of the emergency spillway), and read across for the desired maximum spillway design discharge (from Step 2). Read the design bottom width of the emergency spillway

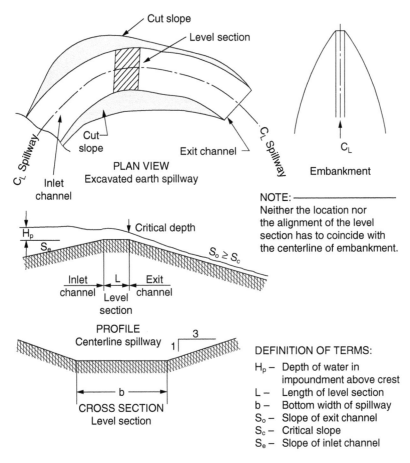

Figure 17.30 Vegetated emergency spillways: typical plan and section. (From *Virginia Stormwater Management Handbook*, 1999, Virginia Department of Conservation and Recreation, Division of Soil and Water Conservation.)

(in feet) at the top of the table, and verify the minimum exit slope (s) and length (x) or, if a maximum bottom width (b) is known because of grading or topographic constraints, enter Table 17.17a at the top with the desired bottom width and read down to find the desired discharge, Q. Then read across to the left to determine the required flow depth, H_p.

5. Add the appropriate H_p and discharge Q values to the stage–storage–discharge table.

Example 17.2

Problem:

Find the width of spillway, b, velocity, v, and depth of water above the spillway crest, H_p.

Given:
Q = 250 cfs (determined from postdeveloped condition hydrology)
S_o = 4% (slope of exit channel)
L = 50 ft (length of level section)

Solution:

Complete Step 1 through Step 5 of design procedure 1 for vegetated emergency spillways by using the given information as follows:

Table 17.17a Headloss Coefficient, K_c, for Square Conduit Flowing Full

$$H_1 = (k_p \text{ or } k_c) L \frac{v^2}{2g}$$

Nomenclature:

a = Cross-sectional area of flow in sq ft
di = Inside diameter of pipe in inches
q = Acceleration of gravity = 32.3 ft per sec
H_1 = Loss of head in feet due to friction in length L
k_c = Head loss coefficient for square conduit flowing full
k_p = Head loss coefficient for circular pipe flowing full
L = Length of conduit in feet
n = Manning's coefficient of roughness (see Table 17.17b)
Q = Discharge or capacity in cu ft per sec
r = Hydraulic radius in feet
v = Mean velocity in ft per sec

Example 1:

Compute the head loss in 300 ft of 24 in. diam. Concrete pipe flowing full and discharging 30 cfs. Assume n = 0.015.

$$v = \frac{Q}{a} = \frac{30}{3.14} = 9.35 \text{ fps}; \quad \frac{(9.35)^2}{64.4} = 1.42 \text{ ft}$$

$$H_1 = k_p L \frac{v^2}{2g} = 0.0165 \times 300 \times 1.42 = 7.03 \text{ ft}$$

Example 2:

Compute the discharge of a 250 ft, 3 × 3 square conduit flowing full if the loss of head is determined to be 2.25 ft. Assume n = 0.014.

Stage (H_P) in feet	Spillway variables	Bottom width (b) in feet																
		8	10	12	14	16	18	20	22	24	26	28	30	32	34	36	38	40
0.5	Q	6	7	8	10	11	13	14	15	17	18	20	21	22	24	25	27	28
	V	2.7	2.7	2.7	2.7	2.7	2.7	2.7	2.7	2.7	2.7	2.7	2.7	2.7	2.7	2.7	2.7	2.7
	S	3.9	3.9	3.9	3.9	3.8	3.8	3.8	3.8	3.8	3.8	3.8	3.8	3.8	3.8	3.8	3.8	3.8
	X	32	33	33	33	33	33	33	33	33	33	33	33	33	33	33	33	33
0.6	Q	8	10	12	14	16	18	20	22	24	26	28	30	32	34	35	37	39
	V	3.0	3.0	3.0	3.0	3.0	3.0	3.0	3.0	3.0	3.0	3.0	3.0	3.0	3.0	3.0	3.0	3.0

		1	2	3	4	5	6	7	8	9	10	11	12	13	14	15	16	17
0.7	S	3.7	3.7	3.7	3.7	3.7	3.7	3.6	3.6	3.6	3.6	3.6	3.6	3.6	3.6	3.6	3.6	3.6
	X	36	36	36	36	36	36	37	37	37	37	37	37	37	37	37	37	37
	Q	11	13	16	18	20	23	25	28	30	33	35	38	41	43	44	46	48
	V	3.2	3.2	3.3	3.3	3.3	3.3	3.3	3.3	3.3	3.3	3.3	3.3	3.3	3.3	3.3	3.3	3.3
0.8	S	3.5	3.5	3.5	3.5	3.5	3.5	3.5	3.5	3.4	3.4	3.3	3.3	3.3	3.4	3.4	3.4	3.4
	X	39	40	40	40	40	41	41	41	41	42	42	42	41	41	41	41	41
	Q	13	16	19	22	26	29	32	35	38	42	45	46	48	51	54	57	60
	V	3.5	3.5	3.5	3.6	3.6	3.6	3.6	3.6	3.6	3.6	3.6	3.6	3.6	3.6	3.6	3.6	3.6
0.9	S	3.3	3.3	3.3	3.4	3.2	3.2	3.2	3.2	3.2	3.2	3.2	3.1	3.1	3.1	3.1	3.1	3.2
	X	44	44	44	44	45	45	45	45	45	45	45	45	49	45	45	45	45
	Q	17	20	24	28	32	35	39	43	47	51	55	57	60	64	68	71	75
	V	3.7	3.8	3.8	3.8	3.8	3.8	3.8	3.8	3.8	3.8	3.8	3.8	3.8	3.8	3.8	3.8	3.8
1.0	S	3.2	3.1	3.1	3.0	3.0	3.0	3.0	3.1	3.1	3.1	3.1	3.0	3.0	3.0	3.0	3.0	3.1
	X	47	47	48	48	48	48	48	48	48	48	49	49	49	49	49	49	49
	Q	20	24	29	33	38	42	47	51	56	61	63	68	72	77	81	86	90
	V	4.0	4.0	4.0	4.0	4.0	4.0	4.0	4.0	4.0	4.0	4.0	4.0	4.0	4.0	4.0	4.0	4.0
1.1	S	3.1	3.1	3.0	3.0	2.9	2.9	2.9	3.0	3.0	3.0	3.0	3.0	3.0	2.9	2.9	3.0	3.0
	X	51	51	51	51	52	52	52	52	52	52	52	52	52	52	52	52	52
	Q	23	28	34	39	44	49	54	60	65	70	74	79	84	89	95	100	105
	V	4.2	4.2	4.2	4.3	4.3	4.3	4.3	4.3	4.3	4.3	4.3	4.3	4.3	4.3	4.3	4.3	4.3
1.2	S	2.9	2.9	2.9	2.9	2.8	2.8	2.8	2.8	2.9	2.9	2.9	2.9	2.9	2.9	2.9	2.9	2.9
	X	55	55	55	55	55	55	55	56	56	56	56	56	56	56	56	56	56
	Q	28	33	40	45	51	58	64	69	75	80	86	92	98	104	110	116	122
	V	4.4	4.4	4.4	4.4	4.4	4.5	4.5	4.5	4.5	4.5	4.5	4.5	4.5	4.5	4.5	4.5	4.5
1.3	S	2.9	2.9	2.8	2.8	2.8	2.8	2.8	2.8	2.8	2.8	2.8	2.7	2.8	2.8	2.8	2.8	2.8
	X	58	58	59	59	59	59	59	59	60	60	60	60	60	60	60	60	60
	Q	32	38	46	53	59	65	73	80	86	91	99	106	112	119	125	133	140
	V	4.5	4.6	4.6	4.6	4.6	4.6	4.7	4.7	4.7	4.7	4.7	4.7	4.7	4.7	4.7	4.7	4.7
1.4	S	2.8	2.8	2.7	2.7	2.7	2.7	2.7	2.7	2.7	2.7	2.7	2.7	2.7	2.7	2.7	2.7	2.7
	X	62	62	62	63	63	63	63	63	63	63	63	64	64	64	64	64	64
	Q	37	44	51	59	66	74	82	90	96	103	111	119	127	134	142	150	158
	V	4.7	4.8	4.8	4.8	4.8	4.8	4.8	4.8	4.8	4.9	4.9	4.9	4.9	4.9	4.9	4.9	4.9
1.5	S	2.8	2.7	2.6	2.6	2.6	2.6	2.6	2.6	2.6	2.6	2.6	2.6	2.6	2.6	2.6	2.6	2.6
	X	65	66	66	66	66	67	67	67	67	67	68	68	68	68	68	68	69
	Q	41	50	58	66	75	85	92	101	108	116	125	133	142	150	160	169	178
	V	4.8	4.9	5.0	5.0	5.0	5.0	5.0	5.0	5.0	5.0	5.0	5.0	5.0	5.0	5.1	5.1	5.1
1.6	S	2.7	2.7	2.6	2.6	2.6	2.6	2.6	2.6	2.6	2.6	2.6	2.6	2.6	2.6	2.5	2.5	2.5
	X	69	69	70	70	71	71	71	71	71	71	71	72	72	72	72	72	72
	Q	46	56	65	75	84	94	104	112	122	132	142	149	158	168	178	187	197

Table 17.17a Headloss Coefficient, K_c, for Square Conduit Flowing Full (continued)

$$H_i = (k_p \text{ or } k_c)L \, \frac{v^2}{2g}$$

Stage (H_p) in feet	Spillway variables	Bottom width (b) in feet																
		8	10	12	14	16	18	20	22	24	26	28	30	32	34	36	38	40
1.7	V	5.0	5.1	5.1	5.1	5.1	5.2	5.2	5.2	5.2	5.2	5.2	5.2	5.2	5.2	5.2	5.2	5.2
	S	2.6	2.6	2.6	2.6	2.5	2.5	2.5	2.5	2.5	2.5	2.5	2.5	2.5	2.5	2.5	2.5	2.5
	X	72	74	74	75	75	76	76	76	76	76	76	76	76	76	76	76	76
	Q	52	62	72	83	94	105	115	126	135	145	156	167	175	187	196	208	217
1.8	V	5.2	5.2	5.2	5.3	5.3	5.3	5.3	5.4	5.4	5.4	5.4	5.4	5.4	5.4	5.4	5.4	5.4
	S	2.6	2.6	2.5	2.5	2.5	2.5	2.5	2.5	2.5	2.5	2.5	2.5	2.5	2.5	2.5	2.5	2.5
	X	76	78	79	80	80	80	80	80	80	80	80	80	80	80	80	80	80
	Q	58	69	81	93	104	116	127	138	150	160	171	182	194	204	214	226	235
1.9	V	5.3	5.4	5.4	5.5	5.5	5.5	5.5	5.5	5.5	5.5	5.5	5.6	5.6	5.6	5.6	5.6	5.6
	S	2.5	2.5	2.5	2.4	2.4	2.4	2.4	2.4	2.4	2.4	2.4	2.4	2.4	2.4	2.4	2.4	2.4
	X	80	82	84	84	84	84	84	84	84	84	84	84	84	84	84	84	84
	Q	64	76	88	102	114	127	140	152	164	175	188	201	213	225	235	249	260
2.0	V	5.5	5.5	5.5	5.6	5.6	5.6	5.7	5.7	5.7	5.7	5.7	5.7	5.7	5.7	5.7	5.7	5.7
	S	2.5	2.5	2.5	2.4	2.4	2.4	2.4	2.4	2.4	2.4	2.4	2.4	2.4	2.4	2.4	2.4	2.4
	X	84	85	86	87	88	88	88	88	88	88	88	88	88	88	88	88	88
	Q	71	83	97	111	125	138	153	164	178	193	204	218	232	245	258	280	293
2.1	V	5.6	5.7	5.7	5.7	5.8	5.8	5.8	5.8	5.8	5.8	5.8	5.8	5.9	5.9	5.9	5.9	5.9
	S	2.5	2.4	2.4	2.4	2.4	2.4	2.4	2.4	2.3	2.3	2.3	2.3	2.3	2.3	2.3	2.3	2.3
	X	88	90	91	91	91	91	92	92	92	92	92	92	92	92	92	92	92
	Q	77	91	107	122	135	148	162	177	192	207	220	234	250	267	276	291	305
2.2	V	5.7	5.8	5.9	5.9	5.9	5.9	5.9	6.0	6.0	6.0	6.0	6.0	6.0	6.0	6.0	6.0	6.0
	S	2.4	2.4	2.4	2.4	2.4	2.4	2.4	2.3	2.3	2.3	2.3	2.3	2.3	2.3	2.3	2.3	2.3
	X	92	95	95	95	95	95	95	95	95	96	96	96	96	96	96	96	96
	Q	84	100	116	131	146	163	177	194	210	224	238	253	269	283	301	314	330
2.3	V	6.0	6.1	6.1	6.1	6.2	6.2	6.2	6.2	6.3	6.3	6.3	6.3	6.3	6.3	6.3	6.3	6.3
	S	2.4	2.4	2.3	2.3	2.3	2.3	2.3	2.3	2.2	2.2	2.2	2.2	2.2	2.2	2.2	2.2	2.2
	X	96	98	99	99	99	99	99	100	100	100	100	100	100	100	100	100	100
	Q	90	108	124	140	158	175	193	208	226	243	258	275	292	306	323	341	354
2.4	V	6.0	6.1	6.1	6.1	6.2	6.2	6.2	6.2	6.3	6.3	6.3	6.3	6.3	6.3	6.3	6.3	6.3
	S	2.4	2.3	2.3	2.3	2.3	2.3	2.3	2.3	2.2	2.2	2.2	2.2	2.2	2.2	2.2	2.2	2.2
	X	100	102	102	103	103	103	104	104	104	105	105	105	105	105	105	105	105
	Q	99	116	136	152	170	183	206	224	241	260	275	294	312	327	346	364	376

V	6.1	6.2	6.2	6.3	6.3	6.3	6.3	6.4	6.4	6.4	6.4	6.4	6.4	6.4	6.4	6.4	6.4	6.4	6.4	6.4
S	2.3	2.3	2.3	2.3	2.2	2.2	2.2	2.2	2.2	2.2	2.2	2.2	2.2	2.2	2.2	2.2	2.2	2.2	2.2	2.2
X	105	105	106	107	108	108	108	108	109	109	109	109	109	109	109	109	109	109	109	109

$$H_l = k_c L \frac{v^2}{2g}; \quad \frac{H_l}{K_c L} = \frac{2.25}{0.00839 \times 2.50} = 1.073 \text{ ft}$$

$$v = \sqrt{64.4 \times 1.073} = 8.31; Q = 9 \times 8.31 = 74.8 \text{ cfs}$$

Source: From *Virginia Stormwater Management Handbook*, 1999, Virginia Department of Conservation and Recreation, Division of Soil and Water Conservation.

Table 17.17b   Design Data for Earth Spillways

$$K_c = \frac{29.16n^2}{r^{4/3}}$$

Conduit size (ft)	Flow area (sq ft)	Manning's coefficient of roughness "n"				
		0.012	0.013	0.014	0.015	0.016
2 × 2	4.00	0.01058	0.01242	0.01440	0.01653	0.01880
2½ × 2½	6.25	0.00786	0.00972	0.01070	0.01228	0.01397
3 × 3	9.00	0.00616	0.00725	0.00839	0.00963	0.01096
3½ × 3½	12.25	0.00582	0.00589	0.00683	0.00784	0.00892
4 × 4	16.00	0.00420	0.00495	0.00572	0.00656	0.00746
4½ × 4½	20.25	0.00359	0.00421	0.00488	0.00561	0.00638
5 × 5	25.00	0.00312	0.00366	0.00425	0.00487	0.00554
5½ × 5½	30.25	0.00275	0.00322	0.00374	0.00429	0.00488
6 × 6	36.00	0.00245	0.00287	0.00333	0.00382	0.00435
6½ × 6½	42.25	0.00220	0.00258	0.00299	0.00343	0.00391
7 × 7	49.00	0.00199	0.00234	0.00271	0.00311	0.00354
7½ × 7½	56.25	0.00182	0.00213	0.00247	0.00284	0.00325
8 × 8	64.00	0.00167	0.00196	0.00227	0.00260	0.00296
8½ × 8½	72.25	0.00154	0.00180	0.00209	0.00240	0.00273
9 × 9	81.00	0.00142	0.00157	0.00194	0.00223	0.00253
9½ × 9½	90.25	0.00133	0.00156	0.00180	0.00207	0.00236
10 × 10	100.00	0.00124	0.00145	0.00168	0.00135	0.00220

Source: From SCS: U.S. Soil Conservation Service (SCS), 1984, *Engineering Field Manual*, USDA, U.S. Department of Agriculture.

1. Peak rate of inflow: given $Q = 250$ cfs.
2. The flow through the riser and barrel at the estimated maximum water surface elevation is calculated at 163 cfs. The desired maximum spillway design discharge is 250 cfs – 163 cfs = 87 cfs, at an H_p value of 1.3 ft.
3. Emergency spillway excavated into undisturbed material. The slope of the exit channel and length and elevation of level section: given, $S_o = 4\%$; $L = 50$ ft; elevation = 100.0 ft.
4. Enter Table 17.17 with the desired H_p value of 1.3 ft and read across to 86 cfs. Then read up to a bottom width of 24 ft at the top of the table. The minimum exit channel slope is 2.7%, which is less than the 4% provided, and the length of exit channel is required to be 63 ft. The velocity within the exit channel is 4.7 ft/sec at an exit channel slope of 2.7%. Because the provided exit channel slope is 4.0%, erosive velocities may warrant special treatment of the exit channel.
5. Add the elevation corresponding to 1.3 ft above the crest of the emergency spillway to the stage–storage–discharge worksheet.

Procedure 2

1. Determine the design peak rate of inflow from the spillway design flood into the basin, using the developed condition hydrology, or determine the allowable design peak release rate, Q, from the basin, based on downstream conditions or watershed requirements.
2. Estimate the maximum water surface elevation and calculate the associated flow through the riser and barrel system for this elevation. Subtract this flow value from the design peak rate of inflow to determine the desired maximum spillway design discharge.
3. Position the emergency spillway on the basin-grading plan at an embankment abutment.
4. Determine the slope, S_o, of the proposed exit channel, and the length, L, and elevation of the proposed level section from the basin grading plan.
5. Classify the natural soils around the spillway as *erosion-resistant* or *easily erodible* soils.
6. Determine the type and height of vegetative cover to be used to stabilize the spillway.
7. Determine the permissible velocity, v, from the appropriate table, based on the vegetative cover, soil classification, and the slope of the exit channel, S_o.

Table 17.18a H_p and Slope Range for Discharge, Velocity, and Crest Length — Retardance A

Max. velocity, v (ft/sec)	Unit discharge, q (cfs/ft)	Depth of water above spillway, H_p (ft) Length of level section, L (ft)				Slope range, S_o (%)	
		25	50	100	200	Min.	Max.
3	3	2.3	2.5	2.7	3.1	1	11
4	4	2.3	2.5	2.8	3.1	1	12
4	5	2.5	2.6	2.9	3.2	1	7
5	6	2.6	2.7	3.0	3.3	1	9
6	7	2.7	2.8	3.1	3.5	1	12
7	10	3.0	3.2	3.4	3.8	1	9
8	12.5	3.3	3.5	3.7	4.1	1	10

Source: U.S. Soil Conservation Service (SCS), 1984, *Engineering Field Manual*, USDA, U.S. Department of Agriculture.

Table 17.18b H_p and Slope Range for Discharge, Velocity, and Crest Length — Retardance B

Max. velocity, v (ft/sec)	Unit discharge, q (cfs/ft)	Depth of water above spillway, H_p (ft) Length of level section, L (ft)				Slope range, S_o (%)	
		25	50	100	200	Min.	Max.
2	1	1.2	1.4	1.5	1.8	1	12
2	1.25	1.3	1.4	1.6	1.9	1	7
3	1.5	1.3	1.5	1.7	1.9	1	12
3	2	1.4	1.5	1.7	1.9	1	8
4	3	1.6	1.7	1.9	2.2	1	9
5	4	1.8	1.9	2.1	2.4	1	8
6	5	1.9	2.1	2.3	2.5	1	10
7	6	2.1	2.2	2.4	2.7	1	11
8	7	2.2	2.4	2.6	2.9	1	12

Source: U.S. Soil Conservation Service (SCS), 1984, *Engineering Field Manual*, USDA, U.S. Department of Agriculture.

8. Determine the retardance classification of the spillway based on the type and height of vegetative cover from the appropriate table.
9. Determine the unit discharge of the spillway, q, in cubic feet per second per foot, from the appropriate table for the selected retardance, the maximum permissible velocity, v, and the slope of the exit channel, S_o.
10. Determine the required bottom width of the spillway, in feet, by dividing the allowable or design discharge, Q, by the spillway unit discharge, q:

$$\frac{Q(cfs)}{q(cfs/ft)} = ft$$

11. Determine the depth of flow, H_p, upstream of the control section based on the length of the level section, L, from Table 17.18a through Table 17.18d.
12. Enter the stage–discharge information into the stage–storage–discharge table.

Example 17.3

Problem:
Find permissible velocity, v, width of spillway, b, and depth of water above the spillway crest, H_p.

Table 17.18c H_p and Slope Range for Discharge, Velocity, and Crest Length — Retardance C

Max. velocity, v (ft/sec)	Unit discharge, q (cfs/ft)	Depth of water above spillway, H_p (ft) Length of level section, L (ft)				Slope range, S_o (%)	
		25	50	100	200	Min	Max
2	0.5	0.7	0.8	0.9	1.1	1	6
2	1	0.9	1.0	1.2	1.3	1	3
3	1.25	0.9	1.0	1.2	1.3	1	6
4	1.5	1.0	1.1	1.2	1.4	1	12
4	2	1.1	1.2	1.4	1.6	1	7
5	3	1.3	1.4	1.6	1.8	1	6
6	4	1.5	1.6	1.8	2.0	1	12
8	5	1.7	1.8	2.0	2.2	1	12
9	6	1.8	2.0	2.1	2.4	1	12
9	7	2.0	2.1	2.3	2.5	1	10
10	7.5	2.1	2.2	2.4	2.6	1	12

Source: U.S. Soil Conservation Service (SCS), 1984, *Engineering Field Manual*, USDA, U.S. Department of Agriculture.

Table 17.18d H_p and Slope Range for Discharge, Velocity, and Crest Length — Retardance D

Max. velocity, v (ft/sec)	Unit discharge, q (cfs/ft)	Depth of water above spillway, H_p (ft) Length of level section, L (ft)				Slope range, S_o (%)	
		25	50	100	200	Min	Max.
2	0.5	0.6	0.7	0.8	0.9	1	6
3	1	0.8	0.9	1.0	1.1	1	6
3	1.25	0.8	0.9	1.0	1.2	1	4
4	1.5	0.8	0.9	1.0	1.2	1	10
4	2	1.0	1.1	1.3	1.4	1	4
5	1.5	0.9	1.0	1.2	1.3	1	12
5	2	1.0	1.2	1.3	1.4	1	9
5	3	1.2	1.3	1.5	1.7	1	4
6	2.5	1.1	1.2	1.4	1.5	1	11
6	3	1.2	1.3	1.5	1.7	1	7
7	3	1.2	1.3	1.5	1.7	1	12
7	4	1.4	1.5	1.7	1.9	1	7
8	4	1.4	1.5	1.7	1.9	1	12
8	5	1.6	1.7	1.9	2.0	1	8
10	6	1.8	1.9	2.0	2.2	1	12

Source: U.S. Soil Conservation Service (SCS), 1984, *Engineering Field Manual*, USDA, U.S. Department of Agriculture.

Given:

Q = 250 cfs (determined from postdeveloped condition hydrology)
S_o = 4% (slope of exit channel)
L = 50 ft (length of level section)
Erosion-resistant soils
Sod forming grass–legume mixture cover, 6 to 10 in. height
Permissible velocity v = 5 ft/sec

Solution:

Complete Step 1 though Step 12 of design procedure 2 for vegetated emergency spillways by using the given information as follows:

1. Peak rate of inflow: given Q = 250 cfs.
2. The flow through the riser and barrel at the estimated maximum water surface elevation is calculated at 163 cfs. The desired maximum spillway design discharge is 250 cfs – 163 cfs = 87 cfs.
3. Emergency spillway excavated into undisturbed material.
4. Slope of exit channel, and length and elevation of level section: given, S_o = 4%, L = 50 ft, elevation = 100.0 ft.
5. Soil classification: given, erosion-resistant soils.
6. Vegetative cover: given, sod-forming grass–legume mixture.
7. Assume permissible velocity v = 5 ft/sec for sod-forming grass–legume mixtures, erosion-resistant soils, and exit channel slope S_o = 4%.
8. Assume retardance classification, C, for sod-forming grass–legume mixtures, expected height = 6 to 10 in.
9. The unit discharge of the spillway q = 3 cfs/ft from Table 17.18c for retardance C; maximum permissible velocity v = 5 ft/sec; and exit channel slope S_o = 4%.
10. The required bottom width b = Q/q = 87 cfs/3 cfs/ft = 29 ft.
11. The depth of flow, H_p, from Table 17.18c for retardance C: enter at q = 3 cfs/ft, find H_p = 1.4 ft for level section L = 50 ft.
12. The stage–discharge relationship: at stage elevation 1.4 ft above the spillway crest (101.4 ft), the discharge is 87 cfs.

Example 17.4

Problem:

Find permissible velocity, v, width of spillway, b, and depth of water above the spillway crest, H_p. Analyze the spillway for stability during the vegetation establishment period and for capacity once adequate vegetation is achieved.

Given:
Q = 175 cfs (determined from postdeveloped hydrology)
S_o = 8% (slope of exit channel)
L = 25 ft (length of level section)
Easily erodible soil
Bahia grass, good stand, 11 to 24 in. expected

Solution:

Complete Step 1 through through Step 12 of the design procedure 2 for vegetated emergency spillways by using the given information as follows:

1. Q = 175 cfs.
2. The flow through the riser and barrel at the estimated maximum water surface elevation is calculated at 75 cfs. The desired maximum spillway design discharge is 175 cfs – 75 cfs = 100 cfs.
3. Emergency spillway in undisturbed ground.
4. S_o = 8%; L = 25 ft, elevation = 418.0 ft (given).
5. Easily erodible soils.
6. Bahia grass, good stand, 11 to 24 in. expected.
7. Assume permissible velocity, v = 5 ft/sec.
8. (a) Retardance used for stability during the establishment period — good stand of vegetation 2 to 6 in.; retardance D.
   (b) Retardance used for capacity — good stand of vegetation 11 to 24 in.; retardance B.
9. Unit discharge q = 2 cfs/ft stability. From Table 17.18d for retardance D, permissible velocity, v = 5 ft/sec and S_o = 8%.
10. Bottom width b = Q/q = 100 cfs/2 cfs/ft = 50 ft (stability).

11. The depth of flow, H_p for capacity: from Table 17.18b for retardance B, enter at $q = 2$ cfs/ft, find $H_p = 1.4$ ft for $L = 25$ ft.

12. The stage–discharge relationship: at stage (elevation) 1.4 ft above the spillway crest (419.4 ft), the discharge, Q, is 100 cfs.

17.7.12 Hydrograph Routing

This section presents the methodology for routing a runoff hydrograph through an existing or proposed stormwater basin. One of the simplest and most commonly used methods, the "level pool" or storage indication routing technique is based on the continuity equation:

$$
\begin{aligned}
I \quad - \quad O \quad &= \quad ds/dt \\
\text{Inflow} \quad - \quad \text{Outflow} \quad &= \quad \text{Change in storage over time}
\end{aligned}
\tag{17.19}
$$

The goal of the routing process is to create an outflow hydrograph that is the result of the combined effects of the outlet device and the available storage. This allows the engineer to evaluate the performance of the outlet device, the basin storage volume, or both. When multiple iterations are required to create the most efficient basin shape, the routing procedure can be time consuming and cumbersome, especially when done by hand using the methods presented in this section. Note that several computer programs are available to help complete the routing procedure.

We present a step-by-step procedure for routing a runoff hydrograph through a stormwater basin. Note that the first four steps are part of the multistage riser design of the previous section. Remember: the water quality volume is not considered and only one design storm is routed — the 2-year storm. Other design frequency storms can be easily analyzed with the same procedure.

Procedure

1. Generate a postdeveloped condition inflow hydrograph. The runoff hydrograph for the 2-year frequency storm, postdeveloped condition as calculated by the SCS TR-20 computer program (Figure 17.9) is used for the inflow hydrograph.

2. Develop the stage–storage relationship for the proposed basin. The hydrologic calculations and the hydrograph analysis mentioned earlier revealed that the storage volume required to reduce the 2-year postdeveloped peak discharge back to the predeveloped rate was 35,820 ft³. Therefore, a preliminary grading plan should have a stormwater basin with this required storage volume (as a minimum) to control the 2-year frequency storm. Figure 17.18 shows the completed storage volume calculations worksheet and Figure 17.19 shows the stage vs. storage curve.

3. Size the outlet device for the design frequency storm and generate the stage–discharge relationship. An outlet device or structure must be selected to define the stage–discharge relationship. This procedure is covered in the multistage riser design. Use the procedure within the procedure as follows:

   a. Approximate the 2-year maximum head, h_{2max}. Enter the stage–storage curve, Figure 17.19, with the 2-year required storage: 35,820 ft³ and read the corresponding elevation: 88.5 ft. Then, $h_{2max} = 88.5$ ft $- 81.0$ ft (bottom of basin) $= 7.5$ ft. Note that this approximation ignores the centerline of the orifice as the point from which the head is measured. The head values can be adjusted when the orifice size is selected.

   b. Determine the maximum allowable 2-year discharge rate, $Q_{2allowable}$. In the predeveloped hydrologic analysis, the 2-year allowable discharge from the basin was set at 8.0 cfs. (This assumes that watershed conditions or local ordinance limits the developed rate of runoff to be less than or equal to the predeveloped rate.)

   c. Calculate the size of the 2-year controlled release orifice. Solve for the area, a, in square feet by inserting the allowable discharge $Q = 8.0$ cfs and $h_{2max} = 7.5$ ft into rearranged orifice Equation 17.13. This results in an orifice diameter of 10 in.:

$$a = \frac{Q}{C\sqrt{2gh}}$$

where:
a = required orifice area, square feet
Q = maximum allowable discharge = 8.0 cfs
C = orifice coefficient = 0.6
g = gravitational acceleration = 32.2 ft/sec
h = maximum 2-year hydraulic head, h_{2max} = 7.5 ft

$$a = \frac{8.0}{0.6\sqrt{(2)(32)(7.5)}}$$

$$a = 0.61 \text{ ft}^2$$

For orifice diameter:

$$a = 0.61 \text{ ft}^2 = \pi \left(\frac{d}{2}\right)^2$$

$$d = 0.88 \text{ ft} = 10.6 \text{ in}$$

Use a 10-in. diameter orifice.

4. Develop the stage–storage–discharge relationship for the 2-year storm. Substituting the 10-in. orifice size into the orifice equation and solving for the discharge, Q, at various stages provides the information needed to plot the stage vs. discharge curve and complete the stage–storage–discharge worksheet:

$$Q = C_o a \sqrt{2gh}$$

where: $a = a_{10 \text{ in.}} = 0.45 \text{ ft}^2$

$$Q = (0.6)(0.545)\sqrt{(2)(32.2)/(h)}$$

$$Q_2 = 2.62 \text{ (h)}^{0.5}$$

where: h = water surface elev. $- (81.0 + 0.83/2) = -81.4$.

Note: The h is measured to the centerline of the 10-in. orifice.

Figure 17.31 shows the result of the calculations: the stage vs. discharge curve and table.

5. Develop the relationship $2S/\Delta t$ vs. O and plot $2S/\Delta t$ vs. O. The plot of the curve $2S/\Delta t$ vs. O is derived from the continuity equation. The continuity equation is rewritten:

$$\frac{I_n + I_{n+1}}{2} - \frac{O_n + O_{n+1}}{2} - \frac{S_{n+1} - S_n}{\Delta t} \tag{17.20}$$

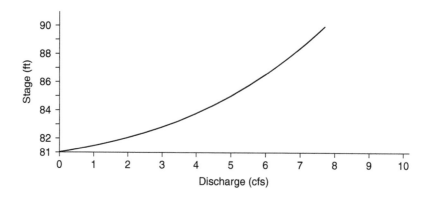

Stage	(h)	(Q)
81.4	0	0
82	0.6	2.0
84	2.6	4.2
86	4.6	5.6
88	6.6	6.7
90	8.6	7.7

Figure 17.31   Stage vs. discharge curve. (From *Virginia Stormwater Management Handbook*, 1999, Virginia Department of Conservation and Recreation, Division of Soil and Water Conservation.)

where:

$I_n + I_{n+1}$ = inflow at time $n = 1$ and time $n = 2$

$O_n + O_{n+1}$ = outflow at time $n = 1$ and time $n = 2$

$S_n + S_{n+1}$ = storage at time $n = 1$ and time $n = 2$

Δt = time interval ($n = 2 - n = 1$)

This equation describes the change in storage over time as the difference between the average inflow and outflow at the given time. Multiplying both sides of the equation by two and rearranging allows the equation to be re-written:

$$I_n + I_{n+1} + \left(\frac{2S_n}{\Delta t} - O_n \right) = \frac{2S_{+1}}{\Delta t} + O_{n+1} \tag{17.21}$$

We know the terms on the left-hand side of the equation from the inflow hydrograph and from the storage and outflow values of the previous time interval. The unknowns on the right-hand side, O_{n+1} and S_{n+1} can be solved interactively from the previously determined stage vs. storage curve, Figure 17.19, and stage vs. discharge curve, Figure 17.31. First, however, the relationship between $2S/\Delta t + O$ and O must be developed. This relationship can best be developed by using the stage vs. storage and stage vs. discharge curves to fill out the worksheet shown in Figure 17.32:

a. Columns 1, 2, and 3 are completed using the stage vs. discharge curve.

b. Columns 4 and 5 are completed using the stage vs. storage curve.

c. Column 6 is completed by determining the time step increment used in the inflow hydrograph. $\Delta t = 1$ h $= 3600$ sec. Δt is in seconds to create units of cubic feet per second for the $2S/\Delta t$ calculation.

d. Adding Columns 3 and 6 completes Column 7. The completed table is presented in Figure 17.33, along with the plotted values from Column 3, O or outflow, and Column 7, $2S/\Delta t + O$.

6. Route the inflow hydrograph through the basin and the 10-in. diameter orifice. The routing procedure is accomplished by use of Figure 17.34, the hydrograph routing worksheet. Note that

1	2	3	4	5	6	7
Elev	Stage	Outflow (0 cfs)	Storage (S cf)	2S (cf)	2S/Δt (cfs)	2S/Δt + O
81	0	0	0	0	0	0
8	0.6	2.0	900	1800	0.50	2.5
84	2.6	4.2	5,940	11,880	3.3	7.5
86	4.6	5.6	14,354	28,708	7.97	13.6
88	6.6	6.7	29,582	59,164	16.4	23.1
90	8.6	7.7	55,564	111,128	30.9	38.6

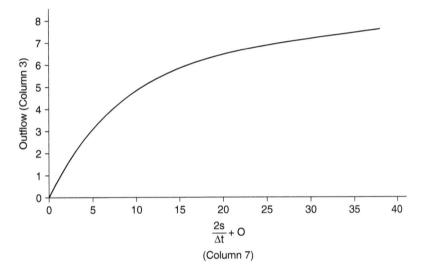

Figure 17.32 Storage indication hydrograph routing (2S/Δt + O) vs. O worksheet. (From *Virginia Stormwater Management Handbook*, 1999, Virginia Department of Conservation and Recreation, Division of Soil and Water Conservation.)

as the work is completed for each value of n, it is necessary to jump to the next row for a value. The table is completed by the following steps:

a. Complete Column 2 and Column 3 for each time n. These values are taken from the inflow hydrograph provided in tabular form in Figure 17.35. This information is taken from the plot of the inflow hydrograph or read directly from the tabular version of the inflow hydrograph (TR-20, TR-55, etc.).

b. Complete Column 4 for each time n by adding two successive inflow values from Column 3. Thus, Column 4_n = Column 3_n + Column 3_{n+1}.

c. Compute the values in Column 6 by adding Column 4 and Column 5 from the previous time step. Note that for $n = 0$, Column 5 through Column 7 are given a value of zero before starting the table. Therefore, Column $6_{n=2}$ = $4_{n=1}$ + Column $5_{n=1}$. (Note that this works down the table and not straight across.)

d. Column 7 is read from the $2S/\Delta t + O$ vs. O curve by entering the curve with the value from Column 6 to obtain the outflow, O.

e. Now backtrack to fill Column 5 by subtracting twice the value of Column 7 (from Step d) from the value in Column 6. Column 5_n = Column 6_n − 2(Column 7_n).

f. Repeat Steps c through e until the discharge (O, Column 7) reaches zero.

The preceding steps are repeated here for the first four time steps and displayed in the completed hydrograph routing worksheet, Figure 17.36.

1. Column 2 and Column 3 are completed for each time step using the inflow hydrograph.
2. Column 4 is completed as follows:

1	2	3	4	5	6	7
Elev.	Stage	Outflow (0 cfs)	Storage (S cf)	2S (cf)	2S/Δt (cfs)	2S/Δt + O
81	0	0	0	0	0	0
8	0.6	2.0	900	1800	0.50	2.5
84	2.6	4.2	5,940	11,880	3.3	7.5
86	4.6	5.6	14,354	28,708	7.97	13.6
88	6.6	6.7	29,582	59,164	16.4	23.1
90	8.6	7.7	55,564	111,128	30.9	38.6

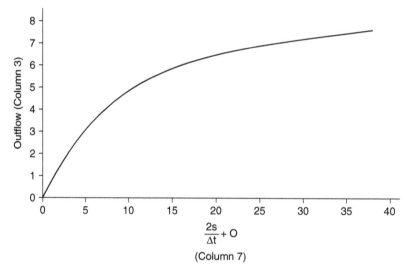

Figure 17.33 Storage indication hydrograph routing (2S/Δt + O) vs. O worksheet, curve and table. (From *Virginia Stormwater Management Handbook*, 1999, Virginia Department of Conservation and Recreation, Division of Soil and Water Conservation.)

1	2	3	4	5	6	7
n	Time (min)	I_n (cfs)	$I_n + I_{n+1}$ (cfs)	$2S_n/\Delta t - O_n$ (cfs)	$2S_{n+1}/\Delta t + O_{n+1}$ (cfs)	O_{n+1} (cfs)
	from hydrograph		Col. 3_n + Col. 3_{n+1}	Col. $6_n - 2$(Col. 7_n)	Col. 4_{n-1} + Col. 5_{n-1}	from chart; use Col. 6_n
0				0	0	0

Figure 17.34 Storage indication hydrograph routing worksheet. (From *Virginia Stormwater Management Handbook*, 1999, Virginia Department of Conservation and Recreation, Division of Soil and Water Conservation.)

Column 4_n = Column 3_n + Column 3_{n+1}
for $n = 1$: Column $4_{n=1}$ = Column $3_{n=1}$ + Column $3_{n=2}$
Column $4_{n=1}$ = 0 + 0.32 = 0.32
for $n = 2$: Column $4_{n=2}$ = 0.3 + 23.9 = 24.2
for $n = 3$: Column $4_{n=3}$ = 23.9 + 4.6 = 28.5

h	Q_i	Q_o	h	Q_i	Q_o	h	Q_i	Q_o
10	0	0	16	1.4	4.8	22	0.7	1.0
11	0.3	0.3	17	1.2	1.8	23	0.7	1.0
12	23.9	6.8	18	1.1	1.5	24	0.6	0.9
13	4.6	7.7	19	1.0	1.4	25	0.0	0.5
14	2.4	7.5	20	0.9	1.3	26	0	0
15	1.6	6.4	21	0.7	1.1			

Inflow and outflow hydrograph values
Q_i taken from TR–20 hydrology computer run
Q_o taken from routing worksheet

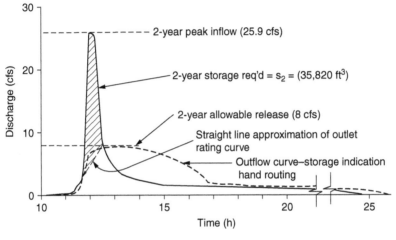

Peak 25.9 cfs, 12.13 h
Calculations =
$(0.40 \text{ in.}^2)(10 \text{ cfs/in.})(2.5 \text{ h/in.})(3600 \text{ sec/h}) = 35{,}820 \text{ ft}^3$
$= 0.82 \text{ acre-ft}$

Figure 17.35 Inflow and discharge hydrographs. (From *Virginia Stormwater Management Handbook*, 1999, Virginia Department of Conservation and Recreation, Division of Soil and Water Conservation.)

for $n = 4$: Column $4_{n=4} = 4.6 + 2.4 = 7.0$
etc.

- $n = 1$

3. Column $6_{n=1} = 0$. $N = 1$ is at time O. The first time step has a value of zero.
4. Column $7_{n=1} = 0$. Entering the $2S/\Delta t$ vs. O curve with a value of zero gives $O = 0$ cfs. (The discharge is always zero at time $t = 0$ unless a base flow exists.)
5. Column $5_{n=1}$ = Column $6_{n=1} - 2$ (Column $7_{n=1}$) Column $5_{n=1} = 0 - 0 = 0$.

- $n = 2$

3. Column $6_{n=2}$ = Column $4_{n=1}$ + Column $5_{n=1}$
   Column $6_{n=2} = 0.3 + 0 = 0.3$.
4. Column $7_{n=2} = 0.3$. Enter the $2S/\Delta t + O$ vs. O curve with $2S/\Delta t + 0.3$ (from Column 6) and read $O = 0.3$.
5. Column $5_{n=2}$ = Column $6_{n=2} - 2$(Column $7_{n=2}$).
   Column $5_{n=2} = 0.3 - 2(0.3) = -0.3 = 0$. (A negative outflow is unacceptable.)

- $n = 3$

1	2	3	4	5	6	7
n	Time (min)	I_n (cfs)	$I_n + I_{n+1}$ (cfs)	$2S_n/\Delta t - Q_n$ (cfs)	$2S_{n-1}/\Delta t + Q_{n+1}$ (cfs)	Q_{n+1} (cfs)
		from hydrograph	Col. 3_n + Col. 3_{n+1}	Col. 6_n − 2(Col. 7_n)	Col. 4_{n-1} + Col. 5_{n-1}	from chart: use Col. 6_n
1	0	0	0.32	0	0	0
2	60	0.32	24.2	0(−0.3)	0.3	0.3
3	120	23.9	28.5	10.6	24.2	6.8
4	180	4.6	7.0	23.7	39.1	7.7
5	240	2.4	4.0	15.7	30.7	7.5
6	300	1.6	3.0	6.9	19.7	6.4
7	360	1.4	2.6	0.3	9.9	4.8
8	420	1.2	2.3	0 (−0.7)	2.9	1.8
9	480	1.1	2.1	0 (−0.7)	2.3	1.5
10	540	1.0	1.9	0 (−0.7)	2.1	1.4
11	600	0.9	1.6	0 (−0.7)	1.9	1.3
12	660	0.7	1.4	0 (−0.6)	1.6	1.1
13	720	0.7	1.4	0 (−0.6)	1.4	1.0
14	780	0.7	1.3	0 (−0.6)	1.4	1.0
15	840	0.6	0.6	0 (−0.5)	1.3	0.9
16	900	0	0	0 (−0.4)	0.6	0.5
17	960	0	0		0	0

Figure 17.36 Storage indication hydrograph routing worksheet. (From *Virginia Stormwater Management Handbook*, 1999, Virginia Department of Conservation and Recreation, Division of Soil and Water Conservation.)

3. Column $6_{n=3}$ = 24.2 + 0 = 24.2
4. Column $7_{n=3}$ = 6.8. Enter $2S/\Delta t + O$ vs. O curve with 24.2, read O = 6.8.
5. Column $5_{n=3}$ = 24.2 − 2(6.8) = 10.6.

- $n = 4$

3. Column $6_{n=4}$ = 28.5 + 10.6 = 39.1
4. Column $7_{n=4}$ = 7.7. Enter $2S/\Delta t = O$ vs. O curve with 39.1, read O = 7.7.
5. Column $5_{n=4}$ = 39.1 − 2(7.7) = 23.7.

- $n = 5$, etc.

This process is continued until the discharge (O, Column 7) equals 0. The values in Column 7 can then be plotted to show the outflow rating curve or discharge hydrograph, as shown in Figure 17.35. The designer should verify that the maximum discharge from the basin is less than the allowable release. If the maximum discharge is greater than or much less than the allowable discharge, the designer should try a different outlet size or basin shape.

17.8 CONCLUSION

With practice, engineers using these methods should be able to design to fit the requirements and needs of their individual facilities and sites.

REFERENCES

Chow, V.T. (1959). *Open Channel Hydraulics.* New York: McGraw-Hill.

Engman, T. (1986). U.S. SCS, Urban Hydrology for Small Watersheds. Technical Release No. 55.

King, H.W. and Brater, E.F. (1976). *Handbook of Hydraulics,* 6th ed. New York: McGraw-Hill.

McGraw-Hill Series: *Hydrology for Engineers,* 3rd ed., 1982. New York: McGraw-Hill.

Morris, H.M. and Wiggert, J.M. (1972). *Applied Hydraulics in Engineering.* New York: John Wiley & Sons, Inc.

Schueler, T. (1987). *Controlling Urban Runoff: A Practical Manual for Planning and Designing.* Urban BMPs. Washington, D.C.: Metropolitan Washington Council of Governments.

Spellman, F.R. and Drinan, J. (2000). *The Drinking Water Handbook.* Lancaster, PA: Technomic Publishing Company.

Tuomari, D.C. and Thompson, S. (2003). "Sherlocks of Stormwater" Effective Investigation Technique for Illicit Connections and Discharge Detection. Accessed April 2004 at: http://www.epa/owow/nps/nat/stormwater03/40Tuomari.pdf

U.S. Bureau of Public Roads (1995). Hydraulic Engineering Circular (H.E.C.) 5. Washington, D.C.: U.S. Department of Transportation.

U.S. DOT, Federal Highway Administration (1984). Hydrology. Hydraulic Engineering Circular No. 19.

U.S. DOT, Federal Highway Administration (2001). *Urban Drainage Design Manual.* Washington, D.C.: Department of Transportation.

U.S. Soil Conservation Service (SCS) (1985). National Engineering Handbook Section 4 — Hydrology.

U.S. Soil Conservation Service (SCS) (1956). National Engineering Handbook, Washington, D.C.: U.S. Department of Agriculture.

U.S. Soil Conservation Service (SCS) (1984). *Engineering Field Manual.* USDA, U.S. Department of Agriculture.

U.S. Soil Conservation Service (SCS) (1986). Urban hydrology for small watersheds. Technical Release No. 55.

U.S. Soil Conservation Services (SCS) (1982). Project formulation — hydrology. Technical Release No. 20.

VDOT. (1999). Drainage Manual. DCR Course "C" Training Notebook. Virginia Department of Transportation.

Virginia Department of Conservation and Recreation, (1992). Virginia *Erosion and Sediment Control Handbook.* Virginia Department of Conservation and Recreation, Division of Soil and Water Conservation.

Virginia Department of Transportation (1999). Virginia Standards 111-55, in *Virginia Stormwater Management Handbook.* Virginia Department of Conservation and Recreation, Division of Soil and Water Conservation.

Walesh, S.G. (1989). *Urban Surface Water Treatment.* New York: John Wiley & Sons, Inc.

Index